Studies in Systems, Decision and Control

Volume 307

Series Editor

Janusz Kacprzyk, Systems Research Institute, Polish Academy of Sciences, Warsaw, Poland

The series "Studies in Systems, Decision and Control" (SSDC) covers both new developments and advances, as well as the state of the art, in the various areas of broadly perceived systems, decision making and control–quickly, up to date and with a high quality. The intent is to cover the theory, applications, and perspectives on the state of the art and future developments relevant to systems, decision making, control, complex processes and related areas, as embedded in the fields of engineering, computer science, physics, economics, social and life sciences, as well as the paradigms and methodologies behind them. The series contains monographs, textbooks, lecture notes and edited volumes in systems, decision making and control spanning the areas of Cyber-Physical Systems, Autonomous Systems, Sensor Networks, Control Systems, Energy Systems, Automotive Systems, Biological Systems, Vehicular Networking and Connected Vehicles, Aerospace Systems, Automation, Manufacturing, Smart Grids, Nonlinear Systems, Power Systems, Robotics, Social Systems, Economic Systems and other. Of particular value to both the contributors and the readership are the short publication timeframe and the world-wide distribution and exposure which enable both a wide and rapid dissemination of research output.

** Indexing: The books of this series are submitted to ISI, SCOPUS, DBLP, Ulrichs, MathSciNet, Current Mathematical Publications, Mathematical Reviews, Zentralblatt Math: MetaPress and Springerlink.

More information about this series at http://www.springer.com/series/13304

Reza Vafashoar · Hossein Morshedlou ·
Alireza Rezvanian · Mohammad Reza Meybodi

Cellular Learning Automata: Theory and Applications

Springer

Reza Vafashoar
Computer Engineering Department
Amirkabir University of Technology
(Tehran Polytechnic)
Tehran, Iran

Alireza Rezvanian
Department of Computer Engineering
University of Science and Culture (USC)
Tehran, Iran

Hossein Morshedlou
Department of Computer Engineering
and Information Technology
Shahrood University of Technology
Shahrood, Iran

School of Computer Science
Institute for Research in Fundamental
Sciences (IPM)
Tehran, Iran

Mohammad Reza Meybodi
Computer Engineering Department
Amirkabir University of Technology
(Tehran Polytechnic)
Tehran, Iran

ISSN 2198-4182 ISSN 2198-4190 (electronic)
Studies in Systems, Decision and Control
ISBN 978-3-030-53140-9 ISBN 978-3-030-53141-6 (eBook)
https://doi.org/10.1007/978-3-030-53141-6

This Springer imprint is published by the registered company Springer Nature Switzerland AG
The registered company address is: Gewerbestrasse 11, 6330 Cham, Switzerland

To my family

Reza Vafashoar

To my family

Hossein Morshedlou

To my family for their lovely support

Alireza Rezvanian

To my family

Mohammad Reza Meybodi

Preface

This book is written for computer scientists, engineers, students, and researchers working on artificial intelligence, machine learning, reinforcement learning, learning automata, and researchers who are studying system engineering, electrical engineering, control, adaptation, and learning. In particular, the reader is assumed to be already familiar with basic statistics, probability, linear algebra, stochastic process, and algorithm. Prior exposure to mathematics, game theory and learning automata is helpful but not necessary. The book in detail describes varieties of cellular learning automata models and their recent developments of applications in optimization, networks, game theory, and cloud computing with detailed mathematical and theoretical perspectives.

This book consists of eight chapters devoted to the theory of cellular learning automata models for computer science and engineering applications. Chapter 1 gives a preliminary introduction and an overview of varieties of cellular learning automata models. Chapter 2 provides a bibliometric analysis of the research studies on cellular learning automata as a systematic review. Chapter 3 is dedicated to the description of learning from multiple reinforcements in cellular learning automata. Chapter 4 is devoted to applications of cellular learning automata and reinforcement learning in global optimization. In Chap. 5, applications of multi-reinforcement cellular learning automata in channel assignments are discussed. Chapter 6 provides cellular learning automata for collaborative loss sharing. Chapter 7 describes cellular learning automata for competitive loss sharing. Finally, Chap. 8 provides a detailed discussion on the cellular learning automata versus multi-agent reinforcement learning.

The authors would like to thank Dr. Thomas Ditzinger, Springer, Editorial Director and Interdisciplinary Applied Sciences, and Ms. Saranya Kalidoss, Project Coordinator and Books Production of Springer Nature for the editorial assistance, excellent support, and cooperative collaboration to produce this important scientific work. We hope that readers will share our pleasure to present this book on the theory of cellular learning automata and will find it useful in their research.

Tehran, Iran
April 2020

Reza Vafashoar
Hossein Morshedlou
Alireza Rezvanian
Mohammad Reza Meybodi

Acknowledgements We are grateful to many people who have contributed to the work presented here and offered critical reviews of prior publication. We thank Springer for their assistance in publishing the book. We are also grateful to our academic supervisor, our family, our parents, and all our friends for their love and support.

Contents

About the Authors

Reza Vafashoar received the B.S. degree in Computer Engineering from Urmia University, Urmia, Iran, in 2007, and the M.S. degree in Artificial Intelligence from Amirkabir University of Technology, Tehran, Iran, in 2010. He also received the Ph.D. degree in Computer Engineering in the Computer Engineering Department from Amirkabir University of Technology (Tehran Polytechnic), Tehran, Iran, in 2019. His research interests include learning systems, cellular learning automata, evolutionary computing, and other computational intelligence techniques.

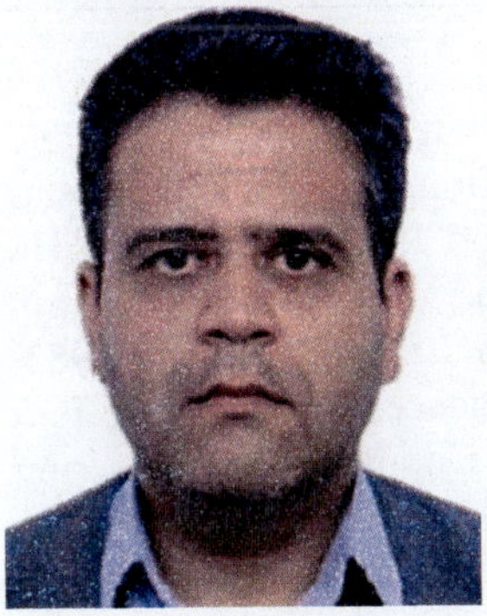

Hossein Morshedlou received the B.Sc. degree in Computer Engineering from Ferdowsi University, Mashhad, Iran, and the M.Sc. degree in Computer Engineering from the Amirkabir University of Technology, Tehran, Iran, in 2005 and 2008, respectively. He also received the Ph.D. degree in Computer Engineering in the Computer Engineering Department from Amirkabir University of Technology (Tehran Polytechnic), Tehran, Iran, in 2017. Currently, he is an Assistant Professor at the Department of Computer Engineering and Information Technology, Shahrood University of Technology, Shahrood, Iran. His research interests include distributed systems, cloud computing, learning automata, reinforcement learning, parallel algorithms, and soft computing.

Alireza Rezvanian received the B.Sc. degree from Bu-Ali Sina University of Hamedan, Iran, in 2007, the M.Sc. degree in Computer Engineering with honors from Islamic Azad University of Qazvin, Iran, in 2010, and the Ph.D. degree in Computer Engineering in the Computer Engineering Department from Amirkabir University of Technology (Tehran Polytechnic), Tehran, Iran, in 2016. Currently, he is an Assistant Professor in the Department of Computer Engineering, University of Science and Culture, Tehran, Iran. He worked from 2016 to 2020 as a researcher at the School of Computer Science from the Institute for Research in Fundamental Sciences (IPM), Tehran, Iran. He has authored or co-authored more than 70 research publications in reputable peer-reviewed journals and conferences including IEEE, Elsevier, Springer, Wiley, and Taylor & Francis. He has been a guest editor of the special issue on new applications of learning automata-based techniques in real-world environments for the journal of computational science (Elsevier). He is an associate editor of both human-centric computing and information sciences (Springer) and CAAI Transactions on Intelligence Technology (IET). His research activities include soft computing, learning automata, complex networks, social network analysis, data mining, data science, machine learning, and evolutionary algorithms.

Mohammad Reza Meybodi received the B.S. and M.S. degrees in Economics from the Shahid Beheshti University in Iran, in 1973 and 1977, respectively. He also received the M.S. and Ph.D. degrees from Oklahoma University, USA, in 1980 and 1983, respectively, in Computer Science. Currently, he is a Full Professor in the Computer Engineering Department, Amirkabir University of Technology, Tehran, Iran. Prior to the current position, he worked from 1983 to 1985 as an Assistant Professor at the Western Michigan University and from 1985 to 1991 as an Associate Professor at Ohio University, USA. His current research interests include learning systems, cloud computing, soft computing, and social networks.

Chapter 1
Varieties of Cellular Learning Automata: An Overview

1.1 Introduction

One of the goals of artificial intelligence is to make a machine thinks/acts like human behavior or thinks/acts rational (Stuart and Peter 2002). This synthetic machine needs to learn to solve problems and adapt to environments. From a psychology perspective, learning is considered as any systematic change in the performance of the system with a particular specific goal (Fu 1970). Learning is defined as any relatively permanent change in behavior resulting from experience. The learning system is characterized by its ability to improve its behavior with time, in some sense tending towards an ultimate goal (Narendra and Thathachar 1974). The concept of learning makes it possible to design systems that can gradually improve their performance during actual operation through learning from past experiences. Every learning task consists of two parts: a learning system and environment. The learning system must learn to act in the environment.

Among the machine learning methods, reinforcement learning as a prominent method for unknown environments is learning from positive and negative rewards (Montague 1999). In standard reinforcement learning, the system is connected to its environment via perception and action. The learning system receives an input at each instant, which is an indication of the current state of the environment. Then the learning system chooses an action to generate as output. This action changes the state of the environment, and the new state of the environment is communicated to the learning system with a scalar reinforcement signal. The reinforcement signal changes the learning system's behavior, and the learning system can learn to choose the best action over time through systematic trial and error. Reinforcement learning tasks can be divided naturally into two types: sequential and non-sequential tasks. In non-sequential tasks, the objective is to learn a mapping from situations to actions that maximize the expected immediate payoff. The non-sequential reinforcement

R. Vafashoar et al., *Cellular Learning Automata: Theory and Applications*, Studies in Systems, Decision and Control 307,
https://doi.org/10.1007/978-3-030-53141-6_1

learning has been studied extensively in the field of learning automata (Kumpati and Narendra 1989). In sequential tasks, the objective is to maximize the expected long-term payoffs. Sequential tasks are difficult, because the action selected may influence future situations and thus future payoffs. The sequential tasks are studied in reinforcement learning techniques based on dynamic programming, an approximation of dynamic programming (Kaelbling et al. 1996).

The theory of learning automaton (LA) as a promising field of artificial intelligence is a powerful source of computational technique that could be exploited for solving many real-world problems. LA is a self-adaptive decision-making unit that interacts with an unknown stochastic environment and is progressively able to make optimal decisions, even if provided with probabilistic wrong hints. Learning automata is especially suitable for modeling, learning, controlling, and solving complex real-world problems where the available information is incomplete; the environment is either noisy or highly uncertain. Thus, LA has made a significant impact in many areas of computer science as well as engineering problems. In the last decade, a wide range of learning automata theories, models, paradigms, simulations, and applications have been published by researchers; the cellular learning automaton (CLA) is the most popular technique. CLA, as a new reinforcement learning approach, is a combination of cellular automata (CA) and learning automata (LA). It is formed by a group of interconnected cells, which are arranged in some regular forms such as a grid or ring, in which each cell contains one or more LAs. A cellular learning automaton can be considered as a distributed model, which inherits both the computational power of cellular automata and the learning ability of learning automata. Accordingly, it is more powerful than a single learning automaton due to the ability to produce more complex patterns of behavior. Also, the learning abilities of cells enable cellular learning automata to produce these complex patterns by understandable behavioral rules rather than complicated mathematical functions commonly used in cellular automata. Owing to these characteristics, cellular learning automata can be successfully employed on modeling, learning, simulating, controlling, and solving challenging problems in uncertain, distributed, and complex environments.

Various studies have been conducted to investigate both the theoretical aspects and the applications of cellular learning automata. From a theoretical point of view, several models have been introduced, and their characteristics, such as their convergence behavior, have been investigated. Also, CLA has been received many attentions in a wide range of application such as channel assignment, image processing, machine vision, data mining, computer networks, scheduling, Peer-to-Peer networks, cellular networks, wireless networks, grid computing, cloud computing, Petri nets, complex network, social network analysis, and optimization, to mention a few.

This chapter's primary focus is to overview the varieties of cellular learning automata (CLA) models. This chapter is organized into three subsections. We give a brief introduction to cellular automata and learning automata models. Then we give a brief description of cellular learning automata definitions and models. Finally, we will give a summarization of the recent successful applications of cellular learning automata.

1.2 Cellular Automata

Cellular automata (CAs) are dynamical systems that exhibit complex global behavior from simple local interaction and computation (Wolfram 1982). Since the inception of cellular automaton (CA) by von Neumann in the 1950 s, it has attracted the attention of many researchers over various fields for modeling different physical, natural, and real-life phenomena. CAs are (typically) spatially and temporally discrete: they are composed of a finite or denumerable set of homogenous and simple units called cells. At each time step, the cells instantiate one of a finite set of states. They evolve in parallel at discrete time steps, following state update functions or dynamical transition rules: the update of a cell state obtains by taking into account the states of cells in its local neighborhood. CAs are computational systems: they can compute functions and solve algorithmic problems. Despite functioning differently from traditional, Turing machine-like devices, CA with suitable rules can emulate a universal Turing machine (Cook 2004), and therefore compute, given Turing's thesis (Copeland 1997), anything computable.

Classically, CAs are uniform. However, non-uniformity has also been introduced in update pattern, lattice structure, neighborhood dependency, and local rule. In dynamic structure CAs, the structure of cells or local rule changes during the time (Bhattacharjee et al. 2018). Non-uniformity also can occur in the parallelism of the updating procedure: cells can be updated asynchronously. CAs depending on their structure can also be classified as regular CAs or irregular CAs. In Irregular CAs, the structure regularity assumption has been relaxed. The irregular CAs and CAs with structural dynamism can be generalized into models known as automata networks in the literature. An automata network is a mathematical system consisting of a network of nodes that evolves according to a set of predetermined rules.

1.3 Learning Automata

Learning automaton (LA) as one of the computational intelligence techniques has been found a very useful tool to solve many complex and real-world problems where a large amount of uncertainty or lacking the information about the environment exists (Thathachar and Sastry 2002; Rezvanian et al. 2018a, 2019a). A learning automaton is a stochastic model operating in the framework of reinforcement learning (Narendra and Thathachar 1974; Thathachar and Sastry 2004; Rezvanian et al. 2019b). This model can be classified under the reinforcement learning schemes in the category of the temporal-difference (*TD*) learning methods. *TD* learning is a combination of the Monte Carlo ideas and dynamic programming ideas. Like Monte Carlo methods, *TD* methods can learn directly from raw experience without a model of the environment's dynamics. Like the dynamic programming, *TD* methods update estimates based in part on the other learned estimates, without waiting for an outcome (Sutton and Barto 1998). Sarsa (Rummery and Niranjan 1994), Q-learning (Watkins 1989), Actor-Critic

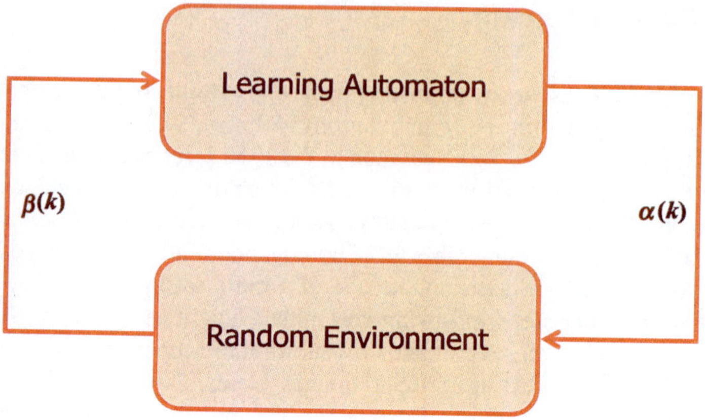

Fig. 1.1 The interaction of a learning automaton and its environment

methods (Barto et al. 1983), and R-learning (Schwartz 1993) are other samples of *TD* methods. Learning automata differ from other *TD* methods in the representation of the internal states and the updating method of the internal states.

The automata approach to learning can be considered as the determination of an optimal action from a set of actions. A learning automaton can be regarded as an abstract object that has a finite number of actions. It selects an action from its finite set of actions and applies it to an environment. The environment evaluates the applied action and sends a reinforcement signal to the learning automaton, as shown in Fig. 1.1. The reinforcement signal provided by the environment is used to update the internal state of the learning automaton. By continuing this process, the learning automaton gradually learns to select the optimal action, which leads to favorable responses from the environment.

Formally, a learning automaton is a quintuple $\langle \alpha, \Phi, \beta, F, G \rangle$ where $\alpha = \{\alpha_1, \alpha_2, \ldots, \alpha_r\}$ is the set of actions that it must choose from, $\Phi = (\Phi_1, \Phi_2, \ldots, \Phi_s)$ is the set of states, $\beta = \{\beta_1, \beta_2, \ldots, \beta_q\}$ is the set of inputs, G: $\Phi \rightarrow \alpha$ is the output map and determines the action taken by the learning automaton if it is in the state Φ_j, and F: $\Phi \times \beta \rightarrow \Phi$ is the transition map that defines the transition of the state of the learning automaton upon receiving an input from the environment.

The selected action at the time instant k, denoted by $\alpha(k)$, serves as the input to the environment which in turn emits a stochastic response, $\beta(k)$, at the time instant k, which is considered as the response of the environment to the learning automaton. Based upon the nature of β, environments could be classified into three classes: P-, Q-, and S-models. The output of a P-model environment has two elements of success or failure. Usually, in P-model environments, a failure (or unfavorable response) is denoted by one, while a success (or a favorable response) is denoted by 0. In Q-model environments, β can take a finite number of values in the interval [0, 1] while in S-model environments β lies in the interval [0, 1]. Based on the response $\beta(k)$, the state of the learning automaton $\Phi(k)$ is updated, and a new action is chosen at the

Algorithm 1-1. Learning automaton

Input: Action-set α
Output: Action probability vector p
Assumptions
Initialize r-dimensional action-set: $\alpha = \{\alpha_1, \alpha_2, \ldots, \alpha_r\}$ with r actions
Initialize r-dimensional action probability vector $p(k) = \{p_1, p_2, \ldots, p_r\} = \left\{\frac{1}{r}, \frac{1}{r}, \ldots, \frac{1}{r}\right\}$ at instant k
Let i denotes the selected action by the automaton
Let j denotes the current checking action
Begin
While (automaton does not converge to any action)
The learning automaton selects an action based on the action probability vector p.
The environment evaluates the selected action and gives the reinforcement signal $\beta \in \{0, 1\}$ to the learning automaton.
//The learning automaton updates its probability vector $p(k)$ using the provided reinforcement signal β
For each action $j \in \{1, \ldots, r\}$ **Do**
If (β=0) // favorable response
The selected action by learning automaton is rewarded according to equation (1-1)
Else If (β=1) // unfavorable response
The selected action by learning automaton is penalized according to equation (1-2)
End If
End For
End While
End Algorithm

Fig. 1.2 Pseudo-code of a learning automaton (LA)

time instant $(k + 1)$. Learning automata can be classified into two main classes: fixed and variable structure learning automata (VSLA) (Kumpati and Narendra 1989). A simple pseudo-code for the behavior of an r-action learning automaton within a stationary environment with $\beta \in \{0, 1\}$ is presented in Fig. 1.2.

1.3.1 Main Characteristics of Learning Automata

Learning automata have several excellent features, which make them suitable for use in many applications. Main features of learning automata are stated below

1. Learning automata can be used without any prior information about the underlying application.
2. Learning automata are very useful for applications with a large amount of uncertainty.
3. Unlike traditional hill-climbing algorithms, hill-climbing in learning automata is done in the expected sense in a probability space.
4. Learning automata require very little and straightforward feedback from their environment.
5. Learning automata are very useful in multi-agent and distributed systems with limited intercommunication and incomplete information.
6. Learning automata are elementary in structure and can be implemented easily in software or hardware.
7. The action set of learning automata can be a set of symbolic or numeric values.

8. Optimization algorithms based on learning automata do not need the objective function to be an analytical function of adjustable parameters.
9. Learning automata can be analyzed by powerful mathematical methodologies. It has been shown that learning automata are optimal in single, hierarchical, and distributed structures.
10. Learning automata require a few mathematical operations at each iteration so it can be used in real-time applications.
11. Learning automata have the flexibility and analytical tractability needed for most applications.

1.3.2 Varieties of Learning Automata

Stochastic learning automata can be categorized into two main classes, namely, finite action-set learning automata (FALA) and continuous action-set learning automata (CALA) (Thathachar and Sastry 2004). The learning automaton is called FALA if it has a finite set of actions, and it is called CALA otherwise. For a FALA with r actions, the action probability distribution is an r-dimensional probability vector. FALA can be categorized into two main families: variable structure learning automata (VSLA), in which the transition and output functions vary in time, and otherwise, fixed structure learning automata (FSLA) (Kumpati and Narendra 1989). Also, learning automata algorithms can be divided into two groups: non-estimator and estimator learning algorithms.

In many applications, there is a need to learn a real-valued parameter. In this situation, the actions of the automaton can be possible values of that parameter. To use a FALA for such an optimization problem, we have to discretize the value space of the parameter to obtain a finite number of actions. However, a fine discretization increases the number of actions, which in turn decreases the convergence speed of the automaton. A natural solution to this problem would be to employ an automaton with a continuous space of actions. Such a model of LA is called continuous action-set learning automata (CALA).

In the finite action-set learning automaton (FALA), learning algorithms can update action probability vectors in discrete or continuous steps. The former is called the discretized learning algorithm, while the latter is called a continuous learning algorithm. The learning algorithms can be divided into two groups: non-estimator and estimator learning algorithms, which are briefly described in the following subsections.

1.3.2.1 Fixed-Structure Learning Automata (FSLA)

The learning automaton is called fixed-structure; either the probability of the transition from one state to another state or the action probability of any action in any state is fixed. An FSLA is a quintuple $\langle \alpha, \beta, \Phi, \mathcal{F}, \mathcal{G} \rangle$, where $\alpha = \{\alpha_1, \alpha_2, \ldots, \alpha_r\}$

is the set of the actions that automaton chooses from, $\beta = \{0, 1\}$ is automaton's set of inputs where it receives a penalty, if $\beta = 1$ and receives a reward otherwise, $\Phi = \{\phi_1, \phi_2, \ldots, \phi_{rN}\}$ indicates its set of states, where N is called the depth of memory of the automaton, $\mathcal{F} : \Phi \times \beta \to \Phi$ illustrates the transition of the state of the automaton on receiving an input from the environment. $\mathcal{F}$ can be stochastic, $\mathcal{G} : \Phi \to \alpha$ is the output function of the automaton. The action taken by the automaton is determined according to its current state. This means that the automaton selects action α_i if it is in any of the states $\{\phi_{(i-1)N+1}, \phi_{(i-1)N+2}, \ldots, \phi_{iN}\}$. The state $\phi_{(i-1)N+1}$ is considered to be the most internal state, and ϕ_{iN} is considered to be the boundary state of action α_i, indicating that the automaton has the most and the least certainty in performing the action α_i, respectively.

The action chosen by the automaton is applied to the environment, which in turn emits a reinforcement signal β. Based on the received signal β, the state of the automaton is updated, and the new action is chosen according to the functions $\mathcal{F}$ and $\mathcal{G}$, respectively. There exist different types of FSLA based on the state transition function $\mathcal{F}$ and the output function $\mathcal{G}$. $L_{2N,2}$, $G_{2N,2}$, and Krinsky (Tsetlin 1962) are some of the essential FSLA types.

1.3.2.2 Variable-Structure Learning Automata (VSLA)

VSLA can be represented by a quadruple $\langle \alpha, \beta, p, T \rangle$, where $\alpha = \{\alpha_1, \alpha_2, \ldots, \alpha_r\}$ indicates the action set from which the automaton chooses, $\beta = \{\beta_1, \beta_2, \ldots, \beta_k\}$ indicates the set of inputs to the automaton, $p = \{p_1, p_2, \ldots, p_r\}$ indicates the action probability vector, such that p_i is the probability of selecting the action α_i, T indicates the learning algorithm that is used to update the action probability vector of the automaton in terms of the environment's response, i.e., $p(t+1) = T[\alpha(t), \beta(t), p(t)]$, where the inputs are the chosen action $\alpha(t)$, the response of the environment $\beta(t)$ and the action probability vector $p(t)$ at time t.

Let $\alpha_i(t)$ be the action selected by the automaton at time t. The action probability vector $p(t)$ is updated as given in Eq. (1.1), if the environment's response is the reward, and $p(t)$ is updated according to Eq. (1.2), if the response of the environment is the penalty.

$$p_j(t+1) = \begin{cases} p_j(t) + a\left[1 - p_j(t)\right] & j = i \\ (1-a)p_j(t) & \forall j \neq i \end{cases} \tag{1.1}$$

$$p_j(t+1) = \begin{cases} (1-b)p_j(t) & j = i \\ \left(\frac{b}{r-1}\right) + (1-b)p_j(t) & \forall j \neq i \end{cases} \tag{1.2}$$

where r denotes the number of actions that can be taken by the automaton, and a and b are the reward and penalty parameters which determine the amount of increases and decreases of the action probabilities, respectively. If $a = b$, the learning algorithm is a linear reward–penalty (L_{R-P}) algorithm, if $a \gg b$, it is a linear reward–ϵ penalty ($L_{R-\epsilon P}$) algorithm, and finally if $b = 0$, it is a linear reward–Inaction (L_{R-I})

algorithm in which the action probability vector remains unchanged when the taken action is penalized by the environment. The reward and penalty parameters a and b influence the speed of convergence as well as how closely the automaton approaches optimal behavior (the convergence accuracy) (Thathachar and Sastry 2002). If a is too small, the learning process is too slow. In contrary, if a is too large, the increments in the action probabilities are too high and the automaton's accuracy in perceiving the optimal behavior becomes low. By choosing the parameter a to be sufficiently small, the probability of convergence to the optimal behavior may be made as close to 1 as desired (Thathachar and Sastry 2002). In the $L_{R-\epsilon P}$ learning algorithm, the penalty parameter b is considered to be small in comparison with the reward parameter a ($b = \epsilon a$, where $0 < \epsilon \ll 1$). In this algorithm, the action probability distribution $p(t)$ of the automaton converges in distribution to a random variable p^* which can be made as close to the optimal vector as desired by choosing ϵ sufficiently small (Thathachar and Ramachandran 1984).

In order to investigate the learning ability of a learning automaton, a pure-chance automaton that always selects its available actions with equal probabilities is used as the standard for comparison (Thathachar and Sastry 2002). Any automaton that is said to learn must perform at least better than such a pure-chance automaton. To make this comparison, one measure can be the average penalty for a given action probability vector. For a stationary random environment with the penalty probability vector $c = \{c_1, c_2, \ldots, c_r\}$, the average penalty probability $M(t)$ received by an automaton is equal to

$$M(t) = E[\beta(t)|p(t)] = \sum_{\alpha_i \in \alpha} c_i p_i(t) \tag{1.3}$$

For a pure-chance automaton, $M(t)$ is a constant and is defined as

$$M(0) = \frac{1}{r} \sum_{\alpha_i \in \alpha} c_i \tag{1.4}$$

An automaton that does better than pure chance must have the average penalty $M(t)$ less than $M(0)$ at least asymptotically as $t \to \infty$. Since $p(t)$ and consequently $M(t)$ are random variables in general, the expected value $E[M(t)]$ is compared with $M(0)$.

Definition 1.1 A learning automaton interacting with a P-, Q-, or S-model environment is said to be *expedient* if

$$\lim_{t \to \infty} E[M(t)] < M(0) \tag{1.5}$$

Expediency means that the average penalty probability decreases when the automaton updates its action probability function. It would be more interesting in determining an updating procedure which would result in $E[M(t)]$ attaining its minimum value. In such a case, the automaton is called *optimal*.

Definition 1.2 A learning automaton interacting with a *P*-, *Q*-, or *S*-model environment is said to be *optimal*

$$\lim_{t\to\infty} E[M(t)] = c_\ell \tag{1.6}$$

where $c_\ell = \min_i c_i$. Optimality implies that asymptotically the action α_ℓ with the minimum penalty probability c_ℓ is chosen with probability one. While optimality is a very desirable property in stationary environments, it may not be possible to achieve it in a given situation. In such a case, one might aim at a suboptimal performance, which is represented by *ε-optimality*.

Definition 1.3 A learning automaton is interacting with a *P*-, *Q*-, or *S*-model environment is said to be *ε-optimal* if the following equation can be obtained for any $\varepsilon > 0$ by a proper choice of the parameters of the learning automaton.

$$\lim_{t\to\infty} E[M(t)] < c_\ell + \varepsilon \tag{1.7}$$

ε-optimality implies that the performance of the automaton can be made as close to the optimal as desired.

1.3.2.3 Variable Action Set Learning Automata

A variable action set learning automaton (also known as a learning automaton with a variable number of actions) is defined as an automaton in which the number of available actions at each instant varies over time (Thathachar and Harita 1987). Such a learning automaton has a finite set of r actions, $\alpha = \{\alpha_1, \alpha_2, \ldots, \alpha_r\}$. At each instant t, the action subset $\hat{\alpha}(t) \subseteq \alpha$ is available for the learning automaton to choose from. Selecting the elements of $\hat{\alpha}(t)$ is made randomly by an external agency. The procedure of choosing an action and updating the action probability vector in this automaton is done as follows.

Let $K(t) = \sum_{\alpha_i \in \hat{\alpha}(t)} p_i(t)$ presents the sum of the probabilities of the available actions in subset $\hat{\alpha}(t)$. Before choosing an action, the available actions probability vector is scaled, as given below.

$$\hat{p}_i(t) = \frac{p_i(t)}{K(t)} \quad \forall \alpha_i \in \hat{\alpha}(t) \tag{1.8}$$

Then, the automaton randomly chooses one of its available actions according to the scaled action probability vector $\hat{p}(t)$. Depending on the reinforcement signal received from the environment, the automaton updates the vector $\hat{p}(t)$. Finally, the available actions probability vector $\hat{p}(t)$ is rescaled according to Eq. (1.9). ε-optimality of this type of LA have been proved in (Thathachar and Harita 1987).

$$p_i(t+1) = \hat{p}_i(t+1).K(t) \quad \forall_{\alpha_i} \in \hat{\alpha}(t) \tag{1.9}$$

Algorithm 1-2. Variable action-set learning automata
Input: Action-set α
Output: Action probability vector p
Assumptions
Initialize r-dimensional action-set: $\alpha = \{\alpha_1, \alpha_2, \ldots, \alpha_r\}$ with r actions
Initialize r-dimensional action probability vector $p(k) = \{p_1, p_2, \ldots, p_r\} = \left\{\frac{1}{r}, \frac{1}{r}, \ldots, \frac{1}{r}\right\}$ at instant k
Let i denotes the selected action by the automaton
Let j denotes the current checking action
Begin
While (automaton does not converge to any action)
Calculate available actions of a learning automaton
Calculate the sum of the probability of available actions
For each action $j \in \{1, \ldots, r\}$ **Do**
If (α_j is available action)
Scale action probability vector $\hat{p}_i(k)$ according to equation (1-8)
End if
End for
Generate map function between available actions and all actions of a learning automaton
The learning automaton selects an action according to the probability vector of available actions $\hat{p}(k)$
The environment evaluates the selected action and gives the reinforcement signal $\beta \in \{0, 1\}$ to the learning automaton.
For each available action $j \in \{1, \ldots, m\}$ **Do**
If (β=0) // favorable response
The selected action by learning automaton is rewarded according to equation (1-1)
Else If (β=1) // unfavorable response
The selected action by learning automaton is rewarded according to equation (1-2)
End if
End for
For each action $j \in [1, \ldots, r]$ do
If (α_j is available action)
Rescale the probability vector of selected available action by equation (1-9)
End
End for
End While
End Algorithm

Fig. 1.3 Pseudo-code of the behavior of a variable action-set learning automaton

The pseudo-code of the behavior of a variable action set learning automaton is shown in Fig. 1.3.

1.3.2.4 Continuous Action-Set Learning Automata (CALA)

In a continuous action-set learning automaton (CALA), the action-set is defined as a continuous interval over the real numbers. This means that each automaton chooses its actions from the real line. In such a learning automaton, the action probability of the possible actions is defined as a probability distribution function. All actions are initially selected with the same probability, that is, the probability distribution function under which the actions are initially selected has a uniform distribution. The probability distribution function is updated depending upon the responses received from the environment.

A continuous action-set learning automaton (CALA) (Thathachar and Sastry 2004) is an automaton whose action-set is the real line, and its action probability distribution is considered to be a normal distribution with mean $\mu(t)$ and standard deviation $\sigma(t)$. At each time instant t, the CALA chooses a real number α at random based on the current action probability distribution $N(\mu(t), \sigma(t))$. The two actions $\alpha(t)$ and $\mu(t)$ are served as the inputs to the random environment. The CALA receives the reinforcement signals $\beta_{\alpha(t)}$ and $\beta_{\mu(t)}$ from the environment for both actions. At last, $\mu(t)$ and $\sigma(t)$ are updated as

$$\mu(t+1) = \mu(t) + \lambda \frac{\left(\beta_{\alpha(t)} - \beta_{\mu(t)}\right)}{\phi(\sigma(t))} \frac{(\alpha(t) - \mu(t))}{\phi(\sigma(t))} \tag{1.10}$$

$$\sigma(t+1) = \sigma(t) + \lambda \frac{\left(\beta_{\alpha(t)} - \beta_{\mu(t)}\right)}{\phi(\sigma(t))} \left[\left(\frac{(\alpha(t) - \mu(t))}{\phi(\sigma(t))} \right)^2 - 1 \right] - \lambda K(\sigma(t) - \sigma_L) \tag{1.11}$$

where,

$$\phi(\sigma(t)) = \begin{cases} \sigma_L & for\, \sigma(t) \leq \sigma_L \\ \sigma(t) & for\, \sigma(t) > \sigma_L \end{cases} \tag{1.12}$$

and $0 < \lambda < 1$ denotes the learning parameter, $K > 0$ is a large positive constant controlling the shrinking of $\sigma(t)$ and σ_L is a sufficiently small lower bound on $\sigma(t)$. Since the updating given for $\sigma(t)$ does not automatically ensure that $\sigma(t) \geq \sigma_L$, the function ϕ provides a projected version of $\sigma(t)$, denoted by $\phi(\sigma(t))$. The interaction with the environment continue until $\mu(t)$ does not change noticeably and $\sigma(t)$ converges close to σ_L.

The objective of CALA is to learn the value of α for which $E\left[\beta_{\alpha(t)}\right]$ attains a minimum. That is, the objective is to make $N(\mu(t), \sigma(t))$ converge to $N(\alpha^*, 0)$, where α^* is a minimum of $E\left[\beta_{\alpha(t)}\right]$. However, we cannot let $\sigma(t)$ converge to zero, since we want the asymptotic behavior of the algorithm to be analytically tractable. Hence, the lower bound $\sigma_L > 0$ is used, and the objective of learning is kept as $\sigma(t)$ converging to σ_L and $\mu(t)$ converging to α^*. By choosing σ_L and λ sufficiently small and K sufficiently large, $\mu(t)$ of the CALA algorithm will be close to a minimum of $E\left[\beta_{\alpha(t)}\right]$ with probability close to unity after a long enough time (Thathachar and Sastry 2004).

1.3.2.5 Non-estimator Learning Algorithms

In non-estimator learning algorithms, the current value of the reinforcement signal is the only parameter that is used to update the action probability vector. Let us assume a finite action-set learning automaton (FALA) with r actions operating in a stationary P-model environment. α_i (for $i = 1, 2, \ldots, r$) denotes the action taken by this automaton at instant n (i.e., $\alpha(n) = \alpha_i$), and $\underline{\beta} = \{0, 1\}$ is the response of the environment to this action. The following is a general learning algorithm for updating the action probability vector proposed in (Aso and Kimura 1979).

$$p_j(n+1) = \begin{cases} p_j(n) - g_{ij}\left[\underline{p}(n)\right] & \text{if } \beta(n) = 0 \\ p_j(n) - h_{ij}\left[\underline{p}(n)\right] & \text{if } \beta(n) = 1 \end{cases} \tag{1.13}$$

for all $j \neq i$. For preserving probability measure, we must have $\sum_{j=1}^{r} p_j(n) = 1$, so we obtain

$$p_i(n+1) = \begin{cases} p_i(n) - g_{ii}\left[\underline{p}(n)\right] & \text{if} \beta(n) = 0 \\ p_i(n) - h_{ii}\left[\underline{p}(n)\right] & \text{if} \beta(n) = 1 \end{cases} \tag{1.14}$$

where $g_{ii}\left(\underline{p}\right) = -\sum_{\substack{j=1 \\ j \neq i}}^{r} g_{ij}\left[\underline{p}(n)\right]$ and $h_{ii}\left(\underline{p}\right) = -\sum_{\substack{j=1 \\ j \neq i}}^{r} h_{ij}\left[\underline{p}(n)\right]$. Functions g_{ij} and h_{ij} (for $i, j = 1, 2, \cdots, r$ are called as reward and penalty functions, respectively.

Non-estimator learning algorithms are divided as linear (Linear Reward-Penalty ($L_{R\text{-}P}$), Linear Reward-Penalty ($L_{R\text{-}P}$), Linear Reward Inaction ($L_{R\text{-}I}$), Linear Inaction-Penalty ($L_{I\text{-}P}$) Linear Reward-Epsilon Penalty ($L_{R\text{-}\varepsilon P}$), Linear Reward-Reward ($L_{R\text{-}R}$)), nonlinear (Meybodi and Lakshmivarahan 1982) and hybrid learning algorithms (Friedman and Shenker 1996).

1.3.2.6 Estimator Learning Algorithms

One of the main difficulties in using LA in many practical applications is the slow rate of convergence. Although several attempts are conducted to increase the rate of convergence, these are not enough. An improvement in the rate of convergence could be to determine the characteristics of the environment as the learning proceeds. This additional information is used when the action probabilities are updated. These algorithms are called *estimator learning algorithms* (Thathachar and Sastry 1985b). Non-estimator learning algorithms update their action probability vectors based on the current response from the environment. In contrast, estimator learning algorithms maintain a running estimate of reward strengths for each action. Then the action probability vector will be updated based on the current response from the environment and the running estimate of reward strengths.

The state of an automaton operating under a P-model environment which is equipped with an estimator learning algorithm is defined as $\left\langle p(n), \hat{d}(n)\right\rangle$ at instance n, where $p(n)$ denotes the action probability vector, $\hat{d}(n) = [\hat{d}_1(n), \hat{d}_2(n), \ldots, \hat{d}_r(n)]^T$ represents the set of reward estimations and $\hat{d}_i(n)$ denotes the estimation of the reward probability d_j at instant n. The estimation of reward probability d_j is defined as the ratio of the number of times that action α_i is rewarded to the number of times α_i is selected (Thathachar and Sastry 1985a).

An estimator learning algorithm operates as follows. An action is initially chosen by the learning automaton (e.g., α_i). The selected action is applied to the environment. The random environment generates response $\beta(n)$ after it evaluates the chosen action. Learning automaton updates the action probability vector by using $\hat{d}(n)$ and $\beta(n)$. $\hat{d}(n)$ is finally updated based on the current response by the following rules.

$$\begin{aligned} R_i(n+1) &= R_i(n) + [1 - \beta(n)] \\ R_j(n+1) &= R_j(n) \quad j \neq i \end{aligned} \tag{1.15}$$

$$\begin{aligned} Z_i(n+1) &= Z_i(n) + 1 \\ Z_j(n+1) &= Z_j(n) \quad j \neq i \end{aligned} \tag{1.16}$$

$$\hat{d}(n) = \frac{R_k(n)}{Z_k(n)} \quad k = 1, 2, \ldots, r, \tag{1.17}$$

where $R_j(n)$ denotes the number of times action α_i is rewarded and $Z_i(n)$ denotes the number of times action α_i is selected. Several well-known estimator learning algorithms are presented in literature including TS stochastic estimator (TSE), Discretized TS estimator algorithm (DTSE) (Lanctot and Oommen 1992), relative strength learning automata (Simha and Kurose 1989), Stochastic estimator learning automata (SELA) (Papadimitriou et al. 1991), S-model ergodic discretized estimator learning algorithm (SEDEL) (Vasilakos and Paximadis 1994), and absorbing stochastic estimator learning algorithm (ASELA) (Papadimitriou et al. 2002).

1.3.2.7 Pursuit Algorithms

From the name of this class of finite action-set learning automata, it can be concluded that in these algorithms, the action probability vector chases after the action, which is most recently estimated as the optimal action. A pursuit learning algorithm is a simplified version of the estimator algorithms inheriting their main characteristics. In pursuit learning algorithms, the choice probability of the action with the maximum rewarding estimation is increased. By this updating method, the learning algorithm always pursues the optimal action. In the following subsections, several well-known pursuit learning algorithms are suggested by researchers, including Pursuit Reward-Penalty ($P_{R\text{-}P}$) (Thathachar and Sastry 1986) and Pursuit Reward-Inaction ($P_{R\text{-}I}$) (John Oommen and Agache 2001).

1.3.2.8 Generalized Pursuit Algorithm

The main disadvantage of the above-mentioned pursuit algorithms is that at each iteration of the algorithm, the choice probability of the action with the highest rewarding estimation has to be increased. This results in the movement of the action probability vector toward the action with the maximum rewarding estimation. This means that the learning automaton may converge to a wrong (non-optimal) action when the action which has the highest rewarding estimation is not the action with the minimum penalty probability. To avoid such a wrong convergence problem, the generalized pursuit algorithm (*GP*) was introduced in (Agache and Oommen 2002). In this algorithm, a set of possible actions with higher estimations than the currently selected action can be pursued at each instant. In (Agache and Oommen 2002), it has been shown that this algorithm is ε-optimal in all stationary environments. Let $K(n)$ denotes the number of actions that have higher estimations than the action selected at instant *n*. Equation (1.18) shows the updating rule of the generalized pursuit algorithm.

$$\begin{aligned} p_j(n+1) &= p_j(n)(1-a) + \frac{a}{K(n)} \quad \forall j (j \neq m) \text{ such that } \hat{d}_j > \hat{d}_i \\ p_j(n+1) &= p_j(n)(1-a) \quad \forall j (j \neq m) \text{ such that } \hat{d}_j \leq \hat{d}_i \\ p_m(n+1) &= 1 - \sum_{j \neq m} p_j(n+1) \end{aligned} \tag{1.18}$$

where α_m denotes the action with the highest rewarding estimation.

A discretized generalized pursuit algorithm (*DGP*) (Agache and Oommen 2002) is a particular case of the generalized pursuit algorithm in which the action probability vector is updated in discrete steps. This algorithm is called pseudo-discretized since the step size of the algorithm varies in different steps. In this algorithm, the choice probability of all the actions with higher rewarding estimations (than the selected action) increases with amount $\frac{\Delta}{K(n)}$, and that of the other actions decreases with amount $\frac{\Delta}{r-K(n)}$. The ε*-optimality* of the discretized generalized pursuit algorithm in all stationary environments has been shown in (Agache and Oommen 2002) *DGP* updates the action probability vector by the following updating rule.

$$\begin{aligned} p_j(n+1) &= \min\left\{ p_j(n) + \frac{\Delta}{K(n)}, 1 \right\} \quad \forall j (j \neq m) \text{ such that } \hat{d}_j > \hat{d}_i \\ p_j(n+1) &= \max\left\{ p_j(n) - \frac{\Delta}{r - K(n)}, 0 \right\} \quad \forall j (j \neq m) \text{ such that } \hat{d}_j \leq \hat{d}_i \\ p_m(n+1) &= 1 - \sum_{j \neq m} p_j(n+1) \end{aligned} \tag{1.19}$$

The pursuit algorithms are ranked in decreasing order of performance as *DGP*, DP_{R-I}, *GP*, DP_{R-P}, P_{R-I}, and P_{R-P}.

1.3.2.9 Interconnected Learning Automata

It seems that the full potential of learning automaton is realized when multiple automata interact with each other. It is shown that a set of interconnected learning automata can describe the behavior of an ant colony capable of finding the shortest path from their nest to food sources and back (Verbeeck et al. 2002). In this section, we study the interconnected learning automata. The interconnected learning automata techniques based on activation type of learning automata for taking action can be classified into three classes: synchronous, sequential, and asynchronous, as follows:

- **Synchronous Model of Interconnected Automata**. In the synchronous model of interconnected automata, at any time instant, all automata are activated simultaneously, choose their actions, then apply their chosen actions to the environment, and finally update their states. Two models of synchronous interconnected learning automata have been reported in the literature: game of automata and synchronous cellular learning automata.
- **Asynchronous Model of Interconnected Automata**. In the asynchronous model of interconnected automata, at any time instant, only a group of the automaton is activated independently. The only proposed model for the asynchronous model is asynchronous cellular learning automata. An asynchronous cellular learning automaton is a cellular automaton in which an automaton (or multiple automata) is assigned to its every cell. The learning automata residing in a particular cell determines its state (action) based on its action probability vector. Like cellular automata, there is a rule that cellular learning automata operate under it. The rule of cellular learning automata and the state of the neighboring cells of any particular cell determine the reinforcement signal to the learning automata residing in that cell. In cellular learning automata, the neighboring cells of any particular cell constitute its environment because they produce the reinforcement signal to the learning automata residing in that cell. This environment is nonstationary. It varies as action probability vectors of a cell vary and called the local environment because it is local to every cell. Krishna proposed an asynchronous cellular learning automaton in which the order to which learning automata is determined is imposed by the environment (Krishna 1993).
- **Sequential Model of Interconnected Automata**. In the sequential model of interconnected automata, at any time instant, only a group of the automaton is activated, and the actions chosen by currently activated automata determine the next automata to be activated. The hierarchical structure learning automata, the network of learning automata, distributed learning automata, and extended distributed learning automata are examples of the sequential model.

1.3.2.10 Hierarchical Structure Learning Automata

When the number of actions for a learning automaton becomes large (more than ten actions), the time taken for the action probability vector to converges it also increases. Under such circumstances, a hierarchical structure of learning automata

(HSLA) can be used. A hierarchical system of automata is a tree structure with a depth of M where each node corresponds to an automaton, and the arcs emanating from that node corresponds to the actions of that automaton. In HSLA, an automaton with r actions is in the first level (the root of the tree), and the kth level has r^{k-1} automata, each with r actions. The root node corresponds to an automaton, which will be referred to as the first-level or top-level automaton. The selection of each action of this automaton leads to activate an automaton at the second level. In this way, the structure can be extended to an arbitrary number of levels.

1.3.2.11 Network of Learning Automata

A network of learning automata (Willianms 1988) is a collection of *LA*s connected as a hierarchical feed-forward layered structure. In this structure, the outgoing link of the *LA*s in the other layers is the input of the *LA*s of the succeeding layers. In this model, the *LA*s (and consequently the layers) are classified into three separate groups. The first group includes the *LA*s located at the first level of the network, called the input *LA*s (input layer). The second group is composed of the *LA*s located at the last level of the network, called the output *LA*s (output layer). The third group includes the *LA*s located between the first and the last layers, called the hidden *LA*s (hidden layer). In a network of *LA*s, the input *LA*s receive the context vectors as external inputs from the environment, and the output *LA*s apply the output of the network to the environment. The difference between the feed-forward neural networks and NLA is that units of neural networks are deterministic.

In contrast, units of NLA are stochastic, and the learning algorithms used in the two networks are different. Since units are stochastic, the output of a particular unit i is drawn from a distribution depending on its input weight vector and output of the units in the other layers. This model operates as follows. The context vector is applied to the input *LA*s. Each input *LA* selects one of its possible actions based on its action probability vector, and the input signals it receives from the environment. The chosen action activates the *LA*s of the next level, which are connected to this *LA*. Each activated *LA* selects one of its actions, as stated before. The actions selected by the output *LA*s are applied to the random environment. The environment evaluates the output action in comparison with the desired output and generates the reinforcement signal. This reinforcement signal is then used by all *LA*s for updating their states.

1.3.2.12 Distributed Learning Automata (DLA)

The hierarchical structure learning automata has a tree structure, in which there exists a unique path between the root of the tree and each of its leaves. However, in some applications, such as routing in computer networks, there may be multiple paths between the source and destination nodes. This system is a generalization of HSLA, which referred to as distributed learning automata (DLA). A Distributed learning automata (DLA) (Beigy and Meybodi 2006) is a network of interconnected learning automata that collectively cooperate to solve a particular problem. The number of actions for a particular LA in DLA is equal to the number of LA's that are connected to this LA. The selection of an action by a LA in DLA activates another LA, which corresponds to this action. Formally, a DLA can be defined by a quadruple $\langle A, E, T, A_0 \rangle$, where $A = \{A_1, A_2, \ldots, A_n\}$ is the set of learning automata, $E \subset A \times A$ is the set of edges where edge (v_i, v_j) corresponds to action α_i^j of automaton A_i, T is the set of learning algorithms with which learning automata update their action probability vectors, and A_1 is the root automaton of DLA at which activation of DLA starts. The DLA was used for solving the several stochastic graph problems (Akbari Torkestani and Meybodi 2010, 2012; Mollakhalili Meybodi and Meybodi 2014; Mostafaei 2015; Rezvanian and Meybodi 2015a, b), cognitive radio networks (Moradabadi and Meybodi 2017a; Fahimi and Ghasemi 2017) and social network analytics problems (Khomami et al. 2016b; Ghavipour and Meybodi 2018a).

1.3.2.13 Extended Distributed Learning Automata (*e*DLA)

An extended distributed learning automata (*e*DLA) (Mollakhalili Meybodi and Meybodi 2014) is a new extension of DLA supervised by a set of rules governing the operation of the LAs. *Mollakhalili-Meybodi* et al. presented a framework based on *e*DLA for solving stochastic graph optimization problems such as stochastic shortest path problem and stochastic minimum spanning tree problem. The eDLA was also applied for solving several social network analytics problems (Ghamgosar et al. 2017).

1.3.3 Recent Applications of Learning Automata

In recent years, learning automata as one of the powerful computational intelligence techniques have been found a very useful technique to solve many real, complex, and dynamic environments where a large amount of uncertainty or lacking the information about the environment exists (Rezvanian et al. 2018a, e). Table 1.1 summarized some recent applications of learning automata.

1.4 Cellular Learning Automata

Cellular learning automaton (*CLA*) is a combination of cellular automaton (*CA*) (Packard and Wolfram 1985) and learning automaton (*LA*) (Kumpati and Narendra 1989). The basic idea of *CLA* is to use *LA* for adjusting the state transition probability of a stochastic *CA*. This model, which opens a new learning paradigm, is superior to *CA* because of its ability to learn and is also superior to single *LA* because it consists of a collection of *LA*s interacting with each other (Beigy and Meybodi 2004). A

Table 1.1 Summary of recent applications of learning automata

Applications	Learning automata model
Big Data	LA (Irandoost et al. 2019a)
Bioinformatics	CLA (Vafaee Sharbaf et al. 2016)
Cellular networks	CLA (Beigy and Meybodi 2010), LA (Rezapoor Mirsaleh and Meybodi 2018a)
Cloud computing	DLA (Hasanzadeh and Meybodi 2014), LA (Jobava et al. 2018), LA (Rahmanian et al. 2018), ICLA (Morshedlou and Meybodi 2018), LA (Morshedlou and Meybodi 2014), FALA (Velusamy and Lent 2018), LA (Qavami et al. 2017), LA (Rasouli et al. 2020), CLA-EC (Jalali Moghaddam et al. 2020)
Cognitive radio network	BLA (Mahmoudi et al. 2020)
Cyber-physical systems	LA (Ren et al. 2018)
Dynamic optimization	LA (Kazemi Kordestani et al. 2020)
Data mining	FALA (Hasanzadeh-Mofrad and Rezvanian 2018), ACLA (Ahangaran et al. 2017), CLA (Sohrabi and Roshani 2017), LA (Savargiv et al. 2020), DLA (Goodwin and Yazidi 2020), LA (Hasanzadeh-Mofrad and Rezvanian 2018)
Graph problems	CLA (Vahidipour et al. 2017a), LA (Rezapoor Mirsaleh and Meybodi 2018b), LA (Mousavian et al. 2013), ICLA (Mousavian et al. 2014), DLA (Soleimani-Pouri et al. 2012), LA (Khomami et al. 2016a), FALA (Vahidipour et al. 2019), DLA (Vahidipour et al. 2019), ICAL (Vahidipour et al. 2019), (Daliri Khomami et al. 2017a), GAPN-LA (Vahidipour et al. 2019), DLA (Rezvanian and Meybodi 2015b), DLA (Rezvanian and Meybodi 2015a), WCLA (Moradabadi and Meybodi 2018b)
Graph sampling	DLA (Rezvanian et al. 2014), DLA (Rezvanian and Meybodi 2017a), EDLA (Rezvanian and Meybodi 2017b), EDLA (Rezvanian and Meybodi 2017a), ICLA (Ghavipour and Meybodi 2017), VSLA (Rezvanian and Meybodi 2017a), FSLA (Ghavipour and Meybodi 2018b), FALA (Khadangi et al. 2016)
Image processing	CLA (Hasanzadeh Mofrad et al. 2015), LA (Damerchilu et al. 2016), LA (Kumar et al. 2015b), CLA (Adinehvand et al. 2017)
Influence maximization	NLA (Daliri Khomami et al. 2018), DGCPA (Ge et al. 2017), DL_{ri} (Huang et al. 2018), CLA (Aldrees and Ykhlef 2014)
Internet of things (IoT)	LA (Di et al. 2018), LA (Sikeridis et al. 2018), LA (Saleem et al. 2020), LA (Deng et al. 2020)
Link prediction	FALA (Moradabadi and Meybodi 2018c), FALA (Moradabadi and Meybodi 2017b), CALA (Moradabadi and Meybodi 2018a), DLA (Moradabadi and Meybodi 2017a), CALA (Moradabadi and Meybodi 2016)
Machine vision	LA (Betka et al. 2020)
MapReduce	GLA (Irandoost et al. 2019b)

(continued)

Table 1.1 (continued)

Applications	Learning automata model
Network security	LA (Krishna et al. 2014), FALA (Di et al. 2019), LA (Farsi et al. 2018), CALA (Kahani and Fallah 2018), FALA (Su et al. 2018), $DP_{R\text{-}P}$ (Seyyedi and Minaei-Bidgoli 2018)
Operations research	LA (Nesi and da Rosa Righi 2020)
Opportunistic networks	CLA (Zhang et al. 2016)
Optimization	FALA (Rezvanian and Meybodi 2010), FALA (Rezvanian and Meybodi 2010), FALA (Mahdaviani et al. 2015), VSLA (Kordestani et al. 2018), CLA (Vafashoar and Meybodi 2016), LA (Rezapoor Mirsaleh and Meybodi 2015), LA (Li et al. 2018), LA (Rezapoor Mirsaleh and Meybodi 2018c), LA (Zarei and Meybodi 2020), PLA (Yazidi et al. 2020), LA (Alirezanejad et al. 2020)
Peer-to-peer networks	LA (Saghiri and Meybodi 2016, 2017a), CLA (Saghiri and Meybodi 2018a), (Amirazodi et al. 2018), ICLA (Rezvanian et al. 2018b), LA (Safara et al. 2020),
Recommender systems	FALA (Krishna et al. 2013), CALA (Ghavipour and Meybodi 2016), CLA (Toozandehjani et al. 2014)
Robotics	LA (Ghosh et al. 2019)
Social network analysis	LA (Amiri et al. 2013), DLA (Khomami et al. 2016b), CALA (Moradabadi and Meybodi 2016), FSLA (Ghavipour and Meybodi 2018b), ICLA (Ghavipour and Meybodi 2017), ICLA (Khomami et al. 2018), ICLA (Zhao et al. 2015), CLA (Aldrees and Ykhlef 2014), FSLA (Roohollahi et al. 2020), LA (Rezvanian et al. 2019a), ICLA (Rezvanian et al. 2019c), LA (Rezvanian et al. 2019d), NLA (Rezvanian and Meybodi 2016a),
Software-defined networks	ICLA (Thakur and Khatua 2019), FALA (Amiri et al. 2013), DLA (Khomami et al. 2016b), ICLA (Khomami et al. 2018), EDLA (Ghamgosar et al. 2017)
Stochastic social networks	DLA (Rezvanian and Meybodi 2016b), LA (Moradabadi and Meybodi 2018c), DLA (Vahidipour et al. 2017b), ICLA (Vahidipour et al. 2019), EDLA (Rezvanian and Meybodi 2017a)
Transportation	DCLA (Ruan et al. 2019), CLA (Chen et al. 2018)
Trust management	DLA (Ghavipour and Meybodi 2018a), DLA (Ghavipour and Meybodi 2018c), FALA (Lingam et al. 2018), CLA (Bushehrian and Nejad 2017), LA (Rezvanian et al. 2019e)
Vehicular environments	LA (Misra et al. 2014; Kumar et al. 2015a), LA (Toffolo et al. 2018)
Wireless mesh networks	LA (Parvanak et al. 2018), (Beheshtifard and Meybodi 2018),
Wireless sensor networks	LA (Han and Li 2019), ICLA (Rezvanian et al. 2018c), FALA (Javadi et al. 2018), DLA (Mostafaei 2018), ICLA (Mostafaei and Obaidat 2018a), DLA (Mostafaei and Obaidat 2018b), GLA (Rahmani et al. 2018)

CLA is a *CA* in which some *LA*s are assigned to every cell. Each *LA* residing in a particular cell determines its action (state) based on its action probability vector. Like *CA*, there is a local rule that the *CLA* operates under. The local rule of the *CLA* and the actions selected by the neighboring *LA*s of any particular *LA* determine the reinforcement signal to that *LA*. The neighboring *LA*s (cells) of any particular *LA* (cell) constitute the local environment of that *LA* (cell). The local environment of an *LA* (cell) is non-stationary due to the fact that the action probability vectors of the neighboring *LA*s vary during the evolution of the *CLA*. The operation of a *CLA* could be described as the following steps (Fig. 1.4): At the first step, the internal state of every cell is determined on the basis of the action probability vectors of the *LA*s residing in that cell. In the second step, the local rule of the *CLA* determines the reinforcement signal to each *LA* residing in that cell. Finally, each *LA* updates its action probability vector based on the supplied reinforcement signal and the chosen action. This process continues until the desired result is obtained.

CLA can be either *synchronous* or *asynchronous*. In a synchronous *CLA*, *LA*s in all cells are activated at the same time synchronously using a global clock. In contrast, in an asynchronous *CLA* (*ACLA*) (Beigy and Meybodi 2008), *LA*s in different cells are activated asynchronously. The *LA*s may be activated in either a time-driven or step-driven manner. In a time-driven *ACLA*, each cell is assumed to have an internal clock that wakes up the *LA*s associated with that cell. In a step-driven *ACLA*, a cell is selected for activation in either a random or a fixed sequence. From another point

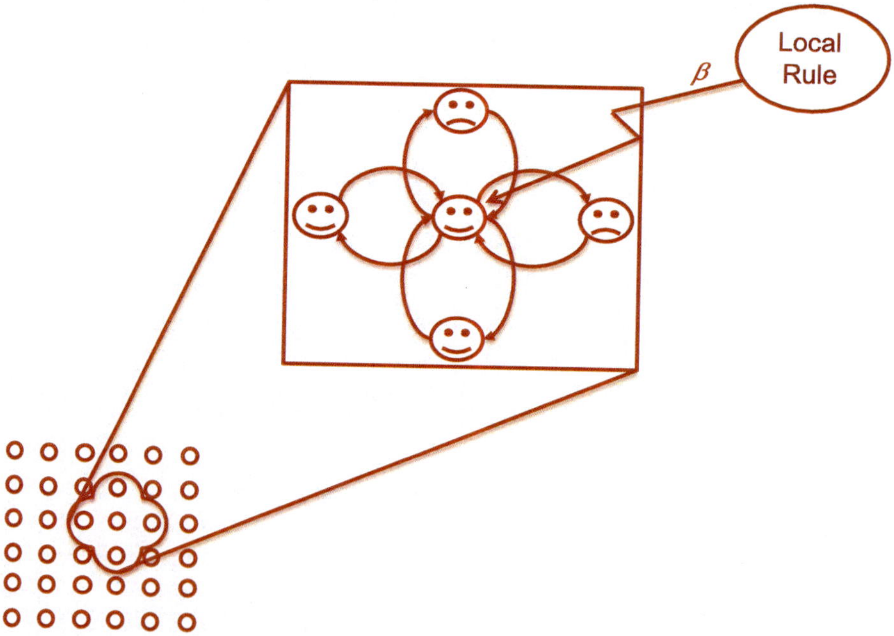

Fig. 1.4 Operation of the cellular learning automaton (CLA)

of view, *CLA* can be either *close* or *open*. In a close *CLA*, the action selection of any particular *LA* in the next iteration of its evolution only depends on the state of the local environment of that *LA* (actions of its neighboring *LA*s).

In contrast, in an open *CLA* (Beigy and Meybodi 2007), this depends on not only the local environment but also on the external environments. In (Beigy and Meybodi 2010), a new type of *CLA*, called *CLA* with multiple *LA*s in each cell, has been introduced. This model is suitable for applications such as channel assignment in cellular networks, in which it is needed that each cell is equipped with multiple *LA*s. In (Beigy and Meybodi 2004), a mathematical framework for studying the behavior of the *CLA* has been introduced. It was shown that, for a class of local rules called commutative rules, different models of *CLA* converge to a globally stable state (Beigy and Meybodi 2004, 2007, 2008, 2010).

Definition 1.4 a *d*-dimensional cellular learning automaton is a framework (Beigy and Meybodi 2004), $\mathcal{A} = \left(Z^d, N, \Phi, A, \mathcal{F}\right)$, where

- $\mathrm{Z^d}$ presents the lattice of d-tuples of integer numbers.
- $N = \{\bar{x}_1, \bar{x}_2, \ldots, \bar{x}_m\}$ is a finite subset of Z^d that is called neighborhood vector, where $\bar{x}_i \in Z^d$.
- Φ denotes the finite set of states. φ_i presents the state of the cell c_i.
- A presents the set of LAs residing in the cells of CLA.
- $F^i : \underline{\Phi_i} \to \underline{\beta}$ defines the local rule of the CLA for each cell c_i, where $\underline{\beta}$ is the set of possible values for the reinforcement signal, and it calculates the reinforcement signal for each LA using the chosen actions of neighboring LAs.

In what follows, a CLA with n cells and neighborhood function $\overline{N}(i)$ is considered. A learning automaton denoted by A_i, which has a finite action set $\underline{\alpha}_i$ is associated with cell i (for $i = 1, 2, \ldots, n$) of CLA. Let cardinality of $\underline{\alpha}_i$ be m_i and the state of CLA represented by $\underline{p} = (\underline{p}'_1, \underline{p}'_2, \ldots, \underline{p}'_n)'$, where $\underline{p}_i = (p_{i1}, \ldots, p_{im_i})'$ is the action probability vector of A_i. The local environment for each learning automaton is the learning automata residing in its neighboring cells. From the repeated application of simple local rules and simple learning algorithms, the global behavior of CLA can be very complicated.

The operation of CLA takes place as the following iterations. At iteration k, each learning automaton chooses an action. Let $\alpha_i \in \underline{\alpha}_i$ be the action chosen by A_i. Then all learning automata receive a reinforcement signal. Let $\beta_i \in \underline{\beta}$ be the reinforcement signal received by A_i. This reinforcement signal is produced by the application of the local rule $\mathcal{F}^i(\alpha_{i+\bar{x}_1}, \alpha_{i+\bar{x}_2}, \ldots, \alpha_{i+\bar{x}_{\bar{m}}}) \to \underline{\beta}$. The higher value of β_i means that the chosen action of A_i is more rewarded. Since each set $\underline{\alpha}_i$ is finite, rule $\mathcal{F}^i(\alpha_{i+\bar{x}_1}, \alpha_{i+\bar{x}_2}, \ldots, \alpha_{i+\bar{x}_{\bar{m}}}) \to \underline{\beta}$ can be represented by a hyper matrix of dimensions $m_1 \times m_2 \times \ldots \times m_{\bar{m}}$. These n hyper matrices together constitute the rule of CLA.

When all of these n hyper matrices are equal, the rule is uniform; otherwise, the rule is nonuniform. For the sake of simplicity in our presentation, the rule $\mathcal{F}^i(\alpha_{i+\bar{x}_1}, \alpha_{i+\bar{x}_2}, \ldots, \alpha_{i+\bar{x}_{\bar{m}}})$ denoted by $\mathcal{F}^i(\alpha_1, \alpha_2, \ldots, \alpha_{\bar{m}})$. Based on set $\underline{\beta}$, the CLA can be classified into three groups: *P*-model, *Q*-model, and *S*-model Cellular learning automata. When $\underline{\beta} = \{0, 1\}$, we refer to CLA as *P-model cellular learning automata,* when $\underline{\beta} = \{b_1, \ldots, b_l\}$, (for $l < \infty$), we refer to CLA as *Q-model cellular learning automata,* and when $\underline{\beta} = [b_1, b_2]$, we refer to CLA as *S-model cellular learning automata.* If learning automaton A_i uses learning algorithm L_i, we denote CLA by the $CLA(L_1, \ldots, L_n)$. If $L_i = L$ for all $i = 1, 2, \ldots, n$, then we denote the CLA by the *CLA(L).* In the following, some definitions and notations are presented.

Definition 1.5 A configuration of CLA is a map $\mathcal{K}: Z^d \to \underline{p}$ that associates an action probability vector with every cell. We will denote the set of all configurations of $\mathcal{A}$ by $\mathcal{K}(\mathcal{A})$ or simply $\mathcal{K}$.

The application of the local rule to every cell allows transforming a configuration into a new one.

Definition 1.6 The global behavior of a CLA is a mapping $\mathcal{G} : \mathcal{K} \to \mathcal{K}$, that describes the dynamics of CLA.

Definition 2.11 The evolution of CLA from a given initial configuration $\underline{p}(0) \in \mathcal{K}$, is a sequence of configurations $\left\{\underline{p}(k)\right\}_{k \geq 0}$, such that $\underline{p}(k+1) = \mathcal{G}\left(\underline{p}(k)\right)$.

CLA may be described as a network of LA assigned to the nodes of a graph (usually a finite and regular lattice). Each node is connected to a set of nodes (its neighbors), and each LA updates its action probability vector at discrete time instants, using a learning algorithm and a rule which depends on its action and that of its neighbors.

Definition 1.7 A CLA is called synchronous if all the learning automata are activated at the same time and asynchronous if at a given time only some LAs are activated.

Definition 1.8 A CLA is called uniform if the neighborhood function and the local rule are the same for all cells of CLA.

Definition 1.9 A rule is called uniform if the rule for all cells is the same; otherwise, it is called nonuniform.

Definition 1.10 A configuration p is called deterministic if the action probability vector of each learning automaton is a unit vector; otherwise, it is called probabilistic. Hence, the set of all deterministic configurations, $\mathcal{K}^*$, set of probabilistic configurations, $\mathcal{K}$, in CLA are

$$\mathcal{K}^* = \left\{ \underline{p} | \underline{p} = (\underline{p}'_1, \underline{p}'_2, \ldots, \underline{p}'_n)', \underline{p}_i = (p_{i1}, \ldots, p_{im_i})', p_{iy} \right.$$
$$\left. = 0 \text{or} 1 \quad \forall y, i, \sum_y p_{iy} = 1 \forall i \right\}$$

and

$$\mathcal{K} = \left\{ \underline{p} | \underline{p} = (\underline{p}'_1, \underline{p}'_2, \ldots, \underline{p}'_n)', \underline{p}_i \right.$$

$$= (p_{i1}, \ldots, p_{im_i})', 0 \le p_{iy} \le 1 \quad \forall y, i, \sum_y p_{iy} = 1 \forall i \Bigg\},$$

Respectively

In the rest of this section, we briefly review the classifications of CLA:

- **Static CLA versus Dynamic CLA**: in static CLAs, the cellular structure of the CLA remains fixed during the evolution of the CLA while in dynamic CLAs one of its aspects such as structure, local rule or neighborhood radius may vary with time (Esnaashari and Meybodi 2011, 2013).
- **Open CLA versus Close CLA**: in close CLAs, the action of each LA depends on the neighboring cells, whereas in open CLAs, the action of each LA depends on the neighboring cells, a global environment that influences all cells, and an exclusive environment for each particular cell (Beigy and Meybodi 2007; Saghiri and Meybodi 2017b).
- **Asynchronous CLA versus Synchronous CLA**: In synchronous CLA, all cells use their local rules at the same time. This model assumes that there is an external clock which triggers synchronous events for the cells. In asynchronous CLA, at a given time, only some cells are activated, and the state of the rest of the cells remains unchanged (Beigy and Meybodi 2007). The LAs may be activated in either time-driven where each cell is assumed to have an internal clock that wakes up the LA associated with that cell or in step-driven where a cell is selected in a fixed or random sequence.
- **Regular CLA versus Irregular CLA**: in regular CLAs the structure of CLA is represented as a lattice of d-tuples of integer numbers while in Irregular CLA (ICLA) the structure regularity assumption is replaced with an undirected graph (Ghavipour and Meybodi 2017; Vahidipour et al. 2017a; Esnaashari and Meybodi 2018).
- **CLAs with one LAs in each cell versus CLAs with multiple LAs in each cell**: in conventional CALAs, each cell equipped with one LA, while in CLAs with multiple LAs in each cell, each cell is equipped with multiple LAs (Beigy and Meybodi 2010).
- **CLAs with a fixed number of LAs in each cell versus CLAs with varying number of LAs in each cell**: in conventional CALAs, the number of LAs in each cell remains fixed during the evolution of CLA, while in CLAs with varying number of LAs in each cell, the number of LAs of each cell changes over time (Saghiri and Meybodi 2017b).
- **CLAs with fixed structure LAs versus CLAs with variable structure LAs**: since LAs can be classified into two leading families; fixed and variable structure (Kumpati and Narendra 1989; Thathachar and Sastry 2004). In CLAs with fixed structure LAs, constituting LAs are of fixed structure type, whereas in CLAs with variable structure LAs, LAs are of variable structure type.

Up to now, various CLA models (Rezvanian et al. 2018d) such as open CLA (Beigy and Meybodi 2007), asynchronous CLA (Beigy and Meybodi 2008), irregular CLA

(Esnaashari and Meybodi 2008), associative CLA (Ahangaran et al. 2017), dynamic irregular CLA (Esnaashari and Meybodi 2018), asynchronous dynamic CLA (Saghiri and Meybodi 2018b), and wavefront CLA (Rezvanian et al. 2019f) are developed and successfully applied on different application domains. We briefly introduce these modes as following subsections.

1.4.1 Open Synchronous Cellular Learning Automata (OCLA)

CLA studied so far are closed, because they do not take into account the interaction between the CLA and the external environments. In this section, a new class of CLA called *open CLA*, which was first introduced by Beigy et al. (Beigy and Meybodi 2007), is presented. In OSCLA, the evolution of CLA is influenced by external environments. Two types of environments can be considered in the open CLA: global environment and exclusive environment. Each CLA has one global environment that influences all cells and an exclusive environment for each particular cell.

The operation of open CLA takes place as iterations of the following steps. At iteration *k*, each learning automaton chooses one of its actions. Let *on* be the action chosen by learning automaton A_i. The actions of all learning automata are applied to their corresponding local environment (neighboring learning automata) as well as the global environment and their corresponding exclusive environment. Then all learning automata receive their reinforcement signal, which is a combination of the responses from local, global, and exclusive environments. These responses are combined using the local rule. Finally, all learning automata update their action probability vectors based on the received reinforcement signal. Note that the local environment for each learning automaton is non-stationary, while global and exclusive environments may be stationary or non-stationary. We now present the convergence result for the open CLA, which ensures convergence to one compatible configuration if the CLA has more than one compatible configurations.

Definition 1.11 A d-dimensional open cellular learning automaton is a structure $\mathcal{A} = \left(Z^d, \Phi, A, E^G, E^E, N, \mathcal{F}\right)$, where

- Z^d is a lattice of d-tuples of integer numbers.
- $\underline{\Phi}$ is a finite set of states.
- A is the set of LAs, each of which is assigned to each cell of the CA.
- E^G is the global environment.
- $E^E = \left\{E_1^E, E_2^E, \ldots, E_n^E\right\}$ is the set of exclusive environments, where E_i^E is the exclusive environment for cell *i*.
- $N = \{\bar{x}_1, \bar{x}_2, \ldots, \bar{x}_{\bar{m}}\}$ is a finite subset of Z^d called neighborhood vector, where $\bar{x}_i \in Z^d$.

- $\mathcal{F} : \underline{\Phi}^{\bar{m}} \times \underline{O(G)} \times \underline{O(E)} \rightarrow \underline{\beta}$ is the local rule of the cellular learning automata, where $\underline{O(G)}$ and $\underline{O(E)}$ are the set of signals of global and exclusive environments, respectively.

In what follows, we consider an open CLA with n cells and neighborhood function $\overline{N}(i)$. A learning automaton denoted by A_i, which has a finite action set $\underline{\alpha}_i$ is associated with cell i (for $i = 1, 2, \ldots, n$) of the open CLA. Let cardinality of $\underline{\alpha}_i$ be m_i and the state of the open CLA represented by $\underline{p} = (\underline{p}_1', \underline{p}_2', \ldots, \underline{p}_n')'$, where $\underline{p}_i = (p_{i1}, \ldots, p_{im_i})'$ is the action probability vector of A_i. The local environment for each learning automaton is the learning automata residing in its neighboring cells. From the repeated application of simple local rules and simple learning algorithms, the global behavior of CLA can be very complicated.

The operation of open synchronous CLA takes place as iterations of the following steps. At iteration *k*, each learning automaton chooses one of its actions. Let α_i be the action chosen by learning automaton A_i. The actions of all learning automata are applied to their corresponding local environment (neighboring learning automata) as well as the global environment and their corresponding exclusive environments. Each environment produces a signal. These signals are then used by the local rule to generate a reinforcement signal to the learning automaton residing in every cell. Finally, all learning automata update their action probability vectors based on the received reinforcement signal. Note that the local environment for each learning automaton is non-stationary, while global and exclusive environments may be stationary or non-stationary. In this study, we assume that global and exclusive environments are stationary. We now present the convergence result for the open synchronous CLA. The result asserts that open synchronous CLA converges to one of its compatible configurations. Based on set $\underline{\beta}$, the OSCLA can be classified into three groups: *P*-model, *Q*-model, and *S*-model Cellular learning automata. When $\underline{\beta} = \{0, 1\}$, we refer to OSCLA as *P-model cellular learning automata,* when $\underline{\beta} = \{b_1, \ldots, b_l\}$, (for $l < \infty$), we refer to OSCLA as *Q-model cellular learning automata,* and when $\underline{\beta} = [b_1, b_2]$, we refer to OSCLA as *S-model cellular learning automata.* If learning automaton A_i uses learning algorithm L_i, we denote OSCLA by the $OSCLA(L_1, \ldots, L_n)$. If $L_i = L$ for all $i = 1, 2, \ldots, n$, then we denote the OSCLA by the *CLA(L).*

1.4.2 Asynchronous Cellular Learning Automata (ACLA)

A cellular learning automaton is called asynchronous if, at a given time, only some LAs are activated independently from each other, rather than all together in parallel. The asynchronous CLA (ACLA) (Beigy and Meybodi 2008) requires the specification of an order in which the learning automata are activated. The learning automata may be activated in either a *time-driven* or *step-driven* manner. In time-driven asynchronous CLA, each cell is assumed to have an internal clock that wakes up the learning automaton associated with that cell while in step-driven asynchronous CLA;

a cell is selected in a fixed sequence or at random. In other words, in step-driven activation methods, an algorithm determines the order of activation of the learning automata.

In contrast, in the time-driven activation methods, an algorithm assigns a specific point in time to every learning automaton at that time, this learning automaton will be activated next. In the ACLA, the trajectory starting from a given initial configuration, in general, depends on the activation order of learning automata. The asynchronous CLA in which cells are selected randomly is of more interest to us because of its application to mobile-cellular networks. Formally a *d*—dimensional asynchronous step-driven CLA is given below.

Definition 1.12 A *d*-dimensional asynchronous step-driven cellular learning automata is a structure $\mathcal{A} = \left(Z^d, \Phi, A, N, \mathcal{F}, \underline{\rho}\right)$, where

- Z^d is a lattice of d-tuples of integer numbers.
- $\underline{\Phi}$ is a finite set of states.
- A is the set of LAs, each of which is assigned to each cell of the CA.
- $N = \{\bar{x}_1, \bar{x}_2, \ldots, \bar{x}_{\bar{m}}\}$ is a finite subset of Z^d called neighborhood vector, where $\bar{x}_i \in Z^d$.
- $\mathcal{F} : \underline{\Phi}^{\bar{m}} \to \underline{\beta}$ is the local rule of the cellular learning automata, where $\underline{\beta}$ is the set of values that the reinforcement signal can take.
- $\underline{\rho}$ is an *n*-dimensional vector called activation probability vector, where ρ_i is the probability that the LA in cell i (for $i = 1, 2, \ldots, n$) to be activated in each stage.

In what follows, we consider CLA with n cells and neighborhood function $\overline{N}(i)$. A learning automaton denoted by A_i, which has a finite action set $\underline{\alpha}_i$ is associated with cell i (for $i = 1, 2, \ldots, n$) of CLA. Let cardinality of $\underline{\alpha}_i$ be m_i and the state of CLA represented by $\underline{p} = (\underline{p}_1', \underline{p}_2', \ldots, \underline{p}_n')'$, where $\underline{p}_i = (p_{i1}, \ldots, p_{im_i})'$ is the action probability vector of A_i. The local environment for each learning automaton is the learning automata residing in its neighboring cells. From the repeated application of simple local rules and simple learning algorithms, the global behavior of CLA can be very complicated.

The operation of asynchronous CLA (ACLA) takes place as the following iterations. At iteration *k*, each learning automaton A_i is activated with probability ρ_i Moreover, activated learning automata choose one of their actions. The activated automata use their current actions to execute the rule (computing the reinforcement signal). The actions of neighboring cells of an activated cell are their most recently selected actions. Let $\alpha_i \in \underline{\alpha}_i$ and $\beta_i \in \underline{\beta}$ be the action is chosen by the activated and the reinforcement signal received by A_i, respectively. This reinforcement signal is produced by the application of local rule $\mathcal{F}^i(\alpha_{i+\bar{x}_1}, \alpha_{i+\bar{x}_2}, \ldots, \alpha_{i+\bar{x}_{\bar{m}}}) \to \underline{\beta}$, where $\mathcal{F}^i$ is the local rule of cell *i*. The higher value of β_i means that the chosen action of A_i is more rewarded. Finally, activated learning automata update their action probability vectors, and the process repeats.

Based on set $\underline{\beta}$, the CLA can be classified into three groups: *P*-model, *Q*-model, and *S*-model Cellular learning automata. When $\underline{\beta} = \{0, 1\}$, we refer to ACLA as

P-model cellular learning automata, when $\underline{\beta} = \{b_1, \ldots, b_l\}$, (for $l < \infty$), we refer to CLA as *Q-model cellular learning automata,* and when $\underline{\beta} = [b_1, b_2]$, we refer to CLA as *S-model cellular learning automata.* If learning automaton A_i uses learning algorithm L_i, we denote CLA by the $CLA(L_1, \ldots, L_n)$. If $L_i = L$ for all $i = 1, 2, \ldots, n$, then we denote the CLA by the *CLA(L).*

Since each set $\underline{\alpha}_i$ is finite, the local rule $\mathcal{F}^i(\alpha_{i+\bar{x}_1}, \alpha_{i+\bar{x}_2}, \ldots, \alpha_{i+\bar{x}_{\bar{m}}}) \to \underline{\beta}$ can be represented by a hyper matrix of dimensions $m_1 \times m_2 \times \cdots \times m_{\bar{m}}$. These n hyper matrices together constitute the rule of ACLA. When all hyper matrices are equal, the rule is uniform; otherwise, the rule is non-uniform. For the sake of simplicity in our presentation, local rule $\mathcal{F}^i(\alpha_{i+\bar{x}_1}, \alpha_{i+\bar{x}_2}, \ldots, \alpha_{i+\bar{x}_{\bar{m}}})$ is denoted by $\mathcal{F}^i(\alpha_1, \alpha_2, \ldots, \alpha_{\bar{m}})$.

1.4.3 *Synchronous CLA with Multiple LA in Each Cell (SLA-MLA)*

In some applications, there is a need for a model of CLA in which each cell is equipped with several LAs, for instance, the channel assignment in mobile cellular networks for which we need to have several decision variables, each of which can be adapted by a learning automaton. Such a CLA, which was first proposed by Beigy and Meybodi (Beigy and Meybodi 2010), is called CLA with multiple learning automata in each cell. In the new model of CLA, LAs may be activated synchronously or asynchronously.

In synchronous CLA with multiple LA in each cell, several LAs are assigned to each cell of CLA, which are activated synchronously. Without loss of generality and for the sake of simplicity, assume that each cell containing s LAs. The operation of a synchronous CLA with multiple learning automata in each cell can be described as follows: At the first step, the internal state of a cell is specified. The state of every cell is determined based on the action probability vectors of all LAs residing in that cell. The initial value of this state may be chosen based on the experience or at random. In the second step, the rule of the CLA determines the reinforcement signal to each LA residing in each cell. The environment for every LA is the set of all LAs in that cell and neighboring cells. Finally, each LA updates its action probability vector based on the supplied reinforcement signal and the chosen action by the cell. This process continues until the desired result is obtained.

Definition 1.13 A d-dimensional synchronous cellular learning automaton with n cells and s LAs in each cell is a structure $\mathcal{A} = \left(Z^d, \Phi, A, N, \mathcal{F}\right)$, where

- Z^d is a lattice of d-tuples of integer numbers.
- $\underline{\Phi}$ is a finite set of states.
- A is the set of LAs assigned to CLA, where A^i is the set of Las assigned to cell i.
- $N = \{\bar{x}_1, \bar{x}_2, \ldots, \bar{x}_{\bar{m}}\}$ is a finite subset of Z^d called neighborhood vector, where $\bar{x}_i \in Z^d$.

- $\mathcal{F}: \underline{\Phi}^{s \times \bar{m}} \rightarrow \underline{\beta}$ is the local rule of the cellular learning automata, where $\underline{\beta}$ is the set of values that can be taken by the reinforcement signal.

The local rule computes the reinforcement signal for each LA based on the actions selected by the neighboring LAs. We assume that there exists a neighborhood function $\overline{N}(u)$ that maps a cell u to the set of its neighbors. Let $\underline{\alpha}_i$ be the set of actions that are chosen by all learning automata in cell i. Hence, the local rule is represented by function $\mathcal{F}^i(\underline{\alpha}_{i+\bar{x}_1}, \underline{\alpha}_{i+\bar{x}_2}, \ldots, \underline{\alpha}_{i+\bar{x}_{\bar{m}}}) \rightarrow \underline{\beta}$. In CLA with multiple LAs in each cell, the configuration of CLA is defined as the action probability vectors of all learning automata. In this model, all learning automata residing in the neighboring cells constitute the local environment of each learning automaton. A configuration is called compatible if no LA in CLA has any reason to change its action. The CLA will be denoted by $CLA(\underline{L}_1, \ldots, \underline{L}_n)$, where $\underline{L}_i = \{L_{i1}, L_{i1}, \ldots, L_{is}\}$ and L_{ij} is the jth learning automaton in cell i. The operation of CLA containing s learning automata in each cell is the same as the CLA with one learning automaton in each cell.

1.4.4 Asynchronous CLA with Multiple LA in Each Cell (ACLA-MLA)

In this model of cellular learning automata, several learning automata are assigned to each cell of cellular learning automata, which are activated asynchronously. Without loss of generality and for the sake of simplicity, assume that each cell containing s learning automata.

Definition 1.14 A d-dimensional step-driven asynchronous cellular learning automata with n cells and s LAs in each cell is a structure $\mathcal{A} = \left(Z^d, \Phi, A, N, \mathcal{F}, \underline{\rho}\right)$, where

- Z^d is a lattice of d-tuples of integer numbers.
- $\underline{\Phi}$ is a finite set of states.
- A is the set of LAs assigned to CLA, where A^i is the set of Las assigned to cell i.
- $N = \{\bar{x}_1, \bar{x}_2, \ldots, \bar{x}_{\bar{m}}\}$ is a finite subset of Z^d called neighborhood vector, where $\bar{x}_i \in Z^d$.
- $\mathcal{F}: \underline{\Phi}^{s \times \bar{m}} \rightarrow \underline{\beta}$ is the local rule of the cellular learning automata, where $\underline{\beta}$ is the set of values that can be taken by the reinforcement signal.

- $\underline{\rho}$ is an $n \times s$-dimensional vector called activation probability vector, where ρ_{ij} is the probability that the LA_j in cell i (for $i = 1, 2, \ldots, n$ and $j = 1, 2, \ldots, s$) to be activated in each stage.

The operation of the asynchronous cellular learning automata with multiple learning automata in each cell can be described as follows: At iteration k, each learning automaton A_{ij} is activated with probability ρ_{ij} and the activated learning automata choose one of their actions. The activated automata use their current actions to execute the rule (computing the reinforcement signal). The actions of neighboring cells of an activated cell are their most recently selected actions. The reinforcement signal is produced by the application of local rule. Finally, activated learning automata update their action probability vectors, and the process repeats.

1.4.5 Continuous Cellular Learning Automata (CCLA)

In this chapter, we considered CLA in which each cell is assigned a finite action-set learning automaton (multiple learning automata). In some applications such as call admission control in cellular mobile networks using limited fractional guard channel policy, the state of each cell is a continuous variable. In such applications, we need to have CLA with continuous action-set learning automata in its cells. We call a CLA with continuous action-set learning automata in each cell as *continuous cellular learning automata* (CCLA). The continuous cellular learning automata can be defined formally as given below.

Definition 1.15 A d-dimensional continuous cellular Learning automaton is a structure $\mathcal{A} = \left(Z^d, \Phi, A, N, \mathcal{F}\right)$, where

- Z^d is a lattice of d-tuples of integer numbers.
- $\underline{\Phi} \in \Re$ is the set of states.
- A is the set of LAs, each of which is assigned to each cell of the CA.
- $N = \{\bar{x}_1, \bar{x}_2, \ldots, \bar{x}_{\bar{m}}\}$ is a finite subset of Z^d called neighborhood vector, where $\bar{x}_i \in Z^d$.
- $\mathcal{F} : \underline{\Phi}^{\bar{m}} \to \underline{\beta}$ is the local rule of the cellular learning automata, where $\underline{\beta}$ is the set of values that can be taken by the reinforcement signal.

The continuous cellular learning automata (CCLA) can be classified in the same manner as the CLA. For example, learning automata may be activated synchronously or asynchronously, or the CCLA may be closed or open

1.4.6 Irregular Cellular Learning Automata (ICLA)

Irregular cellular learning automata (ICLA) (Esnaashari and Meybodi 2008) is a generalization of traditional CLA, which removes the limitation of the rectangular grid structure, as depicted in Fig. 1.5. Such a generalization seems to be necessary since there are many applications such as graph-related applications, social networks, wireless sensor networks, and immune network systems, which cannot be modeled with regular grids (Esnaashari and Meybodi 2008, 2009; Rezapoor Mirsaleh and Meybodi 2016). An ICLA is considered as an undirected graph in which each node is a cell being equipped with a learning automaton, and the neighboring nodes of any particular node constitute the local environment of that cell. The *LA* residing

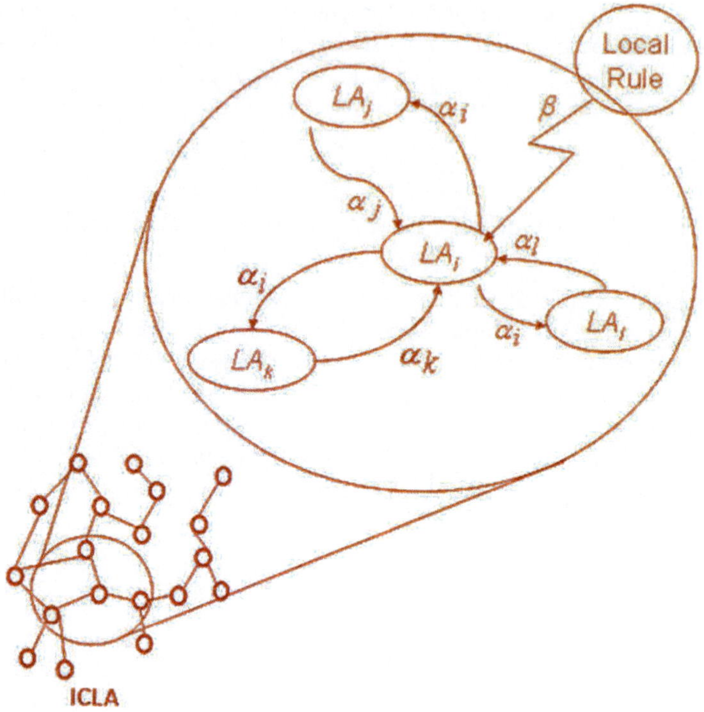

Fig. 1.5 An irregular cellular learning automaton (ICLA)

in a particular cell determines its state (action) according to its action probability vector. Like *CLA*, there is a rule that the *ICLA* operates under. The local rule of the *ICLA* and the actions selected by the neighboring *LA*s of any particular *LA* determine the reinforcement signal to that *LA*. The neighboring *LA*s of any particular *LA* constitute the local environment of that *LA*. The local environment of an *LA* is non-stationary because the action probability vectors of the neighboring *LA*s vary during the evolution of the *ICLA*.

The operation of the *ICLA* is similar to the operation of the *CLA*. In the first step, the internal state of each cell is specified based on the action probability vector of the *LA* residing in that cell. In the second step, the rule of the *ICLA* determines the reinforcement signal to the *LA* residing in each cell. Finally, each *LA* updates its action probability vector based on the supplied reinforcement signal and the internal state of the cell. This process continues until the desired result is obtained. Formally, an *ICLA* is defined as given below.

Definition 1.16 Irregular cellular learning automaton is a structure $\mathcal{A} = (G\langle E, V\rangle, F, A, \mathcal{F})$, where:

- G is an undirected graph, with V as the set of vertices (cells) and E as the set of edges (adjacency relations).

- F is a finite set of states and F_i represents the state of the cell c_i.
- A is the set of *LA*s, each of which is assigned to one cell of the *ICLA*.
- $\mathcal{F} : \underline{\varphi}_i \to \underline{\beta}$ is the local rule of the *ICLA* in each cell c_i, where $\underline{\varphi}_i = \{\varphi_j | \{i, j\} \in E\} \cup \{\varphi_i\}$ is the set of states of all neighbors of c_i and $\underline{\beta}$ is the set of values that the reinforcement signal can take. The local rule gives the reinforcement signal to each *LA* from the current actions selected by the neighboring *LA*s of that *LA*.

Note that in the definition of the *ICLA*, no explicit definition is given for the neighborhood of each cell. It is implicitly defined in the definition of the graph *G*.

In what follows, we consider *ICLA* with *N* cells. The learning automaton LA_i which has a finite action set $\underline{\alpha}_i$ is associated with cell c_i (for $i = 1, 2, \ldots, N$) of the *ICLA*. Let the cardinality of $\underline{\alpha}_i$ be m_i.

The operation of the *ICLA* takes place as the following iterations. At iteration k, each learning automaton selects an action. Let $\alpha_i \in \underline{\alpha}_i$ be the action selected by LA_i. Then all learning automata receive a reinforcement signal. Let $\beta_i \in \underline{\beta}$ be the reinforcement signal received by LA_i. This reinforcement signal is produced by the application of the local rule $\mathcal{F}^i(\varphi_i) \to \underline{\beta}$. Higher values of β_i mean that the selected action of LA_i will receive higher penalties. Then, each LA_i updates its action probability vector based on the supplied reinforcement signal and its selected action α_i.

Like *CLA*, *ICLA* can be either synchronous or asynchronous, and an asynchronous *ICLA* can be either time-driven or step-driven.

Definition 1.17 A configuration of the *ICLA* at step k is denoted by $\underline{p}(k) = \left(\underline{p}_1, \underline{p}_2, \ldots, \underline{p}_N\right)^T$, where $\underline{p}_i$ is the action probability vector of the learning automaton LA_i and T denotes the transpose operator.

Definition 1.18 A configuration $\underline{p}$ is called deterministic if the action probability vector of each learning automaton is a unit vector; otherwise, it is called probabilistic. Hence, the set of all deterministic configurations, K^*, and the set of probabilistic configurations, K, in *ICLA* are

$$K^* = \left\{ \underline{p} \left| \begin{array}{l} \underline{p} = \left(\underline{p}_1, \underline{p}_2, \ldots, \underline{p}_N\right)^T, \underline{p}_i = \left(p_{i1}, \ldots, p_{im_i}\right)^T, \\ p_{iy} = 0 \ or \ 1 \ \forall y, i, \sum_y p_{iy} = 1 \ \forall i \end{array} \right. \right\}, \quad (1.20)$$

and

$$K = \left\{ \underline{p} \left| \begin{array}{l} \underline{p} = \left(\underline{p}_1, \underline{p}_2, \ldots, \underline{p}_N\right)^T, \underline{p}_i = \left(p_{i1}, \ldots, p_{im_i}\right)^T, \\ 0 \le p_{iy} \le 1 \ \forall y, i, \sum_y p_{iy} = 1 \ \forall i \end{array} \right. \right\}, \quad (1.21)$$

respectively.

Lemma 1.1 *K is the convex hull of* K^*.

Proof Proof of this lemma is given in (Beigy and Meybodi 2004). ■

The application of the local rule to every cell allows transforming a configuration into a new one.

Definition 1.19 The global behavior of an *ICLA* is a mapping $\mathcal{G}:\mathcal{K} \rightarrow \mathcal{K}$ that describes the dynamics of the *ICLA*. The evolution of the *ICLA* from a given initial configuration $\underline{p}(0) \in K$ is a sequence of configurations $\left\{\underline{p}(k)\right\}_{k\geq 0}$, such that $\underline{p}(k+1) = G\left(\underline{p}(k)\right)$.

Definition 1.20 Neighborhood set of any particular LA_i, denoted by $N(i)$, is defined as the set of all learning automata residing in the adjacent cells of the cell c_i, that is,

$$N(i) = \left\{LA_j | \{i, j\} \in E\right\}. \tag{1.22}$$

Let N_i be the cardinality of $N(i)$.

Definition 1.21 The average penalty for action r of learning automaton LA_i for configuration $\underline{p} \in K$ is defined as

$$d_{ir}\left(\underline{p}\right) = E\left[\beta_i \middle| \underline{p}, \alpha_i = r\right] = \sum_{y_{j_1},\ldots,y_{jN_i}} F^i\left(y_{j_1}, \ldots, y_{jN_i}, r\right) \prod_{LA_l \in N(i)} p_{ly_{j_l}}, \tag{1.23}$$

and the average penalty for the learning automaton LA_i is defined as

$$D_i\left(\underline{p}\right) = E\left[\beta_i \middle| \underline{p}\right] = \sum_y d_{iy}\left(\underline{p}\right) p_{iy}. \tag{1.24}$$

The above definition implies that if the learning automaton LA_j is not a neighboring learning automaton for LA_i, then $d_{ir}\left(\underline{p}\right)$ does not depend on $\underline{p}_j$. We assume that $d_{ir}\left(\underline{p}\right) \neq 0$ for all i, r and $\underline{p}$, that is, in any configuration, any action has a non-zero chance of receiving a penalty.

Definition 1.22 The total average penalty for the *ICLA* at configuration $\underline{p} \in K$ is the sum of the average penalties for all learning automata in the *ICLA*, that is,

$$D\left(\underline{p}\right) = \sum_i D_i\left(\underline{p}\right). \tag{1.25}$$

1.4.7 Irregular Open Cellular Learning Automata (IOCLA)

An irregular open cellular learning automaton (IOCLA) is a new class of ICLA in which local and external environments influence CLA's evolution. In IOCLA, the neighboring learning automata of any cell constitute its local environment. Two types of external environments can be considered in the IOCLA: global environment and exclusive environment. Each IOCLA has one global environment that influences all cells and an exclusive environment for each particular cell, influencing the evolution of that cell (Rezapoor Mirsaleh and Meybodi 2016). Figure 1.6. shows the interconnection of a typical cell ci and its various types of environments.

The operation of IOCLA could be described as follows: At iteration t, each learning automaton chooses one of its actions. Let α_i be the action chosen by learning automaton LA_i. The actions of all learning automata are applied to their corresponding local environments (neighboring learning automata) as well as the global environment and their corresponding exclusive environments. Each environment produces a signal, which is used by the local rule to generate a reinforcement signal to the learning automaton residing in every cell. Let β_i^l, β_i^e, and β_i^g be the reinforcement signals generated by local, exclusive and global environments of cell c_i, respectively. The local rule of IOCLA determines the reinforcement signal to each learning automaton LA_i using β_i^l, β_i^e, and β_i^g. Finally, all learning automata update their action probability vectors in the basis of supplied reinforcement signal by local rule (β_i^l). This process continues until the desired result is obtained.

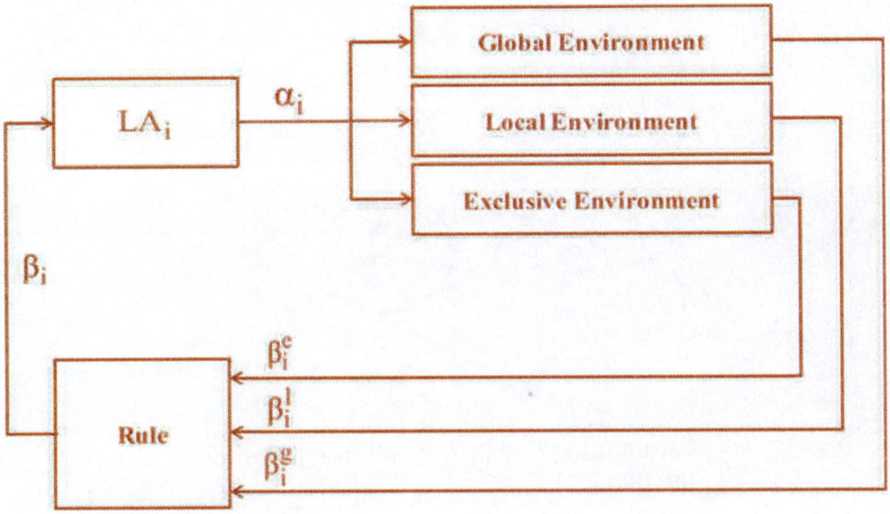

Fig. 1.6 The interconnection of a typical cell in IOCLA with its various environments

1.4.8 Dynamic Irregular Cellular Learning Automata (DICLA)

A *DICLA* is defined as an undirected graph in which each vertex represents a cell, and a learning automaton is assigned to every cell (vertex) (Esnaashari and Meybodi 2018). A finite set of interests is defined for *DICLA*. For each cell of *DICLA*, a tendency vector is defined whose $jtextth$ element shows the degree of tendency of that cell to the jth interest. In *DICLA*, the state of each cell consists of two parts; the action selected by the learning automaton and the tendency vector. Two cells are neighbors in *DICLA* if the distance between their tendency vectors is smaller than or equal to the neighborhood radius. Figure 1.7 gives a schematic of DICLA.

Like *ICLA*, there is a local rule that *DICLA* operates under. The local rule of *DICLA*, the actions selected by the neighboring *LA*s of any particular learning automaton LA_i determine the followings: (1) the reinforcement signal to the learning automaton LA_i, and (2) the restructuring signal to the cell in which LA_i resides. Restructuring signal is used to update the tendency vector of the cell. Dynamicity of *DICLA* is the result of modifications made to the tendency vectors of its constituting cells. Gives a schematic of *DICLA*. A *DICLA* is formally defined below.

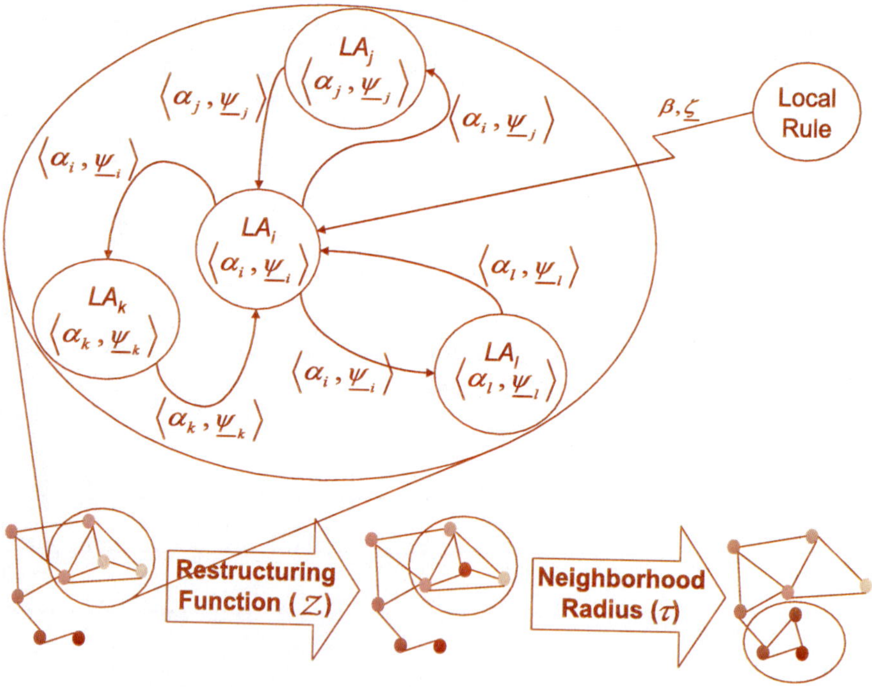

Fig. 1.7 Dynamic Irregular cellular learning automaton (DICLA)

Definition 1.23 Dynamic irregular cellular learning automaton is a structure $\mathcal{A} = \left(G\langle V, E\rangle, \Psi, A, \Phi, \left\langle \alpha, \underline{\psi} \right\rangle, \tau, \mathcal{F}, Z\right)$ where

- G is an undirected graph, with V as the set of vertices (cells) and E as the set of edges (adjacency relations).
- Ψ is a finite set of interests. The cardinality of Ψ is denoted by $|\Psi$.
- A is the set of learning automata, each of which is assigned to one cell of *DICLA*.
- $\Phi\left\langle \alpha, \underline{\psi} \right\rangle$ is the cell state. State of a cell $c_i(\varphi_i)$ consists of two parts; (1) α_i which is the action selected by the learning automaton of that cell, and (2) A vector $\underline{\psi}_i = \left(\psi_{i1}, \psi_{i2}, \ldots, \psi_{i|\psi|}\right)^T$ called the tendency vector of the cell. Each element $\psi_{ik} \in [0, 1]$ in the tendency vector of the cell c_i shows the degree of the tendency of c_i to the interest $\psi_k \in \Psi$.
- τ is the neighborhood radius. Two cells c_i and c_j of *DICLA* are neighbors if $\left|\underline{\psi}_i - \underline{\psi}_j\right| \leq \tau$. In other words, two cells of *DICLA* are neighbors if the distance between their tendency vectors is smaller than or equal to τ.
- $\mathcal{F} : \underline{\varphi}_i \rightarrow \left\langle \underline{\beta}, [0, 1]^{|\Psi|} \right\rangle$ is the local rule of *DICLA* in each cell c_i, where $\underline{\varphi}_i = \left\{\varphi_j \middle| \left|\underline{\psi}_i - \underline{\psi}_j\right| \leq \right\} \cup \{\varphi_i\}$ is the set of states of all neighbors of c_i, $\underline{\beta}$ is the set of values that the reinforcement signal can take, and $[0, 1]^{|\Psi|}$ is a $\Psi|$-dimensional unit hypercube. From the current states of the neighboring cells of each cell c_i, local rule performs the followings: (1) gives the reinforcement signal to the learning automaton LA_i resides in c_i, and (2) produces a restructuring signal $\underline{\zeta}_i = \left(\zeta_{i1}, \zeta_{i2}, \ldots, \zeta_{i|\Psi|}\right)^T$ which is used to change the tendency vector of c_i. Each element ζ_{ij} of the restructuring signal is a scalar value within the close interval $[-1, 1]$.
- $Z : [0, 1]^{|\Psi|} \times [0, 1]^{|\Psi|} \rightarrow [0, 1]^{|\Psi|}$ is the restructuring function which modifies the tendency vector of the cell c_i using the restructuring signal produced by the local rule of the cell.

In what follows, we consider *DICLA* with N cells. The learning automaton LA_i which has a finite action set $\underline{\alpha}_i$ is associated with the cell c_i (for $i = 1, 2, \ldots, N$) of *DICLA*. Let the cardinality of $\underline{\alpha}_i$ be m_i.

The operation of *DICLA* takes place as the following iterations. At iteration k, each learning automaton chooses an action. Let $\alpha_i \in \underline{\alpha}_i$ be the action chosen by LA_i. Then, each *LA* receives a reinforcement signal. Let $\beta_i \in \underline{\beta}_i$ be the reinforcement signal received by LA_i. This reinforcement signal is produced by the application of the local rule $\mathcal{F} : \underline{\varphi}_i \rightarrow \underline{\beta}, [0, 1]^{|\Psi|}$. Higher values of β_i mean that the selected action of LA_i will receive higher penalties. Each *LA* updates its action probability vector based on the supplied reinforcement signal and the action chosen by the cell. Next, each cell c_i updates its tendency vector using the restructuring function $\underline{\psi}_i(k+1) = Z\left(\underline{\psi}_i(k), \underline{\zeta}_i(k)\right)$.

Like *ICLA*, *DICLA* can be either synchronous or asynchronous, and an asynchronous *DICLA* can be either time-driven or step-driven.

Note that most of the definitions used for steady-state analysis of the behavior of the *ICLA* are redefined here for the *DICLA* model.

Definition 1.24 A configuration of *DICLA* at step k is denoted by $\left\langle \underline{p}(k), \underline{\psi}(k) \right\rangle = \left\langle \left(\underline{p}_1, \ldots, \underline{p}_N\right)^T, \left(\underline{\psi}_1, \ldots, \underline{\psi}_N\right)^T \right\rangle$, where $\underline{p}_i$ is the action probability vector of the learning automaton LA_i, $\underline{\psi}_i$ is the tendency vector of the cell c_i, and T denotes the transpose operator.

Definition 1.25 A configuration $\left\langle \underline{p}, \underline{\psi} \right\rangle$ is called unit if the action probability vector of each learning automaton and the tendency vector of each cell are unit vectors; otherwise, it is called general. Hence, the set of all unit configurations, $\langle K^*, Y^* \rangle$, and the set of all general configurations, $\langle K, Y \rangle$, in *DICLA* are

$$\langle K^*, Y^* \rangle = \left\{ \left\langle \underline{p}, \underline{\psi} \right\rangle \middle| \begin{array}{l} \underline{p} = \left(\underline{p}_1, \ldots, \underline{p}_n\right)^T, \underline{p}_i = (p_{i1}, \ldots, p_{im_i})^T, p_{iy} = 0 \; or \; 1 \; \forall i, y, \sum_y p_{iy} = 1 \; \forall i, \\ \underline{\psi} = \left(\underline{\psi}_1, \ldots, \underline{\psi}_n\right)^T, \underline{\psi}_i = (\psi_{i1}, \ldots, \psi_{i|\psi|})^T, \psi_{ij} = 0 \; or \; 1 \; \forall i, j, \sum_j \psi_{ij} = 1 \; \forall i \end{array} \right\}, \tag{1.26}$$

and

$$\langle K, Y \rangle = \left\{ \left\langle \underline{p}, \underline{\psi} \right\rangle \middle| \begin{array}{l} \underline{p} = \left(\underline{p}_1, \ldots, \underline{p}_n\right)^T, \underline{p}_i = (p_{i1}, \ldots, p_{im_i})^T, 0 \le p_{iy} \le 1 \; \forall i, y, \sum_y p_{iy} = 1 \; \forall i, \\ \underline{\psi} = \left(\underline{\psi}_1, \ldots, \underline{\psi}_n\right)^T, \underline{\psi}_i = (\psi_{i1}, \ldots, \psi_{i|\psi|})^T, 0 \le \psi_{ij} \le 1 \; \forall i, j \end{array} \right\} \tag{1.27}$$

respectively. We refer to K as the probability space and to Y as the tendency space of *DICLA* hereafter.

Lemma 1.2 $\langle K, Y \rangle$ is the convex hull of $\langle K^*, Y^* \rangle$.

Proof Proof of this lemma is similar to the proof of lemma 1-1 given in (Beigy and Meybodi 2004). ■

The application of the local rule to every cell allows transforming a configuration into a new one. The local rule of *DICLA* (F) consists of two parts; reinforcement signal generator (F_β) and restructuring signal generator (F_ζ).

Definition 1.26 The global behavior of a *DICLA* is a mapping $\mathcal{G}: \langle \mathcal{K}, \mathcal{Y} \rangle \rightarrow \langle \mathcal{K}, \mathcal{Y} \rangle$ that describes the dynamics of *DICLA*. The evolution of *DICLA* from a given initial configuration $\left\langle \underline{p}(0), \underline{\psi}(0) \right\rangle \in \langle K, Y \langle \rangle \rangle$ is a sequence of configurations $\left\{ \left\langle \underline{p}(k), \underline{\psi}(k) \right\rangle \right\}_{k \ge 0}$, such that $\left\langle \underline{p}(k+1), \underline{\psi}(k+1) \right\rangle = G\left(\left\langle \underline{p}(k), \underline{\psi}(k) \right\rangle \right)$.

Definition 1.27 Neighborhood set of any particular LA_i in configuration $\langle \underline{p}, \underline{\psi} \rangle$, denoted by $N^{\underline{\psi}}(i)$, is defined as the set of all learning automata residing in the adjacent cells of the cell c_i, that is,

$$N^{\underline{\psi}}(i) = \left\{ LA_j \middle| \left\| \underline{\psi}_i - \underline{\psi}_j \right\| \le \tau \right\}. \tag{1.28}$$

Let $N_i^{\underline{\psi}}$ be the cardinality of $N^{\underline{\psi}}(i)$.

Definition 1.28 The average penalty for action r of the learning automaton LA_i for configuration $\langle \underline{p}, \underline{\psi} \rangle \in \langle K, Y\langle\rangle\rangle$ is defined as

$$\begin{aligned} d_{ir}^{\beta}\left(\langle \underline{p}, \underline{\psi} \rangle\right) &= E\left[\beta_i \middle| \langle \underline{p}, \underline{\psi} \rangle, \alpha_i = r\right] \\ &= \sum_{y_{h_1}^{\underline{\psi}}, \ldots, y_{h_{N_i^{\underline{\psi}}}}^{\underline{\psi}}} \left(F_{\beta}^{i}\left(y_{h_1}^{\underline{\psi}}, y_{h_2}^{\underline{\psi}}, \ldots, y_{h_{N_i^{\underline{\psi}}}}^{\underline{\psi}}, r \right) \prod_{\substack{LA_l \in N^{\underline{\psi}}(i) \\ l \ne i}} p_{l y_{h_l}} \right), \end{aligned} \tag{1.29}$$

and the average penalty for the learning automaton LA_i is defined as

$$D_i\left(\langle \underline{p}, \underline{\psi} \rangle\right) = E\left[\beta_i \middle| \langle \underline{p}, \underline{\psi} \rangle\right] = \sum_y d_{iy}^{\beta}\left(\langle \underline{p}, \underline{\psi} \rangle\right) p_{iy}. \tag{1.30}$$

The above definition implies that if the learning automaton LA_j is not a neighboring learning automaton for LA_i, then $d_{ir}^{\beta}\left(\langle \underline{p}, \underline{\psi} \rangle\right)$ does not depend on $\underline{p}_j$. We assume that $d_{ir}^{\beta}\left(\langle \underline{p}, \underline{\psi} \rangle\right) \ne 0$ for all i, r and $\langle \underline{p}, \underline{\psi} \rangle$, that is, in any configuration, any action has a non-zero chance of receiving a penalty.

Definition 1.29 The total average penalty for *DICLA* at configuration $\langle \underline{p}, \underline{\psi} \rangle \in \langle K, Y\langle\rangle\rangle$ is the sum of the average penalties for all learning automata in *DICLA*, that is,

$$D\left(\langle \underline{p}, \underline{\psi} \rangle\right) = \sum_i D_i\left(\langle \underline{p}, \underline{\psi} \rangle\right). \tag{1.31}$$

Definition 1.30 The average restructuring signal for interest j of the cell c_i for configuration $\langle \underline{p}, \underline{\psi} \rangle \in \langle K, Y\langle\rangle\rangle$ is defined as

$$d_{ij}^{\zeta}\left(\left\langle \underline{p}, \underline{\psi} \right\rangle\right) = \sum_{r} \sum_{y_{h_1}^{\underline{\psi}}, \ldots, y_{h_{N_i^{\underline{\psi}}}}^{\underline{\psi}}} \left(\mathcal{F}_{\zeta}^{i}\left(y_{h_1}^{\underline{\psi}}, y_{h_2}^{\underline{\psi}}, \ldots, y_{h_{N_i^{\underline{\psi}}}}^{\underline{\psi}}, r, \underline{\psi}_{*j} \right) \prod_{LA_l \in N^{\underline{\psi}}(i)} p_{l y_{hl}} \right), \tag{1.32}$$

where $\underline{\psi}_{*j} = \left(\psi_{1j}, \psi_{1j}, \ldots, \psi_{nj}\right)^T$.

The above definition implies that all interests are independent of each other since the restructuring signal generator (F_{ζ}) for any interest of any cell depends only on the tendency values of the neighboring cells to that interest.

1.4.9 Heterogeneous Dynamic Irregular Cellular Learning Automata (HDICLA)

As it was indicated, *DICLA* is suitable for modeling problems in the areas with dynamic structure, such as mobile ad hoc and sensor networks. *DICLA* has been successfully used in deploying a wireless sensor network throughout the sensor field using mobile sensor nodes (Esnaashari and Meybodi 2011). In some applications, different cells need different learning parameters. For example, consider the *k*-coverage problem in *WSN*s, where different sub-regions of the sensor field require different degrees of coverage. Here, each sensor node has to consider a value for the parameter *k*, which depends on the location of that node. *DICLA* cannot be used in such applications with heterogeneous learning parameters. Therefore, in this section, we propose heterogeneous dynamic irregular cellular learning automaton (*HDICLA*) as a generalization of *DICLA*, which can support heterogeneous learning parameters in different cells.

We define heterogeneous *DICLA* (*HDICLA*) as an undirected graph in which each vertex represents a cell, and a learning automaton is assigned to every cell (vertex). A finite set of interests and a finite set of attributes are defined for the *HDICLA*. In *HDICLA*, the state of each cell consists of the following three parts:

- *Selected action*: The action selected by the learning automaton residing in the cell.
- *Tendency vector*: The tendency vector of the cell is a vector whose *j*th element shows the degree of the tendency of the cell to the *j*th interest.
- *Attribute vector*: The attribute vector of the cell is a vector whose *j*th element shows the value of the *j*th attribute within the cell's locality.

Two cells are neighbors in *HDICLA* if the distance between their tendency vectors is smaller than or equal to the neighborhood radius.

Like *DICLA*, there is a local rule that *HDICLA* operates under. The rule of *HDICLA*, the actions selected by the neighboring learning automata of any particular learning automaton LA_i, and the attribute vector of the cell in which LA_i resides (c_i)

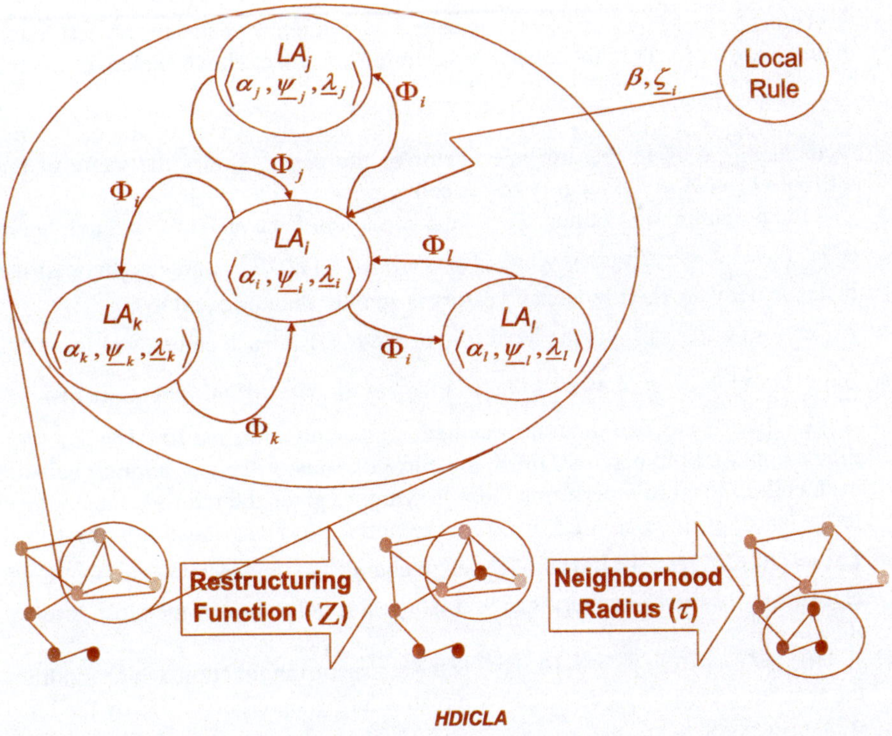

Fig. 1.8 Heterogeneous dynamic irregular cellular learning automaton (*HDICLA*)

determine the followings: (1) the reinforcement signal to the learning automaton LA_i and (2) the restructuring signal to the cell c_i, which is used to update the tendency vector of the cell. Figure 1.8 gives a schematic of *HDICLA*. An *HDICLA* is formally defined below.

Definition 1.31 A heterogeneous dynamic irregular cellular learning automaton (*HDICLA*) is a structure $A = \left(G\langle E, V\rangle, \Psi, \Lambda, A, \Phi\left\langle \alpha, \underline{\psi}, \underline{\lambda}\right\rangle, \tau, F, Z\right)$ where

- G is an undirected graph, with V as the set of vertices and E as the set of edges. Each vertex represents a cell in *HDICLA*.
- Ψ is a finite set of interests. The cardinality of Ψ is denoted by $|\Psi|$.
- Λ is a finite set of attributes. The cardinality of Λ is denoted by $|\Lambda|$.
- A is the set of learning automata, each of which is assigned to one cell of the *HDICLA*.
- $\Phi\left\langle \alpha, \underline{\psi}, \underline{\lambda}\right\rangle$ is the cell state. State of a cell c_i (φ_i) consists of the following three parts:
- α_i which is the action selected by the learning automaton of that cell.

- A vector $\underline{\psi}_i = (\psi_{i1}, \ldots, \psi_{i|\Psi|})^T$ is called the tendency vector of the cell. Each element $\psi_{ik} \in [0, 1]$ in the tendency vector of the cell c_i shows the degree of the tendency of c_i to the interest $\psi_k \in \Psi$.
- A vector $\underline{\lambda}_i = (\lambda_{i1}, \ldots, \lambda_{i|\Lambda|})^T$ is called the attribute vector of the cell. Each element $\lambda_{ik} \in \mathbb{R}$ in the attribute vector of the cell c_i shows the value of the attribute $\lambda_k \in \Lambda$ in the locality of the cell c_i.
- τ is the neighborhood radius. Two cells c_i and c_j of the *HDICLA* are neighbors if $\left\|\underline{\psi}_i - \underline{\psi}_j\right\| \leq \tau$. In other words, two cells of the *HDICLA* are neighbors if the distance between their tendency vectors is smaller than or equal to τ.
- $F : \underline{\varphi}_i \rightarrow \left\langle \underline{\beta}, [0, 1]^{|\Psi|} \right\rangle$ is the local rule of *HDICLA* in each cell c_i, where
- $\underline{\varphi}_i = \left\{\varphi_j \middle| \left\|\underline{\psi}_i - \underline{\psi}_j\right\| \leq \tau \right\} \cup \{\varphi_i\}$ is the set of states of all neighbors of c_i, $\underline{\beta}$ is the set of values that the reinforcement signal can take, and $[0, 1]^{|\Psi|}$ is a $|\Psi|$-dimensional unit hypercube. From the current states of the neighboring cells of each cell c_i, local rule performs the following: 1. gives the reinforcement signal to the learning automaton LA_i resides in c_i, and 2. produces a restructuring signal $\left(\underline{\zeta}_i = (\zeta_{i1}, \ldots, \zeta_{i|\Psi|})^T\right)$, which is used to change the tendency vector of c_i. Each element ζ_{ij} of the restructuring signal is a scalar value within the close interval [0, 1].
- $Z : [0, 1]^{|\Psi|} \times [-1, 1]^{|\Psi|} \rightarrow [0, 1]^{|\Psi|}$ is the restructuring function, which modifies the tendency vector of a cell using the restructuring signal produced by the local rule of the cell.

In what follows, we consider *HDICLA* with *N* cells. The learning automaton LA_i which has a finite action set $\underline{\alpha}_i$ is associated with cell c_i (for $i = 1, 2, \ldots, N$) of *HDICLA*. Let the cardinality of $\underline{\alpha}_i$ being m_i. The state of the *HDICLA* is represented by the triple $\left\langle \underline{p}, \underline{\psi}, \underline{\lambda} \right\rangle$, where

- $\underline{p} = \left(\underline{p}_1, \ldots, \underline{p}_N\right)^T$, where $\underline{p}_i = (p_{i1}, p_{i2}, \ldots, p_{im_i})^T$ is the action probability vector of LA_i.
- $\underline{\psi} = \left(\underline{\psi}_1, \ldots, \underline{\psi}_N\right)^T$, where ψ_i is the tendency vector of the cell c_i.
- $\underline{\lambda} = (\underline{\lambda}_1, \ldots, \underline{\lambda}_N)^T$, where λ_i is the attribute vector of the cell c_i.

The operation of the *HDICLA* takes place as the following iterations. At iteration *n*, each learning automaton chooses an action. Let $\alpha_i \in \underline{\alpha}_i$ be the action chosen by LA_i. Then, each learning automaton receives a reinforcement signal. Let $\beta_i \in \underline{\beta}$ be the reinforcement signal received by LA_i. This reinforcement signal is produced by the application of the local rule $F\left(\underline{\varphi}_i\right) \rightarrow \left\langle \underline{\beta}, [0, 1]^{|\Psi|} \right\rangle$. Each *LA* updates its action probability vector based on the supplied reinforcement signal and the action chosen by the cell. Next, each cell c_i updates its tendency vector using the restructuring function *Z* (Eq. 1.33).

$$\underline{\psi}_i(k+1) = Z\left(\underline{\psi}_i(k), \underline{\zeta}_i(k)\right). \tag{1.33}$$

After the application of the restructuring function, the attribute vector of the cell c_i may change due to the modifications made to its local environment.

Like *DICLA*, *HDICLA* can be either synchronous or asynchronous, and an asynchronous *DICLA* can be either time-driven or step-driven.

Note that most of the definitions used for steady-state analysis of the behavior of the *DICLA* are redefined here for the *HDICLA* model.

Definition 1.32 A configuration of *HDICLA* at step k is denoted by $\left\langle \underline{p}(k), \underline{\psi}(k), \underline{\lambda}(k) \right\rangle = \left\langle \left(\underline{p}_1, \ldots, \underline{p}_N\right)^T, \left(\underline{\psi}_1, \ldots, \underline{\psi}_N\right)^T, \left(\underline{\lambda}_1, \ldots, \underline{\lambda}_N\right)^T \right\rangle$, where $\underline{p}_i$ is the action probability vector of the learning automaton LA_i, $\underline{\psi}_i$ is the tendency vector of the cell c_i, $\underline{\lambda}_i$ is the attribute vector of the cell c_i, and T denotes the transpose operator.

Definition 1.33 A configuration $\left\langle \underline{p}, \underline{\psi}, \underline{\lambda} \right\rangle$ is called unit if the action probability vector of each learning automaton and the tendency vector of each cell are unit vectors; otherwise, it is called general. Hence, the set of all unit configurations, $\langle K^*, Y^*, \Lambda \rangle$, and the set of all general configurations, $\langle K, Y, \Lambda \rangle$, in *HDICLA* are

$$\langle K^*, Y^*, \Lambda \rangle = \left\{ \left\langle \underline{p}, \underline{\psi}, \underline{\lambda} \right\rangle \middle| \begin{array}{l} \underline{p} = \left(\underline{p}_1, \ldots, \underline{p}_N\right)^T, \underline{p}_i = \left(p_{i1}, \ldots, p_{im_i}\right)^T, p_{iy} = 0 \text{ or } 1 \; \forall i, y, \sum_y p_{iy} = 1 \; \forall i, \\ \underline{\psi} = \left(\underline{\psi}_1, \ldots, \underline{\psi}_N\right)^T, \underline{\psi}_i = \left(\psi_{i1}, \ldots, \psi_{i|\psi|}\right)^T, \psi_{ij} = 0 \text{ or } 1 \; \forall i, j, \sum_j \psi_{ij} = 1 \; \forall i, \\ \underline{\lambda} = \left(\underline{\lambda}_1, \ldots, \underline{\lambda}_N\right)^T, \underline{\lambda}_i = \left(\lambda_{i1}, \ldots, \lambda_{i|\Lambda|}\right)^T, \lambda_{ik} \in \Lambda \end{array} \right\} \tag{1.34}$$

and

$$\langle K, Y, \Lambda \rangle = \left\{ \left\langle \underline{p}, \underline{\psi}, \underline{\lambda} \right\rangle \middle| \begin{array}{l} \underline{p} = \left(\underline{p}_1, \ldots, \underline{p}_N\right)^T, \underline{p}_i = \left(p_{i1}, \ldots, p_{im_i}\right)^T, 0 \le p_{iy} \le 1 \; \forall i, y, \sum_y p_{iy} = 1 \; \forall i, \\ \underline{\psi} = \left(\underline{\psi}_1, \ldots, \underline{\psi}_N\right)^T, \underline{\psi}_i = \left(\psi_{i1}, \ldots, \psi_{i|\psi|}\right)^T, 0 \le \psi_{ij} \le 1 \; \forall i, j, \\ \underline{\lambda} = \left(\underline{\lambda}_1, \ldots, \underline{\lambda}_N\right)^T, \underline{\lambda}_i = \left(\lambda_{i1}, \ldots, \lambda_{i|\Lambda|}\right)^T, \lambda_{ik} \in \Lambda \end{array} \right\} \tag{1.35}$$

respectively. We refer to $\mathcal{K}$ as the probability space, to Y as the tendency space, and to Λ as the attribute space of *HDICLA* hereafter.

Lemma 1.3 *$\langle K, Y \rangle$, which represents the probability and tendency spaces of HDICLA, is the convex hull of $\langle K^*, Y^* \rangle$.*

Proof Proof of this lemma is similar to the proof of lemma 1-1 given in (Beigy and Meybodi 2004). ■

The application of the local rule to every cell allows transforming a configuration into a new one. The local rule of *HDICLA* (F) consists of two parts; reinforcement signal generator (F_β) and restructuring signal generator (F_ζ).

Definition 1.34 The global behavior of an *HDICLA* is a mapping $\mathcal{G}: <\mathcal{K}, \Upsilon, \mathcal{A}> \to <\mathcal{K}, \Upsilon, \mathcal{A}>$ that describes the dynamics of *HDICLA*. The evolution of *HDICLA* from a given initial configuration $\left\langle \underline{p}(0), \underline{\psi}(0), \underline{\lambda}(0) \right\rangle \in \langle K, Y, \Lambda \rangle$ is a sequence of configurations $\left\{ \left\langle \underline{p}(k), \underline{\psi}(k), \underline{\lambda}(k) \right\rangle \right\}_{k \geq 0}$, such that $\left\langle \underline{p}(k+1), \underline{\psi}(k+1), \underline{\lambda}(k+1) \right\rangle = G\left(\left\langle \underline{p}(k), \underline{\psi}(k), \underline{\lambda}(k) \right\rangle \right)$.

Definition 1.35 Neighborhood set of any particular LA_i in configuration $\left\langle \underline{p}, \underline{\psi}, \underline{\lambda} \right\rangle$, denoted by $N^{\underline{\psi}}(i)$, is defined as the set of all learning automata residing in the adjacent cells of the cell c_i, that is,

$$N^{\underline{\psi}}(i) = \left\{ LA_j \middle| \left\| \underline{\psi}_i - \underline{\psi}_j \right\| \leq \tau \right\}. \tag{1.36}$$

Let $N_i^{\underline{\psi}}$ be the cardinality of $N^{\underline{\psi}}(i)$.

Definition 1.36 The average penalty for action r of the learning automaton LA_i for configuration $\left\langle \underline{p}, \underline{\psi}, \underline{\lambda} \right\rangle \in \langle K, Y, \Lambda \rangle$ is defined as

$$\begin{aligned} d_{ir}^{\beta}\left(\left\langle \underline{p}, \underline{\psi}, \underline{\lambda} \right\rangle \right) &= E\left[\beta_i \middle| \left\langle \underline{p}, \underline{\psi}, \underline{\lambda} \right\rangle, \alpha_i = r \right] \\ &= \sum_{y_{h_1}^{\psi}, \ldots, y_{h_{N_i^{\psi}}}^{\psi}} \left(F_\beta^i\left(y_{h_1}^{\psi}, y_{h_2}^{\psi}, \ldots, y_{h_{N_i^{\psi}}}^{\psi}, r, \underline{\lambda} \right) \prod_{\substack{LA_l \in N^{\psi}(i) \\ l \neq i}} p_{l y_{h_l}} \right), \end{aligned} \tag{1.37}$$

and the average penalty for the learning automaton LA_i is defined as

$$D_i\left(\left\langle \underline{p}, \underline{\psi}, \underline{\lambda} \right\rangle \right) = E\left[\beta_i \middle| \left\langle \underline{p}, \underline{\psi}, \underline{\lambda} \right\rangle \right] = \sum_y d_{iy}^{\beta}\left(\left\langle \underline{p}, \underline{\psi}, \underline{\lambda} \right\rangle \right) p_{iy}. \tag{1.38}$$

The above definition implies that if the learning automaton LA_j is not a neighboring learning automaton for LA_i, then $d_{ir}^{\beta}\left(\left\langle \underline{p}, \underline{\psi}, \underline{\lambda} \right\rangle \right)$ does not depend on $\underline{p}_j$. We assume that $d_{ir}^{\beta}\left(\left\langle \underline{p}, \underline{\psi}, \underline{\lambda} \right\rangle \right) \neq 0$ for all i, r and $\left\langle \underline{p}, \underline{\psi}, \underline{\lambda} \right\rangle$, that is, in any configuration, any action has a non-zero chance of receiving a penalty.

Definition 1.37 The total average penalty for *HDICLA* at configuration $\langle \underline{p}, \underline{\psi}, \underline{\lambda} \rangle \in \langle K, Y, \Lambda \rangle$ is the sum of the average penalties for all learning automata in *HDICLA*, that is,

$$D\left(\left\langle \underline{p}, \underline{\psi}, \underline{\lambda} \right\rangle\right) = \sum_{i} D_i\left(\left\langle \underline{p}, \underline{\psi}, \underline{\lambda} \right\rangle\right). \tag{1.39}$$

Definition 1.38 The average restructuring signal for interest j of the cell c_i for configuration $\langle \underline{p}, \underline{\psi}, \underline{\lambda} \rangle \in \langle K, Y, \Lambda \rangle$ is defined as

$$d_{ij}^{\zeta}\left(\left\langle \underline{p}, \underline{\psi}, \underline{\lambda} \right\rangle\right) = \sum_{r} \sum_{y_{h_1}^{\psi}, \ldots, y_{N_i^{\psi}}^{\psi}} \left(F_{\zeta}^{i}\left(y_{h_1}^{\psi}, y_{h_2}^{\psi}, \ldots, y_{h_{N_i^{\psi}}}^{\psi}, r, \underline{\psi}_{*j}, \underline{\lambda} \right) \prod_{LA_l \in N^{\underline{\psi}}(i)} p_{l y_{h_l}} \right), \tag{1.40}$$

where $\underline{\psi}_{*j} = (\psi_{1j}, \psi_{1j}, \ldots, \psi_{nj})^T$.

The above definition implies that all interests are independent of each other since the restructuring signal generator (F_{ζ}) for any interest of any cell depends only on the tendency values of the neighboring cells to that interest.

1.4.10 Closed Asynchronous Dynamic Cellular Learning Automata (CADCLA)

In Closed Asynchronous Dynamic Cellular Learning Automata (*CADCLA*), we have a rule called structure updating rule, which can be defined by the designer for controlling the evolution of the cellular structure (Saghiri and Meybodi 2017b).

Definition 1.39 Closed Asynchronous Dynamic Cellular Learning Automaton (*CADCLA*) is a 7-tuple *CADCLA* = (G, A, N, Φ, Ψ, F_1, F_2), where:

- $G = (V, E)$ is an undirected graph which determines the structure of *CADCLA* where
  V = $\{cell_1, cell_2, \ldots, cell_n\}$ is the set of vertices and E is the set of edges.
- A = $\{LA_1, LA_2, \ldots, LA_n\}$ is a set of *LAs,* each of which is assigned to one cell of *CADCLA*. The set of actions of automaton for a cell is the same as the set of states for that cell.
- N = $\{N_1, N_2, \ldots, N_n\}$ where $N_i = \{cell_j \in V | dist(cell_i, cell_j) < \theta_i\}$ where θ_i is the neighborhood radius of $cell_i$ and $dist(cell_i, cell_j)$ is the length of the shortest path between $cell_i$ and $cell_j$ in G. N_i^1 determines the immediate neighbors of $cell_i$.

- $\Phi = \{\Phi_1, \Phi_2, \ldots, \Phi_n\}$ where $\Phi_i = \{(j, \alpha_l) | cell_j \in N_i \; and \; action \; \alpha_l$ has been chosen by $LA_j\}$ denotes the state of $cell_i$. Φ_i^1 determines the state of $cell_i$ when $\theta_i = 1$.
- $\Psi = \{\Psi_1, \Psi_2, \ldots, \Psi_n\}$, $\Psi_i = \{(j, X_j) | cell_j \in N_i\}$ denotes the attribute of $cell_j$ where $X_j \subseteq \{x_1, x_2, \ldots, x_s\}$. $\{x_1, x_2, \ldots, x_s\}$ is the set of allowable attributes. Ψ_i^1 determines the attribute of $cell_i$ when $\theta_i = 1$.
- $F_1 : (\underline{\Phi}, \underline{\Psi}) \rightarrow (\underline{\beta}, \underline{\zeta})$ is the local rule of *CADCLA*. In each cell, the local rule computes the reinforcement signal and the restructuring signal for the cell based on the states and attributes of that cell and its neighboring cells. The reinforcement signal is used to update the learning automaton of that cell.
- $F_2 : (\underline{N}, \underline{\Psi}, \underline{\zeta}) \rightarrow (\underline{N}^1)$ is the structure updating rule. In each cell, the structure updating rule finds the immediate neighbors of the cell based on the restructuring signal computed by the cell, the attributes of the neighbors of the cell, and the neighbors of the cell. For example, in $cell_i$, structure updating rule takes $\langle N_i, \Psi_i, \zeta_i \rangle$ and returns N_i^1.

The internal structure of $cell_i$ and its interaction with local environments is shown in Fig. 1.9.

The main loop for the operation of a *CADCLA* is described in Fig. 1.10. The application determines which cell must be activated. Upon activation of a cell, the cell performs a process containing three phases: **preparation**, **structure updating,** and **state updating**.

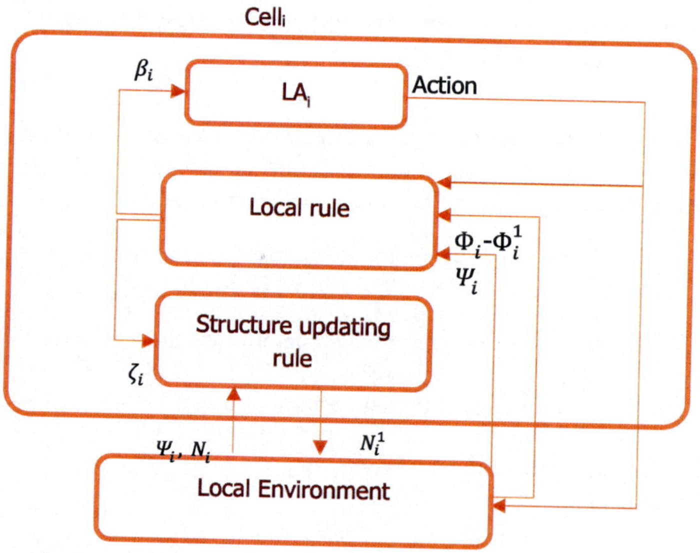

Fig. 1.9 The internal structure of $cell_i$ and its interaction with the local environment

Algorithm 1-3. CADCLA: mainloop()

Begin

Repeat

Choose a cell for activation; // the mechanism of choosing the cells is application-dependent

Activate the selected cell and perform the following phases;

- **Preparation phase:** In this phase, the cell performs the following steps.
  - The cell set its attribute.
  - The cell and its neighboring cells compute its *restructuring signal* using the *local rule* (F_1).
- **Structure updating phase:** In this phase, the cell performs the following steps.
  - The neighborhood structure of the cell is updated using the structure updating rule (F_2) if the value of the *restructuring signal* of that cell is 1.
- **State updating phase:** In this phase, the cell performs the following steps.
  - The *LA* of the cell select one of their actions. The action selected by the *LA* in the cell determines the new state for that cell.
  - The local rule (F_1) is applied, and a *reinforcement signal* is generated
  - The probability vectors of the *LA* of the cell is updated.

Until (there is no cell for activation)

End

Fig. 1.10 The main loop for the operation of *CADCLA*

Algorithm 1-4. DCLA

Begin

Repeat

Choose a $cell_{z_t}$ for activation; // The sequence according to which the cells are activated is application-dependent//

Call *Activate* ($cell_{z_t}$); // The pseudo code for algorithm Activate//

Until (there is no cell to be activated)

End

Fig. 1.11 The pseudo-code for the operation of *CADCLA*

In what follows, Fig. 1.11 shows the operation of the DCLA, and Fig. 1.12 shows the pseudo-code of the process, which is performed upon the activation of a cell of DCLA.

1.4.11 Adaptive Petri Nets Based on Irregular Cellular Learning Automata (APN-ICLA)

In the Adaptive Petri Nets based on Learning Automata (APN-LA) (Vahidipour et al. 2017a), the controllers that control different sets of conflicting transitions in the PN act independently from each other. However, there could be situations where resolving the conflict within a set of conflicting transitions affects another

```
Algorithm 1-5. Activate ()
Input
            cell_i
Notations
            F1 denotes the local rule
            F2 denotes the structure updating rule
            Ψ_i denotes the attribute of cell_i
            Φ_i denotes the state of cell_i
            ζ_i denotes the restructuring signal of cell_i
            N_i denotes the set of neighbors of cell_i
            β_i denotes the reinforcement signal of the learning automata of cell_i
Begin
                        // preparation phase\\
  Set the attribute of the cell_i.
  Compute ζ_i using F1;
Gather the restructuring signals of the Neighboring cells;

                // structure updating phase\\
If (ζ_i is 1) Then
    Compute N_i and Ψ_i using F2;
EndIf

                //state updating phase\\
Each learning automaton of cell_i chooses one of its actions;
Set Φ_i;// set Φ_i to be the set of actions chosen by the set of learning automata in cell_i;
  Compute β_i using F1;
  Update the action probabilities of learning automata of cell_i using β_i;
End
```

Fig. 1.12 The pseudo-code of the process which is executed upon the activation of a cell in CADCLA

set of conflicting transitions by possibly enabling or disabling some of the transitions within that set. In such situations, it seems it is more appropriate to let the controllers within the PN cooperate, instead of operating independently. In other words, if resolving conflicts among two different sets of transitions affects each other, then their controllers must be aware of the decisions made by each other to make better decisions.

Irregular cellular learning automata (ICLA), which is a network of learning automata, is utilized to construct a new adaptive Petri net, called APN-ICLA. The APN-ICLA consists of two layers: PN-layer and ICLA-layer, both of which are constructed according to the problem to be solved. The PN-layer is a Petri net, in which conflicting transitions are partitioned into several clusters. A cluster in the PN-layer is mapped into a cell of ICLA-layer, and hence, the LA residing in that cell, acting as the controller of that cluster. The connections among the cells in the ICLA-layer are determined by the locality defined in the application, for which the APN-ICLA is designed. Two clusters in the PN-layer are said to be neighbors if their corresponding cells in the ICLA-layer are neighbors. We will also define the firing rules for the proposed adaptive Petri net, and hence, the definition of the APN-ICLA will be completed.

An APN-ICLA consists of two layers: PN-layer and an ICLA-layer (Fig. 1.13). To construct the PN-layer, a Petri net is needed. This PN is designed according to the problem to be solved. This Petri net is used to design an algorithm for solving that problem. To create this PN-layer, we first determine all sets $s_i, i = 1, \ldots, n$ of maximal potential conflicts which call them clusters in the PN. Then, the remaining transitions in this PN, which are not in any of the maximal potential conflicts sets $s_i, i = 1, \ldots, n$, form the cluster s_0.

Every cluster $s_i, i = 1, \ldots, n$ in the PN-layer is mapped into a cell in the ICLA-layer. The connections among the cells in the ICLA-layer are determined by the locality defined in the application, for which the APN-ICLA is designed. Two clusters in the PN-layer are said to be neighbors if their corresponding cells in the ICLA-layer are neighbors. The LA resides in a particular cell in the ICLA-layer acts as the controller of the corresponding cluster in the PN-layer.

Like ICLA, there is a local rule that APN-ICLA operates under. The local rule of the APN-ICLA and the actions selected by the neighboring LAs of any particular LA determine the reinforcement signal to that LA. The neighboring LAs of any

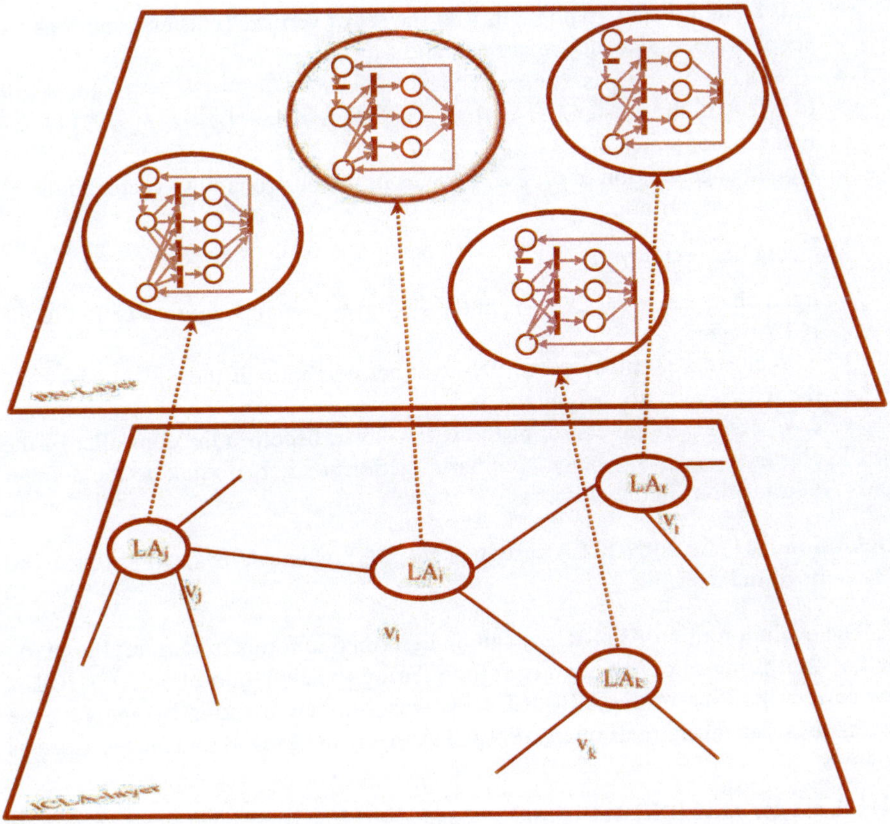

Fig. 1.13 The two-layered model of an APN-ICLA

particular LA constitute the local environment of that LA. The local environment of an LA in the APN-ICLA is non-stationary because the action probability vectors of the neighboring LAs vary during the evolution of APN-ICLA. An APN-ICLA can be formally defined as follows:

Definition 1.40 An Adaptive Petri Net based on Irregular Cellular Learning Automata (APN-ICLA) with n cells is a tuple $\mathcal{N} = \left(PN - layer = \left(\widehat{P}, \widehat{T}, \widehat{W}, S\right), ICLA - layer = (G\langle E, V\rangle, \Phi, A, F)\right)$, where

1. PN-Layer
   - $\widehat{P}, \widehat{T}, \widehat{W}, S$ are defined in PN
2. ICLA-layer
   - Φ is a finite set of states. The state of the cluster $s_i, i = 1, \ldots, n$ is denoted by φ_i,
   - $A = \{LA_1, \ldots .LA_n\}$ is a set of learning automata with varying number of actions,
   - G is an undirected graph, with V as the set of vertices (clusters) and E as the set of edges (adjacency relations), and
   - $\widehat{F} = \{F_1, \cdots, F_n\}$ is the set of local rules. $F_i : \underline{\Phi}_i \rightarrow \beta_i$ is the local rule of ICLA in the cluster $s_i, i = 1, \ldots, n$, where $\underline{\Phi}_i = \left\{\varphi_j | (i, j) \in E\right\} \cup \{\varphi_i\}$ is the set of states of all neighbors of s_i and β_i is the reinforcement signal. Upon the generation of β_i, LA_i updates its action probability vector using the learning algorithm.
3. Mapping between layers
   - Each cluster $s_i, i = 1, \ldots, n$ in the PN-layer is mapped into a cell c_i in the ICLA-layer,
   - Two clusters s_i and s_j in the PN-layer are neighbors if the cells c_i and c_j in the ICLA-layer are neighbors, and
   - LA_i, resides in the cell c_i of the ICLA-layer, becomes the controller of the cluster s_i of the PN-layer. Number of actions of LA_i is equal to the number of transitions in the cluster s_i. □

Definition 1.41 An APN-ICLA system is $(\mathcal{N}, M_0)$, where $\mathcal{N}$ is an APN-ICLA and M_0 is the initial marking. □

The evolution of an APN-ICLA can be described in terms of its initial marking and several firing rules. At any given time during the evolution of an APN-ICLA, the current marking of the APN-ICLA evolves to a new marking by applying the firing rules. The firing rules of an APN-ICLA in any marking M can be described as follows:

(1) A cluster s_i of PN-layer is said to be enabled if at least one transition in s_i is enabled.

(2) If s_0 is enabled, then s_0 is selected as fired with probability $\frac{|t' \in s_0, M[t'\rangle|}{|t \in \widehat{T}, M[t\rangle|}$, where $||$ stands for norm operator; otherwise, an enabled cluster $s_i, i = 1, \ldots, n$ is selected as fired with probability $\frac{|E_M^{s_i}|}{|t \in \widehat{T}, M[t\rangle|}$ (only one enabled cluster can be selected as the fired cluster in each marking).

(3) Only one enabled transition $t \in \widehat{T}$ from the fired cluster can fire in each marking M.

(4) If s_0 is the fired cluster, then an enabled transition t is randomly selected from s_0 for firing; otherwise, the LA in the corresponding cell with the fired cluster is activated. The available action set of the activated LA consists of the actions corresponding to the enabled transitions in the fired cluster. The activated LA selects an action from its available action set according to its action probability vector, and then the corresponding transition is selected for firing.

(5) Firing of a transition $t \in \widehat{T}$ removes $\widehat{W}(p, t)$ tokens from each input place p of t, and adds $\widehat{W}(t, p)$ tokens to each output place p of t.

(6) The local rule $F_i \in \widehat{F}$ is used to generate the reinforcement signal β_i, when the updating transition $t_i^u \in T^u$ fires.

(7) Using β_i, the action probability vector of LA_i is updated by the learning algorithm.

1.4.12 *Closed Asynchronous Dynamic Cellular Learning Automata with Varying Number of LAs in Each Cell (CADCLA-VL)*

The Closed Asynchronous Dynamic Cellular Learning Automata with varying number of LAs in each cell (CADCLA-VL) is a version of *CADCLA* in which the number of LAs of the cells changes over time.

Definition 1.42 Closed Asynchronous Dynamic Cellular Learning Automaton with a varying number of Learning Automata in each cell (*CADCLA-VL*) is a network of cells whose structure changes with time, and each cell contains a set of *LAs* and a set of attributes (Saghiri and Meybodi 2017b). In this model, the connections among cells, and the number of *LAs* of each cell changes during the evolution of *CLA*. This model can be formally defined by a 7-tuples as follows

$$CADCLA - VL = (G, A, \Psi, \Phi, \mathrm{N}, F_1, F_2),$$

where

- $G = (V, E)$ is an undirected graph which determines the structure of *CADCLA-VL* where
  $\mathrm{V} = \{cell_1, cell_2, \ldots, cell_n\}$ is the set of vertices and E is the set of edges.

- $A = \{LA_1, LA_2, \ldots, LA_v\}$ is a set of *LAs*. A subset of set A is assigned to a cell.
- $N = \{N_1, N_2, \ldots, N_n\}$ where $N_i = \{cell_j \in V | dist(cell_i, cell_j) < \theta_i\}$ where θ_i is the neighborhood radius of $cell_i$ and $dist(cell_i, cell_j)$ is the length of the shortest path between $cell_i$ and $cell_j$ in G. N_i^1 determines the immediate neighbors of $cell_i$.
- $\Psi = \{\Psi_1, \Psi_2, \ldots, \Psi_n\}, \Psi_i = \{(j, X_j, C_j) | cell_j \in N_i\}$ denotes the attribute of $cell_j$ where $X_j \subseteq \{x_1, x_2, \ldots, x_s\}$ and $C_j \subseteq A$. $\{x_1, x_2, \ldots, x_s\}$ is the set of allowable attributes. Ψ_i^1 determines the attribute of $cell_i$ when $\theta_i = 1$.
- $\Phi = \{\Phi_1, \Phi_2, \ldots, \Phi_n\}$ where $\Phi_i = \{(j, k, \alpha_l) | cell_j \in N_i \; and \; action \; \alpha_l$ has been chosen by $LA_k \in C_i\}$ denotes the state of $cell_i$. Φ_i^1 determines the state of $cell_i$ when $\theta_i = 1$.
- $F_1 : (\underline{\Phi}, \underline{\Psi}) \rightarrow (\underline{\beta}, \underline{\zeta})$ is the local rule of *CADCLA-VL*. In each cell, the local rule computes the reinforcement signal and the restructuring signal for the cell based on the states and attributes of that cell and its neighboring cells. for example, in $cell_i$, the local rule takes $\langle \Phi_i, \Psi_i \rangle$ and returns $\langle \beta_i, \zeta_i \rangle$. The reinforcement signal is used to update the learning automaton of that cell.
- $F_2 : (\underline{N}, \underline{\Psi}, \underline{\zeta}) \rightarrow (\underline{N}^1, \underline{\Psi}^1)$ is the structure updating rule. In each cell, the structure updating rule finds the immediate neighbors and attributes of that cell. For example, in $cell_i$, structure updating rule takes $\langle N_i, \Psi_i, \zeta_i \rangle$ and returns N_i^1 and Ψ_i^1.

The internal structure of $cell_i$ and its interaction with local environments is shown in Fig. 1.14. In this model, each cell has a set of *LAs* which may vary during the operation of *CLA*.

The main loop for the operation of a *CADCLA* is described in Fig. 1.15.

1.4.13 Wavefront Cellular Learning Automata (WCLA)

A cellular automaton (CA) is made of similar cells where each cell has a state, and transitions over the set of possible states are based on a local rule. A learning automaton (LA) is also an adaptive decision-making unit in unknown random environments. Intuitively, the LA, during its learning mechanism, tries to choose the optimal action from its action set based on the environment's response to the previous action. To improve the local interactions of the LA in complex systems, the CA and LA are merged and proposed in terms of a CLA. The CLA is preferable to a CA because it tries to learn optimal actions, and it is also preferable to an LA because it can improve the learning capability using a set of learning automata that interact with each other.

In many real-world learning problems, the learning process can proceed in stages in a particular order. For example, in pattern classification with decision trees, when a node makes a decision, its decision is transmitted to its successor nodes as a wave. In problems related to sampling social networks, when we label a link or node as a

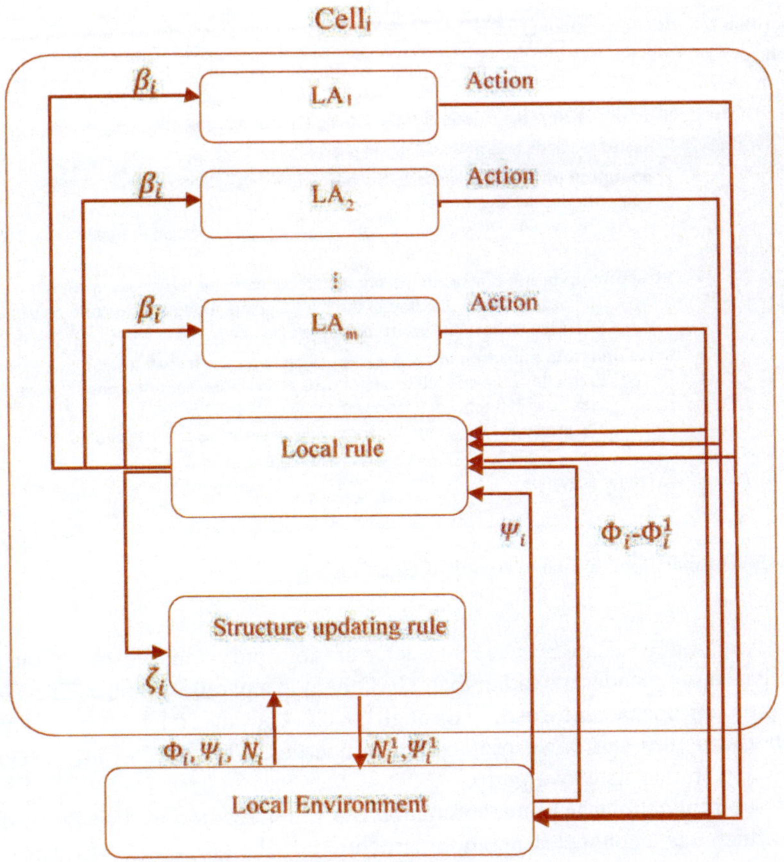

Fig. 1.14 The internal structure of $cell_i$ and its interaction with the local environment in CADCLA-VL

sample instance, this decision can propagate into the network as a wave. Therefore, we consider a CLA with one LA for each region and define the neighbors of each cell to be the cells of successor regions, such that each LA can send a wave to its neighbors if its chosen action is different from the previous action. This model of CLA is called a wavefront CLA (WCLA) (Moradabadi and Meybodi 2018b). In a wavefront CLA, we partition the problem space into stages such that each stage tries to learn the optimal action and helps the neighbors to act accordingly in the environment.

A wavefront CLA (WCLA) (Moradabadi and Meybodi 2018b) is an extension to the asynchronous CLA model, with a connected neighbor structure and propagation property where each LA can send a wave to its neighbors and activate their learning automata to choose new actions if the chosen action is changed from the previous action. Furthermore, this procedure continues. Each cell that receives the wave is activated, and its corresponding LA must choose a new action. The proposed WCLA

Algorithm 1-6. DCLA: mainloop()

Begin

Repeat

Choose a cell for activation;// the mechanism of choosing the cells is application-dependent//

Activate the selected cell and perform the following phases;

- **Preparation phase:** In this phase, the cell performs the following steps.
  - The cell set its attribute.
  - The cell and its neighboring cells compute their *restructuring signals* using the *local rule* (F_1).
- **Structure updating phase:** In this phase, the cell performs the following steps.
  - The neighborhood structure of the cell is updated using the structure updating rule (F_2) if the value of the *restructuring signal* of that cell is 1.
- **State updating phase:** In this phase, the cell performs the following steps.
  - Each *LAs* of the cell select one of their actions. The set of actions selected by the set of *LAs* in the cell determines the new state for that cell.
  - The local rule (F_1) is applied, and a *reinforcement signal* is generated
  - The probability vectors of the *LAs* of the cell are updated.

Until (there is no cell for activation**)**

End

Fig. 1.15 The main loop for the operation of *CADCLA-VL*

is an asynchronous CLA because, at each iteration, only some learning automata are activated independently, rather than all of them in a parallel manner. The WCLA enables us to propagate information through the CLA because it has both a connected neighbor structure and wave propagation properties. This section introduces the WCLA and studies its convergence.

A wavefront cellular learning automaton is a generalization of the asynchronous CLA, which has a connected neighbor structure and the property of diffusion. The main features of a WCLA are that the neighbors of each cell are defined as its successor cells and also that each LA can send a wave to its neighbors and active them if its chosen action is different from the previous action (diffusion property). Each cell that receives the wave is activated, and its corresponding LA must choose a new action. The WCLA operates like a CLA in the following respects. In the initial phase, the initial state of every cell is specified using the initial action probability distribution of the corresponding LA. At instance *k*, one LA called the root LA chooses an action, and if the chosen action is different from the previous action, it sends a wave over the network. Each cell that receives the wave is activated, and its corresponding LA must choose a new action. Then, using the local rule of the WCLA and the selected actions of its neighbors, a reinforcement signal is generated for the LA residing in that cell, and the LA uses this reinforcement signal to update its action probability distribution using the learning algorithm. This procedure continues until a stopping criterion is reached. Figure 1.16 shows the procedure of WCLA in a simple view. In this figure, first, the root learning automata that is labeled as one is activated. This learning automaton chooses an action, and because the new action is different from the previous one, it activates all of its children LAs in the next level (labeled as 2). Then each activated child chooses a new action, and if its action is the same

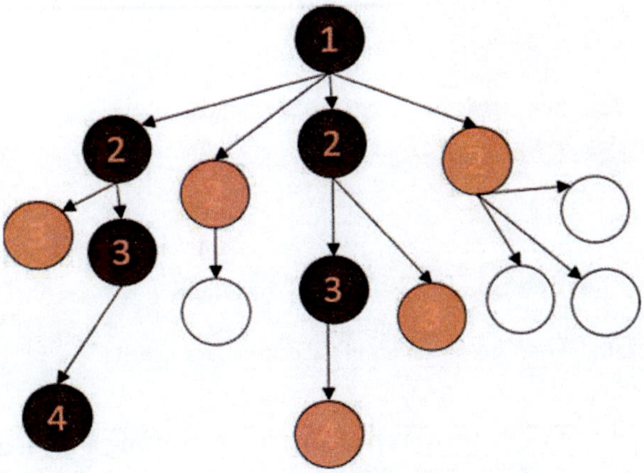

Fig. 1.16 A toy example of a WCLA

as its previous action, then the chain is stopped at that node (colored as gray). If the new action is different from the previous action, it activates its children (colored as red), and this procedure goes on. The proposed WCLA is asynchronous because at each iteration, one LA selects a new action, and therefore, the action selection does not occur in a parallel manner. It is also irregular and closed because it does not use a lattice for its structure and only uses the local environment to update the CLA behavior. Finally, it is static because the structure of the CLA remains fixed.

As mentioned previously, the WCLA utilizes two new concepts compared to the CLA:

(1) Different neighbor structure. In the traditional CLA, the structure must be a lattice, while in the WCLA, the structure can be any connected structure, either a regular structure such as a lattice or an irregular structure such as a graph or tree. The only necessity is that the structure must be connected. The structure of a WCLA is connected when there is a path between every pair of cells. In a connected structure, there are no unreachable cells. The connectivity property of the structure of the WCLA ensures that a wave can reach any cell in the network, and therefore there is a learning capability throughout the network. Figure 1.16 shows some examples of logical structures for a WCLA. Figure 1.17a shows a lattice structure as an example of a regular structure. Figure 1.17b, c show graph and tree structures, respectively, as examples of irregular structures. All the structures are connected.

(2) Propagation property using waves. As previously mentioned, in a traditional CLA, each learning automaton chooses its action in each iteration. Then the local rule of each learning automaton generates the reinforcement signal for the corresponding learning automaton using its chosen action and the chosen actions of its neighbors. Finally, each learning automaton updates its action probability based on its reinforcement signal. As we see in the traditional CLA, there is no propagation and diffusion property. When a cell state is changed, the other cells are not triggered to choose new actions. To overcome this problem, the

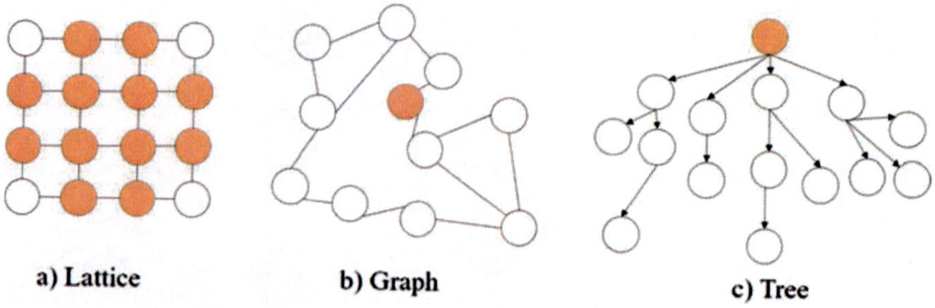

Fig. 1.17 A toy examples of valid structures in a WCLA

WCLA has a propagation property using waves. Each LA can send a wave to its neighbors and activate them if the chosen action is different from the previous action (diffusion property). Each cell that receives the wave is activated, and its corresponding LA must choose a new action. Figure 1.18 shows different waves over different WCLA structures. Each wave starts from a cell in the network and moves to its neighbors. The diffusion path of the wave depends only on the neighboring structure of the WCLA. Because the structure is connected and the waves can move over the entire network, each cell can be activated and improve its state after receiving waves, thus also improving its learning capability. However, in order to control the movement of the waves over the network, we define energy for each wave in the WCLA. The energy of each wave determines the wave's ability to propagate itself. The energy of the wave decreases as it moves through the network until it reaches zero, at which point the wave disappears, and movement stops. In this way, each cell receiving the wave is activated, and its LA is triggered to choose a new action.

Generally, a WCLA model is characterized as follows:

Definition 1.43 A wavefront cellular learning automaton is defined with structure $\mathcal{A} = (S, \Phi, A, \mathcal{F}, W, C)$, where:

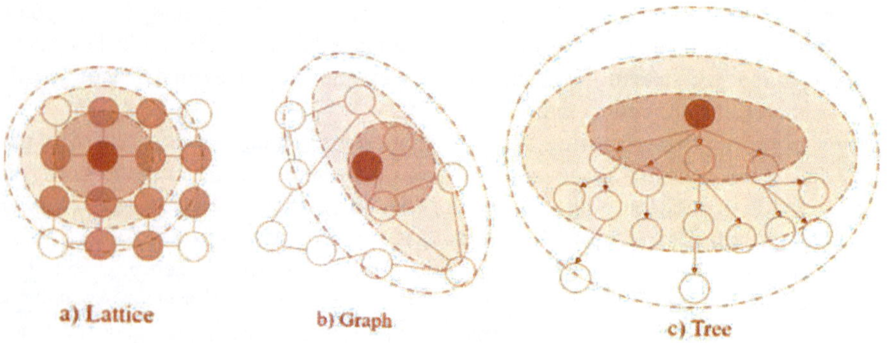

Fig. 1.18 Waves in different structures of WCLA

- S is the structure of the WCLA that is a connected network of learning automata.
- Φ is a finite set of states.
- A represents the set of learning automata residing in the cells of the CLA.
- $F^i : \underline{\Phi_i} \rightarrow \underline{\beta}$ defines the local rule of the WCLA for each cell c_i, where $\underline{\Phi_i}$ is the current state of LA_i and its successors. Also, $\underline{\beta}$ is the set of possible values for the reinforcement signal, and the reinforcement signal is calculated for each LA using the chosen actions of successor learning automata.
- W defines the wave energy and determines when and where the wave stops moving through the network.
- C is a condition where an LA sends a wave over the network and triggers its children to choose a new action. In a WCLA, C is defined according to whether the chosen action of the LA is different from the previous action or not.

As previously mentioned, the WCLA differs in two main respects compared to the CLA. Firstly, in the WCLA the lattice Z^d moreover, the neighborhood vector N is replaced by any connected network. Secondly, in the proposed model, each LA can send a wave to its neighbors, and each cell that receives the wave is activated and triggers its LA to choose a new action. In the rest of this section, we present the learning algorithm we use for updating the WCLA action:

As mentioned above, in the WCLA, we have a set of N learning automata indexed from 1 to N. The ith LA has the action set A_i with r_i actions as follows:

$$A_i = \{a_{i1}, \ldots, a_{ir_i}\} \tag{1.41}$$

Besides, $p_i(k)$ and $\alpha_i(k)$ are its action probability vector and its chosen action in iteration k, respectively. The probability of choosing action a_{ij} at instance k is defined as:

$$Prob\{\alpha_i(k) = a_{ij}\} = p_{ij}(k) \tag{1.42}$$

To update $p_i(k)$ for instance k, we use the linear reward-inaction ($L_{R\text{-}I}$) algorithm according to the following equations:

$$\begin{aligned} p_{ij}(k+1) &= p_{ij}(k) + \lambda\beta(k)\big(1 - p_{ij}(k)\big) && if\ \alpha_i(k) = a_{ij} \\ p_{ij}(k+1) &= (1 - \lambda\beta(k))\, p_{ij}(k) && if\ \alpha_i(k) \neq a_{ij} \end{aligned} \tag{1.43}$$

where $\beta(k)$ is the reinforcement signal in instance k and the learning parameter λ satisfies the relation $0 < \lambda < 1$. Also, in (Moradabadi and Meybodi 2018b), it is shown that WCLA is expedient and converges to global optima.

1.4.14 Cellular Learning Automata-Based Evolutionary Computing (CLA-EC)

CLA-EC model, proposed in (Rastegar and Meybodi 2004; Rastegar et al. 2006), is a combination of *CLA* and the evolutionary computing model (Fig. 1.19). In this model, each cell of the *CLA* is equipped with an *m*-bit binary genome. Each genome has two components; model genome and string genome. The model genome is composed of *m* learning automata. Each *LA* has two actions; 0 and 1. The set of actions selected by the set of *LA*s of a particular cell concatenated to each other to form the second component of the genome, i.e., the string genome. The operation of a CLA-EC, in any particular cell c_i, takes place as follows. Each *LA* residing within the cell c_i randomly selects one of its actions according to its action probability vector. The actions selected by the set of *LA*s of the cell c_i are concatenated to each other to form a new string genome for that cell. The fitness of this newly generated genome is then evaluated. If the newly generated genome is better than the current genome of the cell, then the current genome of the cell c_i is replaced by the newly generated genome. Next, a number of the neighboring cells of the cell c_i, according to the fitness evaluation of their corresponding genomes, are selected for mating. Note that the mating in this context is not reciprocal, i.e., a cell selects another cell for mating but not necessarily vice versa.

The results of the mating process in the cell c_i are some reinforcement signals, one for each *LA* of the cell. The process of computing the reinforcement signal for each *LA* is described in Fig. 1.20. Each *LA* updates its action probability vector based on the supplied reinforcement signal and its selected action. This process continues until a termination condition is satisfied.

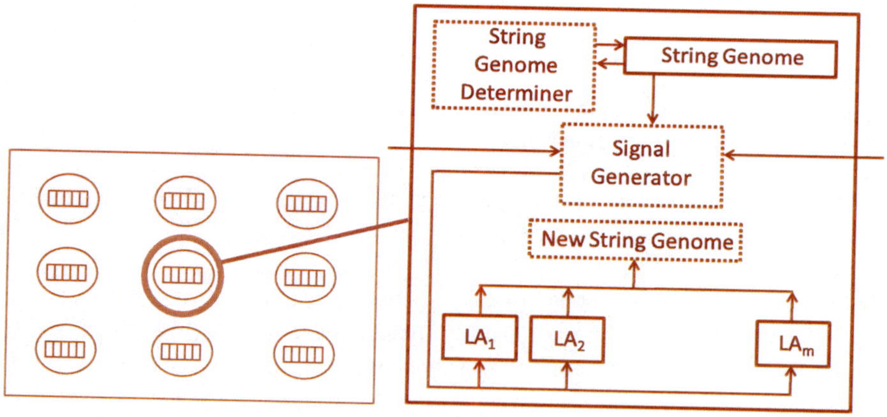

Fig. 1.19 The structure of the CLA-EC model

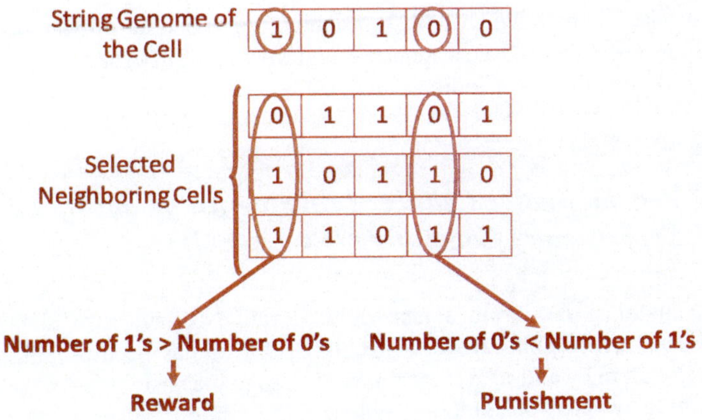

Fig. 1.20 The process of computing the reinforcement signal for each *LA*

1.4.15 Cooperative Cellular Learning Automata-Based Evolutionary Computing (CLA-EC)

The performance of the CLA-EC model is highly dependent on the learning rate of the LAs reside in its constituting cells. Large values of the learning rate decrease the performance of the model, whereas small values of the learning rate decrease its speed of convergence. The idea behind the cooperative CLA-EC model (Masoodifar et al. 2007) is to use two CLA-EC models with different learning rates in cooperation with each other. The operations of the CLA-ECs in the cooperative CLA-EC model are synchronized. At any iteration, each CLA-EC performs its normal operation independent of the other CLA-EC. Any m iteration, the genomes of the corresponding cells in the two CLA-ECs are exchanged, as depicted in Fig. 1.21. This way, one of the CLA-ECs (the one with the high learning rate) explores through the space of the problem, and the other CLA-EC (the one with the small learning rate) exploits and enhances the solutions found thus far.

In (Khaksarmanshad and Meybodi 2010), a different cooperative CLA-EC model has been proposed in which a CLA-EC model cooperates with a self-adaptive genetic

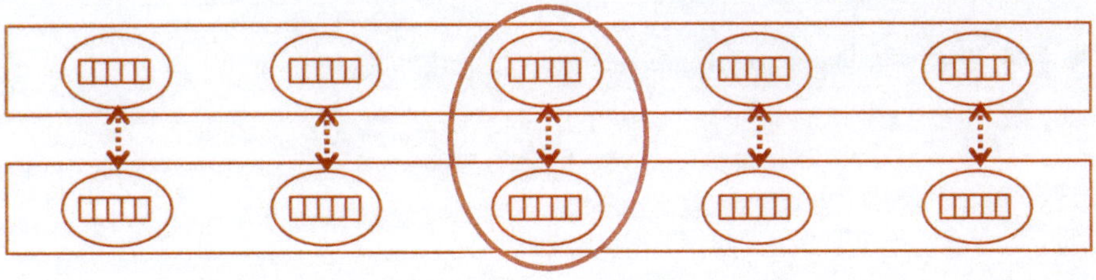

Fig. 1.21 Cooperative CLA-EC model

algorithm. The Cooperative CLA-EC model has been used in function optimization (Masoodifar et al. 2007; Khaksarmanshad and Meybodi 2010) as well as solving the 0/1 knapsack problem (Masoodifar et al. 2007).

1.4.16 Recombinative Cellular Learning Automata-Based Evolutionary Computing (RCLA-EC)

CLA-EC model suffers from a lack of explorative behavior. In (Jafarpour and Meybodi 2007), recombinative CLA-EC (RCLA-EC) has been proposed to overcome this problem by adding a recombination step to the basic steps of the operation of the CLA-EC model. The authors argue that there might be some string genomes with poor fitness, which contain good partial solutions. However, the CLA-EC model has no mechanism for partial string exchange between cells, and thus, it cannot reap the benefits of such good partial solutions. To overcome this problem, in recombinative CLA-EC, a shuffle recombination (Burkowski 1999) step has been added to the basic steps of the operation of the CLA-EC model. In this step, every cell c_i (for even i) recombines its string genome with that of the cell c_{i+1}. Next, the string genomes of the cells c_i and c_{i+1} are replaced by the string genomes resulted from the recombination. The recombination step of the recombinative CLA-EC model is presented in Fig. 1.22.

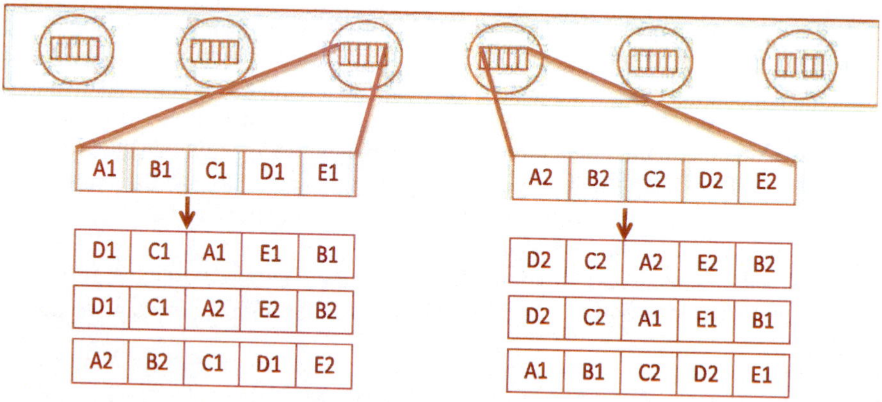

Fig. 1.22 Recombination step of the recombinative CLA-EC model

1.4.17 Combination of Extremal Optimization and CLA-EC (CLA-EC-EO)

Extremal Optimization (EO) (Boettcher and Percus 2002) is an optimization heuristic inspired by the Bak-Sneppen model (Bak and Sneppen 1993) of the self-organized criticality from the field of statistical physics. This heuristic is designed to address the complex combinatorial optimization problems such as the traveling salesman problem and spin glasses. However, the technique has been demonstrated to function in optimization domains. Unlike genetic algorithms that work with a population of candidate solutions, EO evolves a single solution and makes local modifications to the worst components. The technique can be regarded as a fine-grained search that tries to enhance a candidate solution gradually.

The EO technique may lead to a deterministic search for some types of problems, such as the single spin-flip neighborhood for the spin Hamiltonian (Boettcher and Percus 2002). To overcome this problem, the basic EO algorithm has been slightly modified in the τ-EO algorithm (Boettcher and Percus 2002). In this algorithm, unlike the basic EO algorithm in which the worst component is modified at each step, any component of the candidate solution has a chance of being modified at each step. Another drawback of the EO algorithm is that a general definition of the fitness for the individual components may prove ambiguous or even impossible. This problem has been addressed in the generalized EO (GEO) algorithm, proposed by Sousa and Ramos in (de Sousa and Ramos 2002). In this algorithm, the candidate solution is represented as a binary string. The fitness of each component i of the candidate solution (fitness of the ith bit) is assumed to be the difference between the fitness of the candidate solution and the fitness of the candidate solution in which the ith bit is flipped.

CLA-EC-EO model, proposed in (Khatamnejad and Meybodi 2008), is a combination of the CLA-EC and the GEO models. The main idea behind this model is that the GEO model is used to enhance the solutions found by the CLA-EC model. The operation of this model takes place as follows. The CLA-EC model explores throughout the search space until it finds a candidate solution with the fitness above a specified threshold. This candidate solution is then given to the GEO model for enhancement. The enhanced candidate solution is given back to the CLA-EC model. This process continues until a termination condition is met.

1.4.18 Cellular Learning Automata-Based Differential Evolution (CLA-DE)

Differential evolution (DE), proposed by Storn and Price (1997), is an evolutionary algorithm that optimizes a problem by iteratively trying to improve a population of candidate solutions concerning a given measure of quality. DE is used for multi-dimensional real-valued functions but does not use the gradient of the problem being optimized. This means that DE does not require the optimization problem to be differentiable, as is required by classic optimization methods such as gradient descent.

The operation of the basic DE algorithm takes place as follows. First, the best candidate solution in the current population is identified as *BCS*. Next, candidate solutions of the population are selected one by one. For each candidate solution CS_i, three other candidate solutions, namely RCS_{i1}, RCS_{i2}, and RCS_{i3}, are selected randomly. A new candidate solution NCS_i is then created by combining CS_i, RCS_{i1}, RCS_{i2}, and RCS_{i3}, and *BCS* through one of the following formulae

$$NCS_i = RCS_{i1} + F.(RCS_{i2} - RCS_{i3}),\ F \in [0, 2]. \tag{1.44}$$

$$NCS_i = BCS + F.(RCS_{i1} - RCS_{i2}),\ F \in [0, 2]. \tag{1.45}$$

$$NCS_i = CS_i + F.(BCS - CS_i) + F.(RCS_{i1} - RCS_{i2}),\ F \in [0, 2]. \tag{1.46}$$

The new candidate solution NCS_i is then recombined with CS_i using either of the binomial or the exponential crossover operators. If this newly recombined candidate solution is more fitted than the candidate solution CS_i, then it is replaced with CS_i in the current population.

The CLA-DE model is proposed in (Vafashoar et al. 2012) in order to improve the convergence rate and the accuracy of the results obtained from the CLA model. In this model, each cell of the CLA is equipped with m learning automata, where m is the dimension of the problem at hand. The LA assigned to the *i*th dimension is responsible for learning the admissible interval for the *i*th dimension of the candidate solution. Each dimension is divided into some intervals, and the LA assigned to each dimension has to choose an interval among the intervals of that dimension. The operation of the CLA-DE model within a cell c_i takes place as follows. First, each LA selects an interval randomly according to its action probability vector. The candidate solution of the cell is then created by selecting random numbers from the selected intervals in each dimension. Next, the candidate solutions of the neighboring cells are evaluated, and the most fitted candidate solution (BCS) is identified. Three other neighboring cells are selected at random (RCS_{i1}, RCS_{i2}, and RCS_{i3}). Using different combinations of RCS_{i1}, RCS_{i2}, and RCS_{i3}, and by using Eqs. (1.44) and (1.45), a set of candidate solutions is generated. All newly generated candidate solutions are recombined by the candidate solution of the cell. Finally, the most fitted candidate solution among these newly generated candidate solutions, and the current candidate solution of the cell is selected as the new candidate solution of the cell. The action selected by the LA, assigned to the *i*th dimension, is rewarded if one of the following two conditions is met:

1. The interval selected by the LA coincides with the selection of more than half of the neighboring cells.

2. The candidate solution of this cell is better than the candidate solution of more than half of the neighboring cells.

1.4.19 Cellular Learning Automata-Based Particle Swarm Optimization (CLA-PSO)

The particle swarm optimization (PSO) algorithm, introduced by Kennedy and Eberhart (1995), is based on the social behavior metaphor. In this algorithm, a group of particles flies through a D dimensional space, adjusting their positions in the search space according to their own and other particles' experiences. Each particle i is represented with a position vector $p_i = (p_{i1}, p_{i2}, \ldots, p_{iD})^T$ and a velocity vector $v_i = (v_{i1}, v_{i2}, \ldots, v_{iD})^T$. The elements of the vectors p_i and v_i are initially selected randomly. In every time instant k, each particle i updates its velocity and position vectors according to the following equations

$$v_i(k+1) = wv_i(k) + c_1 r_1 (pbest_i(k) - p_i(k)) + c_2 r_2 (gbest - p_i(k)), \quad (1.47)$$

$$p_i(k+1) = p_i(k) + v_i(k+1), \quad (1.48)$$

where w is the inertial weight, and c_1 and c_2 are positive acceleration coefficients used to scale the contribution of the cognitive and the social components, respectively. r_1 and r_2 are uniform random variables in the range [0, 1]. $pbest_i$ is the most suitable position, visited by the particle i during its lifetime. *gbest* is the best position found by the swarm at any time. The standard PSO algorithm is also known as the *gbest* PSO.

Another basic PSO algorithm, which differs from the *gbest* PSO in the size of the particles' neighborhood, is *lbest* PSO. This algorithm defines a neighborhood for each particle. The social component reflects information exchanged within the neighborhood of the particle, reflecting local knowledge of the environment. Regarding the velocity equation, the social contribution to a particle's velocity is proportional to the distance between the particle and the best position found by the neighboring particles of that particle. The velocity is calculated as Eq. (1.49) given below

$$v_i(k+1) = wv_i(k) + c_1 r_1 (pbest_i(k) - p_i(k)) + c_2 r_2 (lbest_i(k) - p_i(k)), \quad (1.49)$$

where $lbest_i$ is the local best position in the neighborhood of particle i.

The main difference between the *gbest* PSO and the *lbest* PSO algorithms is their convergence characteristics (Kiink et al. 2002). Due to the larger particle interconnectivity in the *gbest* PSO, it converges faster than the *lbest* PSO. However, this faster convergence comes at the cost of faster losing the diversity compared to the *lbest* PSO. The larger diversity of the *lbest* PSO makes it less susceptible to being trapped in local minima. Generally, the neighborhood structures, such as the ring topology (Kennedy 1999; Mendes et al. 2003; Peer et al. 2003), and inertia weight strategy

(Nabizadeh et al. 2010; Kordestani et al. 2014, 2016) improve the performance of the PSO in multi-modal environments.

Many real-world problems are dynamic optimization problems where the environment changes over time. Optimization algorithms in such dynamic environments must not only find the optimal solution in a short time but also track it after each change in the environment. In order to design PSO algorithms for dynamic environments, two important points have to be taken into consideration: old memory and diversity loss. Old memory refers to the condition in which memory of the particles, which is the best location visited in the past and its corresponding fitness, may no longer be valid after a change in the environment (Blackwell 2007). Diversity will be lost when all particles of the swarm converge on a few peaks in the landscape. This situation reduces the capability of the swarm to find new peaks if an environmental change occurs (Blackwell 2007).

Cellular PSO (CPSO), recently introduced in (Hashemi and Meybodi 2009), is one of the best PSO-based algorithms which can be used for optimization in dynamic environments. In this algorithm, a cellular automaton (CA) is embedded in the search space and divides it into some partitions; each corresponds to one cell in the CA. Figure 1.23 illustrates a 2-D CA embedded into a two-dimensional search space. CA implicitly divides the swarm into several sub-swarms, each of which resides in a cell. The swarm resides in the cell ci is responsible for finding the highest peak in ci and its neighboring cells. Initially, particles are distributed randomly among the cells of the CA. The particles may migrate from one cell to another in order to provide more searching power for the cells in which peaks may exist with a high probability.

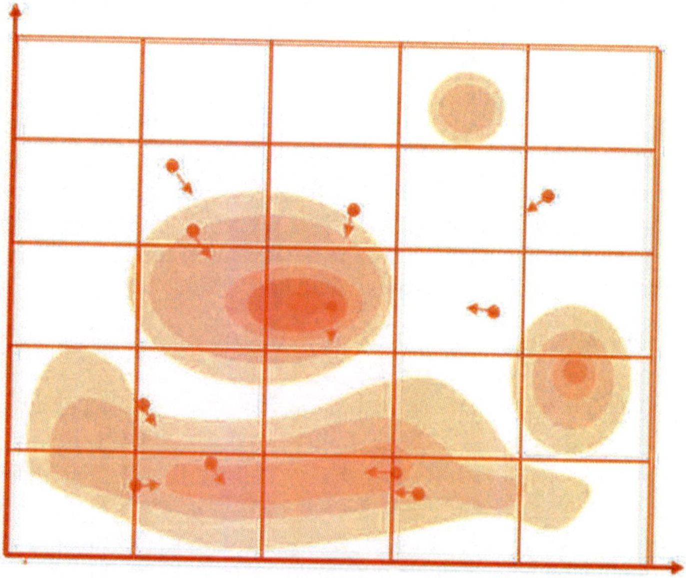

Fig. 1.23 Partitioning of 2-D search space by the CLA

When a cell becomes crowded, some randomly chosen particles move through the CA to some randomly selected cells to search for new areas of the search space. When a change in the environment is detected, the particles around any peak begin a random search around that peak in order to track the changes. The outdated memory problem is solved in the CPSO algorithm by re-evaluating the memory (Eberhart and Shi 2001), and the diversity loss problem is prevented by imposing a limit on the number of particles searching in each cell.

Cell capacity has a considerable effect on the performance of the CPSO algorithm. Therefore, making this parameter adaptive to the environmental changes significantly improves the performance of the CPSO algorithm. To this end, an adaptive version of the CPSO, called CLA-PSO, is proposed in (Hashemi and Meybodi 2009), in which the capacity of each cell is determined using a learning automaton. In this algorithm, instead of a CA, a CLA is embedded in the search space. The LA of each cell is responsible for adjusting the value of the capacity of that cell. Each LA has two actions, namely, INCREASE (the capacity of the cell) and DECREASE (the capacity of the cell). The process of the CLA-PSO in each cell ci takes place as the following steps: First, the particles reside within the cell ci update their velocities and positions according to Eqs. (1.47) and (1.48). Then the LA of the cell c_i selects one of its actions. According to the selected action, the capacity of the cell ci is changed. Afterward, the LA updates its action probability vector, as described below.

If the selected action of the LA is INCREASE, and currently, the particle density of the cell is higher than its capacity, then the LA will be rewarded. Otherwise, it will be penalized. This is because if a cell has attracted too many particles, the cell may contain a peak; thus, it will be a wise choice to increase its capacity. On the other hand, if the particle density in a cell is currently less than its capacity, there is no need to increase the cell capacity. Therefore, in this case, the LA of the cell will be rewarded if it has selected DECREASE action. Otherwise, it will be penalized.

1.4.20 Associative Cellular Learning Automata (Associative CLA)

In associative CLA (Ahangaran et al. 2017), each cell receives two inputs: an input vector from the global environment of the CLA and a reinforcement signal for its performed actions. During each iteration, each cell receives an input form the environment. Then, it selects an action and applies it to the environment. Based on the performed actions, the local rule of the CLA determines the reinforcement signal to each LA. Associative CLA has been applied to applications such as clustering and image segmentation. For instance, clustering can be performed in the following manner. During each iteration, sample data are chosen by the environment and are given to each LA. According to the received input, each LA selects an action using its decision function. The defined action selection works based on the distance of the current state of the cell and the input vector. A LA is rewarded if its chosen action is smaller than those selected in its neighborhood. The learning algorithm of the proposed LA is defined so that upon receiving the reward, a LA update its state and becomes nearer to the input data, while the receiving penalty does not affect the cell's state. Accordingly, the smallest action is chosen by the nearest cell to the input data. So, it is expected that, after some iterations, the state-vectors of cells lie on the centers of the clusters.

Definition 1.44 A d-dimensional associative cellular learning automaton is a structure $\left(Z^d, \Phi, A, \mathrm{X}, N, \mathrm{F}, \mathrm{T}, \mathrm{G}\right)$, where

- Z^d is a lattice of d-tuples of integer numbers,
- Φ is a finite set of states for cells,
- A is the set of LAs, each of which is assigned to one cell of the CLA,
- X is the set of states of the global environment,
- $N = \{x_1, x_2, \ldots, x_k\}$ is the neighborhood vector such that $x_i \in Z^d$,
- $\mathcal{F} : \Phi^k \rightarrow \beta$ is the local rule of the cellular learning automata, where β is the set of values which reinforcement signal can take,
- $T : X \times \Phi \times \beta \rightarrow \Phi$ is a learning algorithm for the learning automata,
- $G : X \times \Phi \rightarrow \alpha$ is the probabilistic decision function. Based on this function, each LA chooses an action from its action set.

In what follows, different components of the Associative CLA are explained.

- **Structure of cellular learning automata**: CLA's cells can be arranged in one dimension (linear), two-dimension (grid), or *n*-dimension (lattice). One can use each of the possible structures, depending on the nature of the problem. For example, in the image processing applications, two dimensional CLA is often used.
- **Input**: Each cell in this model has a learning automaton that gets its input vector from the global environment. The input vector can be the state of the global environment or a feature vector. Also, the input can be discrete or continuous.
- **Learning algorithm**: Learning algorithm of the learning automaton residing in each cell should be an associative algorithm. This family of algorithms updates the cell's state according to the reinforcement signal and the environment's state. A_{R-P} (Barto and Anandan 1985) Moreover, parametrized learning algorithm (PLA) (Thathachar and Phansalkar 1995) are two examples of such algorithms.
- **Environment**: The environment under which the proposed model works can be stationary or non-stationary.
- **Neighborhood**: Each cell in this model has some neighbors. According to the state of neighboring cells, the reinforcement signal of a cell is generated. The neighboring vector of any cell may be fixed or variable.
- **Local rule**: Local rule determines the reinforcement signal for each cell according to the states of neighboring cells.
- **Reinforcement signal**: Reinforcement signal can be binary (zero or one representing punishment or reward, respectively), multi-valued or continuous.

Algorithm 1-7. Associative CLA

```
Initialize the state of each cell in the CLA.
For each cell i in the CLA do
    Give the input vector from the environment to cell i.
    Cell i selects an action according to its decision function.
    Apply the selected action to the environment.
    Apply the local rule and give the reinforcement signal to the cell i.
    Update the state of the cell i according to the reinforcement signal.
End for
```

Fig. 1.24 The pseudo-code for the operation of the Associative CLA

- **Action**: The action set from which LA residing on each cell can select an action. Action set may be a discrete set or a continuous interval.

The Associative CLA is similar to the self-organizing map (SOM). The cells in the Associative CLA are the cells of SOM, and there is a neighborhood concept in both of them. However, the proposed model has stochastic nature in contrast with the SOM. Thus, decisions in the Associative CLA are probabilistic, while in the SOM are deterministic. As a result, the Associative CLA can be regarded as probabilistic SOM. The pseudo-code for the operation of the associative CLA is shown in Fig. 1.24.

1.5 CLA Timeline

Since the introduction of the original model of CLA by in 2004 by Beigy and Meybodi (Beigy and Meybodi 2004), several different models of CLA have been proposed by researchers such as Asynchronous CLA (ACLA), Open Synchronous CLA (OCLA), Synchronous CLA with Multiple LA in Each Cell (SLA-mLA), Asynchronous CLA with Multiple LA in Each Cell (ACLA-mLA), Continuous CLA (CCLA), Irregular CLA (ICLA), Dynamic Irregular CLA (DICLA), Heterogeneous Dynamic Irregular CLA (HDICLA), Closed Asynchronous Dynamic CLA (CADCLA), Closed Asynchronous Dynamic CLA with varying number of LAs in each cell (CADCLA-VL), Wavefront CLA (WCLA), CLA-based Evolutionary Computing (CLA-EC), and Associative CLA. The timeline of appearing in these models is depicted in Fig. 1.25.

1.6 Recent Applications of Cellular Learning Automata

In the recent years, cellular learning automata have been used in distributed, decentralized, dynamic, and unknown environments where a large amount of uncertainty or lacking the information about the environment exists (Beigy and Meybodi 2004;

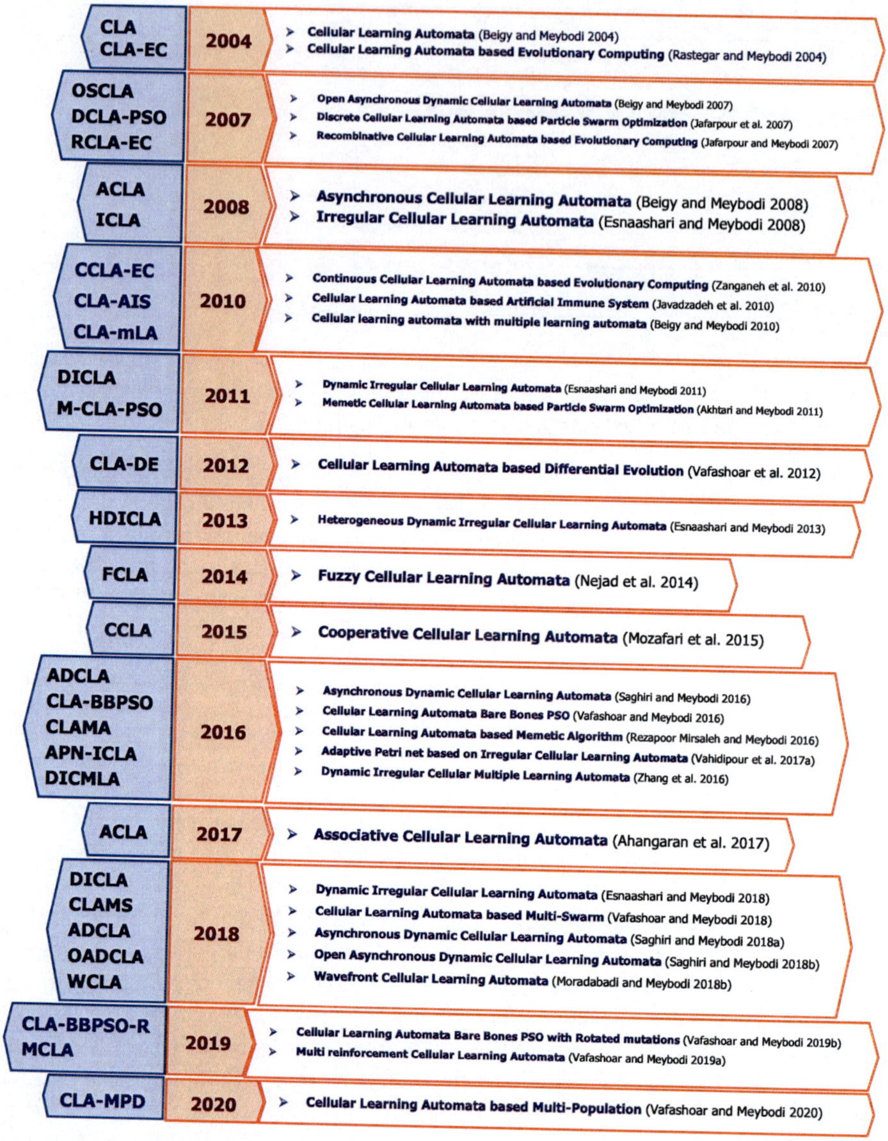

Fig. 1.25 Timeline of cellular learning automata evolution

Rezvanian et al. 2018a, e) and for these reasons, CLA has been successfully applied to a wide range of domains and applications as given in Table 1.2.

Table 1.2 Summary of recent applications of cellular learning automata models

Application	Cellular learning automata model
5G networks	CLA (Qureshi et al. 2019)
Bioinformatics	CLA (Vafaee Sharbaf et al. 2016)
Biomedical	ACO-CLA (Boveiri et al. 2020)
Business process management system	CLA (Saraeian et al. 2019)
Cellular networks	CLA (Beigy and Meybodi 2010)
Channel assignment	CLA (Vafashoar and Meybodi 2019a, b)
Cloud computing	ICLA (Morshedlou and Meybodi 2017), CLA (Kheradmand and Meybodi 2014), CLA-EC (Jalali Moghaddam et al. 2020)
Community detection	ICLA (Zhao et al. 2015; Khomami et al. 2018), ACLA (Motiee and Meybodi 2009)
Data clustering	Associative CLA (Ahangaran et al. 2017), CLA (Rastegar et al. 2005; Hasanzadeh-Mofrad and Rezvanian 2018), CLA-EC (Rastegar et al. 2005)
Data mining	ACLA (Ahangaran et al. 2017), CLA (Sohrabi and Roshani 2017), CLA (Vafaee Sharbaf et al. 2016), (Toozandehjani et al. 2014), CLA-EC (Abad and Derakhshan-Barjoei 2012)
Graph problems	WCLA (Moradabadi and Meybodi 2018b), ICLA (Vahidipour et al. 2017a), ICLA (Rezapoor Mirsaleh and Meybodi 2016), ICLA (Mousavian et al. 2014)
Hardware design	CLA (Hariri et al. 2005), CLA (Zamani et al. 2003)
Hub location	OADCLA (Saghiri and Meybodi 2018b)
Image processing	CLA (Adinehvand et al. 2017), (Arish et al. 2016), CLA (Hasanzadeh Mofrad et al. 2015), (Hadavi et al. 2014), CLA (Gharehchopogh and Ebrahimi 2012)

(continued)

Table 1.2 (continued)

Application	Cellular learning automata model
Intrusion detection	ICLA (FathiNavid and Aghababa 2012), CLA (Aghababa et al. 2012)
Link prediction	ICLA (Khaksar Manshad et al. 2020)
Machine vision	CLA (Mirzamohammad et al. 2014)
Marketing	OCLA (Aldrees and Ykhlef 2014)
Medical imaging	CLA (Hadavi et al. 2014)
Multi-agent systems	CLA (Khani et al. 2017), CLA (Saraeian et al. 2019), CLA (Xue et al. 2019)
Network sampling	ICLA (Ghavipour and Meybodi 2017)
Opportunistic networks	CLA (Zhang et al. 2016)
Optimization	CLA-DE (Vafashoar et al. 2012), CLA (Vafashoar and Meybodi 2016), CLA (Vafashoar and Meybodi 2018), CLA (Mozafari et al. 2015), CLA (Vafashoar and Meybodi 2020),
Peer to peer networks	CLA (Saghiri and Meybodi 2016), ADCLA (Saghiri and Meybodi 2018a), CADCLA (Saghiri and Meybodi 2017b)
Recommender systems	CLA (Talabeigi et al. 2010), CLA (Toozandehjani et al. 2014)
Robotics	CLA (Santoso et al. 2016)
Scheduling	CLA (Abdolzadeh and Rashidi 2010), ACO-CLA (Boveiri et al. 2020)
Social network analysis	ICLA (Ghavipour and Meybodi 2017), ICLA (Khomami et al. 2018), ICLA (Zhao et al. 2015), CLA (Aldrees and Ykhlef 2014), ICLA (Daliri Khomami et al. 2017b)
Software-defined networks	ICLA (Thakur and Khatua 2019)
Task allocation	CLA (Khani et al. 2017)
Transportation	DCLA (Ruan et al. 2019), CLA (Chen et al. 2018)
Wireless networks	ICLA (Mostafaei and Obaidat 2018a), DICLA (Esnaashari and Meybodi 2018), CLA (Assarian et al. 2019), CLA (Torshizi and Sheikhzadeh 2020), CLA (Jahanshahi et al. 2017)

1.7 Chapter Map

The book begins with an introductory chapter providing background tutorials for the material about cellular learning automata models, which many of the later chapters build upon. Chapter 1 is a brief overview to varieties of the CLA models including Asynchronous CLA (ACLA), Open Synchronous CLA (OCLA), Synchronous CLA with Multiple LA in Each Cell (SLA-mLA), Asynchronous CLA with Multiple LA in Each Cell (ACLA-mLA), Continuous CLA (CCLA), Irregular CLA (ICLA), Dynamic Irregular CLA (DICLA), Heterogeneous Dynamic Irregular CLA (HDICLA), Closed Asynchronous Dynamic CLA (CADCLA), Closed Asynchronous Dynamic CLA with varying number of LAs in each cell (CADCLA-VL), Wavefront CLA (WCLA), CLA-based Evolutionary Computing (CLA-EC), and Associative CLA (Associative CLA). Chapter 2 describes the history of the CLA and research trends as bibliometric perspectives. Chapter 3 covers learning from multiple reinforcements in cellular learning automata and its application. Chapter 4 is dedicated to applications of cellular learning automata in optimization and a very recent class of reinforcement learning models for dynamic optimization. Chapter 5 provides a tutorial on applications of multi-reinforcement cellular learning automata in channel assignment as one of the advantages of cellular learning automata.

Then the next set of chapters describes several CLA-based approaches for cloud computing. Chapter 6 provides an introduction to Cellular Learning Automata for Collaborative Loss Sharing for both model and application perspectives. In addition to describing the CLA-based model, the chapter presents some simulation results. Chapter 7 describes Cellular Learning Automata for Competitive Loss Sharing in the application of cloud computing. Finally, Chap. 8, gives a brief description of a Cellular Learning Automata versus Multi-Agent Reinforcement Learning its discussion of models.

1.8 Conclusion

In this chapter, we made a beginning with the basic concepts of cellular automata and learning automata models. We described the definitions and various models of cellular learning automata and also their properties as well. In addition to basic cellular learning automata, recent models of cellular learning automata, like the Irregular CLA, Dynamic Irregular CLA, Heterogeneous Dynamic Irregular CLA, Closed Asynchronous Dynamic CLA, Wavefront CLA, CLA-EC, CLA-DE, and Associative CLA are introduced. We also reported recent applications of cellular learning automata models as well.

References

Abad, M.J.F.H., Derakhshan-Barjoei, P.: Heuristic model of cellular learning automata for fuzzy rule extraction. Res. J. Appl. Sci. Eng. Technol. **4**, 1701–1707 (2012)

Abdolzadeh, M., Rashidi, H.: An approach of cellular learning automata to job shop scheduling problem. Int. J. Simul. Syst. Sci. Technol. **34**, 391–401 (2010)

Adinehvand, K., Sardari, D., Hosntalab, M., Pouladian, M.: An efficient multistage segmentation method for accurate hard exudates and lesion detection in digital retinal images. J. Intell. Fuzzy Syst. **33**, 1639–1649 (2017). https://doi.org/10.3233/JIFS-17199

Agache, M., Oommen, B.J.: Generalized pursuit learning schemes: new families of continuous and discretized learning automata. IEEE Trans. Syst. Man Cybern. Part B Cybern. **32**, 738–749 (2002). https://doi.org/10.1109/TSMCB.2002.1049608

Aghababa, A.B., Fathinavid, A., Salari, A., Zavareh, S.E.H.: A novel approach for malicious nodes detection in ad-hoc networks based on cellular learning automata. In: 2012 World Congress on Information and Communication Technologies. IEEE, pp. 82–88 (2012)

Ahangaran, M., Taghizadeh, N., Beigy, H.: Associative cellular learning automata and its applications. Appl. Soft Comput. **53**, 1–18 (2017). https://doi.org/10.1016/j.asoc.2016.12.006

Akbari Torkestani, J., Meybodi, M.R.: Learning automata-based algorithms for finding minimum weakly connected dominating set in stochastic graphs. Int. J. Uncertain., Fuzziness Knowl.-Based Syst. **18**, 721–758 (2010). https://doi.org/10.1142/S0218488510006775

Akbari Torkestani, J., Meybodi, M.R.: A learning automata-based heuristic algorithm for solving the minimum spanning tree problem in stochastic graphs. J. Supercomput. **59**, 1035–1054 (2012). https://doi.org/10.1007/s11227-010-0484-1

Akhtari, M., Meybodi, M.R.: Memetic-CLA-PSO: a hybrid model for optimization. In: 2011 UkSim 13th International Conference on Computer Modelling and Simulation. IEEE, pp. 20–25 (2011)

Aldrees, M., Ykhlef, M.: A seeding cellular learning automata approach for viral marketing in social network. In: Proceedings of the 16th International Conference on Information Integration and Web-Based Applications & Services—iiWAS'14. ACM Press, New York, New York, USA, pp. 59–63 (2014)

Alirezanejad, M., Enayatifar, R., Motameni, H., Nematzadeh, H.: GSA-LA: gravitational search algorithm based on learning automata. J. Exp. Theor. Artif. Intell. 1–17 (2020). https://doi.org/10.1080/0952813X.2020.1725650

Amirazodi, N., Saghiri, A.M., Meybodi, M.: An adaptive algorithm for super-peer selection considering peer's capacity in mobile peer-to-peer networks based on learning automata. Peer-to-Peer Netw. Appl. **11**, 74–89 (2018). https://doi.org/10.1007/s12083-016-0503-y

Amiri, F., Yazdani, N., Faili, H., Rezvanian, A.: A novel community detection algorithm for privacy preservation in social networks. In: Intelligent Informatics, pp. 443–450 (2013)

Arish, S., Javaherian, M., Safari, H., Amiri, A.: Extraction of active regions and coronal holes from EUV images using the unsupervised segmentation method in the Bayesian framework. Sol. Phys. **291**, 1209–1224 (2016). https://doi.org/10.1007/s11207-016-0883-4

Aso, H., Kimura, M.: Absolute expediency of learning automata. Inf. Sci. (Ny) **17**, 91–112 (1979). https://doi.org/10.1016/0020-0255(79)90034-3

Assarian, A., Khademzadeh, A., Hosseinzadeh, M., Setayeshi, S.: An efficient resource allocation scheme in a dense RFID network based on cellular learning automata. Int. J. Commun. Syst. **32**, e3835 (2019). https://doi.org/10.1002/dac.3835

Bak, P., Sneppen, K.: Punctuated equilibrium and criticality in a simple model of evolution. Phys. Rev. Lett. **71**, 4083–4086 (1993). https://doi.org/10.1103/PhysRevLett.71.4083

Barto, A.G., Anandan, P.: Pattern-recognizing stochastic learning automata. IEEE Trans. Syst. Man Cybern. SMC-15:360–375 (1985). https://doi.org/10.1109/TSMC.1985.6313371

Barto, A.G., Sutton, R.S., Anderson, C.W.: Neuronlike adaptive elements that can solve difficult learning control problems. IEEE Trans. Syst. Man Cybern. SMC-13:834–846 (1983). https://doi.org/10.1109/TSMC.1983.6313077

Beheshtifard, Z., Meybodi, M.R.: An adaptive channel assignment in wireless mesh network: the learning automata approach. Comput. Electr. Eng. **72**, 79–91 (2018). https://doi.org/10.1016/j.compeleceng.2018.09.004

Beigy, H., Meybodi, M.R.: A mathematical framework for cellular learning automata. Adv. Complex Syst. **07**, 295–319 (2004). https://doi.org/10.1142/S0219525904000202

Beigy, H., Meybodi, M.R.: Utilizing distributed learning automata to solve stochastic shortest path problems. Int. J. Uncertain., Fuzziness Knowl.-Based Syst. **14**, 591–615 (2006). https://doi.org/10.1142/S0218488506004217

Beigy, H., Meybodi, M.R.: Open synchronous cellular learning automata. Adv. Complex Syst. **10**, 527–556 (2007)

Beigy, H., Meybodi, M.R.: Asynchronous cellular learning automata. Automatica **44**, 1350–1357 (2008)

Beigy, H., Meybodi, M.R.R.: Cellular learning automata with multiple learning automata in each cell and its applications. IEEE Trans. Syst. Man, Cybern. Part B **40**, 54–65 (2010). https://doi.org/10.1109/TSMCB.2009.2030786

Betka, A., Terki, N., Toumi, A., Dahmani, H.: Grey wolf optimizer-based learning automata for solving block matching problem. Signal, Image Video Process. **14**, 285–293 (2020). https://doi.org/10.1007/s11760-019-01554-w

Bhattacharjee, K., Naskar, N., Roy, S., Das, S.: A survey of cellular automata: types, dynamics, non-uniformity and applications. Nat. Comput. (2018). https://doi.org/10.1007/s11047-018-9696-8

Blackwell, T.: Particle Swarm Optimization in Dynamic Environments, pp. 29–49 (2007)

Boettcher, S., Percus, A.G.: Optimization with extremal dynamics. Complexity **8**, 57–62 (2002). https://doi.org/10.1002/cplx.10072

Boveiri, H.R., Javidan, R., Khayami, R.: An intelligent hybrid approach for task scheduling in cluster computing environments as an infrastructure for biomedical applications. Expert Syst. https://doi.org/10.1111/exsy.12536 (2020)

Burkowski, F.J.: Shuffle crossover and mutual information. In: Proceedings of the 1999 Congress on Evolutionary Computation-CEC99 (Cat. No. 99TH8406). IEEE, pp. 1574–1580 (1999)

Bushehrian, O., Nejad, S.E.: Health-Care Pervasive Environments: A CLA Based Trust Management, pp. 247–257 (2017)

Montague, P.R.: Reinforcement learning: an introduction, by Sutton, R.S. and Barto, A.G. Trends Cogn. Sci. **3**, 360 (1999). https://doi.org/10.1016/S1364-6613(99)01331-5

Chen, Y., He, H., Zhou, N.: Traffic flow modeling and simulation based on a novel cellular learning automaton. In: 2018 IEEE International Conference of Intelligent Robotic and Control Engineering (IRCE). IEEE, pp. 233–237 (2018)

Cook, M.: Universality in elementary cellular automata. Complex Syst. **15**, 1–40 (2004)

Copeland, B.: The church-turing thesis (1997)

Daliri Khomami, M.M., Haeri, M.A., Meybodi, M.R., Saghiri, A.M.: An algorithm for weighted positive influence dominating set based on learning automata. In: 2017 IEEE 4th International Conference on Knowledge-Based Engineering and Innovation (KBEI). IEEE, pp. 0734–0740 (2017a)

Daliri Khomami, M.M., Rezvanian, A., Bagherpour, N., Meybodi, M.R.: Irregular cellular automata based diffusion model for influence maximization. In: 2017 5th Iranian Joint Congress on Fuzzy and Intelligent Systems (CFIS). IEEE, pp. 69–74 (2017b)

Daliri Khomami, M.M., Rezvanian, A., Bagherpour, N., Meybodi, M.R.: Minimum positive influence dominating set and its application in influence maximization: a learning automata approach. Appl. Intell. **48**, 570–593 (2018). https://doi.org/10.1007/s10489-017-0987-z

Damerchilu, B., Norouzzadeh, M.S., Meybodi, M.R.: Motion estimation using learning automata. Mach. Vis. Appl. **27**, 1047–1061 (2016). https://doi.org/10.1007/s00138-016-0788-0

de Sousa, F.L., Ramos, F.M.: Function optimization using extremal dynamics. In: Icipe'02. Brazil, pp. 115–119 (2002)

Deng, X., Jiang, Y., Yang, L.T., et al.: Learning automata based confident information coverage barriers for smart ocean internet of things. IEEE Internet Things J. 1–1 (2020). https://doi.org/10.1109/JIOT.2020.2989696

Di, C., Su, Y., Han, Z., Li, S.: Learning Automata Based SVM for Intrusion Detection, pp. 2067–2074 (2019)

Di, C., Zhang, B., Liang, Q., et al.: Learning automata based access class barring scheme for massive random access in machine-to-machine communications. IEEE Internet Things J. 1–1 (2018). https://doi.org/10.1109/JIOT.2018.2867937

Eberhart, R.C., Shi, Y.: Tracking and optimizing dynamic systems with particle swarms. In: Proceedings of the 2001 Congress on Evolutionary Computation (IEEE Cat. No. 01TH8546), pp. 94–100 (2001)

Esnaashari, M., Meybodi, M.R.R.: Dynamic point coverage in wireless sensor networks: a learning automata approach. In: Advances in Computer Science and Engineering. Springer, pp. 758–762 (2009)

Esnaashari, M., Meybodi, M.R.: A cellular learning automata based clustering algorithm for wireless sensor networks. Sens. Lett. **6**, 723–735 (2008)

Esnaashari, M., Meybodi, M.R.M.: A cellular learning automata-based deployment strategy for mobile wireless sensor networks. J. Parallel Distrib. Comput. **71**, 988–1001 (2011)

Esnaashari, M., Meybodi, M.R.: Deployment of a mobile wireless sensor network with k-coverage constraint: a cellular learning automata approach. Wirel. Netw. **19**, 945–968 (2013). https://doi.org/10.1007/s11276-012-0511-7

Esnaashari, M., Meybodi, M.R.: Irregular cellular learning automata. IEEE Trans. Cybern. **45**, 1622–1632 (2018). https://doi.org/10.1016/j.jocs.2017.08.012

Fahimi, M., Ghasemi, A.: A distributed learning automata scheme for spectrum management in self-organized cognitive radio network. IEEE Trans. Mob. Comput. **16**, 1490–1501 (2017). https://doi.org/10.1109/TMC.2016.2601926

Farsi, H., Nasiripour, R., Mohammadzadeh, S.: Eye gaze detection based on learning automata by using SURF descriptor. J. Inf. Syst. Telecommun. **21**, 1–10 (2018). https://doi.org/10.7508/jist.2018.21.006

FathiNavid, A., Aghababa, A.B.: Irregular cellular learning automata-based method for intrusion detection in mobile ad hoc networks. In: 51st International FITCE (Federation of Telecommunications Engineers of the European Community), pp. 1–6 (2012)

Friedman, E., Shenker, S.: Synchronous and Asynchronous Learning by Responsive Learning Automata (1996)

Fu, K.-S.: Learning control systems–review and outlook. IEEE Trans. Automat. Contr. **15**, 210–221 (1970). https://doi.org/10.1109/TAC.1970.1099405

Ge, H., Huang, J., Di, C., et al.: Learning automata based approach for influence maximization problem on social networks. In: 2017 IEEE Second International Conference on Data Science in Cyberspace (DSC). IEEE, pp. 108–117 (2017)

Ghamgosar, M., Khomami, M.M.D., Bagherpour, N., Meybodi, M.R.: An extended distributed learning automata based algorithm for solving the community detection problem in social networks. In: 2017 Iranian Conference on Electrical Engineering (ICEE). IEEE, pp. 1520–1526 (2017)

Gharehchopogh, F.S., Ebrahimi, S.: A novel approach for edge detection in images based on cellular learning automata. Int. J. Comput. Vis. Image Process **2**, 51–61 (2012). https://doi.org/10.4018/ijcvip.2012100105

Ghavipour, M., Meybodi, M.R.: An adaptive fuzzy recommender system based on learning automata. Electron. Commer. Res. Appl. **20**, 105–115 (2016). https://doi.org/10.1016/j.elerap.2016.10.002

Ghavipour, M., Meybodi, M.R.: Irregular cellular learning automata-based algorithm for sampling social networks. Eng. Appl. Artif. Intell. **59**, 244–259 (2017). https://doi.org/10.1016/j.engappai.2017.01.004

Ghavipour, M., Meybodi, M.R.: Trust propagation algorithm based on learning automata for inferring local trust in online social networks. Knowl.-Based Syst. **143**, 307–316 (2018a). https://doi.org/10.1016/j.knosys.2017.06.034

Ghavipour, M., Meybodi, M.R.: A streaming sampling algorithm for social activity networks using fixed structure learning automata. Appl. Intell. **48**, 1054–1081 (2018b). https://doi.org/10.1007/s10489-017-1005-1

Ghavipour, M., Meybodi, M.R.: A dynamic algorithm for stochastic trust propagation in online social networks: learning automata approach. Comput. Commun. **123**, 11–23 (2018c). https://doi.org/10.1016/j.comcom.2018.04.004

Ghosh, L., Ghosh, S., Konar, D., et al.: EEG-induced error correction in path planning by a mobile robot using learning automata. In: Soft Computing for Problem Solving, pp. 273–285 (2019)

Goodwin, M., Yazidi, A.: Distributed learning automata-based scheme for classification using novel pursuit scheme. Appl. Intell. (2020). https://doi.org/10.1007/s10489-019-01627-w

Hadavi, N., Nordin, M.J., Shojaeipour, A.: Lung cancer diagnosis using CT-scan images based on cellular learning automata. In: 2014 International Conference on Computer and Information Sciences (ICCOINS). IEEE, pp. 1–5 (2014)

Han, Z., Li, S.: Opportunistic Routing Algorithm Based on Estimator Learning Automata, pp. 2486–2492 (2019)

Hariri, A., Rastegar, R., Zamani, M.S., Meybodi, M.R.: Parallel hardware implementation of cellular learning automata based evolutionary computing (CLA-EC) on FPGA. In: 13th Annual IEEE Symposium on Field-Programmable Custom Computing Machines (FCCM'05). IEEE, pp. 311–314 (2005)

Hasanzadeh, M., Meybodi, M.R.: Grid resource discovery based on distributed learning automata. Computing **96**, 909–922 (2014). https://doi.org/10.1007/s00607-013-0337-x

Hasanzadeh Mofrad, M., Sadeghi, S., Rezvanian, A., Meybodi, M.R.: Cellular edge detection: combining cellular automata and cellular learning automata. AEU—Int. J. Electron. Commun. **69**, 1282–1290 (2015). https://doi.org/10.1016/j.aeue.2015.05.010

Hasanzadeh-Mofrad, M., Rezvanian, A.: Learning automata clustering. J. Comput. Sci. **24**, 379–388 (2018). https://doi.org/10.1016/j.jocs.2017.09.008

Hashemi, A.B., Meybodi, M.R.: Cellular PSO: a PSO for dynamic environments. In: Cai, Z. (ed.) Lecture Notes in Computer Science (including subseries Lecture Notes in Artificial Intelligence and Lecture Notes in Bioinformatics). Springer, Berlin, Heidelberg, Tehran, Iran, pp. 422–433 (2009)

Huang, J., Ge, H., Guo, Y., et al.: A Learning Automaton-Based Algorithm for Influence Maximization in Social Networks, pp. 715–722 (2018)

Irandoost, M.A., Rahmani, A.M., Setayeshi, S.: Learning automata-based algorithms for MapReduce data skewness handling. J. Supercomput. **75**, 6488–6516 (2019a). https://doi.org/10.1007/s11227-019-02855-0

Irandoost, M.A., Rahmani, A.M., Setayeshi, S.: A novel algorithm for handling reducer side data skew in MapReduce based on a learning automata game. Inf. Sci. (Ny) **501**, 662–679 (2019b). https://doi.org/10.1016/j.ins.2018.11.007

Jafarpour, B., Meybodi, M.R., Shiry, S.: A hybrid method for optimization (discrete PSO + CLA). In: 2007 International Conference on Intelligent and Advanced Systems, ICIAS 2007, pp. 55–60 (2007)

Jafarpour, B., Meybodi, M.R.: Recombinative CLA-EC. In: Proceedings—International Conference on Tools with Artificial Intelligence, ICTAI. IEEE, pp. 415–422 (2007)

Jahanshahi, M., Dehghan, M., Meybodi, M.R.: A cross-layer optimization framework for joint channel assignment and multicast routing in multi-channel multi-radio wireless mesh networks. Int. J. Comput. Math. **94**, 1624–1652 (2017). https://doi.org/10.1080/00207160.2016.1227431

Jalali Moghaddam, M., Esmaeilzadeh, A., Ghavipour, M., Zadeh, A.K.: Minimizing virtual machine migration probability in cloud computing environments. Cluster Comput. (2020). https://doi.org/10.1007/s10586-020-03067-5

Javadi, M., Mostafaei, H., Chowdhurry, M.U., Abawajy, J.H.: Learning automaton based topology control protocol for extending wireless sensor networks lifetime. J. Netw. Comput. Appl. **122**, 128–136 (2018). https://doi.org/10.1016/j.jnca.2018.08.012

Javadzadeh, R., Afsahi, Z., Meybodi, M.R.: Hybrid Model Base on Artificial Immune System and Cellular Learning Automata (CLA-AIS). In: IASTED Technology Conferences/705: ARP/706: RA/707: NANA/728: CompBIO. ACTAPRESS, Calgary, AB, Canada (2010)

Jobava, A., Yazidi, A., Oommen, B.J., Begnum, K.: On achieving intelligent traffic-aware consolidation of virtual machines in a data center using learning automata. J. Comput. Sci. **24**, 290–312 (2018). https://doi.org/10.1016/j.jocs.2017.08.005

John Oommen, B., Agache, M.: Continuous and discretized pursuit learning schemes: Various algorithms and their comparison. IEEE Trans. Syst. Man Cybern. Part B Cybern. **31**, 277–287 (2001). https://doi.org/10.1109/3477.931507

Kaelbling, L.P., Littman, M.L., Moore, A.W.: Reinforcement learning: a survey. J. Artif. Intell. Res. **4**, 237–285 (1996). https://doi.org/10.1613/jair.301

Kahani, N., Fallah, M.S.: A reactive defense against bandwidth attacks using learning automata. In: Proceedings of the 13th International Conference on Availability, Reliability and Security—ARES 2018. ACM Press, New York, New York, USA, pp. 1–6 (2018)

Kazemi Kordestani, J., Meybodi, M.R., Rahmani, A.M.: A two-level function evaluation management model for multi-population methods in dynamic environments: hierarchical learning automata approach. J. Exp. Theor. Artif Intell. 1–26 (2020). https://doi.org/10.1080/0952813X.2020.1721568

Kennedy, J., Eberhart, R.: Particle swarm optimization. In: Proceedings of ICNN'95—International Conference on Neural Networks. IEEE, pp. 1942–1948 (1995)

Kennedy, J.: Small worlds and mega-minds: effects of neighborhood topology on particle swarm performance. In: Proceedings of the 1999 Congress on Evolutionary Computation-CEC99 (Cat. No. 99TH8406). IEEE, pp. 1931–1938 (1999)

Khadangi, E., Bagheri, A., Shahmohammadi, A.: Biased sampling from facebook multilayer activity network using learning automata. Appl. Intell. **45**, 829–849 (2016). https://doi.org/10.1007/s10489-016-0784-0

Khaksar Manshad, M., Meybodi, M.R., Salajegheh, A.: A new irregular cellular learning automata-based evolutionary computation for time series link prediction in social networks. Appl. Intell. (2020). https://doi.org/10.1007/s10489-020-01685-5

Khaksarmanshad, M., Meybodi, M.R.: Designing optimization algorithms based on CLA-EC for dynamic environments. In: The 4th Iran Data Mining Conference (IDMC'10), pp. 1–6 (2010)

Khani, M., Ahmadi, A., Hajary, H.: Distributed task allocation in multi-agent environments using cellular learning automata. Soft. Comput. (2017). https://doi.org/10.1007/s00500-017-2839-5

Khatamnejad, A., Meybodi, M.R.: A hybrid method for optimization (CLA-EC + extremal optimization). In: 13th Annual CSI Computer Conf. of Iran, pp. 1–6 (2008)

Kheradmand, S., Meybodi, M.R.: Price and QoS competition in cloud market by using cellular learning automata. In: 2014 4th International Conference on Computer and Knowledge Engineering (ICCKE). IEEE, pp. 340–345 (2014)

Khomami, M.M.D., Bagherpour, N., Sajedi, H., Meybodi, M.R.: A new distributed learning automata based algorithm for maximum independent set problem. In: 2016 Artificial Intelligence and Robotics (IRANOPEN). IEEE, Qazvin, Iran, Iran, pp. 12–17 (2016a)

Khomami, M.M.D., Rezvanian, A., Meybodi, M.R.: Distributed learning automata-based algorithm for community detection in complex networks. Int. J. Mod. Phys. B **30**, 1650042 (2016). https://doi.org/10.1142/S0217979216500429

Khomami, M.M.D., Rezvanian, A., Meybodi, M.R.: A new cellular learning automata-based algorithm for community detection in complex social networks. J. Comput. Sci. **24**, 413–426 (2018). https://doi.org/10.1016/j.jocs.2017.10.009

Kiink, T., Vesterstroem, J.S., Riget, J.: Particle swarm optimization with spatial particle extension. In: IEEE Congress on Evolutionary Computation, pp. 1474–1479 (2002)

Kordestani, J.K., Rezvanian, A., Meybodi, M.R.: An efficient oscillating inertia weight of particle swarm optimisation for tracking optima in dynamic environments. J. Exp. Theor. Artif. Intell. **28**, 137–149 (2016). https://doi.org/10.1080/0952813X.2015.1020521

Kordestani, J.K., Rezvanian, A., Meybodi, M.R.: CDEPSO: a bi-population hybrid approach for dynamic optimization problems. Appl. Intell. **40**, 682–694 (2014). https://doi.org/10.1007/s10489-013-0483-z

Kordestani, J.K., Firouzjaee, H.A., Meybodi, M.R.: An adaptive bi-flight cuckoo search with variable nests for continuous dynamic optimization problems. Appl. Intell. **48**, 97–117 (2018). https://doi.org/10.1007/s10489-017-0963-7

Krishna, K.: Cellular learning automata: a stochastic model for adaptive controllers. Master's thesis, Department of Electrical Engineering, Indian Institue of Science Banglore, India (1993)

Krishna, P.V., Misra, S., Joshi, D., Obaidat, M.S.: Learning automata based sentiment analysis for recommender system on cloud. In: 2013 International Conference on Computer, Information and Telecommunication Systems (CITS). IEEE, pp. 1–5 (2013)

Krishna, P.V., Misra, S., Joshi, D., et al.: Secure socket layer certificate verification: a learning automata approach. Secur. Commun. Netw. **7**, 1712–1718 (2014). https://doi.org/10.1002/sec.867

Kumar, N., Misra, S., Obaidat, M.S.M.S.M.S.M.S.: Collaborative learning automata-based routing for rescue operations in dense urban regions using vehicular sensor networks. IEEE Syst. J. **9**, 1081–1090 (2015a). https://doi.org/10.1109/JSYST.2014.2335451

Kumar, N.N., Lee, J.H.J.-H., Rodrigues, J.J.P.C.: Intelligent mobile video surveillance system as a Bayesian coalition game in vehicular sensor networks: learning automata approach. IEEE Trans. Intell. Transp. Syst. **16**, 1148–1161 (2015b). https://doi.org/10.1109/TITS.2014.2354372

Kumpati, S., Narendra, M.A.L.T.: Learning Automata: An Introduction. Prentice-Hall (1989)

Lanctot, J.K., Oommen, B.J.: Discretized Estimator Learning Automata. IEEE Trans. Syst. Man Cybern. **22**, 1473–1483 (1992). https://doi.org/10.1109/21.199471

Li, W., Ozcan, E., John, R.: A learning automata based multiobjective hyper-heuristic. IEEE Trans. Evol. Comput. 1–1 (2018). https://doi.org/10.1109/TEVC.2017.2785346

Lingam, G., Rout, R.R., Somayajulu, D.: Learning automata-based trust model for user recommendations in online social networks. Comput. Electr. Eng. **66**, 174–188 (2018). https://doi.org/10.1016/j.compeleceng.2017.10.017

Mahdaviani, M., Kordestani, J.K., Rezvanian, A., Meybodi, M.R.: LADE: learning automata based differential evolution. Int. J. Artif. Intell. Tools **24**, 1550023 (2015). https://doi.org/10.1142/S0218213015500232

Mahmoudi, M., Faez, K., Ghasemi, A.: Defense against primary user emulation attackers based on adaptive Bayesian learning automata in cognitive radio networks. Ad Hoc Netw. **102**, 102147 (2020). https://doi.org/10.1016/j.adhoc.2020.102147

Masoodifar, B., Meybodi, M.R., Hashemi, M.: Cooperative CLA-EC. In: 12th Annual CSI Computer Conference of Iran, pp. 558–559 (2007)

Mendes, R., Kennedy, J., Neves, J.: Watch thy neighbor or how the swarm can learn from its environment. In: Proceedings of the 2003 IEEE Swarm Intelligence Symposium. SIS'03 (Cat. No. 03EX706). IEEE, pp. 88–94 (2003)

Meybodi, M.R., Lakshmivarahan, S.: ε-optimality of a general class of learning algorithms. Inf. Sci. (Ny) **28**, 1–20 (1982). https://doi.org/10.1016/0020-0255(82)90029-9

Mirzamohammad, M., Ahmadi, A., Mirzarezaee, M.: Moving object detection and tracking in video by cellular learning automata and gradient method in fuzzy domain. J. Intell. Fuzzy Syst. **27**, 929–935 (2014). https://doi.org/10.3233/IFS-131052

Misra, S., Interior, B., Kumar, N., et al.: Networks of learning automata for the vehicular environment: a performance analysis study. IEEE Wirel. Commun. **21**, 41–47 (2014). https://doi.org/10.1109/MWC.2014.7000970

Mollakhalili Meybodi, M.R., Meybodi, M.R.: Extended distributed learning automata. Appl. Intell. **41**, 923–940 (2014). https://doi.org/10.1007/s10489-014-0577-2

Moradabadi, B., Meybodi, M.R.: Link prediction based on temporal similarity metrics using continuous action set learning automata. Phys. A Stat. Mech. Its Appl. **460**, 361–373 (2016). https://doi.org/10.1016/j.physa.2016.03.102

Moradabadi, B., Meybodi, M.R.: Link prediction in fuzzy social networks using distributed learning automata. Appl. Intell. **47**, 837–849 (2017a). https://doi.org/10.1007/s10489-017-0933-0

Moradabadi, B., Meybodi, M.R.: A novel time series link prediction method: learning automata approach. Phys. A Stat. Mech. Its Appl. **482**, 422–432 (2017b). https://doi.org/10.1016/j.physa.2017.04.019

Moradabadi, B., Meybodi, M.R.: Link prediction in weighted social networks using learning automata. Eng. Appl. Artif. Intell. **70**, 16–24 (2018a). https://doi.org/10.1016/j.engappai.2017.12.006

Moradabadi, B., Meybodi, M.R.: Wavefront cellular learning automata. Chaos **28**, 21101 (2018b). https://doi.org/10.1063/1.5017852

Moradabadi, B., Meybodi, M.R.: Link prediction in stochastic social networks: learning automata approach. J. Comput. Sci. **24**, 313–328 (2018c). https://doi.org/10.1016/j.jocs.2017.08.007

Morshedlou, H., Meybodi, M.R.: Decreasing impact of SLA violations: a proactive resource allocation approach for cloud computing environments. IEEE Trans. Cloud Comput. **2**, 156–167 (2014). https://doi.org/10.1109/TCC.2014.2305151

Morshedlou, H., Meybodi, M.R.: A new local rule for convergence of ICLA to a compatible point. IEEE Trans. Syst. Man, Cybern. Syst. **47**, 3233–3244 (2017). https://doi.org/10.1109/TSMC.2016.2569464

Morshedlou, H., Meybodi, M.R.: A new learning automata based approach for increasing utility of service providers. Int. J. Commun. Syst. **31**, e3459 (2018). https://doi.org/10.1002/dac.3459

Mostafaei, H.: Stochastic barrier coverage in wireless sensor networks based on distributed learning automata. Comput. Commun. **55**, 51–61 (2015)

Mostafaei, H., Obaidat, M.S.: A distributed efficient algorithm for self-protection of wireless sensor networks. In: 2018 IEEE International Conference on Communications (ICC). IEEE, pp 1–6 (2018b)

Mostafaei, H.: Energy-efficient algorithm for reliable routing of wireless sensor networks. IEEE Trans. Ind. Electron 1–1 (2018). https://doi.org/10.1109/TIE.2018.2869345

Mostafaei, H., Obaidat, M.S.: Learning automaton-based self-protection algorithm for wireless sensor networks. IET Netw. **7**, 353–361 (2018). https://doi.org/10.1049/iet-net.2018.0005

Motiee, S., Meybodi, M.R.: Identification of web communities using cellular learning automata. In: 2009 14th International CSI Computer Conference. IEEE, pp. 553–563 (2009)

Mousavian, A., Rezvanian, A., Meybodi, M.R.: Cellular learning automata based algorithm for solving minimum vertex cover problem. In: 2014 22nd Iranian Conference on Electrical Engineering (ICEE). IEEE, pp. 996–1000 (2014)

Mousavian, A., Rezvanian, A., Meybodi, M.R.: Solving minimum vertex cover problem using learning automata. In: 13th Iranian Conference on Fuzzy Systems (IFSC 2013), pp. 1–5 (2013)

Mozafari, M., Shiri, M.E., Beigy, H.: A cooperative learning method based on cellular learning automata and its application in optimization problems. J. Comput. Sci. **11**, 279–288 (2015). https://doi.org/10.1016/j.jocs.2015.08.002

Nabizadeh, S., Faez, K., Tavassoli, S., Rezvanian, A.: A novel method for multi-level image thresholding using particle swarm optimization algorithms. In: ICCET 2010—2010 International Conference on Computer Engineering and Technology, Proceedings. pp. V4-271–V4-275 (2010)

Narendra, K.S., Thathachar, M.A.L.: Learning automata—a survey. IEEE Trans. Syst. Man Cybern. SMC-4, 323–334 (1974). https://doi.org/10.1109/TSMC.1974.5408453

Nejad, H.C., Azadbakht, B., Adenihvand, K., et al.: Fuzzy cellular learning automata for lesion detection in retina images. J. Intell. Fuzzy Syst. **27**, 2297–2303 (2014). https://doi.org/10.3233/IFS-141194

Nesi, L.C., da Rosa Righi, R.: H2-SLAN: A hyper-heuristic based on stochastic learning automata network for obtaining, storing, and retrieving heuristic knowledge. Expert Syst. Appl. **153**, 113426 (2020). https://doi.org/10.1016/j.eswa.2020.113426

Packard, N.H., Wolfram, S.: Two-dimensional cellular automata. J. Stat. Phys. **38**, 901–946 (1985). https://doi.org/10.1007/BF01010423

Papadimitriou, G.I., Pomportsis, A.S., Kiritsi, S., Talahoupi, E.: Absorbing stochastic estimator learning algorithms with high accuracy and rapid convergence. In: Proceedings ACS/IEEE International Conference on Computer Systems and Applications. IEEE Comput. Soc, pp. 45–51 (2002)

Papadimitriou, G.I., Vasilakos, A.V., Papadimitriou, G.I., Paximadis, C.T.: A new approach to the design of reinforcement schemes for learning automata: stochastic estimator learning algorithms. In: Conference Proceedings 1991 IEEE International Conference on Systems, Man, and Cybernetics. IEEE, pp. 1387–1392 (1991)

Parvanak, A.R., Jahanshahi, M., Dehghan, M.: A cross-layer learning automata based gateway selection method in multi-radio multi-channel wireless mesh networks. Computing (2018). https://doi.org/10.1007/s00607-018-0648-z

Peer, E.S., van den Bergh, F., Engelbrecht, A.P.: Using neighbourhoods with the guaranteed convergence PSO. In: Proceedings of the 2003 IEEE Swarm Intelligence Symposium, SIS'03 (Cat. No. 03EX706). IEEE, pp. 235–242 (2003)

Qavami, H.R., Jamali, S., Akbari, M.K., Javadi, B.: A learning automata based dynamic resource provisioning in cloud computing environments. In: 2017 18th International Conference on Parallel and Distributed Computing, Applications and Technologies (PDCAT). IEEE, pp. 502–509 (2017)

Qureshi, M.N., Tiwana, M.I., Haddad, M.: Distributed self optimization techniques for heterogeneous network environments using active antenna tilt systems. Telecommun. Syst. **70**, 379–389 (2019). https://doi.org/10.1007/s11235-018-0494-5

Rahmani, P., Javadi, H.H.S., Bakhshi, H., Hosseinzadeh, M.: TCLAB: a new topology control protocol in cognitive MANETs based on learning automata. J. Netw. Syst. Manag. **26**, 426–462 (2018). https://doi.org/10.1007/s10922-017-9422-3

Rahmanian, A.A., Ghobaei-Arani, M., Tofighy, S.: A learning automata-based ensemble resource usage prediction algorithm for cloud computing environment. Fut. Gener. Comput. Syst. **79**, 54–71 (2018). https://doi.org/10.1016/j.future.2017.09.049

Rasouli, N., Razavi, R., Faragardi, H.R.: EPBLA: energy-efficient consolidation of virtual machines using learning automata in cloud data centers. Cluster Comput. (2020). https://doi.org/10.1007/s10586-020-03066-6

Rastegar, R., Meybodi, M.R.: A new evolutionary computing model based on cellular learning automata. In: IEEE Conference on Cybernetics and Intelligent Systems, 2004. IEEE, pp. 433–438 (2004)

Rastegar, R., Rahmati, M., Meybodi, M.R.: A clustering algorithm using cellular learning automata based evolutionary algorithm. Adaptive and Natural Computing Algorithms, pp. 144–150. Springer, Vienna (2005)

Rastegar, R., Meybodi, M.R., Hariri, A.: A new fine-grained evolutionary algorithm based on cellular learning automata. Int. J. Hybrid Intell. Syst. **3**, 83–98 (2006). https://doi.org/10.3233/HIS-2006-3202

Ren, J., Wu, G., Su, X., et al.: Learning automata-based data aggregation tree construction framework for cyber-physical systems. IEEE Syst. J. **12**, 1467–1479 (2018). https://doi.org/10.1109/JSYST.2015.2507577

Rezapoor Mirsaleh, M., Meybodi, M.R.: A learning automata-based memetic algorithm. Genet. Program Evol. Mach. **16**, 399–453 (2015). https://doi.org/10.1007/s10710-015-9241-9

Rezapoor Mirsaleh, M., Meybodi, M.R.: A new memetic algorithm based on cellular learning automata for solving the vertex coloring problem. Memet. Comput. **8**, 211–222 (2016). https://doi.org/10.1007/s12293-016-0183-4

Rezapoor Mirsaleh, M., Meybodi, M.R.: Assignment of cells to switches in cellular mobile network: a learning automata-based memetic algorithm. Appl. Intell. **48**, 3231–3247 (2018a). https://doi.org/10.1007/s10489-018-1136-z

Rezapoor Mirsaleh, M., Meybodi, M.R.: A Michigan memetic algorithm for solving the vertex coloring problem. J. Comput. Sci. **24**, 389–401 (2018b). https://doi.org/10.1016/j.jocs.2017.10.005

Rezapoor Mirsaleh, M., Meybodi, M.R.: Balancing exploration and exploitation in memetic algorithms: a learning automata approach. Comput. Intell. **34**, 282–309 (2018c). https://doi.org/10.1111/coin.12148

Rezvanian, A., Meybodi, M.R.: Stochastic Social Networks: Measures and Algorithms. LAP LAMBERT Academic Publishing (2016b)

Rezvanian, A., Meybodi, M.R.: Finding minimum vertex covering in stochastic graphs: a learning automata approach. Cybern. Syst. **46**, 698–727 (2015a). https://doi.org/10.1080/01969722.2015.1082407

Rezvanian, A., Moradabadi, B., Ghavipour, M., et al.: Introduction to learning automata models. In: Learning Automata Approach for Social Networks. Springer, pp. 1–49 (2019b)

Rezvanian, A., Moradabadi, B., Ghavipour, M., et al.: Learning Automata Approach for Social Networks. Springer International Publishing (2019a)

Rezvanian, A., Moradabadi, B., Ghavipour, M., et al.: Social networks and learning systems: a bibliometric analysis. In: Learning Automata Approach for Social Networks. Springer, pp. 75–89 (2019d)

Rezvanian, A., Moradabadi, B., Ghavipour, M., et al.: Social recommender systems. In: Learning Automata Approach for Social Networks. Springer, pp. 281–313 (2019c)

Rezvanian, A., Moradabadi, B., Ghavipour, M., et al.: Social trust management. In: Learning Automata Approach for Social Networks. Springer, pp. 241–279 (2019e)

Rezvanian, A., Moradabadi, B., Ghavipour, M., et al.: Wavefront cellular learning automata: a new learning paradigm. In: Learning Automata Approach for Social Networks. Springer, pp. 51–74 (2019f)

Rezvanian, A., Saghiri, A.M., Vahidipour, S.M., et al.: Learning automata for cognitive Peer-to-Peer networks. In: Recent Advances in Learning Automata, pp. 221–278 (2018b)

Rezvanian, A., Saghiri, A.M., Vahidipour, S.M., et al.: Learning automata for wireless sensor networks. In: Recent advances in learning automata. pp. 91–219 (2018c)

Rezvanian, A., Saghiri, A.M., Vahidipour, S.M., et al.: Recent Advances in Learning Automata. Springer (2018a)

Rezvanian, A., Meybodi, M.R.: Tracking extrema in dynamic environments using a learning automata-based immune algorithm. Communications in Computer and Information Science, pp. 216–225. Springer, Berlin, Heidelberg (2010)

Rezvanian, A., Meybodi, M.R.: Finding maximum clique in stochastic graphs using distributed learning automata. Int. J. Uncertain., Fuzziness Knowl.-Based Syst. **23**, 1–31 (2015). https://doi.org/10.1142/S0218488515500014

Rezvanian, A., Meybodi, M.R.: Stochastic graph as a model for social networks. Comput. Hum. Behav. **64**, 621–640 (2016). https://doi.org/10.1016/j.chb.2016.07.032

Rezvanian, A., Meybodi, M.R.: Sampling algorithms for stochastic graphs: a learning automata approach. Knowl.-Based Syst. **127**, 126–144 (2017a). https://doi.org/10.1016/j.knosys.2017.04.012

Rezvanian, A., Meybodi, M.R.: A new learning automata-based sampling algorithm for social networks. Int. J. Commun. Syst. **30**, e3091 (2017b). https://doi.org/10.1002/dac.3091

Rezvanian, A., Rahmati, M., Meybodi, M.R.: Sampling from complex networks using distributed learning automata. Phys. A Stat. Mech. Its Appl. **396**, 224–234 (2014). https://doi.org/10.1016/j.physa.2013.11.015

Rezvanian, A., Vahidipour, S.M., Esnaashari, M.: New applications of learning automata-based techniques in real-world environments. J. Comput. Sci. **24**, 287–289 (2018a). https://doi.org/10.1016/j.jocs.2017.11.012

Rezvanian, A., Saghiri, A.M., Vahidipour, S.M., et al.: Cellular Learning Automata, pp 21–88 (2018d)

Roohollahi, S., Bardsiri, A.K., Keynia, F.: Using an evaluator fixed structure learning automata in sampling of social networks. J. AI Data Min. **8**, 127–148 (2020). https://doi.org/10.22044/JADM.2019.7145.1842

Ruan, X., Jin, Z., Tu, H., Li, Y.: Dynamic cellular learning automata for evacuation simulation. IEEE Intell. Transp. Syst. Mag. **11**, 129–142 (2019). https://doi.org/10.1109/MITS.2019.2919523

Rummery, G.A.A., Niranjan, M.: On-line Q-Learning Using Connectionist Systems. University of Cambridge, Department of Engineering (1994)

Wolfram, S.: Cellular automata as simple self-organizing systems. Caltech Prepr CALT-68-938 5: (1982)

Safara, F., Souri, A., Deiman, S.F.: Super peer selection strategy in peer-to-peer networks based on learning automata. Int. J. Commun. Syst. **33**, e4296 (2020). https://doi.org/10.1002/dac.4296

Saghiri, A.M., Meybodi, M.R.: An approach for designing cognitive engines in cognitive peer-to-peer networks. J. Netw. Comput. Appl. **70**, 17–40 (2016). https://doi.org/10.1016/j.jnca.2016.05.012

Saghiri, A.M., Meybodi, M.R.: A distributed adaptive landmark clustering algorithm based on mOverlay and learning automata for topology mismatch problem in unstructured peer-to-peer networks. Int. J. Commun. Syst. **30**, e2977 (2017a). https://doi.org/10.1002/dac.2977

Saghiri, A.M., Meybodi, M.R.: A closed asynchronous dynamic model of cellular learning automata and its application to peer-to-peer networks. Genet. Program Evol. Mach. **18**, 313–349 (2017b). https://doi.org/10.1007/s10710-017-9299-7

Saghiri, A.M., Meybodi, M.R.: An adaptive super-peer selection algorithm considering peers capacity utilizing asynchronous dynamic cellular learning automata. Appl. Intell. **48**, 271–299 (2018a). https://doi.org/10.1007/s10489-017-0946-8

Saghiri, A.M., Meybodi, M.R.: Open asynchronous dynamic cellular learning automata and its application to allocation hub location problem. Knowl.-Based Syst. **139**, 149–169 (2018b). https://doi.org/10.1016/j.knosys.2017.10.021

Saleem, A., Afzal, M.K., Ateeq, M., et al.: Intelligent learning automata-based objective function in RPL for IoT. Sustain. Cities Soc. **59**, 102234 (2020). https://doi.org/10.1016/j.scs.2020.102234

Santoso, J., Riyanto, B., Adiprawita, W.: Dynamic path planning for mobile robots with cellular learning automata. J. ICT Res. Appl. **10**, 1–14 (2016). https://doi.org/10.5614/itbj.ict.res.appl.2016.10.1.1

Saraeian, S., Shirazi, B., Motameni, H.: Optimal autonomous architecture for uncertain processes management. Inf. Sci. (Ny) **501**, 84–99 (2019). https://doi.org/10.1016/j.ins.2019.05.095

Savargiv, M., Masoumi, B., Keyvanpour, M.R.: A new ensemble learning method based on learning automata. J. Ambient Intell. Humaniz. Comput (2020). https://doi.org/10.1007/s12652-020-01882-7

Schwartz, A.: A reinforcement learning method for maximizing undiscounted rewards. Mach. Learn. Proc. **1993**, 298–305 (1993)

Seyyedi, S.H., Minaei-Bidgoli, B.: Estimator learning automata for feature subset selection in high-dimensional spaces, case study: email spam detection. Int. J. Commun Syst **31**, e3541 (2018). https://doi.org/10.1002/dac.3541

Sikeridis, D., Tsiropoulou, E.E., Devetsikiotis, M., Papavassiliou, S.: Socio-physical energy-efficient operation in the Internet of multipurpose things. In: 2018 IEEE International Conference on Communications (ICC). IEEE, pp. 1–7 (2018)

Simha, R., Kurose, J.F.: Relative reward strength algorithms for learning automata. IEEE Trans. Syst. Man Cybern. **19**, 388–398 (1989). https://doi.org/10.1109/21.31041

Sohrabi, M.K., Roshani, R.: Frequent itemset mining using cellular learning automata. Comput. Hum. Behav. **68**, 244–253 (2017). https://doi.org/10.1016/j.chb.2016.11.036

Soleimani-Pouri M, Rezvanian A, Meybodi MR (2012) Solving maximum clique problem in stochastic graphs using learning automata. In: 2012 Fourth International Conference on Computational Aspects of Social Networks (CASoN). IEEE, pp. 115–119

Storn, R.M., Price, K.V.: Differential evolution—a simple and efficient heuristic for global optimization over continuous spaces. J. Glob. Optim. **11**, 341–359 (1997). https://doi.org/10.1023%2FA%3A1008202821328

Stuart, R., Peter, N.: Artificial Intelligence: A Modern Approach, 3rd edn. Printice-Hall (2002)

Su, Y., Qi, K., Di, C., et al.: Learning automata based feature selection for network traffic intrusion detection. In: 2018 IEEE Third International Conference on Data Science in Cyberspace (DSC). IEEE, pp. 622–627 (2018)

Sutton, R.S., Barto, A.G.: Reinforcement Learning: An Introduction. MIT Press, Cambridge (1998)

Talabeigi, M., Forsati, R., Meybodi, M.R.: A hybrid web recommender system based on cellular learning automata. In: 2010 IEEE International Conference on Granular Computing. IEEE, pp. 453–458 (2010)

Thakur, D., Khatua, M.: Cellular Learning Automata-Based Virtual Network Embedding in Software-Defined Networks, pp. 173–182 (2019)

Thathachar, M.A.L., Harita, B.R.: Learning automata with changing number of actions. IEEE Trans. Syst. Man Cybern. **17**, 1095–1100 (1987). https://doi.org/10.1109/TSMC.1987.6499323

Thathachar, M., Sastry, P.: Estimator algorithms for learning automata. In: Proceedings of the Platinum Jubilee Conference on Systems and Signal Processing, Bengalore, India (1986)

Thathachar, M.A.L., Ramachandran, K.M.: Asymptotic behaviour of a learning algorithm. Int. J. Control **39**, 827–838 (1984). https://doi.org/10.1080/00207178408933209

Thathachar, M.A.L., Sastry, P.S.: A class of rapidly converging algorithms for learning automata. IEEE Trans. Syst. Man Cybern. SMC-15, 168–175 (1985b)

Thathachar, M.A.L., Sastry, P.S.: A new approach to the design of reinforcement schemes for learning automata. IEEE Trans. Syst. Man Cybern. SMC-15, 168–175 (1985a). https://doi.org/10.1109/TSMC.1985.6313407

Thathachar, M.A.L., Phansalkar, V.V.: Learning the global maximum with parameterized learning automata. IEEE Trans. Neural Netw. **6**, 398–406 (1995). https://doi.org/10.1109/72.363475

Thathachar, M.A.L., Sastry, P.S.: Varieties of learning automata: An overview. IEEE Trans. Syst. Man Cybern. Part B Cybern. **32**, 711–722 (2002). https://doi.org/10.1109/TSMCB.2002.1049606

Thathachar, M.A.L., Sastry, P.S.: Networks of learning automata. Springer, Boston, MA (2004)

Toffolo, T.A.M., Christiaens, J., Van Malderen, S., et al.: Stochastic local search with learning automaton for the swap-body vehicle routing problem. Comput. Oper. Res. **89**, 68–81 (2018). https://doi.org/10.1016/j.cor.2017.08.002

Toozandehjani, H., Zare-Mirakabad, M.-R., Derhami, V.: Improvement of recommendation systems based on cellular learning automata. In: 2014 4th International Conference on Computer and Knowledge Engineering (ICCKE). IEEE, pp. 592–597 (2014)

Torshizi, M., Sheikhzadeh, M.J.: Optimum K-coverage in wireless sensor network with no redundant node by cellular learning automata. Wirel. Pers. Commun. **110**, 545–562 (2020). https://doi.org/10.1007/s11277-019-06741-z

Tsetlin, M.L.: On the behavior of finite automata in random media. Autom. Remote Control **22**, 1210–1219 (1962)

Vafaee Sharbaf, F., Mosafer, S., Moattar, M.H.: A hybrid gene selection approach for microarray data classification using cellular learning automata and ant colony optimization. Genomics **107**, 231–238 (2016). https://doi.org/10.1016/j.ygeno.2016.05.001

Vafashoar, R., Meybodi, M.R.: Multi swarm bare bones particle swarm optimization with distribution adaption. Appl. Soft Comput. J. **47**, 534–552 (2016). https://doi.org/10.1016/j.asoc.2016.06.028

Vafashoar, R., Meybodi, M.R.: Multi swarm optimization algorithm with adaptive connectivity degree. Appl. Intell. **48**, 909–941 (2018). https://doi.org/10.1007/s10489-017-1039-4

Vafashoar, R., Meybodi, M.R.: Reinforcement learning in learning automata and cellular learning automata via multiple reinforcement signals. Knowl.-Based Syst. **169**, 1–27 (2019a). https://doi.org/10.1016/j.knosys.2019.01.021

Vafashoar, R., Meybodi, M.R.: Cellular learning automata based bare bones PSO with maximum likelihood rotated mutations. Swarm Evol. Comput. **44**, 680–694 (2019b). https://doi.org/10.1016/j.swevo.2018.08.016

Vafashoar, R., Meybodi, M.R.: A multi-population differential evolution algorithm based on cellular learning automata and evolutionary context information for optimization in dynamic environments. Appl. Soft Comput. **88**, 106009 (2020). https://doi.org/10.1016/j.asoc.2019.106009

Vafashoar, R., Meybodi, M.R., Momeni Azandaryani, A.H.: CLA-DE: a hybrid model based on cellular learning automata for numerical optimization. Appl. Intell. **36**, 735–748 (2012). https://doi.org/10.1007/s10489-011-0292-1

Vahidipour, S.M., Meybodi, M.R., Esnaashari, M.: Adaptive Petri net based on irregular cellular learning automata with an application to vertex coloring problem. Appl. Intell. **46**, 272–284 (2017a). https://doi.org/10.1007/s10489-016-0831-x

Vahidipour, S.M., Meybodi, M.R., Esnaashari, M.: Finding the shortest path in stochastic graphs using learning automata and adaptive stochastic Petri nets. Int. J. Uncertain., Fuzziness Knowl.-Based Syst. **25**, 427–455 (2017b). https://doi.org/10.1142/S0218488517500180

Vahidipour, S.M., Esnaashari, M., Rezvanian, A., Meybodi, M.R.: GAPN-LA: a framework for solving graph problems using Petri nets and learning automata. Eng. Appl. Artif. Intell. **77**, 255–267 (2019). https://doi.org/10.1016/j.engappai.2018.10.013

Vasilakos, A.V., Paximadis, C.T.: Faulttolerant routing algorithms using estimator discretized learning automata for high-speed packet-switched networks. IEEE Trans. Reliab. **43**, 582–593 (1994). https://doi.org/10.1109/24.370222

Velusamy, G., Lent, R.: Dynamic cost-aware routing of web requests. Fut. Internet **10**, 57 (2018). https://doi.org/10.3390/fi10070057

Verbeeck, K., Nowé, A., Nowe, A.: Colonies of learning automata. IEEE Trans. Syst. Man Cybern. Part B Cybern. **32**, 772–780 (2002). https://doi.org/10.1109/TSMCB.2002.1049611

Watkins, C.C.J.H.: Learning from Delayed Rewards (1989)

Willianms, R.J.: Toward a Theory of Reinforcement-Learning Connectionist Systems. Northeastern University (1988)

Xue, L., Sun, C., Wunsch, D.C.: A game-theoretical approach for a finite-time consensus of second-order multi-agent system. Int. J. Control Autom. Syst. **17**, 1071–1083 (2019). https://doi.org/10.1007/s12555-017-0716-8

Yazidi, A., Bouhmala, N., Goodwin, M.: A team of pursuit learning automata for solving deterministic optimization problems. Appl. Intell. (2020). https://doi.org/10.1007/s10489-020-01657-9

Zamani, M.S., Mehdipour, F., Meybodi, M.R.: Implementation of cellular learning automata on reconfigurable computing systems. In: CCECE 2003—Canadian Conference on Electrical and Computer Engineering. Toward a Caring and Humane Technology (Cat. No. 03CH37436). IEEE, pp. 1139–1142 (2003)

Zanganeh, S., Meybodi, M.R., Sedehi, M.H.: Continuous CLA-EC. In: 2010 Fourth International Conference on Genetic and Evolutionary Computing. IEEE, pp. 186–189 (2010)

Zarei, B., Meybodi, M.R.: Improving learning ability of learning automata using chaos theory. J. Supercomput (2020). https://doi.org/10.1007/s11227-020-03293-z

Zhang, F., Wang X, Li, P., Zhang, L.: An energy aware cellular learning automata based routing algorithm for opportunistic networks. Int. J. Grid Distrib. Comput. **9**, 255–272 (2016). https://doi.org/10.14257/ijgdc.2016.9.2.22

Zhao, Y., Jiang, W., Li, S., et al.: A cellular learning automata based algorithm for detecting community structure in complex networks. Neurocomputing **151**, 1216–1226 (2015). https://doi.org/10.1016/j.neucom.2014.04.087

Chapter 2
Cellular Learning Automata: A Bibliometric Analysis

2.1 Introduction

In this chapter, before describing the methodology for systematic bibliometric analysis on cellular learning automata (CLA), we give a brief introduction to CLA and an overview of applications of cellular learning automata models in Sect. 2.2. The material and method are presented in Sect. 2.3, the results of the bibliometric on the CLA is analyzed in Sect. 2.4, and finally, Sect. 2.5 concludes the chapter.

Cellular learning automaton (CLA) (Beigy and Meybodi 2004; Rezvanian et al. 2018b), as depicted in Fig. 2.1 is a combination of cellular automata (Wolfram 1986) and learning automata (Narendra and Thathachar 1989; Rezvanian et al. 2018a). In this model, each cell of a cellular automaton (CA) contains one or more learning automata (LA). The LA or LAs residing in a particular cell define the state of the cell. Like CA, a CLA rule controls the behavior of each cell. At each time step, the local rule defines the reinforcement signal for a particular LA based on its selected action and the actions chosen by its neighboring LAs.

Consequently, the neighborhood of each LA is considered to be its local environment. A formal definition of CLA is provided by Beigy and Meybodi (Beigy and Meybodi 2004). The authors also investigated the asymptotic behavior of the model and provided some theoretical analysis on its convergence.

The structure of a CLA can be represented by a graph, where each vertex denotes a CLA cell. Each edge of this graph defines a neighborhood relation between its two incident nodes. We represent the action set of the ith LA by $A_i = \{a_{i1}, \ldots, a_{im_i}\}$, where m_i is the number of actions of the LA. The probability vector of the ith LA, which determines the selection probability of its actions, is denoted by p_i. This probability vector defines the internal state of the LA, and the probability vectors of all learning automata in the CLA define the configuration of the CLA at each step k. The configuration of a CLA at step k is denoted by $P(k) = (p_1(\mathrm{k}), p_2(\mathrm{k}), \ldots, p_n(k))^{\mathrm{T}}$,

R. Vafashoar et al., *Cellular Learning Automata: Theory and Applications*, Studies in Systems, Decision and Control 307,
https://doi.org/10.1007/978-3-030-53141-6_2

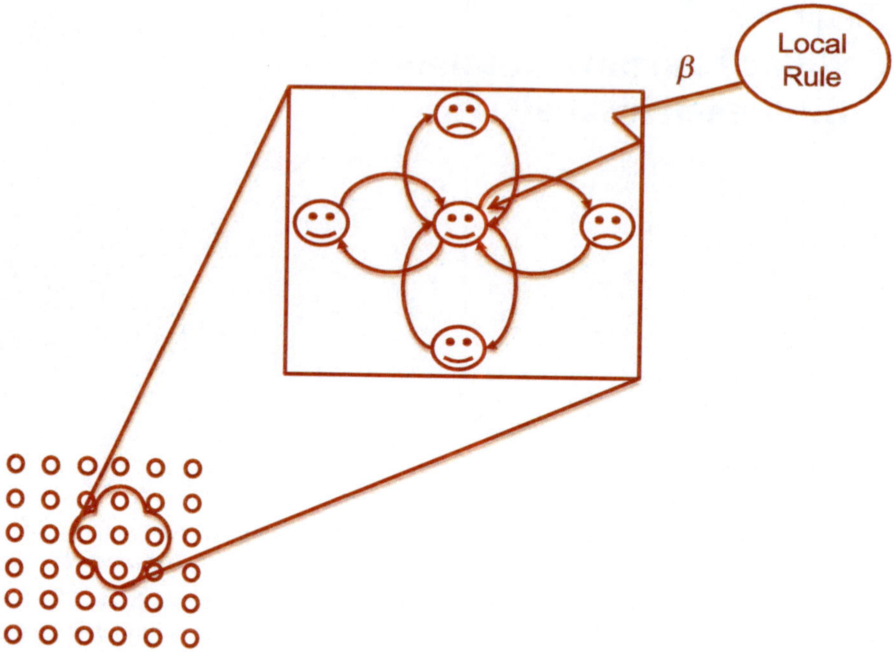

Fig. 2.1 Cellular learning automaton (Beigy and Meybodi 2004)

where n is the number of cells. The operation of a CLA is governed by a set of local rules. At each step k, the local rule of the ith cell determines the reinforcement signal to the learning automaton in the cell as follows:

$$F_i : \varphi_i \rightarrow \underline{\beta}$$
$$with$$
$$\varphi_i = \prod_{j=0}^{n_i} A_{N_i(j)} \tag{2.1}$$

where N_i is the neighborhood function of the ith cell, and $N_i(j)$ returns the index of the jth neighboring cell to cell i with $N_i(0)$ defined as $N_i(0) = i$. n_i denotes the number of cells in the neighborhood of the ith cell. Finally, $\underline{\beta}$ is the set of values that the reinforcement signal can take. The expected reward of each LA at each step can be obtained based on its associated local rule and the CLA configuration. The expected reward of action like $a_r \in A_i$ in configuration P is denoted by $d_{ir}(P)$. A configuration P is called compatible if the following condition holds for any configurations Q and any cell i in the CLA.

$$\sum_r d_{ir}(P) \times p_{ir} \geq \sum_r d_{ir}(Q) \times q_{ir} \tag{2.2}$$

A CLA starts from some initial state (it is the internal state of every cell); at each stage, each automaton selects an action according to its probability vector and performs it in its local environment. Next, the rule of the CLA determines the reinforcement signals (the responses) to the LAs residing in its cells, and based on the received signals; each LA updates its internal probability vector. This procedure continues until a termination condition is satisfied.

2.2 Varieties of Cellular Learning Automata Models

Up to now, various CLA models such as open CLA (Beigy and Meybodi 2007), asynchronous CLA (Beigy and Meybodi 2008), irregular CLA (Esnaashari and Meybodi 2018), associative CLA (Ahangaran et al. 2017), dynamic irregular CLA (Esnaashari and Meybodi 2018), asynchronous dynamic CLA (Saghiri and Meybodi 2018b), and wavefront CLA (Moradabadi and Meybodi 2018; Rezvanian et al. 2019a) are developed and successfully applied on different application domains. Table 2.1 summarizes some application areas for these models.

2.2.1 *Open Cellular Learning Automata*

The basic CLA can be considered as closed since it does not consider the interaction between the CLA and the external environment. In closed models, the behavior of each cell is affected only by its neighboring cells. However, some applications can be easily modeled by CLA if an external environment is also taken into account. This external environment also influences the behavior of the cells. Based on this idea, Beigy and Meybodi introduced open CLA (Beigy and Meybodi 2007), in which in addition to the neighboring environment of a cell, two new environments are also considered: global environment and group environment. The global environment affects all of the LAs of the CLA, and the group environment affects a group of LAs in the CLA.

2.2.2 *Asynchronous Cellular Learning Automata*

Most CLA models are synchronous. In synchronous models, all cells receive their reinforcement signals simultaneously and, based on the received signals, update their internal action probabilities, also, at the same time. The synchronous models can be extended into asynchronous ones. In asynchronous models, the LAs can be updated at different times. In these models, cells can be updated in a time-driven or step driven manner. In step driven manner, cells are updated in some fixed or random orders (Beigy and Meybodi 2008), while in a time-driven manner, each cell

Table 2.1 Some applications of different CLA models

CLA model	Application
Associative CLA	Clustering (Ahangaran et al. 2017; Hasanzadeh-Mofrad and Rezvanian 2018), classification (Ahangaran et al. 2017), image segmentation (Ahangaran et al. 2017)
Asynchronous CLA	Edge detection (Bohlool and Meybodi 2007), super-peer selection (Saghiri and Meybodi 2018a), hub allocation problem (Saghiri and Meybodi 2018b), topology mismatch problem in unstructured peer-to-peer networks (Saghiri and Meybodi 2017), web recommendation (Talabeigi et al. 2010)
Asynchronous dynamic CLA	Topology mismatch problem in unstructured peer-to-peer networks (Saghiri and Meybodi 2017), super-peer selection (Saghiri and Meybodi 2018a), hub allocation (Saghiri and Meybodi 2018b)
Basic CLA	Image processing (Hasanzadeh Mofrad et al. 2015), Frequent itemset mining (Sohrabi and Roshani 2017), gene selection (Vafaee Sharbaf et al. 2016), text summarization (Abbasi-ghalehtaki et al. 2016), skin segmentation (Abin et al. 2008), Lung cancer diagnosis (Hadavi and Nordin 2014), skin detector (Abin et al. 2009), stock market (Mozafari and Alizadeh 2013), edge detection (Sinaie et al. 2009)
CLA-DE	Optimization (Vafashoar et al. 2012)
CLA-EC	Hardware design (Hariri et al. 2005a), FPGA (Hariri et al. 2005b), clustering (Rastegar et al. 2005), optimization (Rastegar et al. 2006), dynamic optimization (Manshad et al. 2011), link prediction in social networks (Khaksar Manshad et al. 2020)
CLA-PSO	Optimization (Akhtari and Meybodi 2011)
Cellular Petri-net	Vertex coloring problem (Vahidipour et al. 2017a), graph problems (Vahidipour et al. 2019)
Closed asynchronous dynamic CLA	Peer-to-peer networks (Saghiri and Meybodi 2017)
Cooperative CLA-EC	Optimization (Masoodifar et al. 2007; Mozafari et al. 2015)
Dynamic irregular CLA	Deployment of mobile wireless sensor networks (Esnaashari and Meybodi 2011), transportation (Ruan et al. 2019)
Irregular CLA	Channel assignment (Morshedlou and Meybodi 2017), sampling social networks (Ghavipour and Meybodi 2017), community detection in complex social networks (Zhao et al. 2015; Khomami et al. 2018), dynamic point coverage in wireless Sensor (Esnaashari and Meybodi 2010), clustering in wireless sensor networks (Esnaashari and Meybodi 2008), link prediction in social network (Khaksar Manshad et al. 2020)

(continued)

Table 2.1 (continued)

CLA model	Application
Open CLA	Hub allocation problem (Saghiri and Meybodi 2018b), edge detection (Bohlool and Meybodi 2007), optimization (Saghiri and Meybodi 2018a)
Re-combinative CLA-EC	Optimization (Jafarpour and Meybodi 2007)
Wavefront CLA	Online social network problems such as prediction and sampling (Moradabadi and Meybodi 2018)

possesses an internal clock that determines its activation times. Asynchronous CLA has been employed in applications such as data mining and peer-to-peer networks.

2.2.3 *Irregular Cellular Learning Automata*

In basic cellular learning automaton (CLA), cells are arranged in some regular forms like a gird or ring. However, there exist some applications which require irregular arrangements of the cells. Esnaashari and Meybodi have proposed an irregular CLA (Esnaashari and Meybodi 2018) for clustering the nodes in a wireless sensor network. Irregular CLA is modeled as an undirected graph in which each vertex represents a cell of the CLA, and its adjacent vertices determine its neighborhood. Up to now, various applications of Irregular CLA are reported in the literature. The application areas of this model include, for instance, (but not limited to) wireless sensor networks (Rezvanian et al. 2018c), graph problems (Vahidipour et al. 2017b), cloud computing (Morshedlou and Meybodi 2017), complex networks (Khomami et al. 2018), and social network analysis (Ghavipour and Meybodi 2017).

2.2.4 *Dynamic Cellular Learning Automata*

In a dynamic CLA, one of its aspects, such as structure, local rule, attributes, or neighborhood radius, may change over time. It should be noted that CLA can have various categories. For instance, it can be jointly categorized as irregular, asynchronous, and open. In this regard, dynamic CLA can be either closed or open. Even it can be synchronous or asynchronous. Several models of dynamic CLA have been investigated in the literature. Esnaashari and Meybodi introduced an interest-based dynamic irregular CLA in which two cells are considered to be neighbors if they share similar interests. In this regard, a set of interests is defined for dynamic irregular CLA. Each cell represents its tendency for each interest using a vector called tendency vector. Two cells are considered to be neighbors if their tendency vectors are close enough. In addition to reinforcement signals, their proposed dynamic irregular CLA uses a second kind of signal, which is called restructuring signal. This latter signal is used to update the tendency vectors of the cells, which in turn may change the neighborhood structures. Later, Saghiri and Meybodi have introduced other models of dynamic CLA, such as asynchronous dynamic CLA and asynchronous dynamic CLA with varying number of LAs in each cell. These later CLA models have proven to be quite successful in applications such as landmark clustering in peer-to-peer networks, and adaptive super-peer selection.

2.2.5 *Wavefront Cellular Learning Automata*

Wavefront CLA (WCLA) (Moradabadi and Meybodi 2018; Rezvanian et al. 2019a) is an asynchronous CLA with a diffusion property. The activation sequence of cells in WCLA is controlled through waves. Each LA in WCLA receiving a wave is activated and selects an action according to its probability vector. If the newly chosen action of the LA is different from its previous action, the LA propagates the wave to its neighbors. A wave propagates through cells in WCLA for some time until one of the following conditions holds: all LAs that have newly received the wave choose similar actions as their previous actions, or the energy of the wave drops to zero. The energy of each wave determines the wave's ability to propagate itself. This energy decreases as it moves through the network until it reaches zero. WCLA has been utilized for solving online social network problems such as prediction methods and sampling.

2.2.6 *Associative Cellular Learning Automata*

In associative CLA (Ahangaran et al. 2017), each cell receives two inputs: an input vector from the global environment of the CLA and a reinforcement signal for its performed actions. During each iteration, each cell receives an input form the environment. Then, it selects an action and applies it to the environment. Based on the performed actions, the local rule of the CLA determines the reinforcement signal to each LA. Associative CLA has been applied to applications such as clustering and image segmentation. For instance, clustering can be performed in the following manner. During each iteration, sample data are chosen by the environment and are given to each LA. According to the received input, each LA selects an action using its decision function. The defined action selection works based on the distance of the current state of the cell and the input vector. An LA is rewarded if its chosen action is smaller than those selected in its neighborhood. The learning algorithm of the proposed LA is defined so that upon receiving the reward, an LA update its state and become nearer to the input data while the receiving penalty does not affect the cell's state. Accordingly, the smallest action is chosen by the nearest cell to the input data. So, it is expected that, after some iterations, the state-vectors of cells lie on the centers of the clusters.

2.3 Material and Method

We adopted a methodology for preparing a literature review similar methodology used in (Rezvanian et al. 2019b) in five steps, including document acquisition, creating notes, structuring the literature review, writing the literature review, and building the bibliography. In this study, similar to this methodology, the research documents from Scopus and Web of Science (WoS) are collected, the results are processed and refined. The refined results are structured as key materials (e.g., top contributing authors, institutes, publications, countries, research topics, and applications), and finally, some insights into current research interests, trends, and future directions are provided. The network visualizations and bibliometric representations are made by Gephi (Bastian et al. 2009) and VOSviewer (van Eck and Waltman 2010, 2013). Gephi is an open-source network exploration, manipulation, analysis, and visualization software package written in Java on the NetBeans platform. VOSviewer provides network visualization on co-authorship, co-citation, and citation concerning authors, organizations, and countries and also co-occurrence concerning keywords.

2.3.1 Data Collection and Initial Results

The data was collected from both Scopus and WoS. The WoS (previously known as Web of Knowledge) is an online subscription-based scientific citation indexing service that provides a comprehensive citation search for many different academic disciplines which covers more than 90 million records during 1900 to present from peer-reviewed journals belonging to publishing houses such as Elsevier, Springer, Wiley, IEEE, ACM, and Taylor & Francis, to name a few. It was initially produced by the Institute for Scientific Information (ISI) and is currently maintained by Clarivate Analytics (previously the Intellectual Property and Science business of Thomson Reuters) (Analytics and Clarivate 2017). Scopus is Elsevier's abstract and citation database which covers more than 69 million records consist of nearly 36,377 titles (22,794 working titles and 13,583 inactive titles) from approximately 11,678 publishers, of which 34,346 are peer-reviewed journals in top-level subject fields such as life sciences, social sciences, physical sciences, and health sciences. All journals covered in the Scopus database, regardless of whom they are published under, are reviewed each year to ensure high-quality standards are maintained.

For data collection, we searched for the ("cellular learning automata" OR "cellular learning automaton") as the two main keywords in the topic of papers belonging to both WoS and Scopus. The initial search resulted in 109 papers for WoS and also 143 papers for Scopus from 2003 until April 2020.

2.3.2 Refining the Initial Results

To refine the search results, the non-English language results such as Chinese (4) and Persian (1) for Scopus results are excluded during the data purification process. Thus, the restriction on these results produced 138 papers for Scopus results. The paper types of the search results, as presented in Table 2.2, can be categorized as conference proceedings papers and journal papers as 34 and 75 papers for WoS, and

Table 2.2 Types of documents about cellular learning automata and related research

Document type	WoS	Scopus
Conference proceedings papers	34	61
Journal papers	75	77
Total	113	138

61 and 77 papers for Scopus, respectively. The final search results consist of 113 and 138 for WoS and Scopus, respectively.

2.4 Analyzing the Results

We performed data analysis in two steps. In the first step, statistical analysis is extracted from the resulted search, networks of entities are extracted using Gephi (Bastian et al. 2009), and bibliometric analysis is performed using VOSviewer (van Eck and Waltman 2013). The network visualization and bibliometric representation, including co-authorship, co-citation, and the citation for authors, organizations, and countries, and also co-occurrence to keywords. Thus, in this section, several statistics, results, and analyses are reported for the research related to the topic of "learning and social networks" based on the final resulted search.

2.4.1 Initial Result Statistics

In Fig. 2.2, the number of papers published during the time-period 2003–2020 is shown whose demonstrates the changing pattern of publications in the research

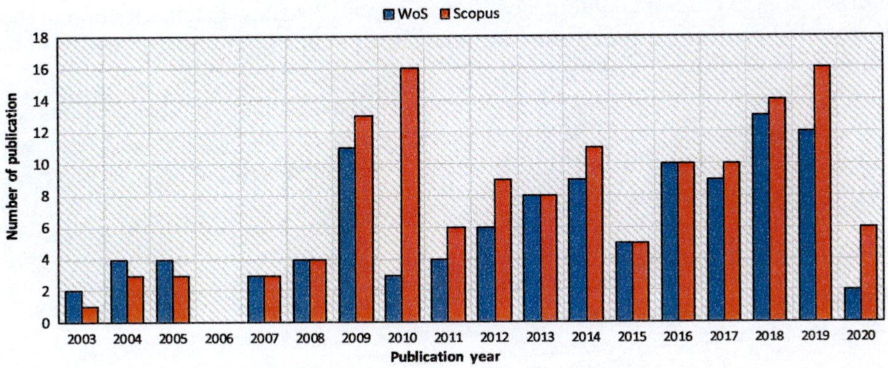

Fig. 2.2 Distribution of papers published during the time 2003–2020

Table 2.3 Distribution of the top 10 journal titles with the most publication related to the topic of "cellular learning automata"

	Journal name	Publisher	No. publications
1	Journal of Computational Science	Elsevier	6
2	Applied Intelligence	Springer Nature	5
3	Ad Hoc and Sensor Wireless Networks	Old City Publishing Inc	3
4	Applied Soft Computing Journal	Elsevier	3
5	Journal of Intelligent and Fuzzy Systems	IOS Press	3
6	Advances in Complex Systems	World Scientific	2
7	IEEE Transactions on Cybernetics	IEEE	2
8	Knowledge-Based Systems	Elsevier	2
9	Swarm and Evolutionary Computation	Elsevier	2
10	Neurocomputing	Elsevier	2

community increasingly. It can be seen from the figure that the number of publications on the research related to the topic of "cellular learning automata" was dramatically increasing from 2008. However, since then, it has been increasing slightly.

2.4.2 *Top Journals*

In order to understand the role of the different scientific journals and also conference series, we identified the top 10 journals and conference titles appearing in the data with the most published in this field of research related to the topic of "cellular learning automata." It was found that these papers have published by 53 different journals and 33 different conferences. Tables 2.3 and 2.4 show the distribution of the journal and conference series titles with the most publication for research related to the topic of study, respectively.

The network of the citation for the journals with a minimum of three citations for each journal in research related to the topic of "cellular learning automata" is shown in Fig. 2.3, whose node size is proportional to the number of citations received by the papers of that journal.

2.4.3 *Top Researchers*

From the resulted search for the research related to the topic of "cellular learning automata," the top ten contributing authors based on the number of publications in

Table 2.4 Distribution of the top 10 conference series titles for the research related to the topic of "cellular learning automata"

	Conference series title	Conference abbreviation	No. publications
1	International CSI Computer Conference	CSI	3
2	Iranian Conference on Electrical Engineering	ICEE	3
3	International Conference on Image Processing Computer Vision and Pattern Recognition	IPCV	3
4	European Modelling Symposium on Computer Modelling And Simulation	UKSIM	3
5	International Conference on Computer Communication and Informatics	ICCCI	2
6	Iranian Conference on Intelligent Systems	ICIS	2
7	IEEE International Conference of Intelligent Robotic and Control Engineering	IRCE	2
8	International Symposium On Telecommunications	IST	2
9	International Conference on Computer and Knowledge Engineering	ICCKE	2
10	International Conference On Hybrid Intelligent Systems	HIS	2

particular for this research topic of study are extracted. These results are summarized in Table 2.5, in which in the second column is the author's name, the third column reflects the number of publications in this topic, and the last column shows the highly cited article in this topic for each author. From the results reported in Table 2.5, it can be observed that Meybodi, M. R. with 21 articles dominates the list, and is followed by Rezvanian, A. each with ten publications.

2.4.3.1 Co-authorship Analysis

Co-authorship analysis can be used in authors and/or publications in order to track and study the relationship between authors, institutions, or countries. If applied to authors, co-authorship analysis reveals the structure of the social relationships between authors (Chen et al. 2010). To conduct co-authorship analysis, we used VOSviewer to the visualization of the co-authorship network based on Scopus data. Then the network clustering (community detection) in this co-authorship network is also applied in order to show the authors that probability works on similar topics

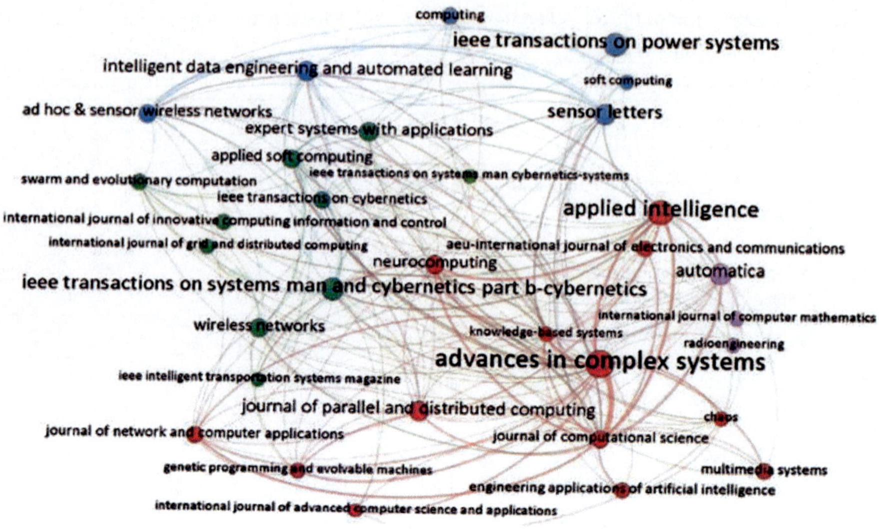

Fig. 2.3 Citation network of key conference series for the research related to the topic of "cellular learning automata"

Table 2.5 Top 10 authors based on the number of publications for "cellular learning automata" related research

	Author	No. publications in WoS	No. publications in Scopus	Highly cited paper
1	Meybodi, M. R.	54	59	Beigy and Meybodi (2004)
2	Esnaashari, M.	11	12	Esnaashari and Meybodi (2008)
3	Beigy, H.	10	9	Beigy and Meybodi (2004)
4	Rastegar, R.	5	4	Rastegar and Meybodi (2004)
5	Saghiri, A. M.	5	6	Saghiri and Meybodi (2016)
6	Vafashoar, R.	5	6	Vafashoar et al. (2012)
7	Rezvanian, A.	4	8	Hasanzadeh Mofrad et al. (2015)
8	Esmaeilpour, M.	4	4	Abbasi-ghalehtaki et al. (2016)
9	Akbari Torkestani, J.	4	2	Akbari Torkestani and Meybodi (2011)
10	Ghavipour, M.	2	5	Ghavipour and Meybodi (2017)

in each community. The resulted network of co-authorship in the topic of "cellular learning automata" consists of 192 nodes (authors) and nine significant communities with a minimum of 5 nodes. This network with its extracted communities is depicted in Fig. 2.4. In Fig. 2.4, the size of each node is proportional to the number of published articles by each author, and the community structures are revealed by both color and position of which node.

This Co-authorship network, with its revealed community structures, is also colored based on the publication year as overlay visualization from 2003 through 2020 in Fig. 2.5. In Fig. 2.5, the size of each node is proportional to the number of published articles by each author, and the color of each node represents the number of publications per year by each author.

In order to study the relationship between authors concerning citing each other papers, the co-citation network of authors is extracted. In the co-citation network

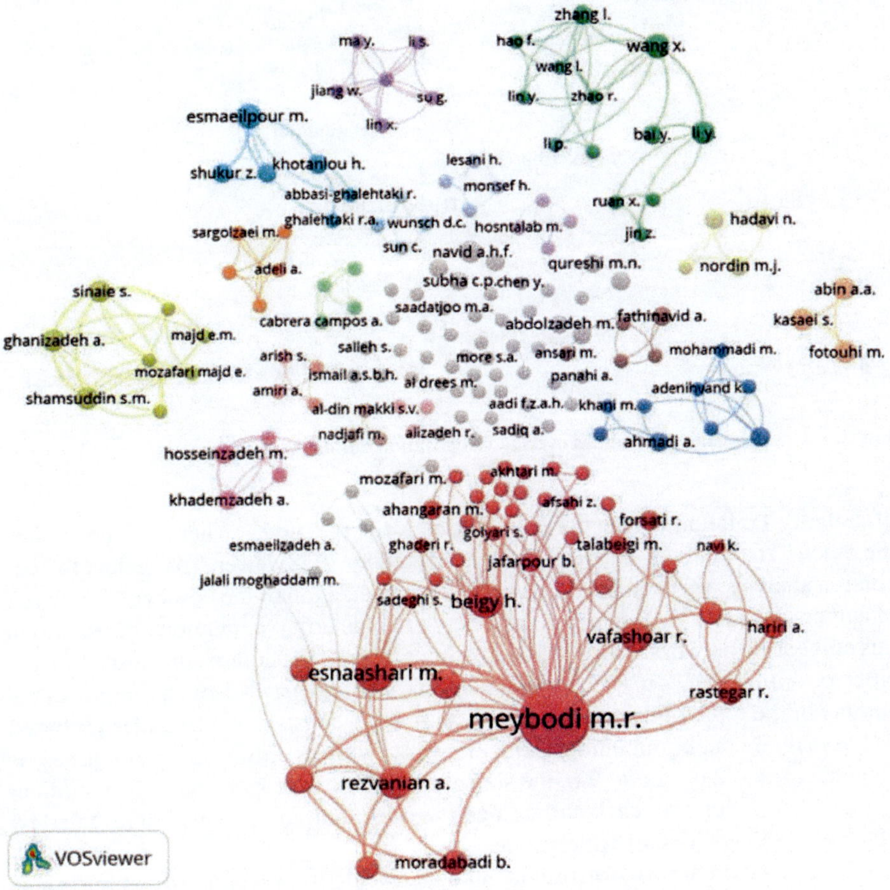

Fig. 2.4 Co-authorship network with its revealed community structures

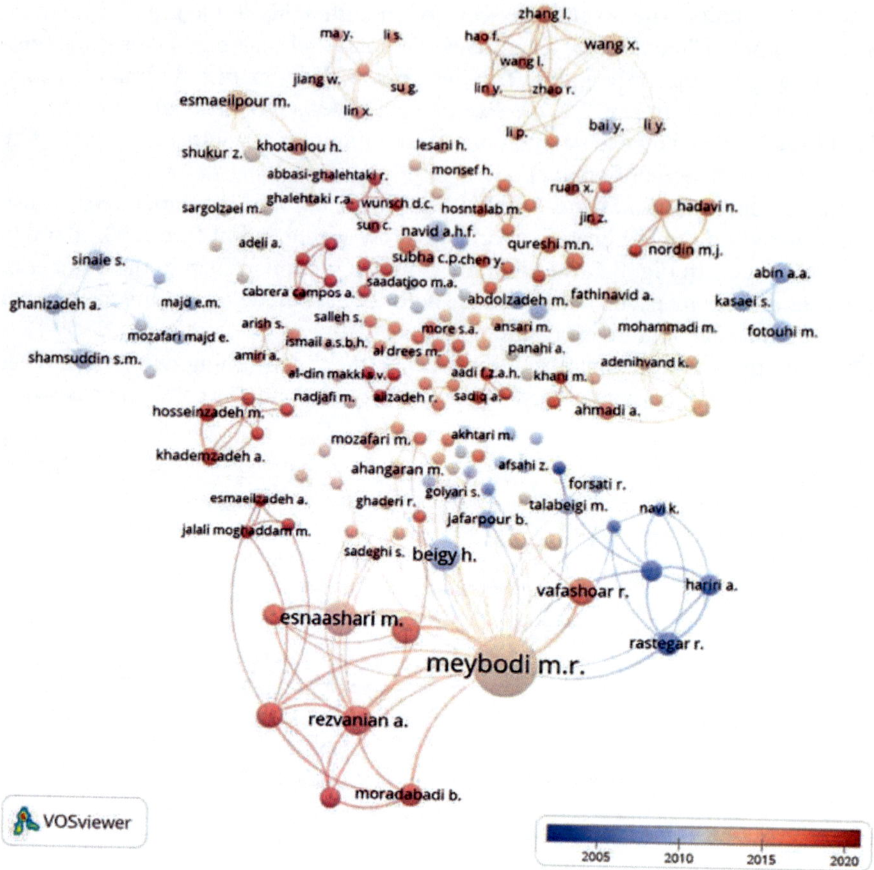

Fig. 2.5 Co-authorship network as overlay visualization during 2003–2020

of authors, a citation connection is a link between two nodes where one node cites the other. Citation links are treated as undirected by VOSviewer. To conduct the co-citation analysis, we used VOSviewer to the visualization of the co-citation network of authors based on WoS data. Then the network clustering (community detection) in this co-citation network is also applied in order to show the authors that probability is citing on similar topics in each community. The resulted network of co-citation of the authors in the topic of "cellular learning automata" consists of 146 nodes (authors) and six significant communities. This network, with its extracted communities, is depicted in Fig. 2.6. In Fig. 2.6, the size of each node is proportional to the number of published articles by each author, and the community structures are revealed by both color and position of which node.

The co-citation network of authors with its revealed community structures is also colored based on the publication year as overlay visualization between 2003 and 2020 in Fig. 2.7. In Fig. 2.7, the size of each node is proportional to the number of

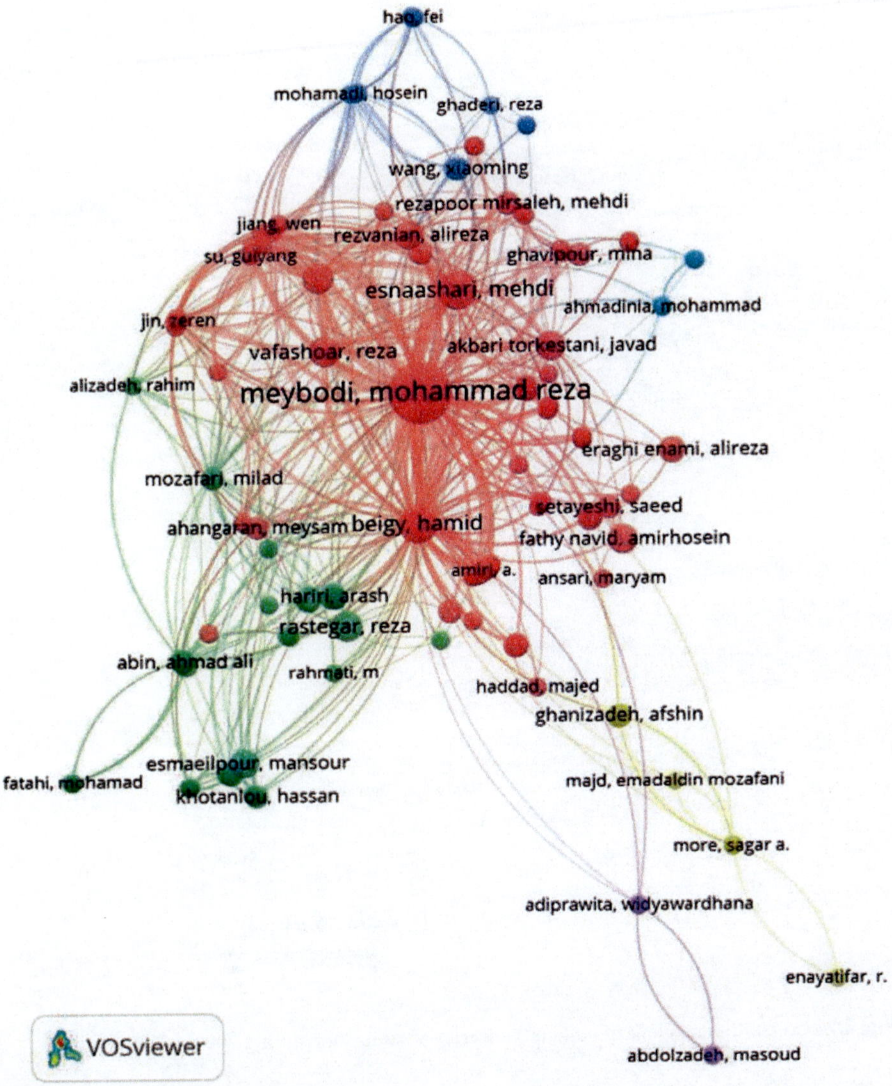

Fig. 2.6 Co-citation network of authors with its revealed community structures

published articles by each author, and the color of each node represents the number of publications per year by each author.

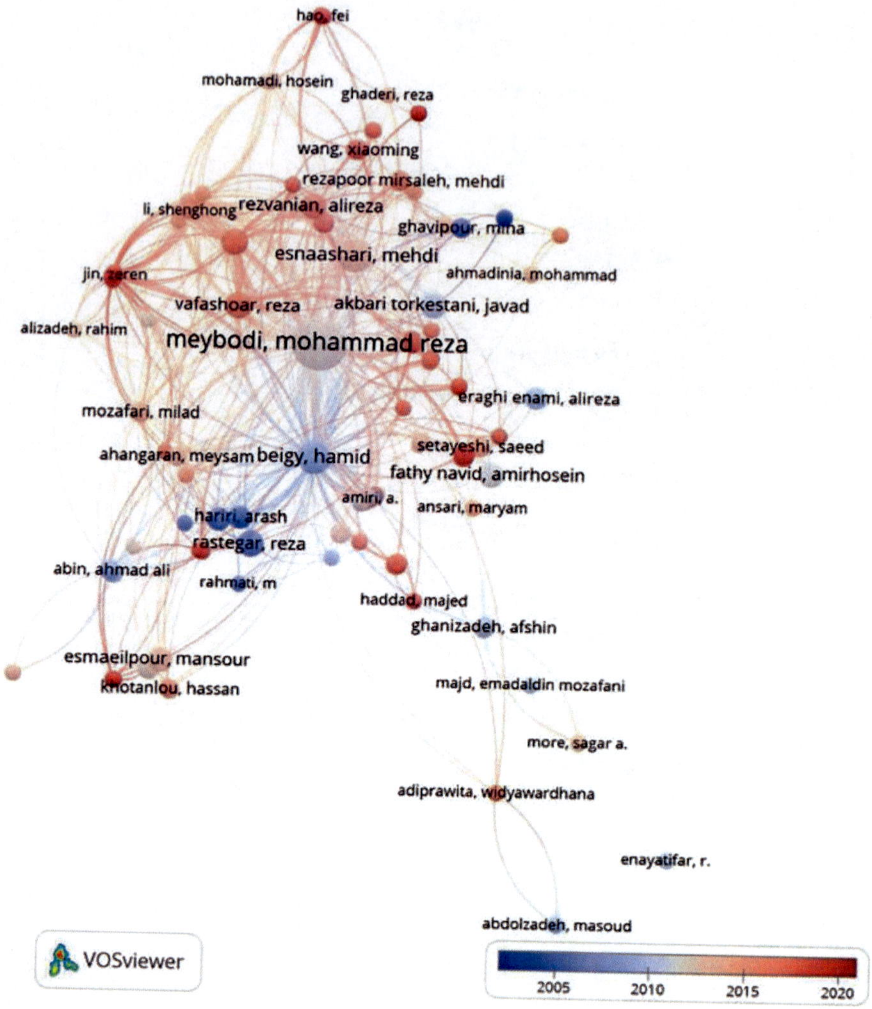

Fig. 2.7 Co-citation network of authors as overlay visualization during 2003–2020

2.4.4 Top Papers

The top articles for the research related to the topic of "cellular learning automata" for highly cited papers (the most number of citations received by each paper) are reported in Table 2.6 in which the paper titles are given in the second column. The year of publication, the number of citations in WoS, the number of citations in Scopus, the average number of citations per year in WoS and the average number of citation per year in WoS are provided in the third, fourth, fifth, sixth and the last column, respectively. From the results reported in Table 2.6, it can be observed that the seminal paper of CLA by Beigy et al. (Sen et al. 2008) with 74 citations and

Table 2.6 Highly cited articles for the research related to the topic of "cellular learning automata."

	Article title (reference)	Year of pub	No. cit. in WoS	No. cit. in Scopus	Avg. No. cit. per year in WoS	Avg. No. cit. per year in Scopus
1	A mathematical framework for cellular learning automata (Beigy and Meybodi 2004)	2004	74	102	4.63	6.38
2	Cellular learning automata with multiple learning automata in each cell and its applications (Beigy and Meybodi 2010)	2010	60	69	6.00	6.90
3	A multi-objective PMU placement method considering measurement redundancy and observability value under contingencies	2013	59	64	8.43	9.14
4	A cellular learning automata based clustering algorithm for wireless sensor networks (Esnaashari and Meybodi 2008)	2008	47	53	3.92	4.42
5	Asynchronous cellular learning automata (Beigy and Meybodi 2008)	2008	42	48	3.50	4.00
6	A cellular learning automata-based deployment strategy for mobile wireless sensor networks	2011	38	41	4.22	4.56

(continued)

Table 2.6 (continued)

	Article title (reference)	Year of pub	No. cit. in WoS	No. cit. in Scopus	Avg. No. cit. per year in WoS	Avg. No. cit. per year in Scopus
7	A cellular learning automata-based algorithm for solving the vertex coloring problem	2011	36	38	4.00	4.22
8	A hybrid gene selection approach for microarray data classification using cellular learning automata and ant colony optimization	2016	33	48	8.25	12.00
9	Deployment of a mobile wireless sensor network with k-coverage constraint: a cellular learning automata approach	2013	30	35	4.29	5.00
10	CLA-DE: a hybrid model based on cellular learning automata for numerical optimization	2012	28	34	3.50	4.25

102 citations in WoS and Scopus, respectively dominates the list. A research study presented by Sharbaf et al. (2016) in the discipline of bioinformatics and also the other research work published by Mazhari et al. in the discipline of electric power systems dominate the top-ten list with 8.25 (12.00), and 8.43 (9.14) values for this measure concerning an average number of citations received each paper in WoS (Scopus) per year, respectively. Since this phenomenon has appeared in these two fields of applications (bioinformatics and electric power systems), this reason may be for the practical applicability of the CLA for applied disciplines.

The network of citations for articles is demonstrated in Fig. 2.8 that the size of each node is proportional to the number of citations that each articled received. In this network, nodes with dark red color represent the recent polished articles, and nodes with dark blue color represent the more past article publication.

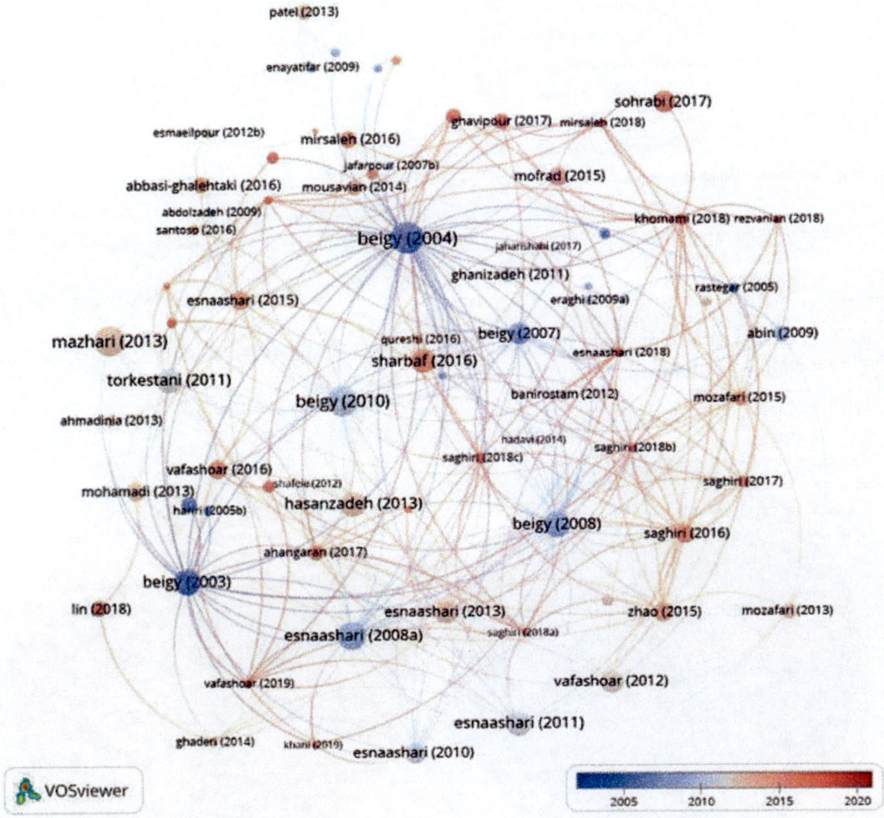

Fig. 2.8 Citation network for papers

2.4.5 Top Affiliations

The plot of the top ten institutions with highly published articles is shown in Fig. 2.9 in which most papers are published by researchers from Amirkabir University of Technology (Tehran Polytechnic) with 56 and 65 papers in WoS and Scopus, respectively. In the list, it is followed by the researchers from Islamic Azad University with 36 and 63 papers in WoS and Scopus, respectively.

The corresponding to each affiliation, the country in which the institution is situated, was taken out for further analysis, and this result is shown in Fig. 2.10. As shown in Fig. 2.10, it can be seen that institutions in Iran (479 papers in WoS and 105 papers

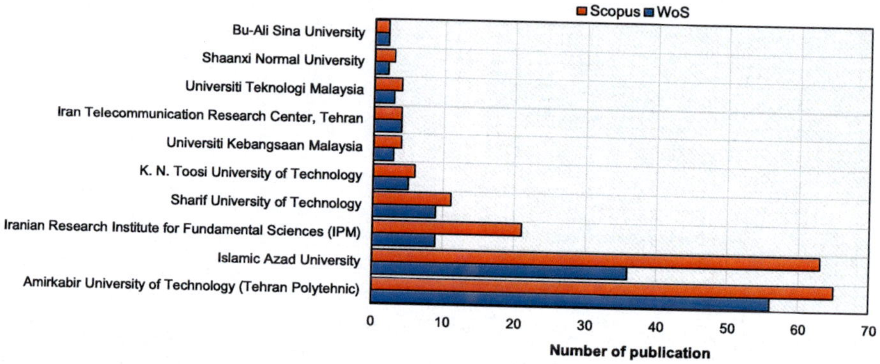

Fig. 2.9 Top 10 institutions with highly published papers for the research related to the topic of "cellular learning automata"

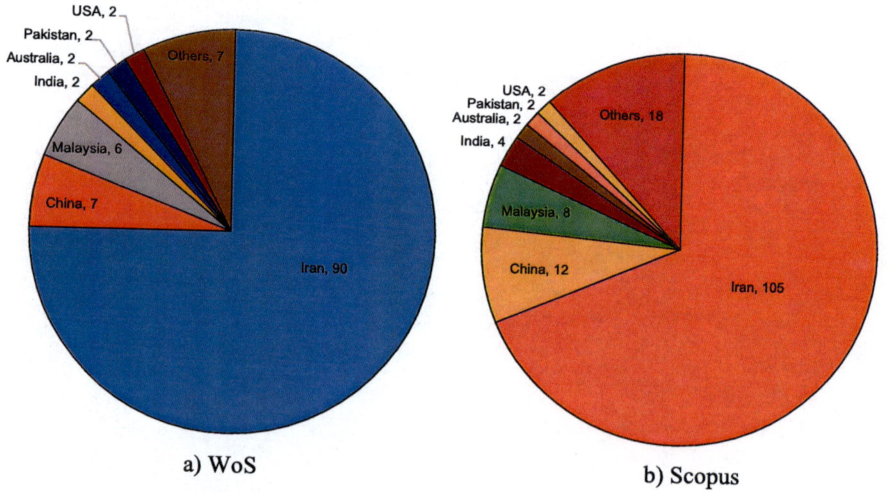

Fig. 2.10 Top 10 contributing countries for the research related to the topic of "cellular learning automata"

in Scopus), China (7 papers in WoS and 12 papers in Scopus) and Malaysia (6 papers in WoS and eight papers in Scopus), are the major contributors. Researchers across the world, in particular in Asia, are attracted to the area of cellular learning automata.

The network of co-authorships for institutions of contributing authors is shown in Fig. 2.11, including 51 institutions in which each node size is proportional to the number of papers published by authors of those institutions. The network of co-authorships for countries of authors' institutions is shown in Fig. 2.12 that each node size is proportional to the number of citations received by each article published by authors of each countries' institutions. This network consists of 18 nodes represented as countries and ten links represented as connections between the institutions of these countries.

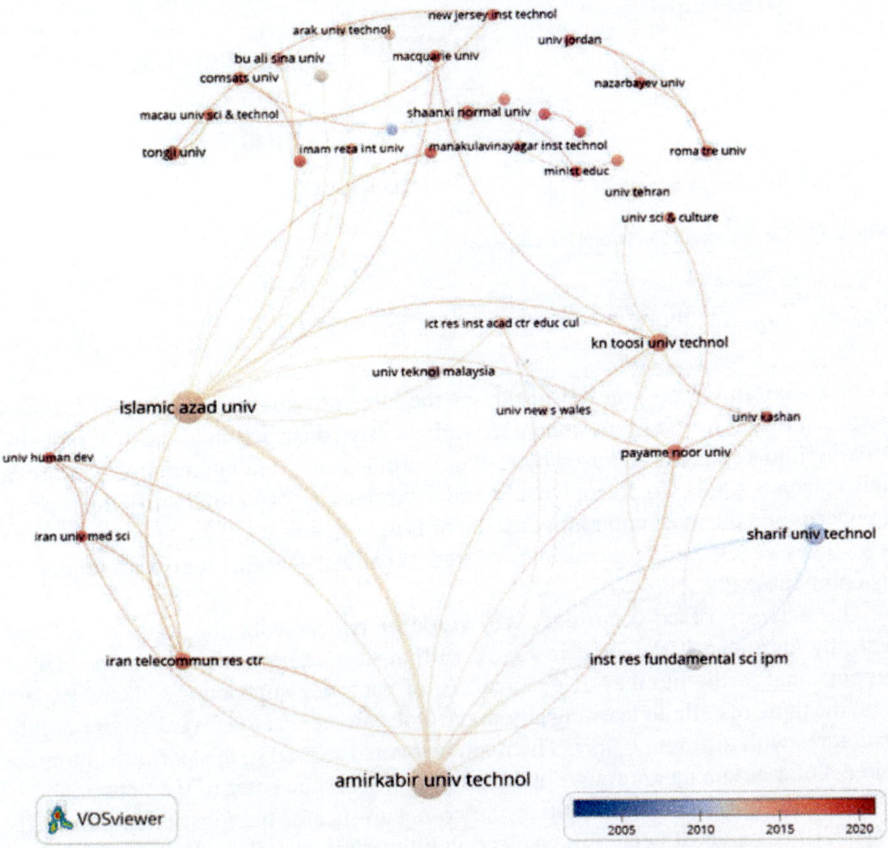

Fig. 2.11 Co-authorship network for institutions for the research related to the topic of "learning and social networks"

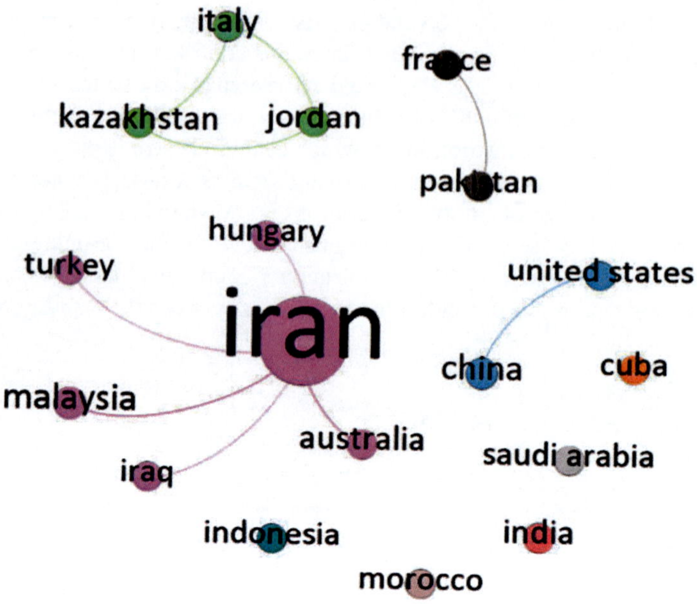

Fig. 2.12 Co-authorship network for countries

2.4.6 *Top Keywords*

In this section, we present the results of the keyword analysis. Such a discussion assists in revealing the intellectual core and identity construction of the discipline by looking into keywords and used by research articles and their aggregation (Sidorova and Valacich 2008). To do so, we adopted a similar approach to identify the top-ten keywords and the most commonly used words in both article titles and abstracts. The top 10 keywords used in the articles related to cellular learning automata topics are listed in Table 2.7.

The network of co-occurrence keywords of papers with the topic of cellular learning automata is depicted in Fig. 2.13 that, in this network, each node size is proportional to the number of occurrences of each keyword among all the papers with the topic of cellular learning automata. One can see that the revealed community structures with different colors. The main keywords related to the learning automata and cellular automata are found in the community at the center of the network.

The density visualization map of the co-occurrence of the most common words in both titles and abstracts of the papers with the topic of cellular learning automata based on the Scopus is depicted in Fig. 2.14. From the result, one can observe that the main words related to the learning automata, cellular learning automata, and cellular learning automata topics are mainly positioned in the center of the map.

Table 2.7 Top 10 keywords of papers related to the "cellular learning automata" topic

	Word	Frequency in WoS	Frequency in Scopus
1	Cellular learning automata	75	130
2	Learning automata	35	57
3	algorithm	31	39
4	Cellular automata	17	32
5	Optimization	13	31
6	Wireless sensor networks	13	30
7	Particle swarm optimization	7	25
8	Clustering	6	21
9	Edge detection	5	12
10	Genetic algorithm	4	9

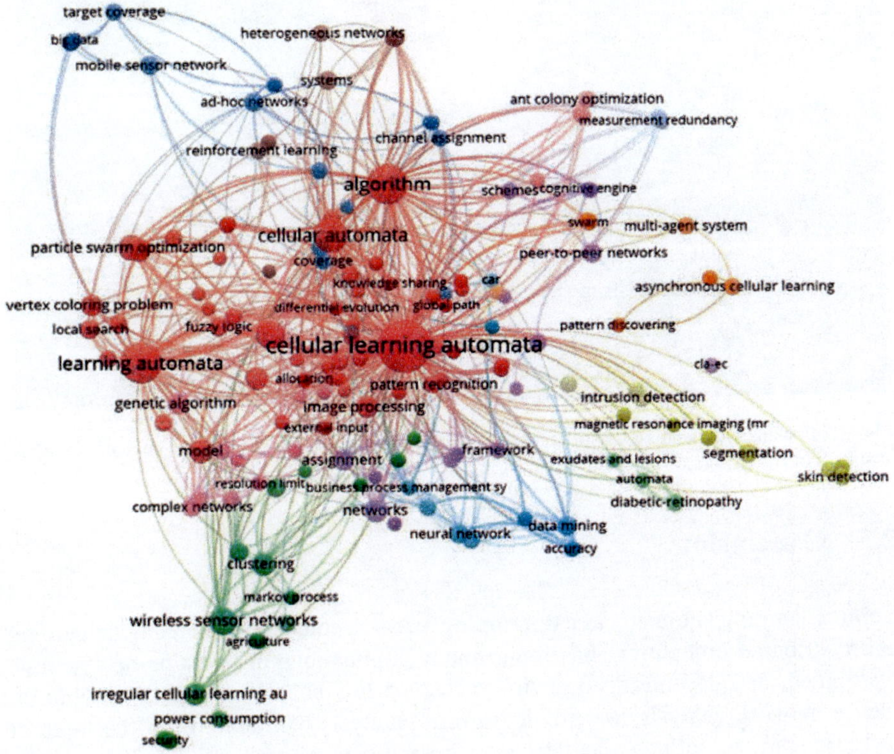

Fig. 2.13 Co-occurrence network of keywords

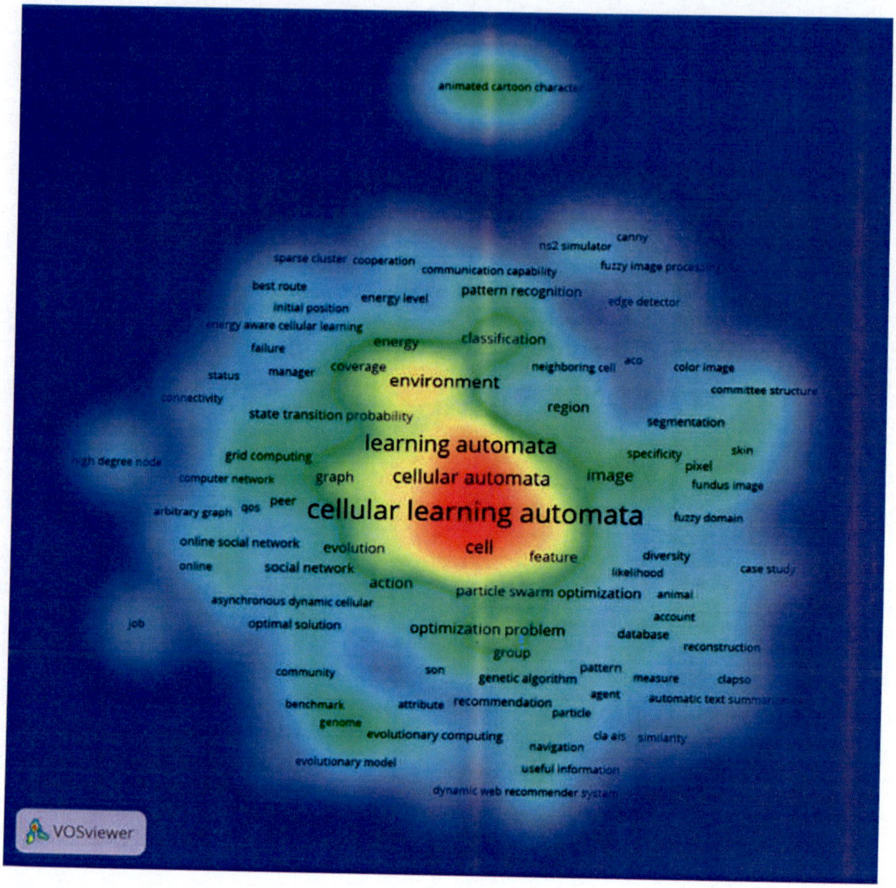

Fig. 2.14 Density visualization of the most common words belonging to the text of titles and abstracts of the papers

2.5 Conclusion

Cellular learning automata, as a developing research study of reinforcement learning and distributed computing, has found many applications in domains of dynamic, distributed, and decentralized environments. In this chapter, based on the bibliometric network analysis, we provide a brief analysis of literature on the topic of Cellular learning automata and related research over 18 years (2003–2020). Also, we provided some insights about the contributing key scientific journals, researchers, institutes, countries, papers, keywords and most common words in the text of title and abstracts of the papers towards advancing cellular learning automata related research as bibliometric network analysis perspective.

References

Abbasi-ghalehtaki, R., Khotanlou, H., Esmaeilpour, M.: Fuzzy evolutionary cellular learning automata model for text summarization. Swarm Evol. Comput. **30**, 11–26 (2016). https://doi.org/10.1016/j.swevo.2016.03.004

Abin, A.A., Fotouhi, M., Kasaei, S.: Skin segmentation based on cellular learning automata. In: Proceedings of the 6th International Conference on Advances in Mobile Computing and Multimedia—MoMM '08, p. 254. ACM, Austria (2008)

Abin, A.A., Fotouhi, M., Kasaei, S.: A new dynamic cellular learning automata-based skin detector. Multimed. Syst. **15**, 309–323 (2009). https://doi.org/10.1007/s00530-009-0165-1

Ahangaran, M., Taghizadeh, N., Beigy, H.: Associative cellular learning automata and its applications. Appl Soft Comput **53**, 1–18 (2017). https://doi.org/10.1016/j.asoc.2016.12.006

Akbari Torkestani, J., Meybodi, M.R.: A cellular learning automata-based algorithm for solving the vertex coloring problem. Expert Syst. Appl. **8**, 9237–9247 (2011)

Akhtari, M., Meybodi, M.R.: Memetic-CLA-PSO: a hybrid model for optimization. In: 2011 UkSim 13th International Conference on Computer Modelling and Simulation, pp. 20–25. IEEE (2011)

Bastian, M., Heymann, S., Jacomy, M.: Gephi: an open source software for exploring and manipulating networks. In: Third international AAAI conference on weblogs and social media, pp. 361–362 (2009)

Beigy, H., Meybodi, M.R.: A mathematical Framework for cellular learning Automata. Adv. Complex Syst. **07**, 295–319 (2004). https://doi.org/10.1142/S0219525904000202

Beigy, H., Meybodi, M.R.: Open synchronous cellular learning automata. Adv. Complex Syst. **10**, 527–556 (2007)

Beigy, H., Meybodi, M.R.: Asynchronous cellular learning automata. Automatica **44**, 1350–1357 (2008)

Beigy, H., Meybodi, M.R.R.: Cellular learning automata with multiple learning automata in each cell and its applications. IEEE Trans. Syst. Man, Cybern. Part B **40**, 54–65 (2010). https://doi.org/10.1109/TSMCB.2009.2030786

Bohlool, M., Meybodi, M.R.: Edge detection using open and asynchronous cellular learning automata. In: 4th Iranian Conference on Machine Vision and Image Processing, pp 1–6 (2007)

Chen, C., Ibekwe-SanJuan, F., Hou, J.: The structure and dynamics of cocitation clusters: a multiple-perspective cocitation analysis. J. Am. Soc. Inf. Sci. Technol. **61**, 1386–1409 (2010). https://doi.org/10.1002/asi.21309

Clarivate Analytics: Acquisition of the Thomson reuters intellectual property and Science business by Onex and Baring Asia Completed (2017). www.prnewswire.com

Esnaashari, M., Meybodi, M.R.: A cellular learning automata based clustering algorithm for wireless sensor networks. Sens. Lett. **6**, 723–735 (2008)

Esnaashari, M., Meybodi, M.R.: Dynamic point coverage problem in wireless sensor networks: a cellular learning automata approach. Ad hoc Sens. Wirel Netw. **10**, 193–234 (2010)

Esnaashari, M., Meybodi, M.R.M.: A cellular learning automata-based deployment strategy for mobile wireless sensor networks. J. Parallel Distrib. Comput. **71**, 988–1001 (2011)

Esnaashari, M., Meybodi, M.R.: Irregular cellular learning automata. IEEE Trans. Cybern. **45**, 1622–1632 (2018). https://doi.org/10.1016/j.jocs.2017.08.012

Ghavipour, M., Meybodi, M.R.: Irregular cellular learning automata-based algorithm for sampling social networks. Eng. Appl. Artif. Intell. **59**, 244–259 (2017). https://doi.org/10.1016/j.engappai.2017.01.004

Hadavi, N., Nordin, M.J., Shojaeipour, A.: Lung cancer diagnosis using CT-scan images based on cellular learning automata. In: 2014 International Conference on Computer and Information Sciences (ICCOINS), pp. 1–5. IEEE (2014)

Hariri A, Rastegar R, Zamani MS, Meybodi MR (2005b) Parallel Hardware Implementation of Cellular Learning Automata Based Evolutionary Computing (CLA-EC) on FPGA. In: 13th Annual IEEE Symposium on Field-Programmable Custom Computing Machines (FCCM'05). IEEE, pp 311–314

Hariri A, Rastegar R, Navi K, et al (2005a) Cellular Learning Automata based Evolutionary Computing (CLA-EC) for Intrinsic Hardware Evolution. In: 2005 NASA/DoD Conference on Evolvable Hardware (EH'05). IEEE, pp 294–297

Hasanzadeh Mofrad, M., Sadeghi, S., Rezvanian, A., Meybodi, M.R.: Cellular edge detection: combining cellular automata and cellular learning automata. AEU—Int. J. Electron. Commun. **69**, 1282–1290 (2015). https://doi.org/10.1016/j.aeue.2015.05.010

Hasanzadeh-Mofrad, M., Rezvanian, A.: Learning automata clustering. J. Comput. Sci. **24**, 379–388 (2018). https://doi.org/10.1016/j.jocs.2017.09.008

Jafarpour, B., Meybodi, M.R.: Recombinative CLA-EC. In: Proceedings—International Conference on Tools with Artificial Intelligence, ICTAI, , pp. 415–422. IEEE (2007)

Khaksar Manshad, M., Meybodi, M.R., Salajegheh, A.: A new irregular cellular learning automata-based evolutionary computation for time series link prediction in social networks. Appl. Intell. (2020). https://doi.org/10.1007/s10489-020-01685-5

Khomami, M.M.D., Rezvanian, A., Meybodi, M.R.: A new cellular learning automata-based algorithm for community detection in complex social networks. J. Comput. Sci. **24**, 413–426 (2018). https://doi.org/10.1016/j.jocs.2017.10.009

Manshad, M.K., Manshad, A.K., Meybodi, M.R.: Memory/search RCLA-EC: A CLA-EC for moving parabola problem. In: 2011 6th International Conference on Computer Sciences and Convergence Information Technology (ICCIT), pp. 1–6 (2011)

Masoodifar, B., Meybodi, M.R., Hashemi, M.: Cooperative CLA-EC. In: 12th Annual CSI Computer Conference of Iran, pp. 558–559 (2007)

Moradabadi, B., Meybodi, M.R.: Wavefront cellular learning automata. Chaos **28**, 21101 (2018). https://doi.org/10.1063/1.5017852

Morshedlou, H., Meybodi, M.R.: A New local rule for convergence of ICLA to a compatible Point. IEEE Trans. Syst. Man, Cybern. Syst. **47**, 3233–3244 (2017). https://doi.org/10.1109/TSMC.2016.2569464

Mozafari, M., Alizadeh, R.: A cellular learning automata model of investment behavior in the stock market. Neurocomputing **122**, 470–479 (2013). https://doi.org/10.1016/j.neucom.2013.06.002

Mozafari, M., Shiri, M.E., Beigy, H.: A cooperative learning method based on cellular learning automata and its application in optimization problems. J. Comput. Sci. **11**, 279–288 (2015). https://doi.org/10.1016/j.jocs.2015.08.002

Narendra, K.S., Thathachar, M.A.L.: Learning automata: an introduction. Prentice-Hall (1989)

Rastegar, R., Meybodi, M.R.: A new evolutionary computing model based on cellular learning automata. In: IEEE Conference on Cybernetics and Intelligent Systems, 2004, pp. 433–438. IEEE (2004)

Rastegar, R., Rahmati, M., Meybodi, M.R.: A clustering algorithm using cellular learning automata based evolutionary algorithm. In: Adaptive and Natural Computing Algorithms, pp. 144–150. Springer-Verlag, Vienna (2005)

Rastegar, R., Meybodi, M.R., Hariri, A.: A new fine-grained evolutionary algorithm based on cellular learning automata. Int. J. Hybrid Intell. Syst. **3**, 83–98 (2006). https://doi.org/10.3233/HIS-2006-3202

Rezvanian, A., Saghiri, A.M., Vahidipour, S.M., et al.: Recent Advances in Learning Automata. Springer (2018a)

Rezvanian, A., Saghiri, A.M., Vahidipour, S.M., et al.: Cellular Learning Automata, pp 21–88 (2018b)

Rezvanian, A., Saghiri, A.M., Vahidipour, S.M., et al.: Learning automata for wireless sensor networks. In: Recent Advances in Learning Automata, pp. 91–219 (2018c)

Rezvanian, A., Moradabadi, B., Ghavipour, M., et al.: Wavefront cellular learning automata: a new learning paradigm. In: Learning Automata Approach for Social Networks, pp. 51–74. Springer (2019a)

Rezvanian, A., Moradabadi, B., Ghavipour, M., et al.: Social networks and learning systems: a bibliometric analysis. In: Learning Automata Approach for Social Networks, pp. 75–89. Springer (2019b)

Ruan, X., Jin, Z., Tu, H., Li, Y.: Dynamic cellular learning automata for evacuation simulation. IEEE Intell. Transp. Syst. Mag. **11**, 129–142 (2019). https://doi.org/10.1109/MITS.2019.2919523

Saghiri, A.M., Meybodi, M.R.: An approach for designing cognitive engines in cognitive peer-to-peer networks. J. Netw. Comput. Appl. **70**, 17–40 (2016). https://doi.org/10.1016/j.jnca.2016.05.012

Saghiri, A.M., Meybodi, M.R.: A closed asynchronous dynamic model of cellular learning automata and its application to peer-to-peer networks. Genet. Program Evolvable Mach. **18**, 313–349 (2017). https://doi.org/10.1007/s10710-017-9299-7

Saghiri, A.M., Meybodi, M.R.: An adaptive super-peer selection algorithm considering peers capacity utilizing asynchronous dynamic cellular learning automata. Appl. Intell. **48**, 271–299 (2018a). https://doi.org/10.1007/s10489-017-0946-8

Saghiri, A.M., Meybodi, M.R.: Open asynchronous dynamic cellular learning automata and its application to allocation hub location problem. Knowledge-Based Syst. **139**, 149–169 (2018b). https://doi.org/10.1016/j.knosys.2017.10.021

Sen, P., Namata, G., Bilgic, M., et al.: Collective classification in network data. AI Mag. **29**, 93 (2008). https://doi.org/10.1609/aimag.v29i3.2157

Sidorova, E., Valacich, R.: Uncovering the intellectual core of the information systems discipline. MIS Q. **32**, 467 (2008). https://doi.org/10.2307/25148852

Sinaie, S., Ghanizadeh, A., Majd, E.M., Shamsuddin, S.M.: A hybrid edge detection method based on fuzzy set theory and cellular learning automata. In: 2009 International Conference on Computational Science and Its Applications, pp. 208–214. IEEE (2009)

Sohrabi, M.K., Roshani, R.: Frequent itemset mining using cellular learning automata. Comput. Human Behav. **68**, 244–253 (2017). https://doi.org/10.1016/j.chb.2016.11.036

Talabeigi, M., Forsati, R., Meybodi, M.R.: A hybrid web recommender system based on cellular learning automata. In: 2010 IEEE International Conference on Granular Computing, pp. 453–458. IEEE (2010)

Vafaee Sharbaf, F., Mosafer, S., Moattar, M.H.: A hybrid gene selection approach for microarray data classification using cellular learning automata and ant colony optimization. Genomics **107**, 231–238 (2016). https://doi.org/10.1016/j.ygeno.2016.05.001

Vafashoar, R., Meybodi, M.R., Momeni Azandaryani, A.H.: CLA-DE: a hybrid model based on cellular learning automata for numerical optimization. Appl. Intell. **36**, 735–748 (2012). https://doi.org/10.1007/s10489-011-0292-1

Vahidipour, S.M., Meybodi, M.R., Esnaashari, M.: Cellular adaptive Petri net based on learning automata and its application to the vertex coloring problem. Discret. Event Dyn. Syst. **27**, 609–640 (2017a). https://doi.org/10.1007/s10626-017-0251-z

Vahidipour, S.M., Meybodi, M.R., Esnaashari, M.: Adaptive Petri net based on irregular cellular learning automata with an application to vertex coloring problem. Appl. Intell. **46**, 272–284 (2017b). https://doi.org/10.1007/s10489-016-0831-x

Vahidipour, S.M., Esnaashari, M., Rezvanian, A., Meybodi, M.R.: GAPN-LA: A framework for solving graph problems using Petri nets and learning automata. Eng. Appl. Artif. Intell. **77**, 255–267 (2019). https://doi.org/10.1016/j.engappai.2018.10.013

van Eck, N.J., Waltman, L.: Software survey: VOSviewer, a computer program for bibliometric mapping. Scientometrics **84**, 523–538 (2010). https://doi.org/10.1007/s11192-009-0146-3

van Eck, N.J., Waltman, L.: VOSviewer manual (2013)

Wolfram, S.: (1986) Theory and Applications of Cellular Automata. World Scientific Publication (1986)

Zhao, Y., Jiang, W., Li, S., et al.: A cellular learning automata based algorithm for detecting community structure in complex networks. Neurocomputing **151**, 1216–1226 (2015). https://doi.org/10.1016/j.neucom.2014.04.087

Chapter 3
Learning from Multiple Reinforcements in Cellular Learning Automata

3.1 Introduction

Learning is a crucial aspect of intelligence, and machine learning has emerged as a vibrant discipline with the avowed objective of developing machines with learning capabilities. Learning has been recognized as an essential aspect of intelligent behavior. Over the last few decades, the process of learning, which was studied earlier mostly by psychologists, has become a topic of much interest to engineers as well, given its role in machine intelligence. Psychologists or biologists, who conduct learning experiments on animals or human subjects, try to create models of behavior through analysis of experimental data. Engineers are more interested in studying learning behavior for helping them to synthesize intelligent machines. While the above two goals are distinctly different, the two endeavors are nonetheless interrelated, because success in one helps to improve our abilities in the other.

Learning is defined as any relatively permanent change in behavior resulting from experience and learning system is characterized by its ability to improve its behavior with time, in some sense tending towards an ultimate goal (Narendra 1989). The concept of learning makes it possible to design systems that can gradually improve their performance during actual operation through a process of learning from past experiences. Every learning task consists of two parts: a learning system and environment. The learning system must learn to act in the environment. The primary learning tasks can be classified into supervised learning, semi-supervised learning, active learning, unsupervised learning, and reinforcement learning.

Cellular learning automaton (CLA) as a new reinforcement learning approach is a combination of cellular automata (CA) (Wolfram 1986) and learning automata (LA) (Rezvanian et al. 2018b, 2019). It is formed by a group of interconnected cells, which are arranged in some regular forms such as a grid or ring, in which each cell contains one or more LAs. A cellular learning automaton can be considered as a

R. Vafashoar et al., *Cellular Learning Automata: Theory and Applications*, Studies in Systems, Decision and Control 307,
https://doi.org/10.1007/978-3-030-53141-6_3

distributed model, which inherits both the computational power of cellular automata and the learning ability of learning automata. Accordingly, it is more powerful than a single learning automaton due to the ability to produce more complex patterns of behavior. Also, the learning abilities of cells enable cellular learning automata to produce these complex patterns by understandable behavioral rules rather than complicated mathematical functions commonly used in cellular automata. Owing to these characteristics, cellular learning automata can be successfully employed on modeling, learning, simulating, controlling, and solving challenging problems in uncertain, distributed, and complex environments.

Since many problems require that a learning automaton (LA) learn the optimal subsets of its actions, the standard LA algorithms can deal with this problem by considering all combinations of their possible actions as new action sets. However, this approach is only applicable to small action spaces. Indeed, sometimes the decision-making entity receives individual feedback from the environment for each one of the actions in its chosen subset of actions.

In this chapter, we aim to extend some standard learning automaton algorithms so that they can efficiently learn the optimal subset of their actions through parallel reinforcements. These parallel reinforcements represent the favorability of each action in the performed subset of actions. The cellular learning automaton (CLA) can model several locally connected decision-making entities. All the introduced learning automaton models in this chapter can be utilized in a cellular learning automaton. Moreover, this chapter investigates the convergence properties of a CLA model in which each decision-making entity selects a subset of its actions. It is worth to note that the main part of this chapter is adopted from (Vafashoar and Meybodi 2019), interested readers could refer to (Vafashoar and Meybodi 2019) for a detailed description and theoretical analysis.

3.2 Learning to Choose the Optimal Subset Using Multiple Reinforcements

This section introduces learning automaton (LA) models (Rezvanian et al. 2018a) that can learn an optimal subset of their actions and update their internal probability vectors according to multiple environmental responses (Vafashoar and Meybodi 2019). In order to distinguish between a vector action and its comprising components, the terms super-action and simple-action are adopted in this chapter. It is assumed that each of the proposed LAs has an action set like $A = \{a_1,\ldots, a_r\}$ simple-actions and aims at learning the optimal super-action which consists of w simple-actions ($1 \leq w < r$). In each time step, the LA selects a super-action $\alpha = \{\alpha_1,\ldots,\alpha_w\}$ consisting of w simple-actions. Then, it performs the combination of these simple-actions as a single super-action in the environment. The environment provides a w-dimensional reinforcement vector in response to the selected super-action. Each element of the reinforcement vector represents the favorability of one of the selected

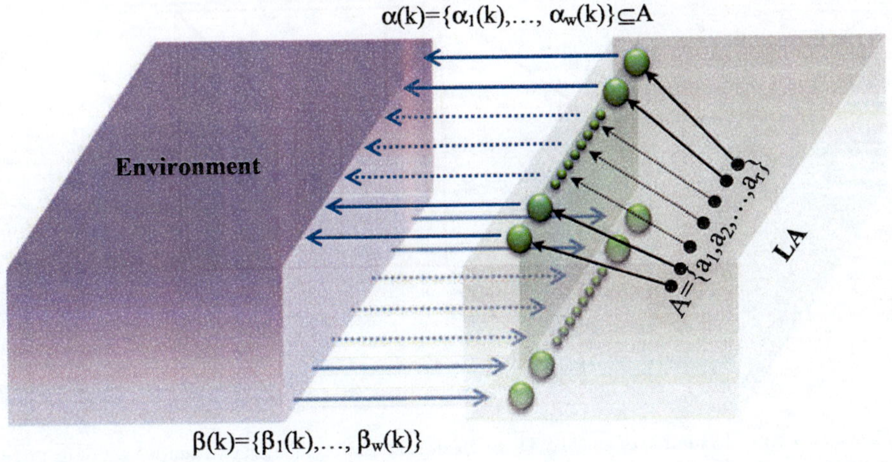

Fig. 3.1 Learning from multiple reinforcements

simple-actions in the combined super-action, as depicted in Fig. 3.1. Accordingly, each selected simple-action like $\alpha_j(k)$ receives an independent reinforcement signal like $\beta_j(k)$. The combined reinforcement for the selected super-action at step k can be considered as the sum of all elements in the reinforcement vector, i.e. $\sum_{l=1}^{w} \beta_l(k)$. The goal of the learning algorithm is to learn the super-action with the highest expected combined reinforcement value.

3.2.1 *Multi-reinforcement Learning Automaton Type I*

The internal state of an LA is ordinarily modeled by a probability vector. During the learning process, this probability vector is adapted to the environment, and the selection probability of the most reward receiving action is gradually increased. However, using this method for subset selection would be inefficient as the number of super-actions grows exponentially with the number of simple-actions. Additionally, the reinforcements received by a typical simple-action depend on the effectiveness of other simple-actions in the experienced super-actions, which adversely affects the learning process. Representing the internal state as a probability vector over simple-actions would also be unsuitable. As the selection probability of a particular simple-action approaches to one, the selection probabilities of all other simple-actions converge to zero.

The first multi-reinforcement learning automaton introduced in this section, called multi-reinforcement learning automaton type I (MLATI), considers an energy level for its actions. The internal state of an MLATI at step k is an r-dimensional vector x, where $x_i \in [0,1]$, as presented in Fig. 3.2. Each component x_i of x defines the energy

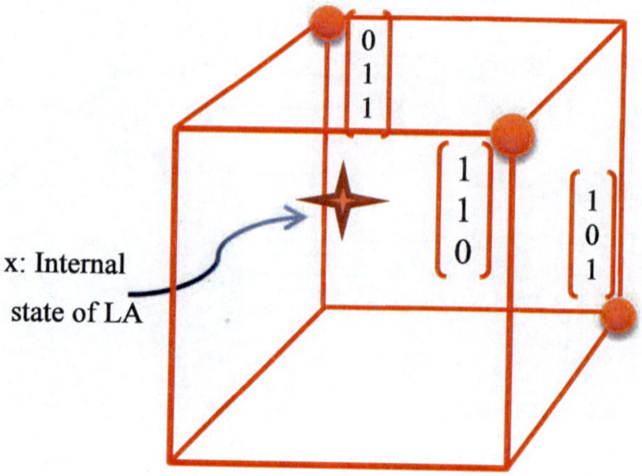

Fig. 3.2 A typical example of an MLATI which aims at finding the best two actions out of its three available actions. The energy vector of the MLATI changes within an r-dimensional unit hypercube

level of the corresponding action a_i. Actions with higher energy levels are selected with higher probability values. The automaton in Fig. 3.2 has three actions and selects a super-action consisting of two simple-actions in each step. These super-actions correspond to the hypercube corners, which are represented by the black circles in the figure. The objective of the proposed LA is to move its energy vector toward the corner of the energy hypercube with the highest expected payoff.

The operation of MLATI follows. At each stage, a dummy super-action is selected based on the energy vector $x(k)$: each simple-action a_i chosen with probability $x_i(k)$ in the dummy super-action. This super-action corresponds to one of the corners of the LA hypercube and may consist of less/more than w simple-actions. After this stage, the algorithm randomly selects one of the closest w-dimensional super-actions to the selected dummy super-action. Accordingly, the final selected super-action (which is denoted by $\alpha(k)$) consists of exactly w simple-actions. It should be noted that if we consider each element of $x(k)$ as the energy level of a simple-action, the final super-action ($\alpha(k)$) is selected based on the energy levels of the simple-actions. After action selection, the selected super-action is carried out in the environment. Then, the LA receives a w-dimensional reinforcement signal, which represents the favorability of each one of the selected simple-actions. Based on this signal, the estimated rewards of the simple-actions are updated according to Eq. (3.1). The algorithm maintains two additional vectors $Z_i(k)$ and $\eta_i(k)$, in order to compute the estimated rewards of the simple-actions. $Z_i(k)$ represents the total received reinforcement by the i-th action, and $\eta_i(k)$ represents the number of times that the i-th action is selected until step k.

$$Z_i(k) = \begin{cases} Z_i(k-1) + \beta_i(k) & if\ a_i \in \alpha(k) \\ Z_i(k-1) & otherwise \end{cases}$$

$$\eta_i(k) = \begin{cases} \eta_i(k-1) + 1 \; if \; a_i \in \alpha(k) \\ \eta_i(k-1) \quad otherwise \end{cases}$$
$$\widehat{d}_i(k) = \frac{Z_i(k)}{\eta_i(k)}, i = 1, \ldots, r \tag{3.1}$$

where $\widehat{d}_i(k)$ represents the estimated reward for the i-th simple-action at step k. Additionally, for simplicity, the reinforcement vector is represented by an r-dimensional vector; however, only w elements of this vector corresponding to the selected simple-actions have valid information.

The next step of the algorithm is to find w simple-actions with the highest estimated rewards. This set of simple-actions corresponds to the super-action with the highest expected reward. Then, the algorithm moves the LA energy vector toward the corner of the LA hypercube associated with this super-action. Consequently, the energy levels of the promising simple-actions are increased. Figure 3.3 represents the pseudocode of the proposed MLATI. It should be noted that the algorithm can be simply implemented without actually calculating C (in Fig. 3.3). Consequently, all of its steps can be efficiently implemented with computational complexities not exceeding o(r).

Convergence analysis of the multi-reinforcement learning automaton type I.

In what follows, we discuss the convergence properties of MLATI. In this regard, the approach given in (Thathachar and Sastry 2004) will be employed, and it will be shown that the LA energy vector converges to the corner with the highest expected reward in the LA hypercube. In order to proceed, we assume that $\lambda(k) = 1 - \gamma^{1/k}$ with $\gamma \in (e^{-1}, 1)$.

Lemma 3.1 Given any $\delta \in (0,1]$ and $N \in \mathbb{Z}^+$, for each simple-action a_i, there exists $K_1(N,\delta)$ such that

$$\Pr\{\eta_i(k) < N\} < \delta \, \forall k > K_1(N, \delta) \tag{3.2}$$

Equation (3.2) is equivalent to $\sum_{s=1}^{N} \Pr\{\eta_i(k) = s\} \leq \delta$, which is satisfied if

$$\Pr\{\eta_i(k) = s\} < \frac{\delta}{N} \forall s \in \{1, \ldots, N\} \tag{3.3}$$

Considering line 12 of the proposed algorithm in Fig. 3.3, the energy of each simple-action can decrease with a rate not more significant than $(1 - \lambda(k))$ at each step k. Considering line 2 of the algorithm, each action is selected according to its energy level in the first phase of the action selection and is added to I. The selection probability of any action like a_i in this phase is x_i. However, I may contain more

Algorithm 3-1. MLATI
Input
r: number of actions
$w<r$: number of actions to be selected.
λ: learning rate.
Initialization:
$Z_i(0) = c_1\ \forall i \in \{1,\ldots,r\}$, where c_1 is a constant value.
$\eta_i(0) = c_2\ \forall i \in \{1,\ldots,r\}$, where c_2 is a constant value.
$\hat{d}_i(0) = 0\ \forall i \in \{1,\ldots,r\}$.
Set $x(0)=[0.5,\ldots,0.5]^T$, where x has r dimensions, and $x_i \in [0,1]$ represents the energy of the ith action.
Let I be an r-dimensional Boolean vector, $I_i \in \{0,1\}$.
Let $C=\{C_1,C_2,\ldots,C_o\}$ be the set of corners (vertices) of the LA hypercube with exactly w 1s and r-w 0s, where $o = |C| = \binom{r}{w}$, $C_{ij} \in \{0,1\}$, and $\sum_j C_{ij} = w\ \forall i \in \{1,\ldots,o\}$.
Set $k = 0$.
repeat
1. Set $k = k+1$ and $\alpha(k)=\{\}$.
2. Select a random set of simple-actions according to the energy level of the LA (i.e., according to x): each action like a_i is selected with probability x_i.
3. For each $i \in \{1,\ldots,r\}$, set $I_i=1$ if action a_i is chosen in step 2, otherwise set $I_i=0$.
4. Let C' be the subset of vertices in C, which are in the minimum distance to I. Randomly set I to one of the vertices in C'.
5. For each $i \in \{1,\ldots,r\}$, add simple-action a_i to $\alpha(k)$ if $I_i=1$.
6. Apply $\alpha(k)$ to the environment.
7. Receive an r-dimensional reinforcement signal β, where β_i is undefined if $a_i \notin \alpha(k)$.
8. Update LA information as follows:
$$Z_i(k) = \begin{cases} Z_i(k-1) + \beta_i(k) & \text{if } a_i \in \alpha(k) \\ Z_i(k-1) & \text{otherwise} \end{cases}$$
$$\eta_i(k) = \begin{cases} \eta_i(k-1) + 1 & \text{if } a_i \in \alpha(k) \\ \eta_i(k-1) & \text{otherwise} \end{cases}$$
$$\hat{d}_i(k) = \frac{Z_i(k)}{\eta_i(k)}, i = 1,\ldots,r$$
9. Update x as follows:
10. Find w actions with the largest estimated average reinforcements represent them by I:
$$\left(I_j \in \{0,1\}\ \forall j \in \{1,\ldots,r\}\right) \text{ and } \sum_j I_j = w$$
and $\forall i,j\left(\left(I_i = 1 \text{ and } I_j = 0\right) \Rightarrow \hat{d}_i \geq \hat{d}_j\right)$
11. Let C_i be the vertex in C corresponding to I.
12. $x(k) = x(k\text{-}1) + \lambda(C_i\text{-}x)$.
until some stopping condition is satisfied

Fig. 3.3 Pseudocode of multi-reinforcement learning automaton type I

than w actions, and accordingly, some of them may be removed randomly according to line 4 of the algorithm. The removal probability of any selected action at this step is less than $(r-1)/r$. Hence

$$\Pr\{a_i \notin \alpha(k)\} \leq \left(1 - \frac{x_i(0)\prod_{t=1}^{k}(1-\lambda(t))}{r}\right) \tag{3.4}$$

Accordingly,

$$\Pr\{\eta_i(k)=s\} < k^s\left(1-\frac{x_i(0)\prod_{t=1}^{k}(1-\lambda(t))}{r}\right)^{k-s} \tag{3.5}$$

Now the lemma can be proved if there exist K_1 such that

$$k^s\left(1-\frac{x_i(0)\prod_{t=1}^{k}(1-\lambda(t))}{r}\right)^{k-s} < \frac{\delta}{N},\ 1\le s\le N \quad \text{for all } k > K_1(N,\delta) \tag{3.6}$$

Using $\lambda(k)=1-\gamma^{1/k}$, it can be observed that

$$\lim_{k\to\infty} f(k) = k^s\left(1-\frac{x_i(0)\prod_{t=1}^{k}(1-\lambda(t))}{r}\right)^{k-s} = 0 \tag{3.7}$$

To verify Eq. (3.7), we can use the harmonic sum formula $\sum\limits_{t=1}^{k}\frac{1}{t}=\ln(k)+\xi+o(1/k)$.

By substituting $c = x_i(0)/r$ and using the harmonic sum, Eq. (3.7) can be restated as

$$\lim_{k\to\infty} f(k) = k^s\left(1-ck^{\ln(\gamma)}\gamma^{\xi}\gamma^{o(1/k)}\right)^{k-s} \tag{3.8}$$

Consequently,

$$\lim_{k\to\infty} f(k) = \lim_{k\to\infty} e^{s\ln(k)+(k-s)\ln\left(1-ck^{\ln(\gamma)}\gamma^{\xi}\gamma^{o(1/k)}\right)} = 0 \tag{3.9}$$

According to Eq. (3.7), for sufficiently large values of k, $f(k)$ becomes small enough and is guaranteed to be less than δ/N.

Lemma 3.2 Given any ε, $\delta\in(0,1]$, for all $i\in\{1,\ldots,r\}$, there exists K_2 satisfying

$$\Pr\left\{\left|\hat{d}_i(k)-d_i\right| > \varepsilon\right\}\langle\delta\forall k\rangle K_2(\varepsilon,\delta) \tag{3.10}$$

where $d_i = \mathrm{E}[\beta_i(k)]$.

Considering the algorithm in Fig. 3.3, the estimated reward for the i-th simple-action (a_i) can be expressed as:

$$\hat{d}_i(k) = \frac{\sum_{s=1}^{\eta_i(k)}\beta_i\left(m_s^i\right)}{\eta_i(k)} \tag{3.11}$$

where m_s^i represents the time instant at which a_i is chosen for the s-th time.

For each i, the sequence $\{\beta_i(m_s^i)\}$ is iid, and each of its members is bounded above by a constant like M. Considering Hoeffding's inequality and substituting $\eta_i(k)$ with N:

$$\Pr\left\{\left|\frac{\sum_{s=1}^{N}\beta_i\left(m_s^i\right)}{N}-d_i\right|>\varepsilon\right\}<2\exp\left(-\frac{2N\varepsilon^2}{M^2}\right) \tag{3.12}$$

Now we define two events A and B as:

$$A=\left\{\left|\widehat{d}_i(k)-d_i\right|>\varepsilon\right\},$$
$$B=\{\eta_i(k)>N\} \tag{3.13}$$

The probability of event A can be obtained as follows:

$$\Pr(A)=\Pr(A|B)\Pr(B)+\Pr(A|\bar{B})\Pr(\bar{B})\langle\Pr(A|B)+\Pr(\bar{B}) \tag{3.14}$$

By considering $N=\left\lceil\frac{M^2}{2\varepsilon^2}\ln\frac{4}{\delta}\right\rceil$, it follows that $\Pr(A|B) < \delta/2$, and by lemma 3.1

$$\Pr(\bar{B})<\frac{\delta}{2},\forall k>K_2(\varepsilon,\delta)\text{ and }\lambda(k)=1-\gamma^{1/k} \tag{3.15}$$

where $K_2(\varepsilon,\delta)=K_1(N,\delta/2)$.

Accordingly, it fallows that $\Pr(A)<\delta/2+\delta/2=\delta$.

Theorem 3.1 A learning automaton using the algorithm in Fig. 3.3 with λ defined as $\lambda(k)=1-\gamma^{1/k}$ and $\gamma\in(e^{-1},1)$ has the following property.

Given any ε, $\delta\in(0, 1)$, there exists $K^* > 0$ such that

$$\Pr\{|x-C_L|<\varepsilon\}>1-\delta \tag{3.16}$$

where C_L corresponds to the corner of the LA hypercube with w best simple-actions.

Proof Considering the set of optimal simple-actions as C_L, we can define the following three events:

$$E_1(k)=\{|C_L-x(k)|<\varepsilon\} \tag{3.17}$$

$$E_2(k)=\left\{\max_i\left|\widehat{d}_i(k)-d_i\right|<\theta\right\} \tag{3.18}$$

$$E_3(k)=\left\{\sup_{s\geq k}\max_i\left|\widehat{d}_i(s)-d_i\right|<\theta\right\} \tag{3.19}$$

where $\theta = \min_l\{d_l|C_{Ll}=1\}$-$\max_i\{d_i|C_{Li}=0\}$.

Using these events, for any k and K, the following inequality holds:

$$\Pr(E_1(k+K)) \geq \Pr(E_1(k+K)|E_3(K))\Pr(E_3(K)) \tag{3.20}$$

Moreover, there exists K_3 such that $\Pr(E_1(k+K)|E_3(K)) = 1$ for all $k > K_3$:

Assuming $x(0) = [1/2,\ldots,1/2]^T$ and defining $D_i(k) = |C_{Li} - x_i(k)|$, the energy vector can move at most K times in the opposite direction of C_L in each dimension. Consequently,

$$D_i(K) \leq \left(1 - D_i(0)\prod_{s=1}^{K}(1-\lambda(s))\right) = \left(1 - 0.5\prod_{s=1}^{K}(1-\lambda(s))\right) < 1 \tag{3.21}$$

After step K, since $E_3(K)$ is satisfied, the LA energy vector approaches C_L in each time step. Accordingly, after K_3 iterations, the following holds:

$$D_i(K+K_3) \leq D_i(K)\prod_{s=1}^{K_3}(1-\lambda(s+K)) \tag{3.22}$$

Using $\lambda(k) = 1 - \gamma^{1/k}$,

$$D_i(K+K_3) \leq D_i(K)\prod_{s=1}^{K_3}\gamma^{\frac{1}{(s+K)}} \tag{3.23}$$

In Eq. (3.23),

$$\prod_{s=1}^{K_3}\gamma^{\frac{1}{(s+K)}} = \gamma^{\sum_{s=1}^{K_3+K}\frac{1}{s} - \sum_{s=1}^{K}\frac{1}{s}} \tag{3.24}$$

By defining $\kappa = D_i(K)\Big/\left(\gamma^{\sum_{s=1}^{K}\frac{1}{s}}\right)$, Eq. (3.22) can be stated as

$$D_i(K+K_3) \leq \kappa\gamma^{\sum_{s=1}^{K_3+K}\frac{1}{s}} \tag{3.25}$$

Then, it follows that

$$D_i(K+K_3) \leq \kappa\gamma^{\ln(K_3+K)} \tag{3.26}$$

$D_i(K+K_3)$ approaches zero as $K_3 \to \infty$ for all $i \in \{1,\ldots,r\}$. Consequently, there exists K_3 such that $\Pr(E_1(k+K)|E_3(K)) = 1$ for all $k > K_3$.

Hence, the proof of Theorem 3.1 can be completed if there exists K_4 such that $\Pr(E_3(k)) \geq 1 - \delta$ for all $k > K_4$.

The probability of event $E_3(k)$ can be defined as follows:

$$\Pr(E_3(k)) = \Pr\left(\bigcap_{s=k}^{\infty} E_2(k)\right) = 1 - \Pr\left(\bigcup_{s=k}^{\infty} \bar{E}_2(k)\right) \tag{3.27}$$

By defining $m(k)$ as $m(k) = argmax\left(\left|\hat{d}_i(k) - d_i(k)\right|\right)$ and $\eta'(k) = \eta_{m(k)}(k)$, it follows that

$$\Pr\left(\bigcup_{s=k}^{\infty} \bar{E}_2(k)\right) \leq \sum_{s=k}^{\infty} \Pr\left(\bar{E}_2(k)\right) = \sum_{s=k}^{\infty}\sum_{t=0}^{s}\left(\Pr\left(\bar{E}_2(s)|\eta'(s) = t\right)\Pr\left(\eta'(s) = t\right)\right) \tag{3.28}$$

Using Hoeffding's inequality,

$$\Pr\left(\bar{E}_2(s)|\eta'(s) = t\right) \leq 2e^{-2t\theta^2/M^2} \tag{3.29}$$

Consequently,

$$\begin{aligned}\Pr\left(\overset{\infty}{\underset{s=k}{\cup}} \bar{E}_2(k)\right) &\leq \sum_{s=k}^{\infty}\sum_{t=0}^{s}\left(\Pr\left(\bar{E}_2(s)|\eta'(s) = t\right)\Pr\left(\eta'(s) = t\right)\right) \\ &\leq \sum_{s=k}^{\infty} 2\sum_{t=0}^{s}\left(e^{-2t\theta^2/M^2}\Pr\left(\eta'(s) = t\right)\right)\end{aligned} \tag{3.30}$$

The RHS of Eq. (3.30) can be restated as follows:

$$\sum_{s=k}^{\infty} 2\sum_{t=0}^{s}\left(e^{-2t\theta^2/M^2}\Pr\left(\eta'(s) = t\right)\right) = \sum_{s=k}^{\infty} 2E\left[e^{\left(-2\theta^2/M^2\right)\eta'(s)}\right] \tag{3.31}$$

Define $\eta'(k) = \sum_{s=1}^{k} \chi(s)$, where $\chi(s) \in \{0,1\}$, and $\chi(s) = 0,1$ indicates whether or not the simple-action corresponding to $m(k)$ is selected at step s. By defining $\tau = -2\theta^2/M^2$ it follows that

$$\begin{aligned}\sum_{s=k}^{\infty} 2E\left[e^{\tau\eta'(s)}\right] &= \sum_{s=k}^{\infty} 2E\left[e^{\tau(\chi(1)+\chi(2)+\ldots+\chi(s))}\right] \\ &= \sum_{s=k}^{\infty} 2\prod_{t=1}^{s}\left(1 - \Pr(a_{m(s)} \in \alpha(t)) + \Pr(a_{m(s)} \in \alpha(t))e^{\tau}\right)\end{aligned}$$

$$= \sum_{s=k}^{\infty} 2 \prod_{t=1}^{s} (1 - \Pr(a_{m(s)} \in \alpha(t))(1 - e^{\tau})) \tag{3.32}$$

Similar to lemma 3.1, $\Pr(a_{m(s)} \in \alpha(t))$ can be replaced by a smaller term:

$$\begin{aligned}
&\sum_{s=k}^{\infty} 2 \prod_{t=1}^{s} (1 - \Pr(a_{m(s)} \in \alpha(t))(1 - e^{\tau})) \\
&\leq \sum_{s=k}^{\infty} 2 \prod_{t=1}^{s} \left(1 - \left(\frac{x_{m(s)}(0)}{r}\right)(1 - e^{\tau})\gamma^{\sum_{n=1}^{s} 1/n}\right) \\
&= \sum_{s=k}^{\infty} 2 \left(1 - \left(\frac{x_{m(s)}(0)}{r}\right)(1 - e^{\tau})\gamma^{\sum_{n=1}^{s} 1/n}\right)^{s}
\end{aligned} \tag{3.33}$$

The RHS of Eq. (3.33) can be approximated as follows:

$$\sum_{s=k}^{\infty} 2 \left(1 - \left(\frac{x_{m(s)}(0)}{r}\right)(1 - e^{\tau})\gamma^{\sum_{n=1}^{s} 1/n}\right)^{s} \approx 2 \sum_{s=k}^{\infty} \left(1 - \left(\frac{x_{m(s)}(0)}{r}\right)(1 - e^{\tau})\gamma^{\ln(s)}\right)^{s} \tag{3.34}$$

The series in the RHS of Eq. (3.34) is convergent. Consequently, following the convergence of partial sums, for any δ there exists a K_4 such that for $s > K_4$, the series ends up less than $\delta/2$.

We can verify that the series in Eq. (3.34) is convergent as follows.

Define $a = (x_{m(s)}(0)/r)(1-\exp(\tau))$ and $b = -\ln(\gamma)$. The RHS of Eq. (3.34) can be stated as follows:

$$f(s) = \sum_{s=1}^{\infty} \left(1 - \frac{a}{s^b}\right)^{s} \tag{3.35}$$

where $a, b \in (0, 1)$.

As $\left(1 - \frac{a}{s^b}\right)^{s^b} \leq e^{-a/b}$,

$$f(s) \leq \sum_{s=1}^{\infty} \left(e^{-a/b}\right)^{s^{1-b}} \tag{3.36}$$

The RHS of Eq. (3.36) converges like a geometric series.

3.2.2 Multi-reinforcement Learning Automaton Type II

The previously proposed MLATI is based on the pursuit algorithm. It has two main drawbacks, which are due to the characteristics inherent in the pursuit algorithm. In order to illustrate these drawbacks, we consider the following scenario for the pursuit algorithm. Consider a learning automaton with four actions a_1, a_2, a_3, and a_4 with the expected payoffs d_1, d_2, d_3, and d_4, respectively. Assume $d_1 > d_2 > d_3 > d_4$. Also, assume that during some early iterations of the pursuit algorithm, actions a_3 and a_4 collect larger rewards in comparison to a_1 and a_2. Accordingly, the following relations would hold for the estimated rewards and action probability values. $\widehat{d}_4(k_1) > \widehat{d}_3(k_1) > \widehat{d}_1(k_1) > \widehat{d}_2(k_1)$ and $p_4(k_1) > p_3(k_1) > p_1(k_1) > p_2(k_1)$ for some k_1. Assume that during some iterations after step k_1, a_1 receives the highest payoff from the environment, and consequently, $\widehat{d}_1(k_2) > \widehat{d}_4(k_2) > \widehat{d}_3(k_2) > \widehat{d}_2(k_2)$ with $p_1(k_2) > p_4(k_2) > p_3(k_2) > p_2(k_2)$ for some k_2. After step k_2, the received reinforcements are such that the following is always satisfied $\widehat{d}_1(k_3) = \max\left(\widehat{d}_i(k_3)\right)$ for any $k_3 > k_2$. Accordingly, after time step k_2, the selection probability of a_1 always increases, which is favorable; however, the relative order of the selection probabilities of other actions remains unaltered (i.e., $p_4(k_3) > p_3(k_3) > p_2(k_3)$ for any $k_3 > k_2$) regardless of their average received payoffs. The second issue with the pursuit algorithm is closely related to the first one. Consider again in the previous scenario that after step k_1, the average received payoff for action a_1 starts increasing. Define k_5 as the minimum number of steps such that the following holds $\widehat{d}_1(k_1 + k_5) = \max_i\left(\widehat{d}_i(k_1 + k_5)\right)$. During these k_5 steps, the probability of action a_4 always increases regardless of the received rewards of a_4. After these k_5 steps, the probability of action a_1 starts increasing. However, it may take a long time until $p_1 > p_4$. These two issues may result in a slow convergence rate of the pursuit algorithm.

According to the first observation, the selection probabilities of actions other than the one with the highest estimated reward may be unrelated to their obtained rewards. Accordingly, it would not affect the learning process if LA selects these actions with equal probabilities. The second issue can be solved if the action with the highest estimated reward is always selected with the highest probability.

Figure 3.4 summarizes the steps of the proposed multi-reinforcement learning automaton type II (MLATII). The algorithm performs action selection in three manners: greedy, exploratory, and mutated greedy. During step k, the algorithm selects the super-action with the highest estimated reward with probability 1-$\lambda(k)$. We assume that $(\lambda(k))_{k \in \mathbb{N}}$ is a decreasing sequence of numbers, which converges to zero. Hence, it is ensured that the super-action with the highest estimated reward is ultimately selected almost surely in every iteration.

Additionally, each simple-action should be selected for an adequate number of times in order to ensure that the estimated rewards converge to their expected values (as discussed in lemma 3.2). In this regard, with probability $\lambda(k)/2$, the proposed

Algorithm 3-2. MLATII

Input

r: number of simple-actions

$A = \{a_1, a_2, \ldots, a_r\}$: actions set

$w < r$: number of simple-actions to be selected.

$(\lambda(k))_{k \in \mathbb{N}}$: learning rate.

$(\eta(k))_{k \in \mathbb{N}}$: averaging factor.

Initialization:

Initialize the estimated rewards for the simple-actions:

$Q_i(0) = c_1 \; \forall i \in \{1, \ldots, r\}$, where c_1 is some constant value.

repeat

1. With probability $1-\lambda(k)$: // *Greedy selection*
    a. Select the w simple-actions with the highest expected payoffs as $\alpha(k)$.
2. Otherwise with probability $\lambda(k)/2$: // *exploratory*
    a. Select a random super-action as $\alpha(k)$.
3. Otherwise: // *mutated greedy*
    a. Select the w simple-actions with the highest expected payoffs as $\alpha(k)$.
    b. Select a random action $a_i \notin \alpha(k)$ according to its probability value defined as:
       $p_i = \frac{Q_i(k)}{\sum_{j \in I} Q_j(k)} \; \forall i \in I = \{l \mid a_l \notin \alpha(k) \text{ and } a_l \in A\}$
    c. Select a random action $a_j \in \alpha(k)$ according to its probability value:
       $p_l = \frac{1-q_l}{\sum_{j \in I} 1-q_j} \; \forall l \in I$
       Where $q_h = \frac{Q_h(k)}{\sum_{j \in I} Q_j(k)} \; \forall h \in I$ and $I = \{s \mid a_s \in \alpha(k)\}$
    d. Replace a_j with a_i in $\alpha(k)$.
4. Perform $\alpha(k)$ in the environment.
5. Receive a vector reinforcement signal β.
6. For each action a_i in $\alpha(k)$ do:
    a. $Q_i(k+1) = Q_i(k) + \eta(k)(\beta_i(k) - Q_i(k))$

until some stopping condition is satisfied.

Fig. 3.4 Pseudocode of multi-reinforcement learning automaton type II

method randomly selects a super-action at line 2, which guarantees that each simple-action is chosen for an adequate number of times. Finally, with probability $\lambda(k)/2$, the algorithm can utilize the mutated greedy selection strategy in each iteration. The mutated greedy selection has the benefits of both exploratory and greedy selection strategies. It greedily selects a slightly perturbed version of the best-estimated super-action. Consequently, by using mutated greedy selection, the algorithm can frequently select and exploit estimated better simple-actions, while experiencing all simple-actions for an adequate number of times. After action selection, the selected super-action is carried out in the environment, and the estimated rewards are updated based on the received reinforcement signals.

It is easy to show that by using a decreasing sequence $(\lambda(k))_{k \in \mathbb{N}}$ which satisfies $\lambda(k) = \kappa \mu(t)$ with $\sum_k \mu(k) = \infty$ and $\kappa > 0$, each simple-action is selected infinitely often. Also, if the averaging factor satisfies the two conditions $\sum_k \eta(k) = \infty$ and

$\sum_k (\eta(k))^2 < \infty$, the estimated reward of each simple-action converges to the actually expected reward of the simple-action. Hence as $\lambda(k) \rightarrow 0$ with $k \rightarrow \infty$, the selection probability of the most promising super-action converges to 1. Based on these observations, the proposed algorithm is ε-optimal.

3.2.3 Multi-reinforcement Learning Automata Type III

The multi-reinforcement LA proposed in the previous section is not absolutely expedient. In order to illustrate this issue, consider $\underline{a}_i$ as the optimal super-action. Also, assume that during some iteration like k, for some super-action like $\underline{a}_j$, the following holds $\sum_l I_l Q_l(k) < \sum_l J_l Q_l(k)$, where I and J are Boolean vectors representing the members of $\underline{a}_i$ and $\underline{a}_j$ respectively. Accordingly, $\underline{a}_j$ will be selected with a higher probability for some iterations after step k. During these subsequent iterations, $\sum_l J_l Q_l(k)$ starts decreasing as $\underline{a}_j$ has a lesser expected payoff in comparison to $\underline{a}_i$. However, the selection probability of $\underline{a}_j$ would increase for some iterations as $\lambda(k)$ is decreasing. This increase in the selection probability of $\underline{a}_j$ continues regardless of the received payoffs until the following condition is satisfied $\sum_l I_l Q_l(k) > \sum_l J_l Q_l(k)$. Absolute expediency is an advantageous property, which, under appropriate conditions, guarantees the convergence of a group of learning automata in a CLA structure (Beigy and Meybodi 2004, 2010; Rezvanian et al. 2018c).

One of the absolutely expedient learning automata algorithms is L_{RI} (Narendra 1989; Rezvanian et al. 2018a). In this section, another multi-reinforcement LA model is introduced, which is based on the L_{RI} algorithm. Like MLATI, each simple-action has an associated energy and is selected according to its energy level in each iteration. The internal state of the learning automaton is modeled as an energy vector which changes in an r-dimensional unit hypercube. However, in contrast to MLATI, the total energy of the LA (energy sum of all simple-actions) is always kept constant and equal to w. Accordingly, an increase in the energy level of a particular simple-action should result in a decrease in the energy levels of the others. In this regard, whenever a selected simple-action receives a favorable reinforcement, its energy level increases, and this increased amount is subtracted from the total energy level of the unselected simple-actions.

A multi-reinforcement learning automaton type III (MLATIII) operates as follows. At each iteration k, a w-dimensional super-action is selected in the same way as MLATI. Then, this super-action is performed in the environment. The environment provides the learning system with a w-dimensional feedback signal, which represents the favorability of each one of the selected simple-actions in the super-action. If the associated signal to a selected simple-action like a_i is favorable, i.e. $\beta_i(k) = 1$, the energy level of a_i is increased according to Eq. (3.37). This increased amount is subtracted from the energy level of unselected actions in order to keep the total energy level of the LA constant (Eq. (3.38)).

$$x_i = x_i + \lambda\beta(k)(1 - x_i) \tag{3.37}$$

$$x_j = x_j - \frac{\lambda\beta_i(k)(1 - x_i)}{\sum_l x_l(1 - I_l)}(x_j)\forall j \in \{h|a_h \notin \alpha(k)\, and\, a_h \in A\} \tag{3.38}$$

Here, λ is the learning rate parameter, and similar to MLATI, the selected set of simple-actions at step k is represented by $\alpha(k)$. I is an r-dimensional Boolean vector that represents the selected simple-actions. Figure 3.5 illustrates the proposed multi-reinforcement learning automaton algorithm.

3.2.4 Performance Evaluation on Subset Selection Problems

This section illustrates the performance of the proposed multi-reinforcement LA models. We assume that there is a hypothetical stationary environment which can be in one of its states at each time step. The environment contains r devices that are

Algorithm 3-3. MLATIII

Input

r: number of simple-actions

$A = \{a_1,\ldots,a_r\}$

$w<r$: number of simple-actions to be selected.

λ: learning rate.

Initialization:

Set $x(0)=[w/r, w/r,\ldots, w/r]^T$, where $x_i \in [0,1]$ represents the energy of the ith simple-action.

Consider an r-dimensional Boolean vector I: $I_i \in \{0,1\}\ \forall i \in \{1, \ldots, r\}$.

Let $C=\{C_1,C_2,\ldots,C_o\}$ be the set of corners (vertices) of the LA hypercube with exactly w 1s and $(r-w)$ 0s, where $o = |C| = \binom{r}{w}$, $C_{ij} \in \{0,1\}$, and $\sum_j C_{ij} = w\ \forall i \in \{1,\ldots,o\}$.

Set $k = 0$.

Repeat

1. Set $k = k+1$ and $\alpha(k) = \{\}$.
2. Select a random set of simple-actions according to the energy level of the LA (i.e., according to x): each action like a_i is selected with probability x_i.
3. For each $i \in \{1, \ldots, r\}$, set $I_i = 1$ if action a_i is chosen in step 2, otherwise set $I_i = 0$.
4. Let C' be the subset of vertices in C, which are in the minimum distance to I. Randomly set I to one of the vertices in C'.
5. For each $i \in \{1, \ldots, r\}$, add simple-action a_i to $\alpha(k)$ if $I_i = 1$.
6. Apply $\alpha(k)$ to the environment.
7. Receive an r-dimensional reinforcement signal β, where β_j is undefined if $a_j \notin \alpha(k)$ (i.e. $I_j = 0$).
8. Update LA information as follows:
   For each action a_i such that $I_i = 1$ and $\beta_i(k) > 0$ do:
   $$x_j = x_j - \frac{\lambda\beta_i(k)(1-x_i)}{\sum_l x_l(1-I_l)}(x_j)\ \forall j \in \{h\,|\,a_h \notin \alpha(k)\ and\ a_h \in A\}$$
   $$x_i = x_i + \lambda\beta_i(k)(1-x_i)$$

until some stopping condition

Fig. 3.5 Pseudocode of Multi-reinforcement learning automata type III

Table 3.1 10 instances of the decision-making problem. r and w respectively represent the number of available devices and the number of devices to be selected by the agent in each step. d_i is the activation probability of device i

Prob.	r,w	d1	d2	d3	d4	d5	d6	d7	d8	d9	d10
1	6,3	0.7296	0.2987	0.3741	0.5187	0.1398	0.7555				
2	6,3	0.4780	0.1300	0.5339	0.5144	0.1665	0.6029				
3	7,3	0.7150	0.7125	0.9987	0.9173	0.3971	0.9228	0.1271			
4	7,3	0.9093	0.5233	0.2607	0.7955	0.9818	0.2673	0.1049			
5	8,3	0.8126	0.5963	0.0153	0.0855	0.0245	0.1393	0.2593	0.7124		
6	8,3	0.8095	0.3593	0.8131	0.5743	0.9497	0.5437	0.3369	0.0266		
7	9,5	0.5745	0.8156	0.3232	0.9225	0.8017	0.8098	0.0074	0.0617	0.6051	
8	9,5	0.0671	0.3588	0.6395	0.0236	0.0899	0.0505	0.5532	0.7165	0.2128	
9	10,4	0.6954	0.1900	0.3978	0.1777	0.5689	0.6425	0.1005	0.7337	0.6832	0.2827
10	10,4	0.6414	0.3879	0.1581	0.7081	0.1114	0.6702	0.3283	0.6734	0.8162	0.4271

hidden from the outside world. During each time step, each device i can be active or inactive with a fixed probability d_i independent from other devices. An agent is interacting with the environment. At each time step, the agent selects w devices from the environment to use for its purposes. However, the agent can only employ active devices. After using the chosen devices, the agent releases its selected devices to the environment, and the environment activates or deactivates some of its devices for the next period. This procedure continues for an infinite number of iterations. The objective of the agent is to utilize as many devices as possible to fulfill its assigned tasks in time. Accordingly, it needs to find w devices with the highest activation probabilities. Table 3.1 gives some instances of the introduced problem, which are used in the experiments of this section.

In order to solve this problem with a multi-reinforcement LA, we consider an LA with r actions. At each time step, the LA selects w actions according to its internal state. Each selected action receives a reward if the corresponding device is active.

3.2.4.1 The Sensitivity of the Proposed Models to the Learning Rate

In this section, the sensitivity of the proposed models to variations of their learning rates is investigated. In this regard, each model is examined with various settings for its learning rate on several instances of the described problem. The expected error of the learning agent is used as a performance metric in this section. First, we define the expected reward of each super-action $\underline{a}_i$ as follows:

$$R_i = \sum_{j \in I} d_j \text{ with } I = \left\{ j | a_j \in \underline{a}_i \right\}. \tag{3.39}$$

If we represent the super-action set of the LA with $\underline{A}$, we can obtain the expected error of the LA at iteration k as follows:

$$\chi(k) = \max_{i \in I}(R_i) - \sum_{i \in I} q_i(k) R_i,$$

with

$$I = \{1, \ldots, |\underline{A}|\}, \tag{3.40}$$

where $q_i(k)$ is the selection probability of super-action $\underline{a}_i$ at step k.

Experiments 1: multi-reinforcement LA type III

Figure 3.6 shows the obtained average results for MLATIII over 20 independent runs. Two major parameters affect the difficulty of each learning problem: the dimensionality of the problem and the closeness of the reward probabilities of different simple-actions. The difference between the third and the fourth-highest reward probabilities is considerable in problem 5 (Table 3.1). As the LA needs to select three simple-actions in this problem instance, it can easily converge to the optimal super-action in a few iterations. In contrast, the difference between the third and the fourth reward probabilities is very small in problem 3. Accordingly, despite its smaller action set, problem 3 requires more learning steps in comparison to problem 5.

As the learning rate increases from 0.001 to 0.1, the convergence speed of the algorithm increases significantly. However, the algorithm can examine fewer actions with a high learning rate, which may result in a premature convergence. Consequently, hard problems or problems with larger action sets require lower learning rates. For instance, using a high learning rate such as 0.1 resulted in the premature convergence of the algorithm in some of its executed runs on difficult problems 3 and 9. Based on the obtained results, a learning rate between 0.05 and 0.1 seems to be appropriate for most of the tested instances as the algorithm can find optimal or near-optimal results in the designated number of iterations.

Experiment 2: multi-reinforcement LA type I

The same experiments, as in the previous section, are conducted using MLATI. As it is evident from the achieved results in Fig. 3.7 MLATI demonstrates rather high convergence rates in comparison to MLATIII on all of the tested problems. Besides, problems 2 and 7, it can converge to the optimal super-action with all of the tested learning rates. Accordingly, MLATI seems to be a better choice for random stationary environments in comparison to MLATIII.

Experiment 3: multi-reinforcement LA type II

This section illustrates the behavior of the proposed MLATII. In the experiments of this section, $\eta(k)$ is kept constant at 0.01, and an increasing sequence with an initial value of 0.1 is used as $(\lambda(k))_{k \in \mathbb{N}}$. Herein, we use a simple equation for defining the values of this parameter as follows $\lambda(k) = \lambda(k-1) + \delta(1 - \lambda(k-1)$. The average expected errors of MLATII on different problems with different settings for δ are provided in Fig. 3.8. The obtained results demonstrate the effectiveness of the proposed learning scheme in finding the optimal subset of actions for the introduced problems. It can be observed that MLATII is capable of finding the optimal super-actions on different problems in less than 1000 iterations with the settings of δ in the range [0.005,0.1]. Additionally, the algorithm can achieve expected errors very close to zero in a short period when $\delta \in$[0.025,0.1].

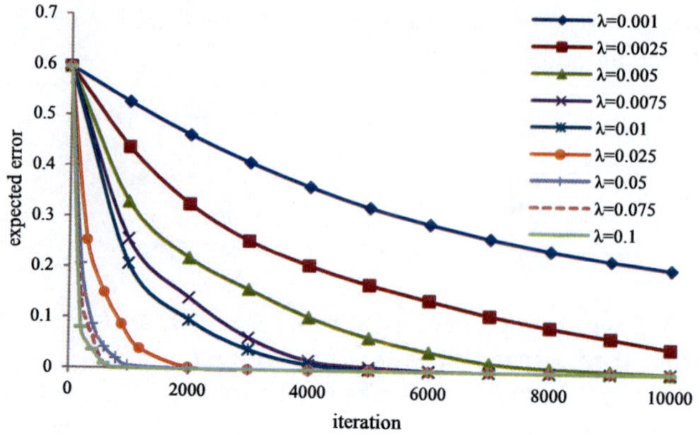

a) Problem 1

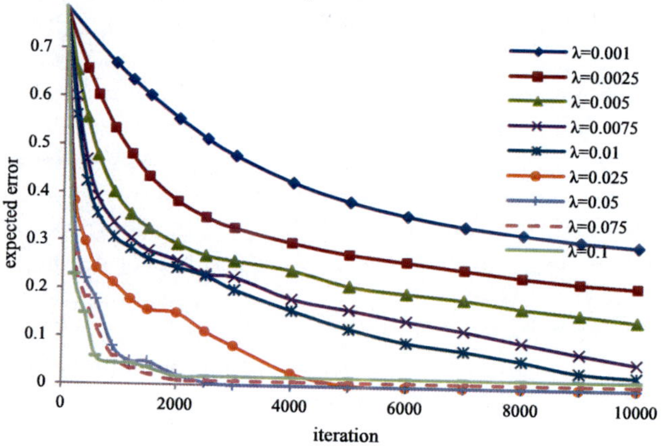

b) Problem 3

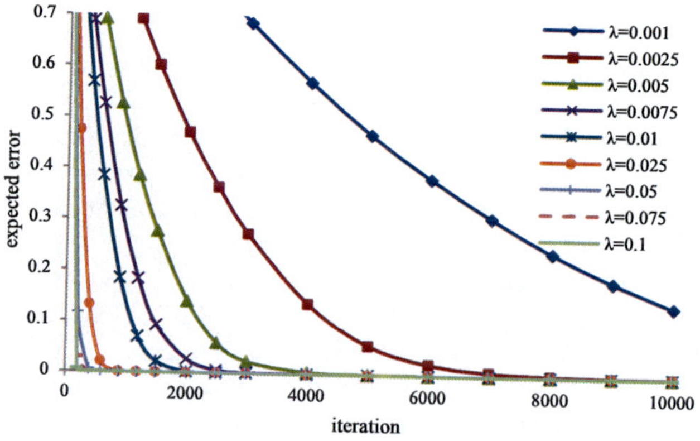

c) Problem 5

Fig. 3.6 The average expected error curves of MLATIII on different problems from Table 3.1. The average expected errors of the algorithm on each problem instance are obtained using different settings of the learning rate (λ)

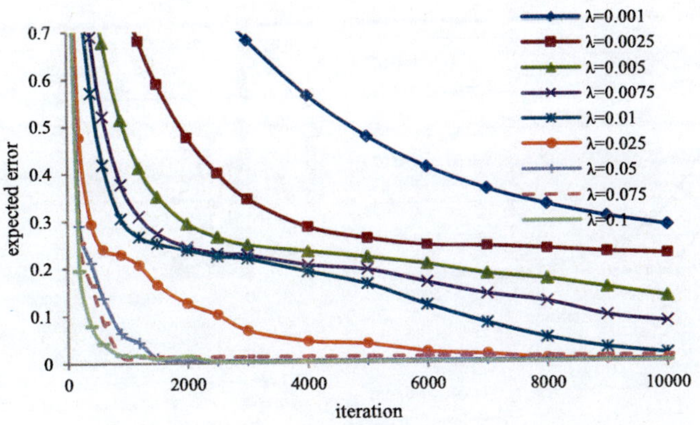

d) Problem 7

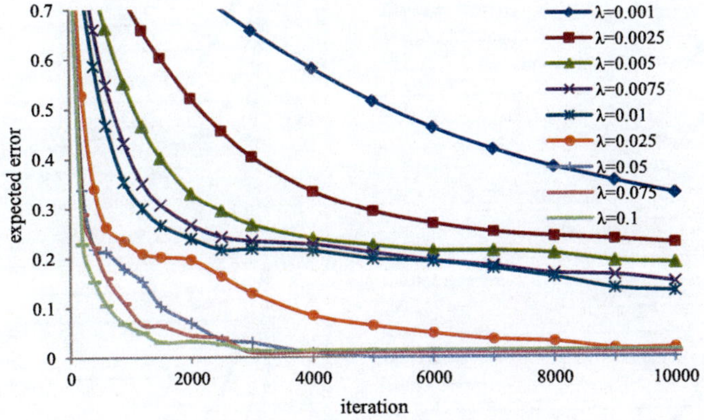

e) Problem 9

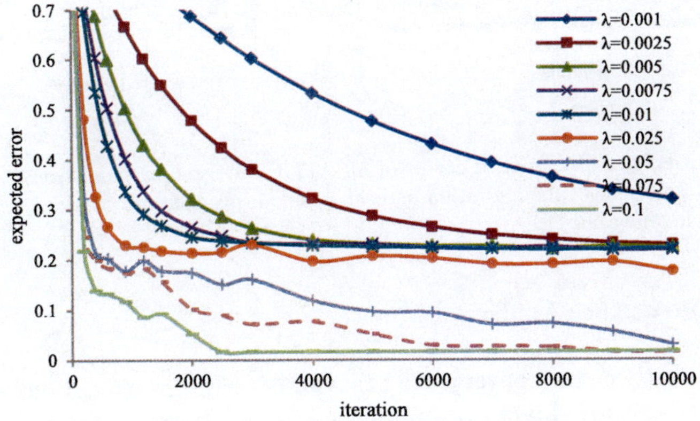

f) Problem 10

Fig. 3.6 (continued)

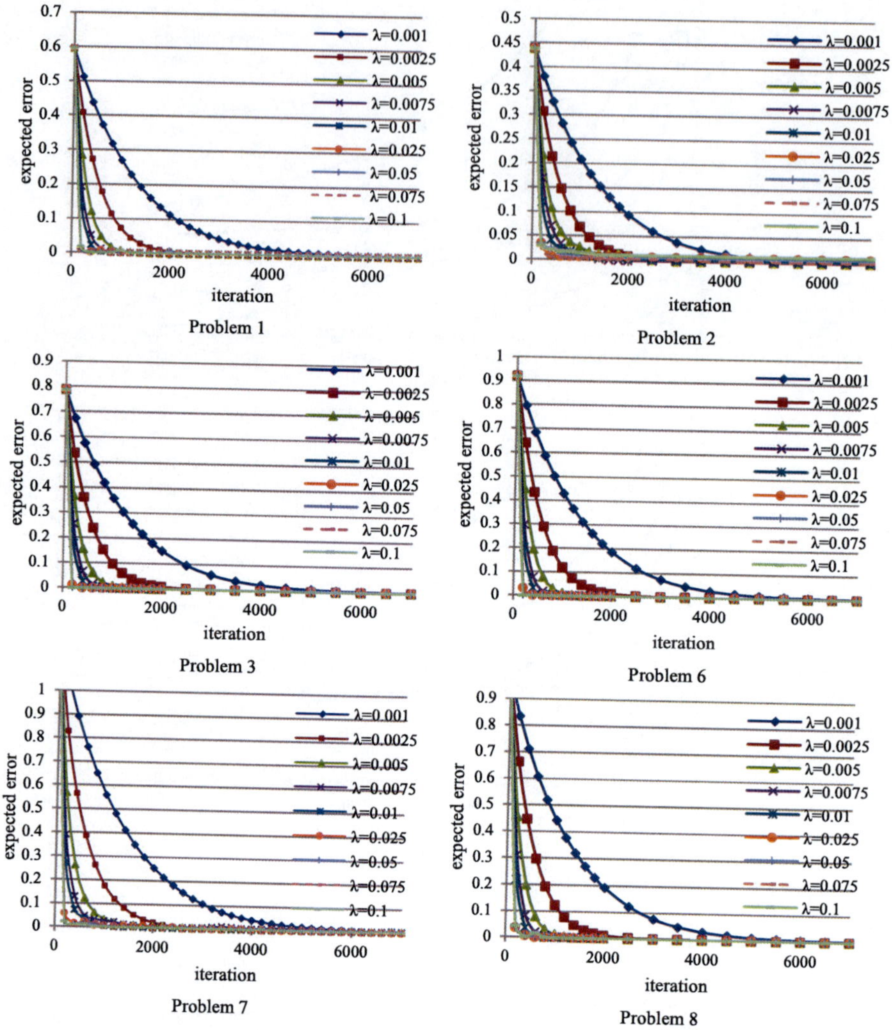

Fig. 3.7 The average expected error curves of MLATI on different problems from Table 3.1. The average expected errors of the algorithm on each problem instance are obtained using different settings of the learning rate (λ)

3.2.4.2 Convergence Analysis

In order to investigate the convergence behavior of the proposed LA models, we use the concept of entropy, which was introduced in the context of information theory by Shannon (2001). If we consider the selected super-action of an LA during step k as a random variable, we can define its entropy using the following equation:

$$H(k) = \sum_{i=1}^{\binom{r}{w}} q(i,k).\ln(q(i,k)), \tag{3.41}$$

where $q(i, k)$ is the selection probability of the k-th super-action at time step k. $H(k)$ can be considered as a measure of uncertainty associated with the LA at step k; larger values of $H(k)$ represent more uncertainty in the decisions of the LA. The

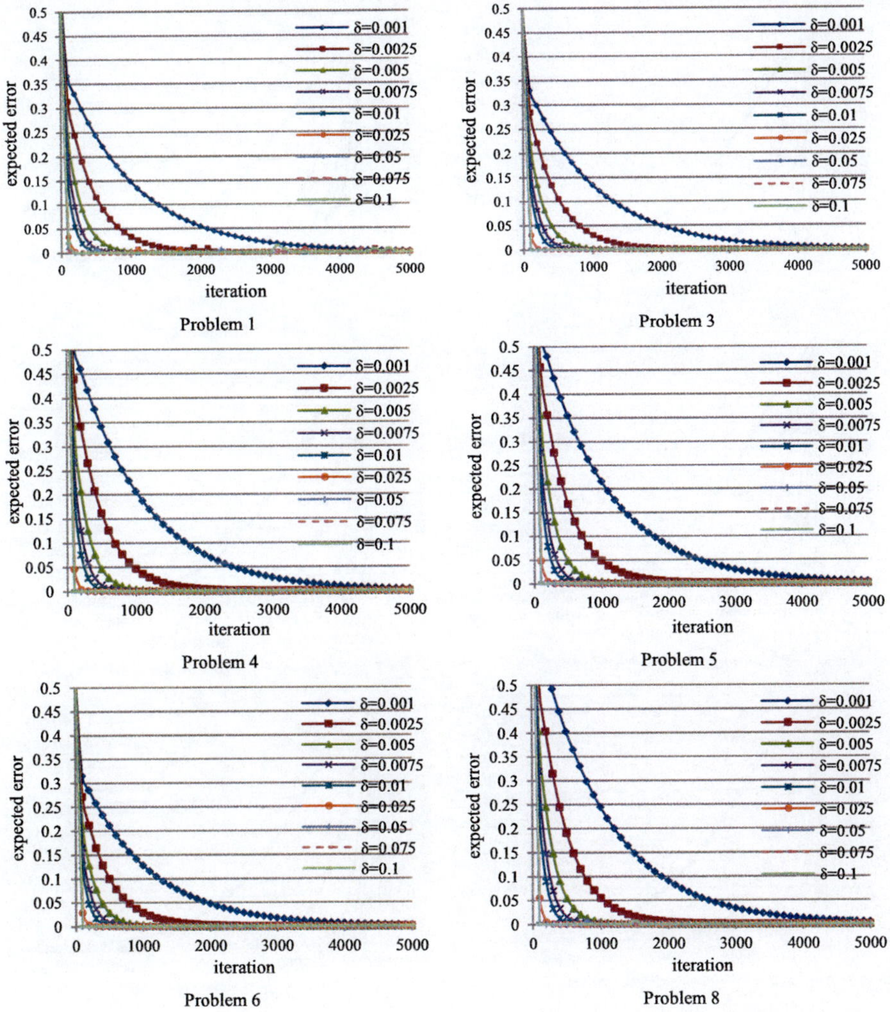

Fig. 3.8 The average expected error curves of MLATI on different problems from Table 3.1. The average expected errors of the algorithm on each problem instance are obtained using different settings of δ

experiments of this section are conducted using MLATI on different problems with different settings for the learning rate. Figure 3.9 represents the average Entropy of the LA over 20 independent runs.

Considering Fig. 3.9, it can be observed that the entropy is high at the beginning of the algorithm, but gradually decreases during the learning procedure. In all experiments, the average entropy converges to 0, meaning that the LA ultimately selects a particular super-action in every step. As demonstrated in the previous section, the expected error of the algorithm converges to 0 as well. Accordingly, it can be

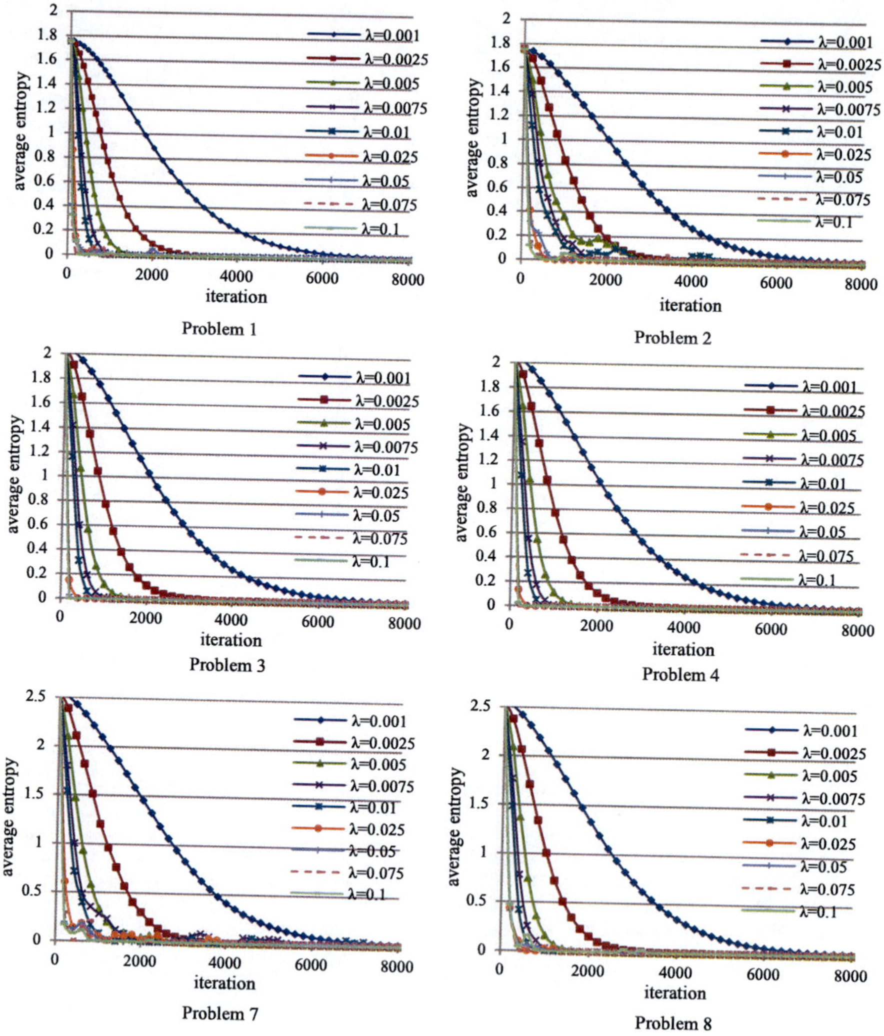

Fig. 3.9 The average entropy of LA type I under different settings for the learning rate (λ) during 1E + 4 iterations. The results are obtained over 20 independent runs

concluded that the proposed multi-reinforcement LA scheme always learns to select the optimal super-action with probability one.

3.2.4.3 Comparison of Different Introduced MLAs for Subset Selection

In this section, we compare the average final errors of the three proposed LA methods. Table 3.2 provides the average expected errors of the introduced LA models on the given problems after 1E + 4 learning steps. The learning rate of all models is set to 0.075. Wilcoxon's rank-sum test at a 5% significance level is also conducted to compare the algorithms (Janez Demšar 2006). The results corresponding to the conducted Wilcoxon's rank-sum tests between MLATI and MLATII are given in the columns associated with MLATII, where '+' represents that MLATI is statistically better than MLATII on the tested problem instance. When there is no significant difference between the two approaches, '=' is utilized, and '−' denotes that MLATII is statistically better than MLATI. Similarly, the Wilcoxon's rank-sum test results in the columns associated with MLATIII, compare MLATI, and MLATII with MLATIII. It should be noted that the results smaller than 1E-8 are considered zero in the Table.

Considering the obtained results, the performances of different proposed approaches are very close. On all of the tested problems, the median of the final achieved results is zero, which means that the presented LA models were successful in finding the optimal super-action on most of their executed runs. However, on some problems like P3, P7, P9, and P10, some of the LA models were unsuccessful in some of their executed runs. We can resolve this issue by slightly decreasing the learning rate of the algorithms. Lowering the learning rate would prolong the convergence time of the algorithms but can guarantee the acquisition of optimal solutions.

3.2.4.4 Comparison Between MLATIII and L_{RI}

In order to illustrate how using multiple reinforcements can improve the performance of a learning approach, this section compares MLATIII with L_{RI} on the introduced decision-making problems. Figure 3.10 provides the average expected errors for the L_{RI} algorithm under different settings of its learning rate. As can be seen from the achieved results, L_{RI} achieves its best results in the given time when it is using a learning rate near 0.0075. The algorithm suffers from premature convergence on the tested problems when its learning rate is set to a value higher than 0.0075. Moreover, it is occasionally unable to approach the optimal solutions with appropriate precisions within 10,000-time steps. This issue is more apparent when the learning rate of the algorithm is lower than 0.0025.

Figure 3.11 compares MLATIII with L_{RI} on a set of problem instances from Table 3.1. We compared the algorithms under their best settings for their learning rates. It can be observed from the achieved results that MLATIII demonstrates superior convergence rates in comparison to L_{RI}. The obtained expected errors of the proposed approach in 2000 iterations are much lower than those that are obtained

Table 3.2 Comparison results of different proposed LA schemes on the problem set of Table 3.1. Wilcoxon rank-sum = WRS

Prob.		LA-I	LA-II	LA-III
P1	Mean	0.00E + 00	0.00E + 00	0.00E + 00
	Median	0.00E + 00	0.00E + 00	0.00E + 00
	Std	0.00E + 00	0.00E + 00	0.00E + 00
	WRS		(=)	(= ,=)
P2	Mean	0.00E + 00	1.19E-02	1.73E-03
	Median	0.00E + 00	0.00E + 00	0.00E + 00
	Std	0.00E + 00	1.91E-02	7.94E-03
	WRS		(+)	(= ,-)
P3	Mean	0.00E + 00	0.00E + 00	0.00E + 00
	Median	0.00E + 00	0.00E + 00	0.00E + 00
	Std	0.00E + 00	0.00E + 00	0.00E + 00
	WRS		(=)	(= ,=)
P4	Mean	0.00E + 00	0.00E + 00	0.00E + 00
	Median	0.00E + 00	0.00E + 00	0.00E + 00
	Std	0.00E + 00	0.00E + 00	0.00E + 00
	WRS		(=)	(= ,=)
P5	Mean	0.00E + 00	0.00E + 00	0.00E + 00
	Median	0.00E + 00	0.00E + 00	0.00E + 00
	Std	0.00E + 00	0.00E + 00	0.00E + 00
	WRS		(=)	(= ,=)
P6	Mean	0.00E + 00	0.00E + 00	0.00E + 00
	Median	0.00E + 00	0.00E + 00	0.00E + 00
	Std	0.00E + 00	0.00E + 00	0.00E + 00
	WRS		(=)	(= ,=)
P7	Mean	1.53E-03	1.53E-03	1.07E-02
	Median	0.00E + 00	0.00E + 00	0.00E + 00
	Std	6.84E-03	6.84E-03	1.50E-02
	WRS		(=)	(+,+)
P8	Mean	0.00E + 00	0.00E + 00	0.00E + 00
	Median	0.00E + 00	0.00E + 00	0.00E + 00
	Std	0.00E + 00	0.00E + 00	0.00E + 00
	WRS		(=)	(= ,=)
P9	Mean	0.00E + 00	0.00E + 00	3.68E-03
	Median	0.00E + 00	0.00E + 00	0.00E + 00
	Std	0.00E + 00	0.00E + 00	1.65E-02
	WRS		(=)	(+,+)
P10	Mean	0.00E + 00	0.00E + 00	1.23E-02
	Median	0.00E + 00	0.00E + 00	0.00E + 00
	Std	0.00E + 00	0.00E + 00	1.55E-02
	WRS		(=)	(+,+)

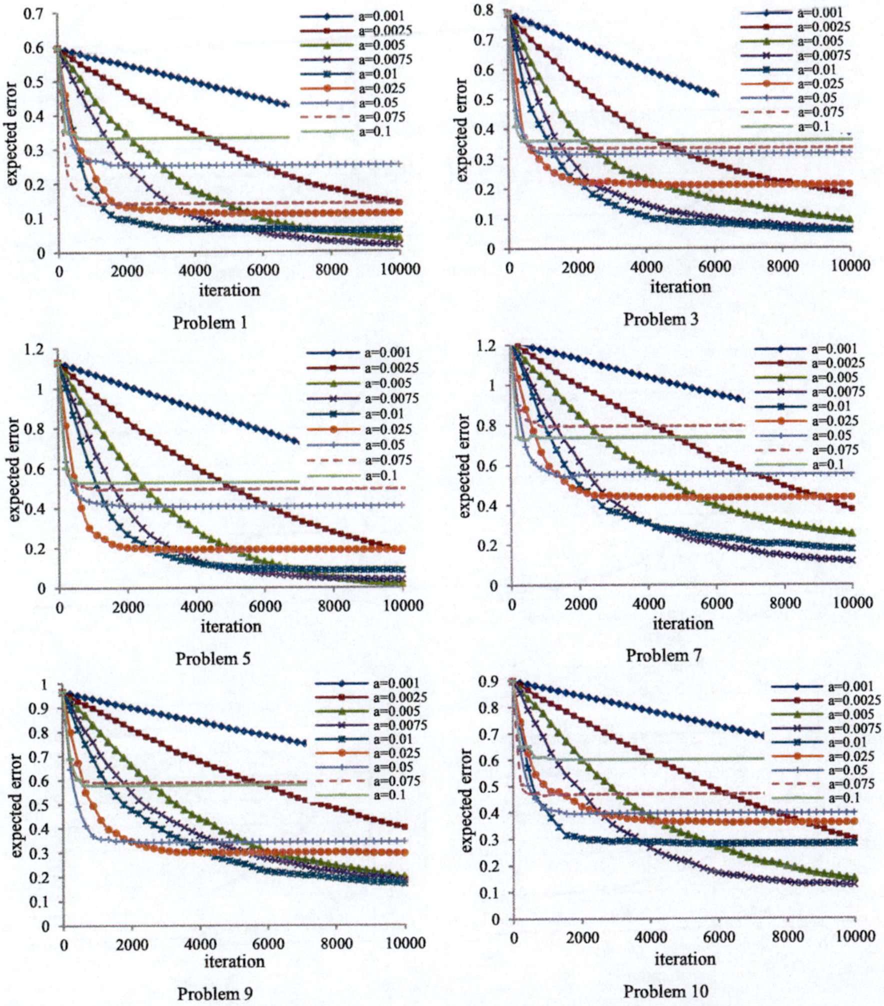

Fig. 3.10 The average expected error curves of L_{RI} on different problems from Table 3.1. The average expected errors of the algorithm on each problem instance are obtained using different settings of the learning rate (**a**)

by the L_{RI} algorithm after 10,000 iterations. On problems such as P1, MLATIII even obtains superior results in comparison to L_{RI} in less than 1000 iterations. MLATIII also demonstrates superior performance in terms of the accuracy of its obtained final solutions. Its achieved expected errors are almost zero on all of the tested instances, while the obtained expected errors of L_{RI} on problems such as P9 are considerable.

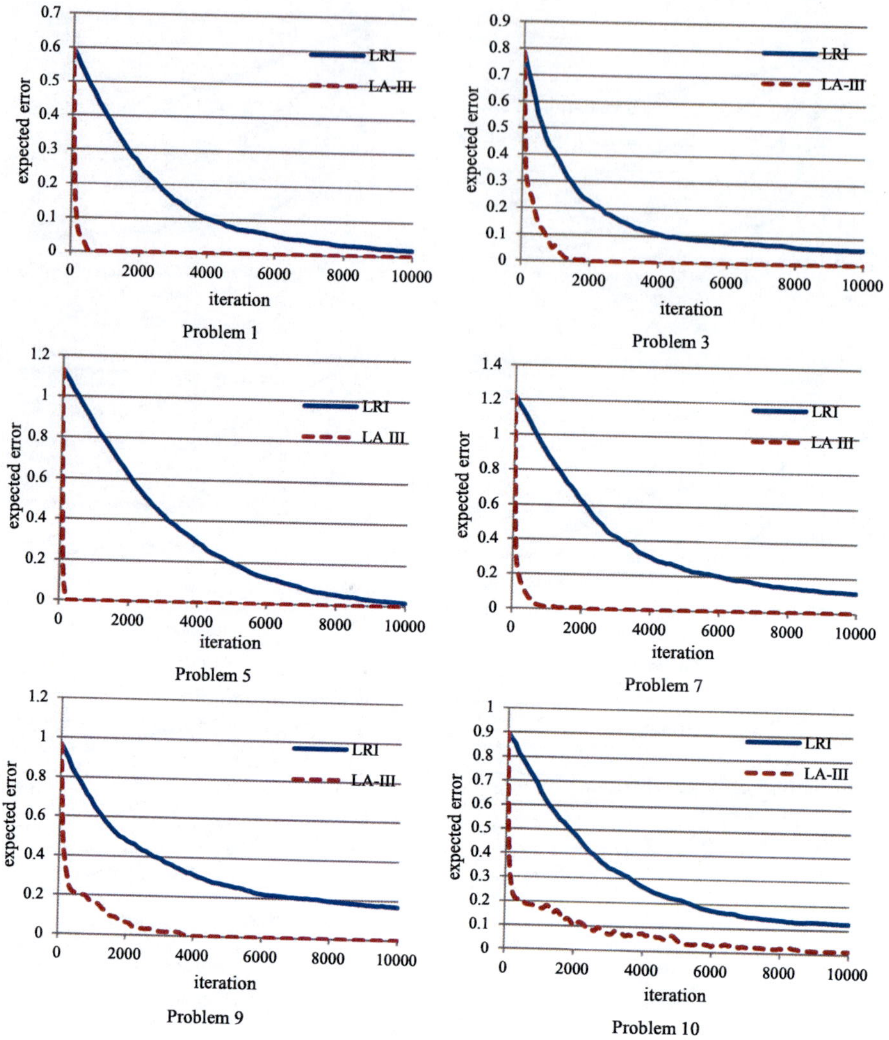

Fig. 3.11 The average expected error curves of MLATIII and L_{RI} during 1E + 4 iterations on a set of problems from Table 3.1

3.3 Cellular Learning Automata Models with Multiple Reinforcements

The three multi-reinforcement LA models introduced in Sect. 3.2 can be directly integrated into the CLA model. The CLA models based on the three introduced multi-reinforcement LAs will be investigated empirically in this chapter (Beigy and Meybodi 2004). In this section, we will investigate a CLA model that aims at maximizing the expected reward for each LA. It will be shown that the proposed model converges to a compatible point (Morshedlou and Meybodi 2017) when the local rules obey the characteristics of a potential function. Many distributed problems, such as channel assignment in mesh networks, can be modeled as potential games (Duarte et al. 2012). At each step k, the local rule of the *i*-th cell determines the reinforcement signal to the learning automaton in the cell as follows:

$$F_i : \varphi_i \to \underline{\beta} \text{ with } \varphi_i = \prod_{j=0}^{n_i} A_{N_i(j)} \quad (3.42)$$

where N_i is the neighborhood function of the *i*-th cell, and $N_i(j)$ returns the index of the *j*-th neighboring cell to cell *i* with $N_i(0)$ defined as $N_i(0) = i$. n_i denotes the number of cells in the neighborhood of the *i*-th cell.

Finally, $\underline{\beta}$ is the set of values that the reinforcement signal can take. We can extend the local rules of a CLA as follows:

$$T_i(\alpha_1, \ldots, \alpha_n) = F_i\left(\alpha_i, \alpha_{N_i(1)}, \ldots, \alpha_{N_i(n_i)}\right) \forall \alpha_j \in A_j \quad (3.43)$$

A local rule will be called a potential local rule if there exists a function like ψ, called a potential function, for which the condition in Eq. (3.44) is satisfied. The characteristics of local utility functions in potential games have been investigated in several recent works, such as (Ortiz 2015; Babichenko and Tamuz 2016).

$$\begin{aligned} &T_i(\alpha_1, \ldots, \alpha_i, \ldots, \alpha_n) - T_i\left(\alpha_1, \ldots, \alpha_i', \ldots, \alpha_n\right) \\ &= \psi(\alpha_1, \ldots, \alpha_i, \ldots, \alpha_n) - \psi\left(\alpha_1, \ldots, \alpha_i', \ldots, \alpha_n\right) \forall \alpha_j \in A_j, \forall \alpha_i' \in A_i \end{aligned} \quad (3.44)$$

3.3.1 Multi-reinforcement Cellular Learning Automata with the Maximum Expected Rewards

This section introduces a multi-reinforcement CLA model, which will be called MCLA for short (Vafashoar and Meybodi 2019). A schematic representation of multi-reinforcement CLA is presented in Fig. 3.12. The model of multi-reinforcement CLA considered in this section aims at maximizing the expected reward for each LA. This CLA is closely related to the model presented in (Morshedlou and Meybodi 2017) and also will be discussed in Chap. 7. In the model presented (Morshedlou and Meybodi 2017), each LA learns a probability distribution over its actions. This distribution leads to the maximum expected reward in response to the empirical joint distribution of actions that are performed by other learning automata in the neighborhood. Consequently, each LA converges to a probability distribution, which may have non zero probabilities on several actions. However, some applications require each LA to learn a single optimal action, or super-action. In this regard, the

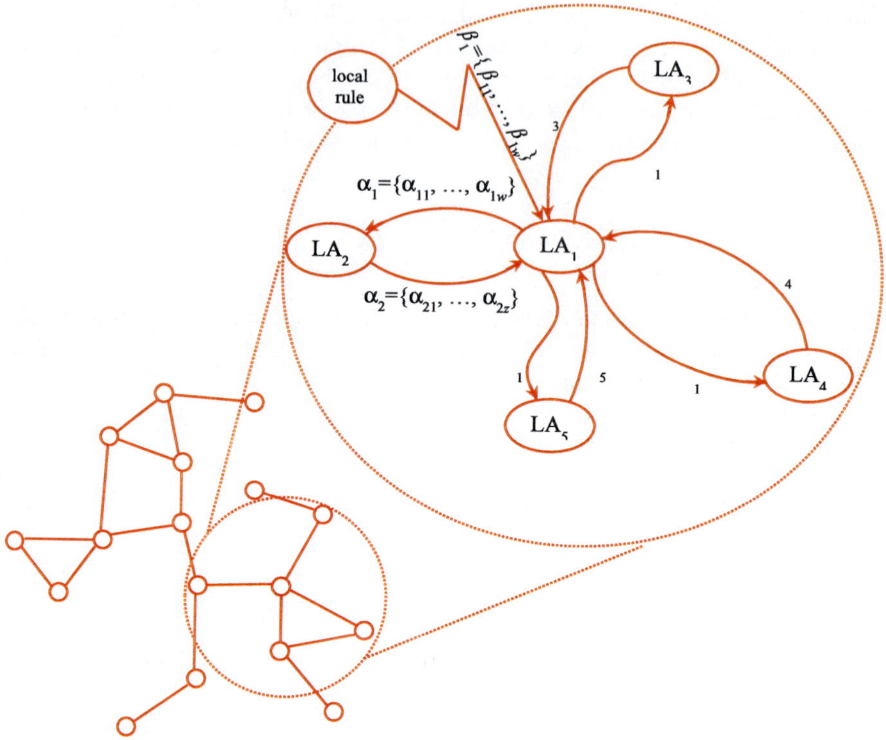

Fig. 3.12 Schematic representation of MCLA

proposed multi reinforcement CLA estimates the payoff for each simple-action in response to the performed actions in the neighborhood. Then, it gradually increases the selection probability of the most favorable super-action, the one consisting of simple-actions with the highest expected rewards. The three introduced LA models can be generalized into a CLA scheme, which maximizes the expected reward of each LA according to the distribution of actions in the neighborhood. For instance, in MLATI, each LA of the CLA selects a super-action as described in Fig. 3.3. Next, it observes the selected actions in its neighborhood and updates its empirical distributions. Then, the LA receives feedback from the environment for each one of its selected simple actions. Based on the empirical probability distributions and the received feedbacks, the estimated payoff of each combination of actions is updated. Finally, similar to Fig. 3.3, the energies of simple actions with w_i highest expected rewards are increased. In what follows, an MCLA model based on MLATII will be investigated. In order to proceed, the required notations are summarized in Table 3.3. Additionally, some basic equations and relations can be derived as follows:

Empirical action selection probability of the i-th LA in step k: $\widehat{p}_i(k) = \left[\widehat{p}_{i1}(k), \widehat{p}_{i2}(k), \ldots, \widehat{p}_{iz_i}(k)\right]^T$

Table 3.3 Summary of necessary notations and definitions used in MCLA

Definition	Notation
Number of actions of the *i*-th LA	m_i
Number of actions to be selected by the *i*-th LA	w_i
The set of all mixed strategies associated with $\underline{A}_i$	Δ_i
The number of neighbors of the *i*-th cell (LA)	n_i
Estimated selection probability of $\underline{a}_{ij}$ in step *k*	$\widehat{p}_{ij}(k)$
Action set of the *i*-th LA	$A_i = \{a_{i1}, \ldots, a_{im_i}\}$
Super-action set of the *i*-th LA	$A_i = \{\underline{a}_{i1}, \ldots, \underline{a}_{iz_i}\}$
Number of Super-actions of the *i*-th LA	$z_i = \binom{m_i}{w_i}$
A mixed strategy over $\underline{A}_i$ (a probability distribution over the elements $\underline{A}_i$)	σ
Index of cells in the neighborhood of the cell *i*	$N_i(1), N_i(2), \ldots, N_i(n_i)$
Average reinforcement of a simple-action like a_{ij} for the neighbors' joint actions until step *k*	$Q_i\left(k, a_{ij}, \underline{a}_{N_i(1)l_1}, \ldots, \underline{a}_{N_i(n_i)l_{n_i}}\right)$

Estimated action selection probabilities in the neighborhood of cell *i*: $\widehat{pn}_i(k) = \left[\widehat{p}_{N_i(1)}(k), \ldots, \widehat{p}_{N_i(n_i)}(k)\right]$

The expected reward for action like a_{ij} at time step *k*:

$$Er\left(a_{ij}, \widehat{pn}_i(k)\right) = \sum_{l_1=1}^{z_{N_i(1)}} \cdots \sum_{l_{n_i}=1}^{z_{N_i(n_i)}} \left(\prod_{s=1}^{n_i} \widehat{p}_{N_i(s)l_s}(k)\right) Q_i\left(k, a_{ij}, \underline{a}_{N_i(1)l_1}, \ldots, \underline{a}_{N_i(n_i)l_{n_i}}\right) \tag{3.45}$$

Super-action set with the maximum expected reward for the *i*-th LA at time step *k*:

$$B_i\left(k, \widehat{pn}_i(k)\right) = \left\{\underline{a}_{ih} \in \underline{A}_i \mid \sum_{a_{il} \in \underline{a}_{ih}} Er\left(a_{il}, \widehat{pn}_i(k)\right) = \max_{\underline{a}_{ij} \in \underline{A}_i} \left(\sum_{a_{il} \in \underline{a}_{ij}} Er\left(a_{il}, \widehat{pn}_i(k)\right)\right)\right\} \tag{3.46}$$

The actual reward for the mixed strategy σ:

$$F_i\left(\sigma, \widehat{pn}_i(k)\right) = \sum_{u=1}^{z_i} \sigma_u \sum_{a_{ij} \in \underline{a}_{iu}} \left[\sum_{l_1=1}^{z_{N_i(1)}} \cdots \sum_{l_{n_i}=1}^{z_{N_i(n_i)}} \left(\prod_{s=1}^{n_i} \widehat{p}_{N_i(s)l_s}(k)\right) F_i\left(k, a_{ij}, \underline{a}_{N_i(1)l_1}, \ldots, \underline{a}_{N_i(n_i)l_{n_i}}\right)\right] \tag{3.47}$$

Epsilon maximum expected reward set for the i-th LA:

$$b_i^{\varepsilon}\left(k, \widehat{pn}_i(k)\right) = \left\{\sigma \in \Delta_i \mid \max_{\sigma' \in \Delta_i}\left(F_i\left(\sigma', \widehat{pn}_i(k)\right)\right) - F_i\left(\sigma, \widehat{pn}_i(k)\right) \leq \varepsilon\right\} \quad (3.48)$$

Figure 3.13 gives pseudocode for the proposed MCLA model. In this model, each learning automaton learns the empirical distribution of actions that are occurring in its neighborhood. Also, it learns the estimated payoff of each one of its simple actions in combination with different super-actions, which are performed by the neighboring learning automata. Consequently, it can compute the expected reward for each one of its simple-actions. During each iteration k, every LA like LA_i finds its w_i simple-actions with the highest expected rewards. Then, with probability $1-\lambda(k)$, the LA performs these actions in the environment, or with probability $\lambda(k)$, it experiments two other action selection strategies similar to Fig. 3.4. After performing these

Algorithm 3-4. MCLA

Input

r: number of actions

$w<r$: number of actions to be selected.

$(\lambda(k))_{k\in N}$: learning rate.

$(\eta(k))_{k\in N}$: averaging factor.

$(\gamma(k))_{k\in N}$: discount factor.

While the termination condition is not satisfied, do:

I. For each cell like i, synchronously with other cells, do:

1. With probability $\lambda(k)/2$ do: // exploratory selection
    1. Select a random super-action as $\alpha(k)$.
2. With probability $\lambda(k)/2$ do: // mutated greedy selection
    1. Select the maximum expected reward super-action. Let $\alpha(k)$ be this action.
    2. Define I as $I=\{l \mid a_{il} \in \alpha(k)\}$
    3. Select a random simple-action from $\alpha(k)$ according to its probability value defined as:
       $p_l = \frac{1-q_l(k)}{\sum_{j\in I} 1-q_j(k)} \; \forall l \in I,$
       where
       $q_h = \frac{Er(a_{ih}, \hat{p}n_i(k))}{\sum_{a_{il}\in\alpha_i(k)} Er(a_{il}, \hat{p}n_i(k))} \; \forall h \in I$
    4. Select a random simple-action not in $\alpha(k)$ according to the probability values defined as:
       $$p_h = \frac{Er(a_{ih}, \hat{p}n_i(k))}{\sum_{a_{is}\notin\alpha_i(k)} Er(a_{is}, \hat{p}n_i(k))} \; \forall h \in \{l \mid a_{il} \in A_i \& a_{il} \notin \alpha_i(k)\}$$
    5. Replace the action obtained in step 2.3 with the one obtained in step 2.4 in $\alpha(k)$.
3. With probability 1- $\lambda(k)$ do:
    1. Select the maximum expected reward super-action as $\alpha(k)$.
4. Perform the selected super-action, and receive the reinforcement vector $\beta(k)$.
5. $\hat{p}_{ij}(k+1) = \begin{cases}(1-\gamma(k))\hat{p}_{ij}(k) + \gamma(k) & \text{if } \alpha_i(k) = a_{ij} \\ (1-\gamma(k))\hat{p}_{ij}(k) & \text{otherwise}\end{cases} \; \forall j=1, \ldots, z$
6. $Q(k+1,:)=Q(k,:)$
7. $Q_i(k+1, a_{ij}, \alpha_{N_i(1)}(k), \ldots, \alpha_{N_i(n_i)}(k)) =$
   $Q_i(k, a_{ij}, \alpha_{N_i(1)}(k), \ldots, \alpha_{N_i(n_i)}(k)) +$
   $\eta(k)\left(\beta_{ij}(k) - Q_i(k, a_{ij}, \alpha_{N_i(1)}(k), \ldots, \alpha_{N_i(n_i)}(k))\right) \; \forall j\in\{l \mid a_{il}\in \alpha(k)\}$

End for

II. Set $k \leftarrow k+1$.

End while

Fig. 3.13 The pseudocode of MCLA

selected simple-actions, the LA updates the empirical distribution of t he actions of neighboring learning automata and the estimated payoff of its performed actions. It should be noted that $\lambda(k)$ is a decreasing sequence of numbers; accordingly, the selection probability of the most promising super-actions approaches to one during the learning procedure.

3.3.2 Convergence Analysis of MCLA

Lemma 3.3 Assume that the following hold w.p.1: $E\left[F_i\left(a_{ij}, a'\right)\right] = \text{A}$, $E\left[F_i(a_{ij}, a')^2\right] < \infty$, $\sum_{k=1}^{\infty} \eta(k) = \infty$, and $\sum_{k=1}^{\infty} \eta(k)^2 < \infty$. Also, assume that $Q_i\left(., a_{ij}, a'\right)$ is updated infinitely often. Then, $Q_i\left(k, a_{ij}, a'\right)$ converges to A w.p.1 as $k \rightarrow \infty$ (Szepesvári and Littman 1999).

Lemma 3.4 If $\lambda(k) = 2\tau r k^{-1/s}$, where r is the maximum number of super-actions in the LAs of the MCLA and s is the maximum neighborhood size in the MCLA, assuming the conditions of lemma 3.3 is hold, then in the proposed model:

$$\lim_{k \rightarrow \infty} \left| Q_i\left(k, a_{ij}, a'\right) - F_i\left(a_{ij}, a'\right) \right| = 0 \forall i, j, a' \tag{3.49}$$

Proof For a typical LA like LA_i, the selection probability of each super-action is at least $\frac{\lambda(k)}{2\bar{m}_i} \geq \frac{\lambda(k)}{2r}$. Accordingly, the occurrence of each joined super-action in the neighborhood of LA_i is at least $\left(\frac{\lambda(k)}{2r}\right)^{n_i} \geq \left(\frac{\lambda(k)}{2r}\right)^{s}$. With $\lambda(k) = \tau 2 r k^{-1/s}$, the selection probability of each joint action is at least $\tau^s k^{-1}$. $\tau^s > 0$ is constant, and it can be verified that $\sum_{k=1}^{\infty} k^{-1} = \infty$. Accordingly, the average estimated reinforcements converge to the actual ones w.p.1 (Szepesvári and Littman 1999).

Theorem 3.2 An MCLA with potential local rules converges to a set of Nash equilibria under the following conditions.

We assume that the conditions of lemma 3.3 and lemma 3.4 are satisfied and, also, assume that $(\gamma\ (k))_{k \in \mathbb{N}}$ is a sequence satisfying the two requirements $\sum_{k=1}^{\infty} \gamma(k) = \infty$ and $\sum_{k=1}^{\infty} (\gamma(k)) < \infty$. Additionally, the reinforcement signals are bounded and fall in the range $[0, D_{\max}]$.

Following lemma 3.4, $\lim_{k \rightarrow \infty} \left| Q_i\left(k, a_{ij}, a'\right) - F_i\left(a_{ij}, a'\right) \right| = 0 \forall i, j, a'$, and consequently, there is a sequence like $(\mu_i(k))_{k \in \mathbb{N}}$ such that $\mu_i(k) \rightarrow 0$ as $k \rightarrow \infty$ and

$$\left| Q_i\left(k, a_{ij}, a'\right) - F_i\left(a_{ij}, a'\right) \right| < \mu_i(k) \forall i, j, a' \tag{3.50}$$

We first show that, at each iteration k, the super-action of a typical LA like LA_i is selected from the set of its epsilon maximum expected payoff super-action set concerning $\widehat{pn}_i(k)$. To do this, it should be shown that the expected reward of the selected actions of LA_i plus ε is greater than the maximum of the achievable actual reward.

During each time step k, the action of a learning automaton like LA_i is selected according to a mixed strategy like $\pi_i(k)$ (considering three action selection strategies in Fig. 3.13. Also, by representing the expected reward of the i-th cell at time k with $Er_i(k)$ and following the procedure of MCLA in Fig. 3.14, we have

$$
\begin{aligned}
&Er_i \geq (1-\lambda(k))max_{\underline{a}_{ij}\in\underline{A}_i}\left(\sum_{a_{il}\in\underline{a}_{ij}} Er\left(a_{il},\widehat{pn}_i(k)\right)\right)+\\
&\frac{\lambda(k)}{2}\frac{1}{|A_i|}\sum_{\underline{a}_{ij}\in A_i}\sum_{a_{il}\in\underline{a}_{ij}} Er\left(a_{il},\widehat{pn}_i(k)\right)+\\
&\frac{\lambda(k)}{2}\left[\sum_{\underline{a}_{ij}\in B_i(k)}\sum_{a_{il}\in\underline{a}_{ij}}\left(1-\frac{1}{w_i}\right)Er\left(a_{il},\widehat{pn}_i(k)\right)+\frac{1}{m_i-w_i}\sum_{a_{ih}\notin\underline{a}_{ij}} Er\left(a_{ih},\widehat{pn}_i(k)\right)\right]
\end{aligned}
\tag{3.51}
$$

$$
Er_i(k) \geq (1-\lambda(k))max_{\underline{a}_{ij}\in\underline{A}_i}\left(\sum_{a_{il}\in\underline{a}_{ij}} Er\left(a_{il},\widehat{pn}_i(k)\right)\right) \geq
$$

$$
(1-\lambda(k))\left[max_{\underline{a}_{ij}\in\underline{A}_i}\left(\sum_{a_{il}\in\underline{a}_{ij}} F\left(a_{il},\widehat{pn}_i(k)\right)\right)-w_i\mu_i\right] \geq
$$

Algorithm 3-5. finding a feasible n-tuple

Input:

w // maximum number of distinct actions that can be selected

A = {$a_1, a_2, \ldots, a_r$} *// action set*

$d_{n\times r}$ *// estimated feedback of each action for each component of an n-tuple.*

1. $\delta(0)\leftarrow$argmax$_j(d_{ij})$ $\forall i\in 1..n$.
2. $t\leftarrow 0$.
3. for j=1 to r do:
   a. if $j\notin\delta(0)$, then $d_{ij}\leftarrow -\infty$ $\forall i\in 1..n$.
4. while the number of distinct actions in $\delta(t)$ is greater than w do:
   a. $\delta(t+1)\leftarrow\varnothing$.
   b. $c\leftarrow -\infty$.
   c. for each distinct $j\in\delta(t)$ do:
      i. $e\leftarrow d$.
      ii. $e_{ij}\leftarrow -\infty$ $\forall i\in 1..n$.
      iii. $\delta'\leftarrow$argmax$_l(e_{il})$ $\forall i\in 1..n$.*// a new tuple which doesn't include a_j*
      iv. $c'\leftarrow\sum_{l=1}^{n} d_{l\delta'(l)}$. *// the expected feedback sum of new tuple*
      v. if $c'>c$
         - $c\leftarrow c'$.
         - $\delta(t+1)\leftarrow\delta'$.
   d. $t\leftarrow t+1$.
   e. for j=1 to r do:
      i. if $j\notin\delta(t)$, then $d_{ij}\leftarrow -\infty$ $\forall i\in 1..n$.

Fig. 3.14 Estimating the feasible n-tuple with the highest sum of expected feedbacks

$$max_{\underline{a}_{ij} \in \underline{A}_i} \left(\sum_{a_{il} \in \underline{a}_{ij}} F\Big(a_{il}, \widehat{pn}_i(k)\Big) \right) - \lambda(k) w_i D_{max} - w_i \mu_i + \lambda(k) w_i \mu_i =$$

$$max_{\underline{a}_{ij} \in \underline{A}_i} \left(\sum_{a_{il} \in \underline{a}_{ij}} F\Big(a_{il}, \widehat{pn}_i(k)\Big) \right) + \varepsilon_i(k) \tag{3.52}$$

As $k \to \infty$, we have $\mu_i \to 0$ and $\lambda(k) \to 0$; consequently, $\varepsilon_i(k) \to 0$. Therefore, the selected super-action of each LA converges to its epsilon maximum expected payoff super-action set.

Based on the updating rules of the proposed algorithm, $\widehat{p}_i(k+1) = (1 - \gamma_{k+1})\widehat{p}_i(k) + \gamma_{k+1} I_{\alpha_i(k)}$. The expected value of $I_{\alpha_i(k)}$ is $\pi_i(k)$. Consequently $I_{\alpha_i(k)} = \pi_i(k) + M_i(k)$, where $M_i(k)$ is the difference between the expected value $I_{\alpha_i(k)}$ at step k and its actual value. As discussed before $\pi_i(k) \in b_i^{\varepsilon}\Big(k, \widehat{pn}_i(k)\Big)$. Accordingly, $\widehat{p}_i(k+1) \in (1 - \gamma_{k+1})\widehat{p}_i(k) + \gamma_{k+1}\Big(b_i^{\varepsilon}\Big(\widehat{pn}_i(k)\Big) + M_i(k)\Big)$. $M_i(k)$ is a sequence of perturbations such that, with the assumed conditions on $\gamma(k)$,

$$\lim_{k \to \infty} \sup_{t} \left\{ \left\| \sum_{j=k}^{t-1} \gamma_i(j+1) M_i(j+1) \right\| \Big| \sum_{j=k}^{t-1} \gamma_i(j+1) \leq T \right\} = 0 \quad \text{for any } T > 0 \tag{3.53}$$

Accordingly, $\widehat{p}_i(k+1) \in (1 - \gamma_{k+1})\widehat{p}_i(k) + \gamma_{k+1}\Big(b_i^{\varepsilon}\Big(\widehat{pn}_i(k)\Big) + M_i(k)\Big)$ is a GWFP process (Leslie and Collins 2006), and converges to the set of Nash equilibria if local rules satisfy the conditions in Eq. (3.44).

3.3.3 On the Computational Complexity of MCLA

The proposed MCLA has updating steps with exponential computational complexities, which can significantly affect its performance. However, considering the properties of local rules in some applications, the time complexity of the approach can be reduced effectively. The multi-reinforcement LA schemes introduced in this chapter are developed on the basis that the environment can provide each selected simple-action of a super-action with an individual reinforcement signal. Accordingly, different simple-actions of a super-action are learned separately, which significantly reduces the learning time. The same approach can be employed in MCLA. For instance, consider a multi-agent environment where each agent interacts with several other agents in its neighborhood. During each iteration, a typical agent selects a set of actions and applies each action to one of its neighbors. The neighboring agent which receives an action responds to the executor of the action based on its selected actions. Accordingly, a LA can learn the estimated payoff for each simple-action in combination with different other simple-actions rather than joined super-actions. Then, the expected rewards for simple-actions can be calculated based on these approximated payoffs. In the local rule of the application, which is studied in Chap. 5, the expected payoff for each simple-action only depends on the selection of the same simple-action by the neighboring LAs. Hence, its learning procedure can be implemented efficiently. Additionally, we can simply use the LA algorithms proposed in Sect. 3.2 in each cell of the CLA. All of these algorithms have linear time updating steps, and as it will be demonstrated in Chap. 5, they can achieve up-and-coming solutions.

3.4 Extensions of Proposed Models for N-Tuple Selection

This section discusses how the multi-reinforcement technique can be used for optimal n-tuple selection problem. We can easily extend the previous methods so that each LA can select an n-tuple of its actions. As the approaches proposed in this section are very similar to the previously discussed ones, we briefly discuss the required LA extensions.

3.4.1 Reinforcement Learning Using Multiple Reinforcements for Optimal N-Tuple Selection

Multi- reinforcement models previously introduced try to tackle the subset selection problem. Using the same ideas, this section introduces a new learning automaton, which will be called multi-reinforcement n-tuple LA (MNLA). MNLA aims at finding an n-tuple of its actions under some restrictions. Assume that an MNLA has an action set $A = \{a_1, a_2, \ldots, a_r\}$. During each iteration k, the MLA selects an n-tuple of its actions under the restriction that the number of distinct actions that appear in the chosen n-tuple is at most w. Then, the MNLA performs the chosen n-tuple, which is denoted by $\alpha(k) = [\alpha_1(k), \ldots, \alpha_n(k)]^{\mathrm{T}}$, in the environment. In response to the performed n-tuple, the environment triggers an n-dimensional feedback signal $\beta(k)$. Each component $\beta_i(k)$ of the feedback signal represents a separate reinforcement from the environment to the associate action $\alpha_i(k)$. The objective of the MNLA is to find an n-tuple $(a_{j1}, a_{j2}, \ldots, a_{jn})$ with the maximum expected total reinforcement signals, i.e., $E\left[\sum_{j=1}^{n} \beta_j(k)\right]$.

We represent the internal state of the MNLA by an $(n \times r)$-dimensional probability matrix M. Each row i of this probability matrix represents the selection probabilities of different actions in A for the i-th position of the n-tuple. During each iteration k of the algorithm, first, a dummy n-tuple $\grave{}\alpha(k)$ is selected. Each component i of the dummy n-tuple is selected according to the probability vector M_i. Consequently, the generated dummy n-tuple may contain more than w distinct actions. In order to decrease the number of distinct actions in a selected n-tuple, the following procedure will be employed. Two distinct actions $a_i \in \dot{\alpha}(k)$ and $a_j \in \dot{\alpha}(k)$ are selected randomly. Then, the energy level of each of these actions is obtained as follows:

$$e_i = \prod_{l \in I} M_{li} \text{ with } I = \{l | \dot{\alpha}_l(k) = a_i \vee \dot{\alpha}_l(k) = a_j\} \tag{3.54}$$

$$e_j = \prod_{l \in I} M_{lj} \text{ with } I = \{l | \dot{\alpha}_l(k) = a_i \vee \dot{\alpha}_l(k) = a_j\} \tag{3.55}$$

After obtaining the energy levels, random action is chosen among a_i and a_j according to their probability values defined respectively as $q_i = e_i/(e_i + e_j)$ and $q_j = e_j/(e_i + e_j)$. Assume that a_l is selected in this stage. Next, all components of $\dot{\alpha}(k)$ consisting of $a_{i,}$ or a_j are replaced with a_l. The above-described procedure continues until the acquisition of a feasible n-tuple $\alpha(k)$. The chosen feasible n-tuple $\alpha(k)$ is carried out in the environment, and n-dimensional feedback signal $\beta(k)$ is received from the environment. Each component of the feedback signal is a reinforcement for its corresponding component in $\alpha(k)$. Based on the received reinforcements, the probability matrix M is updated. In order to update M, the following three matrices are maintained by the MNLA η, Z, and d. $\eta_{ij}(k)$ represents the number of times that action a_j is selected for the i-th component of the n-tuple till iteration k, and $Z_{ij}(k)$ denotes the total reinforcement obtained by a_j when it was used as the j-th component of the selected n-tuples. Finally, $d_{ij}(k)$ is the estimated feedback for action a_j when it is used as the i-th component of the chosen n-tuples.

Based on the obtained feedback vector, the LA first updates η, d, and Z as follows:

$$Z_{ij}(k+1) = \begin{cases} Z_{ij}(k) + \beta_i(k) \; if \; \alpha_i(k) = a_j \\ Z_{ij}(k) \qquad otherwise \end{cases} \tag{3.56}$$

$$\mu_{ij}(k+1) = \begin{cases} \mu_{ij}(k) + 1 \; if \; \alpha_i(k) = a_j \\ \mu_{ij}(k) \qquad otherwise \end{cases} \tag{3.57}$$

$$d_{ij}(k+1) = \mu_{ij}(k+1)/Z_{ij}(k+1) \tag{3.58}$$

Based on the obtained estimated feedbacks, the LA finds the feasible n-tuple with the highest (or lowest for minimization) sum of estimated feedbacks. The problem of finding this n-tuple is NP-complete. A greedy method for this problem will be described herein. Assume that the set of all feasible n-tuples is represented by B. the n-tuple $\delta \in B$ has the highest estimated feedbacks if the following condition is satisfied.

$$\sum_{s=1}^{n} d_{sI(\delta_s)} \geq \sum_{s=1}^{n} d_{sI(\rho_s)} \forall \rho \in B, \tag{3.59}$$

where I is an index function, and $I(\sigma_s)$ return the index of the action in the s-th component of σ. After obtaining the n-tuple δ with the minimum estimated feedbacks, the probability matrix M is updated as follows:

$$M_{ij}(k+1) = \begin{cases} M_{ij}(k) - \lambda M_{ij}(k) + \lambda \; if \; \delta_i = a_j \\ M_{ij}(k) - \lambda M_{ij}(k) \quad otherwise \end{cases} \tag{3.60}$$

3.4.1.1 A Greedy Method for Finding a Feasible N-Tuple

This section describes a greedy algorithm for finding a feasible n-tuple with the highest sum of estimated feedbacks. Assume that we wish to find a tuple of size n, which consists of at most w distinct actions. The greedy algorithm first chooses the n-tuple $\delta(0)$, which is composed of actions with the highest estimated feedbacks. If this n-tuple consists of more than w distinct actions, the algorithm iteratively will reduce the number of its distinct actions. During each iteration t, the algorithm finds an action $a_i \in \delta(t)$ which can be replaced by other actions in $\delta(t)$ with the smallest decrease in the total feedback sum, and then replaces this action. This procedure continues until the number of distinct actions equals w. The pseudocode for the described algorithm is given in Fig. 3.14.

3.4.2 Modified Multi-reinforcement LA with Weak Memory for N-Tuple Selection

Like the previous section, we represent the internal state of modified multi-reinforcement LA (MMLA) by an $(n \times r)$-dimensional probability matrix M. Each row i of this probability matrix represents the selection probabilities of different actions in A for the i-th position of the n-tuple. During each iteration k of the algorithm, first, a feasible n-tuple is selected similar to the previous section. The chosen feasible n-tuple $\alpha(k)$ is carried out in the environment, and an n-dimensional feedback signal $\beta(k)$ is received from the environment. Each component of the feedback signal is a reinforcement for its corresponding component in $\alpha(k)$. Based on the received reinforcements, the probability matrix M is updated. In order to update the probability matrix, MMLA maintains the last received reinforcements of each selected action in an $n \times r$-dimensional matrix η. During each iteration k, η is updated based on the received reinforcement vector as follows:

$$\eta_{ij}(k) = \begin{cases} \tau\beta_i(k) + (1-\tau)\eta_{ij}(k) \; if \; \alpha_i(k) = a_j \\ \eta_{ij}(k) \quad otherwise \end{cases} \tag{3.61}$$

Here $\alpha_i(k)$ represents the i-th component of the selected action, $\alpha(k)$. τ is the averaging factor, which puts more weight on the recently received reinforcements. Based on the updated last received reinforcements, the MMLA finds the feasible n-tuple with the highest sum of the estimated feedbacks and algorithm in Fig. 3.14 will be used for this purpose. After obtaining the n-tuple δ with the maximum estimated feedbacks, the probability matrix M is updated as follows:

$$M_{ij}(k+1) = \begin{cases} M_{ij}(k) - \lambda M_{ij}(k) + \lambda \; if\; \delta_i = a_j \; and \; \alpha_i(k) = a_j \\ M_{ij}(k) - \lambda M_{ij}(k) \quad if\; \delta_i = a_l \; and \; \alpha_i(k) = a_l \; and \, l \neq j \\ M_{ij}(k) \quad otherwise \end{cases}, \tag{3.62}$$

where λ is the learning rate of the MLA.

3.4.3 Modified Learning Automaton for N-Tuple Selection with Free Actions

In the previous section, a LA is introduced, which is capable of choosing an n-tuple of actions under the restriction that the number of distinct selected actions should not be higher than a constant like w. This section relaxes this constraint and introduces a new MMLA called FMLA. Assume that the action set can be partitioned into two sets A^r and A^f, $A = A^r \cup A^f$ and $A^r \cap A^f = \emptyset$. The FMLA chooses an n-tuple from A during each iteration k with the restriction that the number of distinct actions chosen from A^r is at most w; however, there is no restriction on the number of distinct actions chosen from the set A^f.

In the action selection step of MMLA, described in the previous section, first, a dummy n-tuple $\dot{\alpha}(k)$ is selected by the MMLA. During each step k, the selection probability of each action a_j for the i-th component of the dummy n-tuple is $M_{ij}(k)$. Accordingly, the probability that a_j is not chosen for the i-th component is $1 - M_{ij}(k)$. Hence, the probability that each action a_j is not chosen in the dummy n-tuple is $\prod_{i=1}^{n}(1 - M_{ij})$. We define the elimination probability of each action a_j in the k-th step as follows:

$$p_j(k) = \prod_{i=1}^{n}(1 - M_{ij}(k)) \tag{3.63}$$

In the action selection step of FMLA, first, $|A^r|$-w actions are selected from A^r according to their normalized elimination probability values. Let this set of actions denoted by $A^e(k)$. Then, an n-tuple is selected from the set $A^f \cup (A^r$-$A^e)$ according to the probability matrix of the FMLA. This chosen feasible n-tuple $\alpha(k)$ is carried out in the environment, and n-dimensional feedback signal $\beta(k)$ is received from the

```
Algorithm 3-6. finding a feasible n-tuple with the highest sum of estimated feedbacks in FMLA
Input:
w // maximum number of distinct actions that can be selected form A^r
A = {a_1,a_2,...,a_r} // action set
A = A^f ∪ A^r
η_{n×r} // estimated feedback of each action in each dimension of the selected n-tuples
δ(0)←argmax_j(η_ij) ∀i∈1..n
t←0
η_ij←-∞ ∀i∈1..n ∀j(j∉δ(0) and a_j∈A^r)
while number of distinct actions selected form A^r in δ(t) is greater than w do
        δ(t+1)←∅
        c←-∞
        for each distinct action a_j∈δ(t) such that a_j∈A^r do:
                e←η
                e_ij←-∞ ∀i∈1..n
                δ'←argmax_j(e_ij) ∀i∈1..n
                c'← Σ_{i=1}^{n} η_{iδ'(i)}
                If c'>c
                        c←c'
                        δ(t+1)←δ'
                End if
        End for
        t←t+1
        η_ij←-∞ ∀i∈1..n ∀j(j∉δ(t) and a_j∈A^r)
end while
```

Fig. 3.15 Estimating the feasible n-tuple with the highest sum of expected feedbacks in FMLA

environment. The updating procedure of FMLA is much similar to that of MMLA, and Fig. 3.15 describes a greedy procedure for finding the feasible n-tuple with the highest sum of estimated feedbacks in FMLA.

3.4.4 Experimental Studies

In this section, we experimentally analyze the performance of the proposed multi-reinforcement schemes for optimal n-tuple selection. Similar experiments as of Sect. 3.2.4 can be conducted here; however, we keep the experiments concise for the sake of space economy. Two problem instances are considered for this section, and their properties are summarized in Tables 3.4 and 3.5. As can be seen from the tables, these two problems have a large number of super-actions numerously. Consequently, the classical LA models are inapplicable for solving these problems. Finding the optimal n-tuple is an NP-complete problem; accordingly, the proposed models use a greedy scheme for estimating the optimal solution. The considered problems can be considered as hard optimization problems due to their high dimensionalities and the noisy nature of the feedbacks (in the proposed problems, the feedback

Table 3.4 General properties of tested problems

Prob.	Action set size	n-tuple size	Maximum number of distinct actions	Free actions for FMLA
P_1	12	14	5	$\{a_1, a_2\}$
P_2	7	12	4	$\{a_1\}$

for each simple-action is its expected value from the Table 3.5 with an augmented uniform noise from the interval [−0.25, 0.25]).

3.4.4.1 Investigating the Learning Rate (λ) of the Proposed LA Models

In this section, we investigate the sensitivity of the proposed models to their learning rates. The averaging factor of MMLA and FMLA is set as $\tau = 0.01$. Figures 3.16, 3.17 and 3.18 give the expected sum of the received feedbacks under different settings of λ with $\lambda \in [0.0005, 0.05]$. As can be seen from the obtained results, the proposed LA models require rather small learning rates for effectively exploring the search space of the problems. With large settings for learning rates (like $\lambda = 0.05$), an LA cannot adequately estimate the expected feedback of its different actions and may converge to weak solutions with a high pace. In general, as the number of actions grows, lower learning rates are required for the adequate performance of an LA model. Also, when the expected feedbacks are too close for different actions, and the noise level is high, the learning rate should be set to a small value to allow the LA to explore its different actions sufficiently. Considering the obtained results, a setting near $\lambda = 0.05$ seems to be an appropriate choice for the proposed models on the tested problems.

3.4.4.2 Comparison of MNLA, MMLA, and FMLA

In this section, we experimentally compare the performance of the proposed models. Figure 3.19 gives the expected total payoff for different models on P_1 and P_2. As can be seen from the achieved results, the obtained payoffs for different models are very close. MNLA has a slightly better performance in comparison to MMLA due to the way of calculating Z_{ij} for different actions in Eq. 3.56. MNLA uses the complete history of different actions to estimate their expected payoffs. According to stationary environments, it can achieve better estimates for the expected payoffs.

In contrast, MMLA and FMLA put more weight on the recent history of the performed actions, which is suitable for time-varying and non-stationary environments. Indeed, FMLA should obtain better total payoffs in comparison to MMLA and MNLA, as it puts less restriction on the choice of its actions. However, in the tested problems, the optimal solutions are very close for P_1 and P_2 and their less restricted versions used in FMLA.

Table 3.5 The expected reward of each action a_i when it is chosen for different positions in an n-tuple

Prob.	Pos.	a_1	a_2	a_3	a_4	a_5	a_6	a_7	a_8	a_9	a_{10}	a_{11}	a_{12}
P_1	1	0.6123	0.7624	0.5980	0.3286	0.3739	0.2029	0.0095	0.3738	0.2709	0.5349	0.2667	0.6989
	2	0.4160	0.0764	0.8927	0.0264	0.3422	0.8946	0.6600	0.6224	0.9611	0.1802	0.3535	0.9928
	3	0.4223	0.5036	0.5628	0.0575	0.1347	0.2037	0.4560	0.2835	0.8465	0.4917	0.2114	0.1891
	4	0.4962	0.2232	0.1174	0.3283	0.3994	0.0972	0.9985	0.9284	0.2880	0.1711	0.7781	0.3157
	5	0.6728	0.4905	0.0423	0.7848	0.6194	0.5750	0.2121	0.8601	0.1738	0.8385	0.9383	0.7557
	6	0.0555	0.9075	0.2660	0.2070	0.1622	0.0186	0.9915	0.9848	0.6945	0.1550	0.4625	0.0321
	7	0.0585	0.4302	0.7541	0.8268	0.5846	0.5748	0.2252	0.0131	0.6272	0.8247	0.1520	0.5749
	8	0.0640	0.4560	0.6779	0.6126	0.7957	0.6621	0.5198	0.2017	0.1402	0.3886	0.4407	0.1428
	9	0.2191	0.5940	0.1523	0.2846	0.1012	0.5108	0.2838	0.9497	0.8199	0.8158	0.3442	0.7942
	10	0.5245	0.8216	0.6005	0.1177	0.7822	0.5447	0.2178	0.1872	0.1753	0.6114	0.5336	0.7525
	11	0.8273	0.4255	0.9682	0.7114	0.5423	0.9366	0.8909	0.8592	0.7238	0.6387	0.1475	0.9393
	12	0.5830	0.6458	0.7853	0.4365	0.7582	0.9401	0.7991	0.5727	0.2327	0.6536	0.4696	0.4454
	13	0.0311	0.7265	0.3198	0.2860	0.9386	0.7991	0.9895	0.0371	0.3965	0.3404	0.1438	0.9606
	14	0.2009	0.3419	0.9185	0.7555	0.5938	0.1288	0.5146	0.3592	0.6624	0.9412	0.2582	0.9929
P_2	1	0.7613	0.5766	0.0748	0.0563	0.3864	0.2598	0.0557					
	2	0.2709	0.7843	0.6486	0.9388	0.0966	0.6094	0.8967					
	3	0.2830	0.7145	0.1017	0.1147	0.3447	0.3446	0.8274					
	4	0.8426	0.2305	0.4105	0.6921	0.8771	0.2960	0.3432					
	5	0.9796	0.9582	0.9821	0.4475	0.7908	0.0674	0.7559					
	6	0.8737	0.5906	0.6790	0.6033	0.0981	0.5452	0.3947					
	7	0.7721	0.5230	0.2979	0.4702	0.2958	0.6824	0.5489					

(continued)

Table 3.5 (continued)

Prob.	Pos.	a_1	a_2	a_3	a_4	a_5	a_6	a_7	a_8	a_9	a_{10}	a_{11}	a_{12}
	8	0.8200	0.5491	0.8180	0.2788	0.2473	0.8070	0.5812					
	9	0.0248	0.0767	0.4051	0.4360	0.8422	0.3131	0.0567					
	10	0.0123	0.0361	0.3464	0.2567	0.5573	0.2052	0.9663					
	11	0.1107	0.9785	0.0589	0.4830	0.2357	0.5345	0.2126					
	12	0.4747	0.8748	0.7464	0.0412	0.3919	0.4243	0.5753					

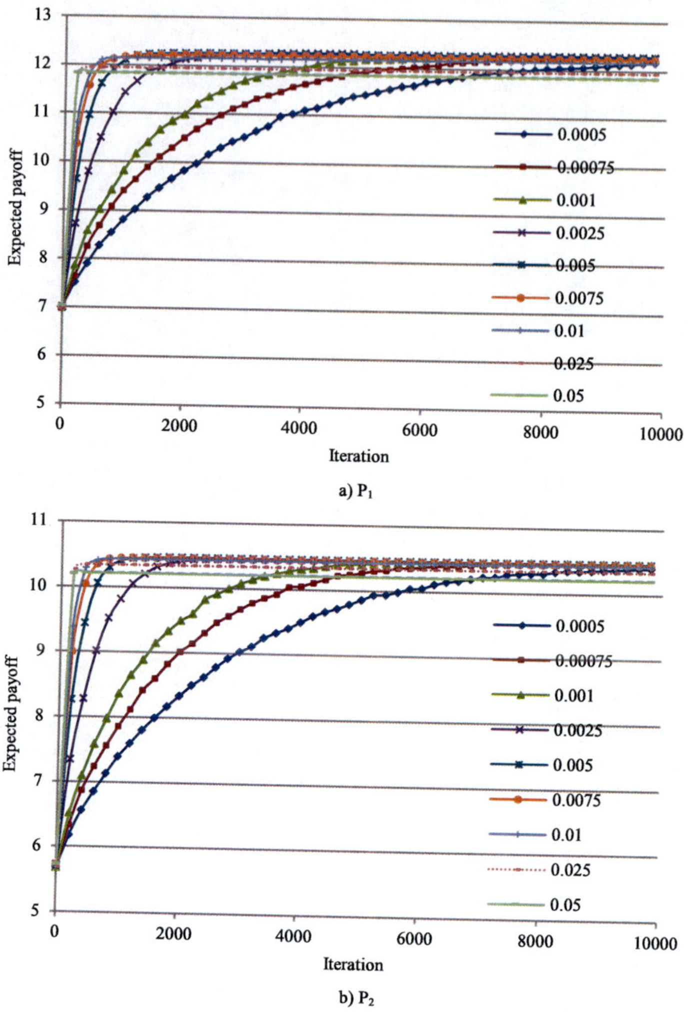

Fig. 3.16 Expected total received feedback of MNLA on P_1 and P_2

3.5 Conclusion

This chapter investigated new learning automata (LA) and cellular learning automata (CLA) models, which are based on multiple reinforcement signals. Each LA in the proposed models aims at learning the optimal subset or n-tuple of its actions using multiple feedbacks from its environment. Using a separate reinforcement for each chosen action enables the LA to learn the effectiveness of its different actions in the chosen super-actions in parallel. This kind of decision making is useful in scenarios like multi-agent environments where each agent has limited resources and actuators and performs several tasks simultaneously via its actuators. The convergence

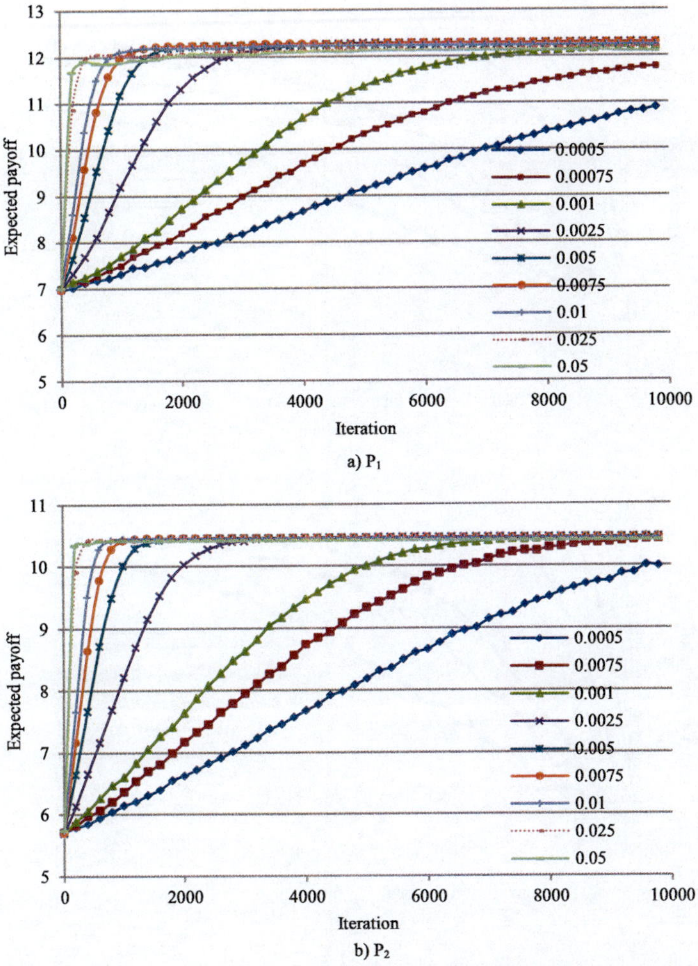

Fig. 3.17 Expected total received feedback of MMLA on P_1 and P_2

behavior of some of the proposed models is theoretically analyzed. It is shown that the proposed CLA model named MCLA, which is based on the idea of maximizing the expected rewards for each LA, converges to a compatible point when it uses potential local rules. Also, it is discussed that two of the proposed LA models are ε-optimal.

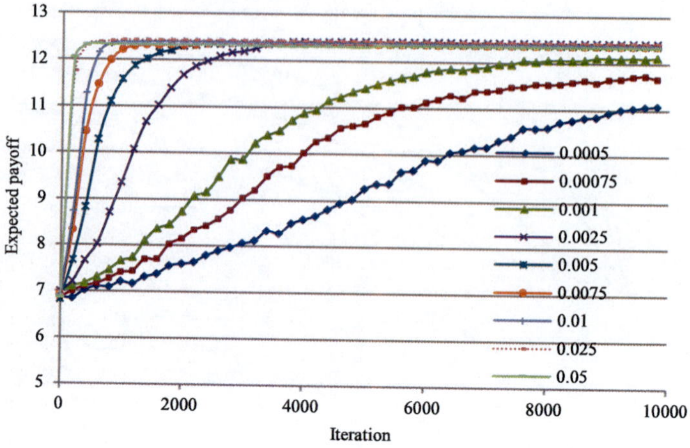

a) P_1

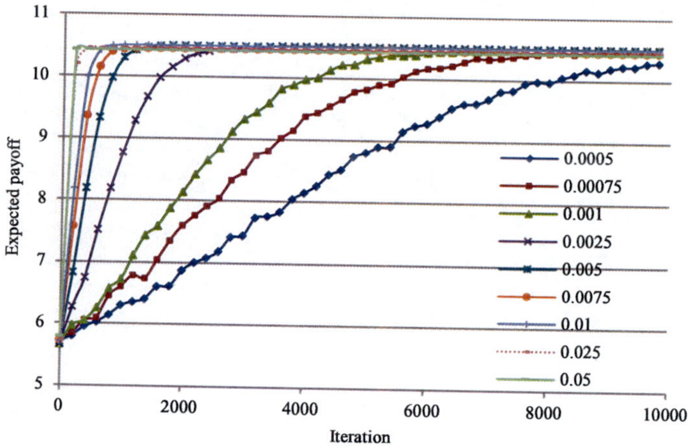

b) P_2

Fig. 3.18 Expected total received feedback of FMLA on P_1 and P_2

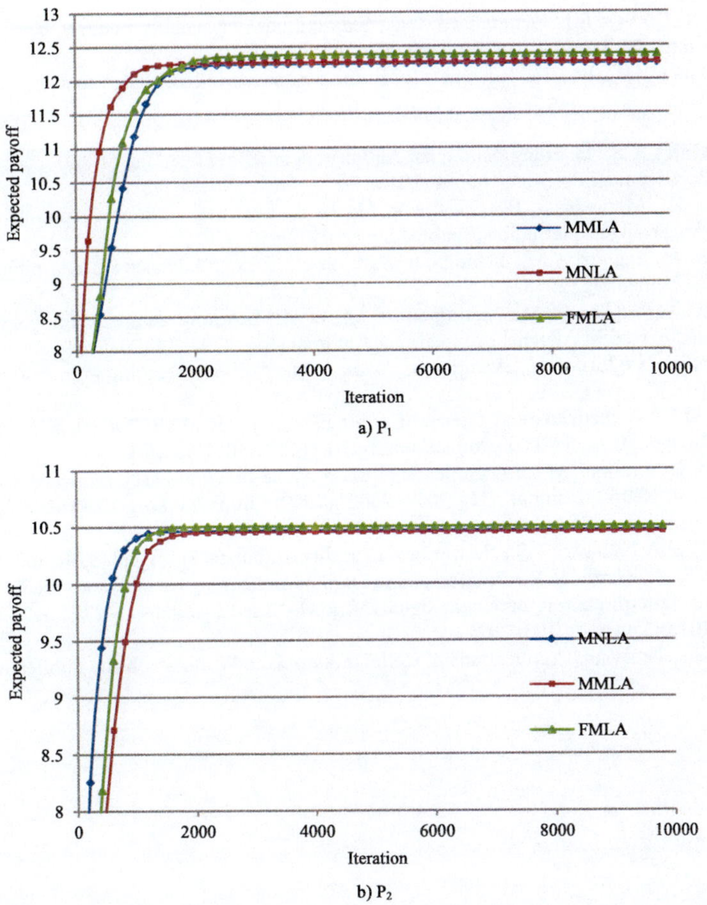

Fig. 3.19 The total payoff for MNLA, MMLA, and FMLA on P1 and P2 in 10,000 iterations

References

Babichenko, Y., Tamuz, O.: Graphical potential games. J. Econ. Theor. **163**, 889–899 (2016). https://doi.org/10.1016/j.jet.2016.03.010

Beigy, H., Meybodi, M.R.: A mathematical framework for cellular learning automata. Adv. Complex Syst. **07**, 295–319 (2004). https://doi.org/10.1142/S0219525904000202

Beigy, H., Meybodi, M.R.R.: Cellular learning automata with multiple learning automata in each cell and its applications. IEEE Trans. Syst. Man Cybern. Part B **40**, 54–65 (2010). https://doi.org/10.1109/TSMCB.2009.2030786

Duarte, P.B.F., Fadlullah, Z.M., Vasilakos, A.V., Kato, N.: On the partially overlapped channel assignment on wireless mesh network backbone: a game theoretic approach. IEEE J. Sel. Areas Commun. **30**, 119–127 (2012). https://doi.org/10.1109/JSAC.2012.120111

Demšar, J.: Statistical comparisons of classifiers over multiple data sets. J. Mach. Learn. Res. **7**, 1–30 (2006)

Leslie, D.S., Collins, E.J.: Generalised weakened fictitious play. Games Econ. Behav. **56**, 285–298 (2006). https://doi.org/10.1016/j.geb.2005.08.005

Morshedlou, H., Meybodi, M.R.: A new local rule for convergence of ICLA to a compatible point. IEEE Trans. Syst. Man Cybern. Syst. **47**, 3233–3244 (2017). https://doi.org/10.1109/TSMC.2016.2569464

Narendra Malt, K.S.: Learning Automata: An Introduction. Prentice-Hall (1989)

Ortiz, L.E.: Graphical Potential Games (2015)

Rezvanian, A., Moradabadi, B., Ghavipour, M., et al.: Learning Automata Approach for Social Networks. Springer International Publishing (2019)

Rezvanian, A., Saghiri, A.M., Vahidipour, S.M., et al.: Recent Advances in Learning Automata. Springer, Heidelberg (2018a)

Rezvanian, A., Saghiri, A.M., Vahidipour, S.M., et al.: Learning automata for complex social networks. In: Recent Advances in Learning Automata, pp. 279–334 (2018b)

Rezvanian, A., Saghiri, A.M., Vahidipour, S.M., et al.: Cellular Learning Automata, pp. 21–88 (2018c)

Shannon, C.E.: A mathematical theory of communication. ACM SIGMOBILE Mob. Comput. Commun. Rev. **5**, 3–55 (2001). https://doi.org/10.1145/584091.584093

Szepesvári, C., Littman, M.L.: A unified analysis of value-function-based reinforcement-learning algorithms. Neural Comput. **11**, 2017–2060 (1999). https://doi.org/10.1162/089976699300016070

Thathachar, M.A.L., Sastry, P.S.: Networks of Learning Automata. Springer US, Boston, MA (2004)

Vafashoar, R., Meybodi, M.R.: Reinforcement learning in learning automata and cellular learning automata via multiple reinforcement signals. Knowl.-Based Syst. **169**, 1–27 (2019). https://doi.org/10.1016/j.knosys.2019.01.021

Wolfram, S.: Theory and Applications of Cellular Automata. World Scientific Publication (1986)

Chapter 4
Applications of Cellular Learning Automata and Reinforcement Learning in Global Optimization

4.1 Introduction

Various scientific and engineering problems can be modeled as optimization tasks. Roughly speaking, optimization problems can be categorized as static problems and dynamic problems. The problem attributes are considered to be constant in a static optimization problem. Accordingly, the optima of the problem do not change during the optimization procedure. However, many optimization problems often contain several uncertain and dynamic factors. Typical examples include traffic-aware routing in wireless networks and air traffic scheduling. Such problems are commonly referred to as dynamic optimization problems. Due to the time-varying characteristics of a dynamic optimization problem, its fitness landscape changes with time. Accordingly, the locations of its optima may change with time as well.

Among different optimization approaches, nature-inspired methods such as evolutionary algorithms (Yu and Gen 2010; Eiben and Smith 2015) have attracted considerable interest in recent years. A nature-inspired algorithm is an iterative process. During each iteration, the algorithm aims at generating new and better candidate solutions to a given problem from the set of current candidate solutions. In order to generate new candidate solutions, nature-inspired optimization algorithms incorporate schemes or operators such as mutation, crossover, and selection.

Much effort has been put into the development of versatile nature-inspired optimization methods. However, it is well accepted that a general-purpose universal optimization algorithm is impossible: no strategy or scheme can outperform the others on all possible optimization problems (Wu et al. 2019). Additionally, a single evolutionary scheme or operator may always follow similar trajectories. Therefore, it would be better to use diverse evolutionary operators to increase the chance of

R. Vafashoar et al., *Cellular Learning Automata: Theory and Applications*, Studies in Systems, Decision and Control 307,
https://doi.org/10.1007/978-3-030-53141-6_4

finding optima. Moreover, the parameter and strategy configuration of an optimization algorithm can significantly affect its performance. On a specific problem, proper configurations can result in high-quality solutions and high convergence speeds.

Considering the discussed issues, the utilization of multiple strategies or techniques along with a suitable adaptation mechanism can significantly enhance a nature-inspired optimization method. Indeed, some recently developed nature-inspired optimization methods incorporate a collection of strategies in some of their algorithmic elements such as mutation, crossover, or neighborhood structure. Such collections are referred to as ensemble components in some literature, such as (Wu et al. 2019). In this chapter, we will show how cellular learning automata (CLA) and reinforcement learning (RL) can be utilized for the adaptive selection of proper strategies in nature-inspired optimization algorithms.

The rest of the chapter is organized as follows: Sect. 4.1 presents some background concerning nature-inspired optimization algorithms. In Sect. 4.2, some hybrid optimization approaches based on nature-inspired methods and reinforcement learning are briefly reviewed. Then, in Sect. 4.3, we describe an optimization algorithm based on CLA and PSO, which is presented in (Vafashoar and Meybodi 2018). Section 4.4 gives an optimization algorithm based on CLA and bare-bones particle swarm optimizer, which is adopted from (Vafashoar and Meybodi 2016). A CLA based dynamic optimization method is given in Sect. 4.5, which is presented in (Vafashoar and Meybodi 2020). Finally, Sect. 4.6 provides a summary of the chapter.

4.2 Evolutionary Algorithms and Swarm Intelligence

Biological and natural principles have long been a source of inspiration for human inventions. The natural evolution can be viewed as an optimization process of spices which is based on learning how to adapt to the environment. Evolutionary algorithms (EAs) are optimization algorithms which are based on the survival of the fittest principle. These algorithms generally have three main characteristics (Yu and Gen 2010):

- Population-based: EAs maintain a group of individuals, termed as a population. Each individual is a candidate solution to the given optimization problem.
- Fitness-oriented: Each individual has a gen representation. Also, EAs has mechanisms to determine the quality of a candidate solution. The quality of an individual is called its fitness value. During their optimization procedure, EAs prefer to preserve high-quality individuals.
- Variation-driven: In order to search the solution space, EAs incorporates some various operations that mimic the genetic gene changes.

Inspired from the natural evolution process, several EAs like a genetic algorithm (GA), evolutionary strategies (ES), differential evolution (DE) are introduced in the literature (Yu and Gen 2010; Eiben and Smith 2015). There are other similar phenomena in nature that can be viewed as an optimization and learning process. A swarm of not smart insects sometimes demonstrates a brilliant behavior. This swarm-level intelligent behavior has inspired the design of algorithms such as particle swarm optimization (PSO). In what follows, we briefly review three well-known nature-inspired optimization algorithms including, DE, PSO, and bare-bones PSO (BBPSO) (Yu and Gen 2010; Eiben and Smith 2015; Vafashoar and Meybodi 2016).

4.2.1 *Differential Evolution Algorithm*

DE evolves a population of *NP* candidate solutions, called individuals, through a process consisting of mutation, crossover, and selection. Each individual $X_i(G) = \{x_i^1(G), \ldots, x_i^D(G)\}$ $\forall i = 1, \ldots, NP$ at generation G represents a D dimensional real vector within the problem search space. The search space is defined using its upper and lower bounds: $X_{\min} = \{x_{\min 1}, x_{\min 2}, \ldots, x_{\min D}\}$ and $X_{\max} = \{x_{\max 1}, x_{\max 2}, \ldots, x_{\max D}\}$. The DE population is initially generated using uniform distribution within these bounds:

$$x_{i,0}^j = x_{\min}^j + \text{rand}(0, 1).(x_{\max}^j - x_{\min}^j),\ j = 1, .., D, \text{ and}, i = 1, ..., NP \quad (4.1)$$

At each generation G, one mutant vector $V_{i,G}$ is generated for each individual using a mutation strategy. Several mutation strategies have been developed in the DE literature; three of the most common ones are listed below:

$$DE/rand/1: \quad V_{i,G} = X_{r1,G} + F \cdot \left(X_{r2,G} - X_{r3,G}\right), \quad (4.2)$$

$$DE/best/1: \\ V_{i,G} = X_{best,G} + F \cdot \left(X_{r1,G} - X_{r2,G}\right), \quad (4.3)$$

$$DE/rand-to-best/1: \\ V_{i,G} = X_{i,G} + F \cdot \left(X_{best,G} - X_{i,G}\right) + F \cdot \left(X_{r1,G} - X_{r2,G}\right), \quad (4.4)$$

where r_1, r_2, and r_3 are three mutually different random integer numbers uniformly taken from the interval [1, *NP*]. F is a constant control parameter which lies in the range [0,2] and scales the differential vectors. $X_{best,G}$ represents the fittest individual of the Gth generation.

After the generation of mutant vectors, a crossover operator is applied to each mutant vector $V_{i,G}$ in order to generate a trail vector $U_i(G) = \{u_i^1(G), \ldots, u_i^D(G)\}$ $\forall i = 1, \ldots, NP$. DE algorithms use two types of crossover operators: exponential and binomial. The binomial operator is the most preferred crossover operator in the DE literature and is defined as follows:

$$U_{i,G}^{j} = \begin{cases} V_{i,G}^{j} & \text{if rand}(0,1) < CR \text{ or } j = irand \\ X_{i,G}^{j} & \text{otherwise} \end{cases} \quad j = 1, \ldots, D, \tag{4.5}$$

where *CR* is a constant parameter controlling the portion of the offspring chosen from the mutant vector. *irand* is a random integer generated uniformly from the interval [1,*N*] for each trial vector to ensure that at least one of its components is chosen from the mutant vector. rand(0,1) is a uniform random real number generator from the interval [0,1].

Each generated trial vector is evaluated using a fitness function; then, either of the parent vector (also called target vector) or its corresponding trial vector is selected for the next generation by using a greedy selection scheme:

$$X_{i,G+1} = \begin{cases} U_{i,G} & \text{if } f\left(U_{i,G}\right) < f\left(X_{i,G}\right) \\ X_{i,G} & \text{otherwise} \end{cases}, \tag{4.6}$$

4.2.2 *Particle Swarm Optimization*

PSO executes its search in a specified space through the accumulation of velocity and position information. In the beginning, each particle is initiated randomly within a *D* dimensional search space. It keeps the information about its best-visited position known as *pbest*, swarm best position known as *gbest*, and its current flying velocity. Canonical PSO maintains a swarm of particles; each particle X_i in the swarm maintains three *D*-dimensional vectors: $x_i = [x_{i1}, x_{i2}, \ldots, x_{iD}]$ representing its current position, $p_i = [p_{i1}, p_{i2}, \ldots, p_{iD}]$ representing the best location previously experienced by the particle, and $v_i = [v_{i1}, v_{i2}, \ldots, v_{iD}]$ representing its flying velocity. The whole swarm keeps the information about its best-experienced position in a *D* dimensional vector like $p_g = [p_{g1}, p_{g2}, \ldots, p_{gD}]$. During the search, at step *k*, the velocity of the particle X_i and its current position is updated according to the following equations:

$$\begin{aligned} v_{id}(k+1) &\leftarrow \omega v_{id}(k) + c_1 r_1 (p_{id} - x_{id}(k)) + c_2 r_2 \left(p_{gd} - x_{id}(k)\right), \\ x_{id}(k+1) &\leftarrow x_{id}(k) + v_{id}(k+1), \end{aligned} \tag{4.7}$$

where $v_{id}(k)$ is the *d*th dimension of the velocity vector of the particle in step *k*; $x_{id}(k)$ and $p_{id}(k)$ are respectively the *d*th dimension of its position and historical best position vectors; $p_{gd}(k)$ represents the *d*th dimension of the historical best position of the whole swarm in step *k*; ω is the inertia weight, which was introduced to bring a balance between the exploration and exploitation characteristics (Shi and Eberhart 1998); c_1 and c_2 are acceleration constants and represent cognitive and social learning weights; and, finally, r_1 and r_2 are two random numbers from the uniform distribution $u(0, 1)$.

After acquiring the new positions of the particles, the historical best position of each particle is updated, which may also affect the historical global best position of the swarm.

4.2.3 Bare Bones Particle Swarm Optimization

BBPSO was originally developed to study the dynamics of the particle swarm optimization algorithm. As in canonical PSO, the spread of the positions visited by a single particle resembles a Gaussian-like distribution, BBPSO replaces the update equations of PSO with explicit probability distributions (Kennedy 2003). Consequently, it is suggested that the new position of a particle like X_i can be generated according to the following equation, in which the velocity term of PSO is eliminated:

$$x_{id}(k+1) \leftarrow \frac{p_{id} + p_{gd}}{2} + N(0, 1) \times \left|p_{id} - p_{gd}\right|, \tag{4.8}$$

where $N(0,1)$ denotes a random number taken from the standard normal distribution, p_g is the global best position of the swarm, and p_i is the personal best position of the particle itself.

Kennedy suggested that preserving some components of the personal best positions would speed up the search process of BBPSO. Accordingly, the following updating rule was recommended:

$$x_{id}(k+1) \leftarrow \begin{cases} \frac{p_{id}+pl_{id}}{2} + N(0, 1) \times |p_{id} - pl_{id}| & \text{if } r < IP \\ p_{id} & \text{otherwise} \end{cases} \tag{4.9}$$

where IP is a parameter which controls the portion of a particle to be remained unchanged, r is a random number from uniform distribution $r \sim (0,1)$, and pl_i is the local best position in the neighborhood of the ith particle.

4.3 A Brief Review of Hybrid Models Based on RL and Nature-Inspired Optimization Methods

Up to now, various hybrid models based on reinforcement learning schemes and nature-inspired methods have been introduced for solving a variety of problems (Table 4.1). A great deal of these methods uses a reinforcement learning scheme as a controller unit to adaptively control the comprising components of an optimization method. This section briefly reviews some of these approaches. Although in most of the works that are reviewed in this section, reinforcement learning is utilized as a mechanism to improve the performance of a nature-inspired method, some works are based on a different idea. Iima and Kuroe introduced some swarm reinforcement

Table 4.1 Some applications of reinforcement learning in population-based optimization methods

Brief description	Application	Heuristic	Reinforcement learning mechanism	Ref.
Parameter/strategy adaptation	Numerical optimization	PSO	LA	Hashemi and Meybodi (2011), Hasanzadeh et al. (2012, 2014), Zhang et al. (2018)
Parameter/ strategy adaptation	Numerical optimization, economic dispatch problem	Quantum-behaved PSO	Q-learning	Sheng and Xu (2015)
Parameter/ strategy adaptation	Numerical optimization	Memetic PSO	Q-learning	Samma et al. (2016)
Parameter/strategy adaptation	Numerical optimization	DE	LA	Sengupta et al. (2012), Kordestani et al. (2014), Mahdaviani et al. (2015)
Parameter/strategy adaptation	Numerical optimization	GA	LA	(Ali and Brohi 2013)
Parameter/strategy adaptation	Numerical optimization	Butterfly algorithm	LA	Arora and Anand (2018)
Parameter/strategy adaptation	Numerical optimization	Harmony search	LA	Enayatifar et al. (2013)
Evaluation budget control	Numerical optimization in noisy environments	PSO	LA	Zhang et al. (2014, 2015)
Optimization entity/local search	Graph isomorphism problem, OneMax problem, object partitioning	Memetic GA	LA, object migration automaton	Rezapoor Mirsaleh and Meybodi (2015, 2018)
Optimization entity/genotype	Numerical optimization	GA	LA	(Howell et al. 2002)
Optimization entity/quantization orthogonal crossover	Numerical multi-objective optimization	Orthogonal EA	LA	Dai et al. (2016)

(continued)

Table 4.1 (continued)

Brief description	Application	Heuristic	Reinforcement learning mechanism	Ref.
Optimization entity/genotype	Traveling salesman problem	GA	Q-learning	Alipour et al. (2018)
Upper-level heuristic/strategy adaptation	Numerical multi-objective optimization, vehicle crashworthiness problem	Hyper-heuristic	LA	Li et al. (2018)
Optimization entity/genotype	Numerical optimization	DE, GA	CLA	Rastegar and Meybodi (2004), Vafashoar et al. (2012), Mozafari et al. (2015)
Scheduler	Dynamic numerical optimization	Multi-population optimizer	LA	Kordestani et al. (2019)
Parameter/strategy adaptation	Dynamic numerical optimization	Cuckoo search	LA	Kordestani et al. (2018)
Parameter/strategy adaptation	Dynamic numerical optimization	Artificial immune system	LA	Rezvanian and Meybodi (2010a, b, c)
Parameter/strategy adaptation	Dynamic numerical optimization	Firefly algorithm	LA	Abshouri et al. (2011)
Parameter/strategy adaptation	Dynamic numerical optimization	PSO	LA	Geshlag and Sheykhzadeh (2012)
Parameter/strategy adaptation	Dynamic software project scheduling	Multi-objective memetic algorithm	Q-learning	Shen et al. (2018)
Parameter/strategy adaptation	Dynamic multi-objective optimization, controlling traffic lights	PSO	Q-learning	El Hatri and Boumhidi (2016)

learning algorithms (Iima and Kuroe 2008). In these algorithms, several Q-learning agents learn concurrently based on two learning strategies: independent learning and cooperative learning. In the independent learning phase, each agent learns according to the ordinary Q-learning algorithm. In the cooperative phase, agents exchange their obtained knowledge according to swarm updating rules of PSO.

4.3.1 *Static Optimization*

One of the earliest uses of a reinforcement learning mechanism in adaptive parameter control of a nature-inspired method is reported by Hashemi and Meybodi (Hashemi and Meybodi 2011). Unified adaptive PSO (UAPSO) and independent adaptive PSO (IAPSO) algorithms use LA-based mechanisms for parameter adaption during the search procedure. During each iteration of UAPSO, a set of learning automata determine the parameter values of the algorithm. In contrast, IAPSO associates a set of learning automata with each particle, and each set determines the parameter values for its associated particle. After the evolution of particles, the learning automata are updated in order to choose better parameter values in future generations (Hashemi and Meybodi 2011). Another, reinforcement learning parameter adaptation mechanism is presented by Sheng and Xu, which adaptively adjusts the parameter values of quantum-behaved PSO by using the Q-learning approach (Sheng and Xu 2015). The presented algorithm is employed for solving the economic dispatch problem and demonstrated some beneficial characteristics.

Reinforcement learning-based memetic particle swarm optimizer (RLMPSO) model uses five well-designed operations consisting of exploration, convergence, high-jump, low-jump, and fine-tuning to update each particle (Samma et al. 2016). To control the application sequence of these five operations, RLMPSO utilizes a Q-learning mechanism. Each state of the Q-learner entity is associated with one specific operation. After the execution of a specific operation on a particle, the Q-learner decides the next operation, which best suits the currently applied operation. The LAPSO algorithm presented by Zhang et al. uses two updating strategies for particles (Zhang et al. 2018). During each generation, an equipped LA chooses the updating mechanism of all particles. Then, using the chosen mechanism, the particles get updated. Next, the probability vector of the LA is updated based on the number of successful position updates.

Similarly, cooperative particle swarm optimization based on learning automata (CPSOLA) hybridizes cooperative and canonical PSO methods into one algorithm to inherit the beneficial characteristics of both methods. It uses a learning automaton to switch between two methods based on the performance of the algorithm (Hasanzadeh et al. 2014). The authors also presented another adaptive cooperative particle swarm optimizer in (Hasanzadeh et al. 2013). In this algorithm, a set of learning automata is associated with the dimensions of the problem. These learning automata aim to find the correlated variables of the search space. By considering the existent dependencies among variables, the proposed PSO can solve complex optimization problems efficiently.

They are historically learning automata are decision-making units designed for unknown, random, and noisy environments. Consequently, they can be well suited for optimization in such environments. One common approach for dealing with noise corrupted fitness values is based on re-evaluation. After re-evaluating a particular candidate solution several times, its fitness can be defined as the mean value of the obtained noise corrupted fitness values. PSOLA and LAPSO, introduced by Zhang et al., utilize LA to allocate re-evaluation budgets to promising particles in an intelligent manner (Zhang et al. 2014, 2015).

Learning automata is also utilized successfully for adaptively controlling the parameter settings and the mutation strategies of differential evolution algorithm (Kordestani et al. 2014; Mahdaviani et al. 2015). In the approach presented in (Sengupta et al. 2012), a population of *NP* candidate solutions is equipped with a set of *NP* learning automata. During each generation, the individuals are ranked according to their fitness values. Then, the *i*th learning automaton chooses the mutation parameter value of the *i*th ranked individual. Next, the individual is updated according the chosen settings and is evaluated by the fitness function. Based on this evaluation, the probability vector of the corresponding learning automaton is updated.

Reinforcement learning has also integrated, in a similar manner as the ones described before, into approaches such as genetic algorithm, harmony search, butterfly algorithm, and firefly algorithm. Ali and Brohi have used fast learning automata to adaptively control the application of three mutation operators, including Cauchy, Gaussian, and Levy, in their proposed genetic algorithm (Ali and Brohi 2013). Similarly, Arora and Anand have used an LA agent to control the behavior of butterflies in the butterfly optimization algorithm. In their approach, a learning automaton is associated with the whole population, which decides whether a butterfly should perform a random local search or move towards the best butterfly (Arora and Anand 2018). The authors have also investigated the applicability of their introduced model on some classical engineering design problems. Another approach for controlling parameter values through a reinforcement learning mechanism is presented in (Enayatifar et al. 2013). This method learns a set of proper parameter values for a variant of the harmony search algorithm.

Rezapoor and Meybodi (2015) introduced a Baldwinian memetic algorithm called MGALA by combining genetic algorithms with learning automata. In this method, LA provides the local search function of the memetic algorithm. Each chromosome of MGALA is an object migration automaton whose states keep information about the history of the local search process. The critical aspect of MGALA is the computation of the fitness function, which is based on the history of the local search and the chromosome's objective value. In another work, the same authors suggested an optimization method composed of two parts: a genetic part and a memetic part. They demonstrated how to use learning automata in order to create a balance between exploration performed by evolution and exploitation performed by local search (Rezapoor Mirsaleh and Meybodi 2018).

Howell et al. used probability strings to represent individuals (Howell et al. 2002). The value of the ith component of a probability string defines the probability of the allele value in a corresponding bit string being one at that position. These probability strings can be regarded as genotypes. At each generation, a population of phonotypes is generated by sampling the probability strings. These generated bit strings are then evaluated by the fitness function, and the probability vectors are updated accordingly.

Additionally, the probability strings can be affected by crossover and mutation operators with the hope of producing better genotypes. An interesting orthogonal evolutionary algorithm based on learning automata is presented by Dai et al. for multi-objective optimization (Dai et al. 2016). In this algorithm, learning automata provide mechanisms for mutation and grouping decision variables for quantization orthogonal crossover.

Q-learning has been reported to be an efficient approach for solving the traveling salesman problem (Alipour et al. 2018). In the approach presented in (Alipour et al. 2018), each city is modeled as a state of a Q-learner. There are several traveling salesmen considered traversing the cities to obtain the best solution. During each iteration, each salesman starts from a random city. Then, it selects an action from the action set of the corresponding state. It follows the action and moves to the next city. This process is repeated until the tour is complete. The best tour fond by all salesmen is used to generate reinforcement signals for the Q-learning system. According, the salesmen can choose better tours as the algorithm proceeds.

The reinforcement learning paradigm can well suit the hyper-heuristic framework of optimization. There are several works reported on employing learning automata to control the application of a set of low-level heuristics. An excellent example of this category of hyper-heuristics has presented by Li et al. for multi-objective optimization (Li et al. 2018). The method is applied for solving vehicle crashworthiness problem; however, it can be used as a general framework for hyper-heuristic design. In this method, the reinforcement learning scheme sits at the core of the meta-heuristic selection process. For each meta-heuristic, it learns the most appropriate meta-heuristic to be applied next. By doing this, a transition matrix is learned by the reinforcement learning mechanism, which governs the application sequence of low-level meta-heuristics during the searching process.

Static optimization based on CLA

Several works have been reported on the use of CLA for solving optimization problems (Rastegar and Meybodi 2004; Vafashoar et al. 2012; Mozafari et al. 2015). In some of these works, in a similar way as (Howell et al. 2002), CLA models the solution space. Usually, a set of learning automata is considered in each cell of the CLA, and each combination of their actions corresponds to a candidate solution to the given problem. The learning automata of each cell give their combination of chosen actions (a sampled candidate solution) to the environment. The environment decides the favorability of the received candidate solution or each of its comprising components and generates reinforcement signals accordingly. The LAs use these reinforcement feedbacks to update their internal probability distributions and generate better candidate solutions in future generations (Rastegar and Meybodi 2004; Vafashoar et al. 2012; Mozafari et al. 2015).

4.3.2 Dynamic Optimization

A general framework for scheduling subpopulations in multi-population dynamic optimization methods is presented in (Kordestani et al. 2019). In this approach, a learning automaton schedules the evolutionary process of subpopulations. The general idea of the presented approach is to allocate more function evaluations to the best performing subpopulations. In another dynamic optimization approach, LA is used to balance the exploration and exploitation characteristics of cuckoos in the cuckoo search algorithm (Kordestani et al. 2018). The proposed method uses two types of Levy distributions with two different β values: one for long jumps and the other for short ones. During each iteration, the search strategy distribution for each egg is adaptively determined through a learning automaton. In a similar approach, Rezvanian and Meybodi proposed to control the mutation probability adaptively in artificial immune systems by learning automata (Rezvanian and Meybodi 2010a, b). Learning automata have been employed for parameter adjustment in the firefly algorithm (Abshouri et al. 2011). In the dynamic optimization approach presented in (Abshouri et al. 2011), each firefly is equipped with three learning automata, and each learning automaton adjusts a specific parameter for its associated firefly. Another approach using learning automata to provide adaptive strategy selection for PSO is presented in (Geshlag and Sheykhzadeh 2012) for dynamic optimization. In this approach, a particle can be update based on two rules. Each particle is associated with a particular LA, and this LA controls which strategy to be utilized by the particle.

Software project scheduling, which allocates employees to tasks in a software project, can be viewed as a dynamic multi-objective optimization problem (Shen et al. 2018). A dynamic multi-objective memetic algorithm based on Q-learning is presented by Shen et al. for solving this scheduling/rescheduling problem (Shen et al. 2018). In this algorithm, Q-learner learns the most appropriate global and local search methods to be used for different software project environment states. A motivating multi-objective optimization approach is presented in (El Hatri and Boumhidi 2016), which can be utilized in static and dynamic environments. The algorithm is tested as a system for controlling traffic lights. It uses Q-learning to adaptively control the parameter settings of each PSO particle based on its state. The state of a particle is defined as its distance from its previous best position and the global best position. Additionally, Q-learning is modified to meet the requirements of multi-objective optimization problems.

4.4 Multi Swarm Optimization Algorithm with Adaptive Connectivity Degree

In the canonical PSO, each particle updates its velocity and position according to its own historical best position and the global best position experienced by the entire swarm. The use of this simple updating scheme is often blamed as a reason for premature convergence and entrapment in local optima. Accordingly, other updating rules and population topologies have been developed to enhance the diversity of the population or balancing its explorative and exploitative behaviors (Mendes et al. 2004; Li and Yang 2009; Montes de Oca et al. 2009; Nabizadeh et al. 2012; Changhe Li et al. 2012; Lim and Mat Isa 2014). Moreover, different problems may require specific updating rules or population topologies according to the properties of their fitness landscape (Li and Yang 2009; Changhe Li et al. 2012). Additionally, the use of different topologies and updating rules according to the state of the search process is also investigated in some literature (Montes de Oca et al. 2009; Lim and Mat Isa 2014).

In all PSO variants, each particle adjusts its flying velocity according to a set of attractor particles. The cardinality of this set is a feature of neighborhood topology, which can greatly influence the exploration and exploitation characteristics of the algorithm (Montes de Oca et al. 2009). In order to investigate this property exclusively, the concept of connectivity degree of particles can be utilized (Vafashoar and Meybodi 2018). We can balance the exploration and exploitation capabilities of the search procedure by appropriately tuning the defined connectivity degree. A CLA is utilized for adaptively controlling the connectivity degree of the particles based on the characteristics of their surrounding fitness landscape. The entire population of the particles is divided into several swarms, and each swarm resides in one cell of a CLA. Each cell of the CLA also contains a group of learning automata. These learning automata are responsible for adjusting the connectivity degrees of their related particles. This task is achieved through periods of learning. In each period, the learning automata realize suitable connectivity degrees for the particles based on their experienced knowledge. Moreover, multi-swarm models maintain population diversity through the isolation of its sub-swarms. This diversity preserving mechanism is beneficial in the optimization of complex multimodal problems.

4.4.1 CLA Based Multi-swarm Optimization Algorithm (CLAMS)

The particles in CLAMS are divided into several swarms. Each particle is updated based on the information obtained from its parent swarm as well as the nearby ones. The global best position of each nearby swarm would be included in the update equation of a particular particle to influence its movement. The swarms or particles that directly affect a single particle in an update step constitute its neighborhood.

The size of this neighborhood controls the convergence speed and the exploration ability of the proposed approach. In the smallest neighborhood, a particle is updated in isolation from its nearby swarms. Hence, it can effectively exploit the promising areas covered by its parent swarm.

On the other hand, the tendency of particles toward their nearby swarms grows as the size of the described neighborhood increases. Increasing the neighborhood size of particles has several effects on the behavior of PSO. First, the convergence speed increases since the particles are utilizing more similar information sources for adjusting their movements. Also, the particles that are trapped in poor local optima may be guided to more appropriate areas; however, they may lose track of the great unexploited regions around their parent swarms. In order to proceed, we first define the following concept to formalize the ideas, as mentioned earlier.

Particle connectivity degree: the connectivity degree of a particle like X during time t is the number of particles that are connected to X at t. These connected particles influence the velocity of X during the time; therefore, X "flies" toward them.

Since the connectivity degree describes the neighborhood topology, the terms particle connectivity degree and particle topology will be used interchangeably in this section. As described earlier, the neighborhood topology of particles has an intense impact on the search behavior of the proposed approach. Tuning the connectivity degree of a particle is not an easy task as it relies on the fitness landscape and the distribution of particles. Also, as the location of particles changes during the search process, different connectivity degrees would be required to achieve an efficient search. The CLAMS uses an approach based on cellular learning automata for dynamically adjusting the connectivity degree of particles. CLA enables the particles to adapt their topology during the search process based on their previous experiences in the fitness landscape.

Each cell of the CLA embeds a swarm of particles and a group of learning automata. There is a one to one correspondence between the learning automata and the particles within each cell. Each cycle of CLAMS starts with topology selection for the particles. In this step, CLA sets the connectivity degree of the particles. Afterward, each particle employs its selected connectivity degree during the search process. The role of a learning automaton is to understand the effectiveness of different topologies for its coupled particle. The learning automaton achieves this purpose by observing the performance of the particle during the search process and employing the gained experience of other nearby learning automata in the CLA. The searching process consists of ξ update steps, ξ is called a learning period. Because of the stochastic nature of the PSO algorithm, an appropriately selected topology probably leads to low fitness positions for a particle or vice versa. Therefore, The CLAMS uses ξ search steps, rather than a single one, to obtain a better understanding of the selected topologies before their evaluation. After this search process, the effectiveness of the adopted topologies will be evaluated by the CLA. Then, the internal probability vectors of the learning automata will be updated accordingly. The block diagram of the proposed approach is illustrated in Fig. 4.1, and its building blocks are thoroughly explained in the subsequent sections.

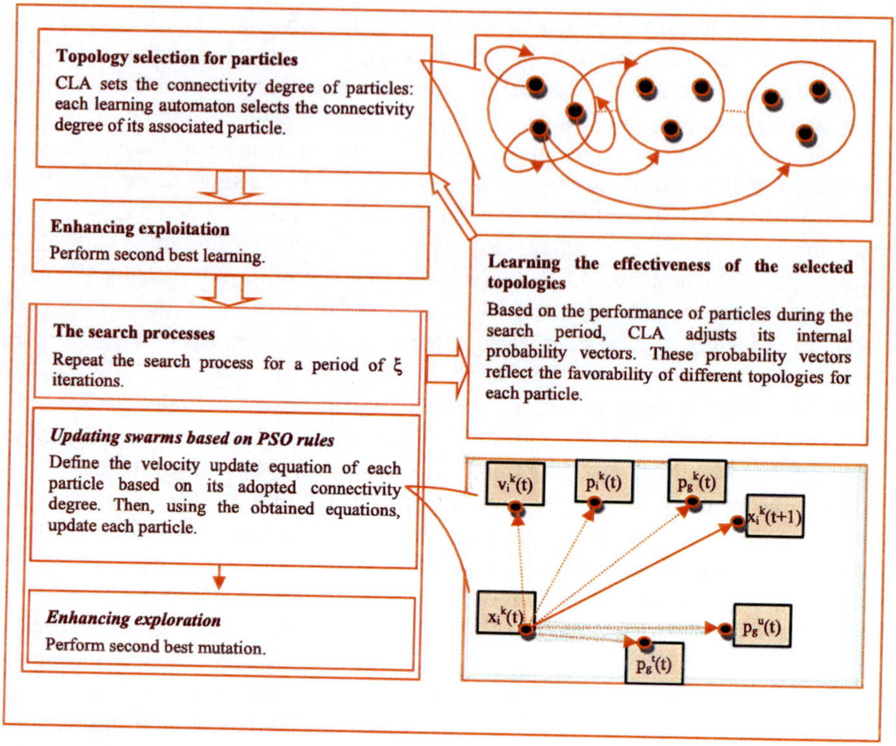

Fig. 4.1 The block diagram of the CLAMS

4.4.1.1 Topology Selection for Particles

In CLAMS, a CLA is responsible for selecting a promising topology for each particle. Each cell of the CLA is composed of two parts: a swarm of the population, which itself holds N particles, and a set of N learning automata, one learning automaton for each particle. Each learning automaton controls the connectivity degree of its associated particle. Figure 4.2 shows a schematic representation of these ideas.

As discussed previously, each particle can "fly" toward the global best position of a nearby swarm. The number of nearby swarms, which will be denoted by γ, defines the maximum connectivity degree of a particle. Each learning automaton has a set of $\gamma + 1$ actions, where each action corresponds to one of the $\gamma + 1$ allowable connectivity degrees. It should be noted that a particle utilizes only its intra-swarm information when its connectivity degree is set to its minimum value (i.e., 2). During the topology selection step, each learning automaton selects an action according to its internal probability vector. The selected action of a learning automaton defines the connectivity degree of its associated particle.

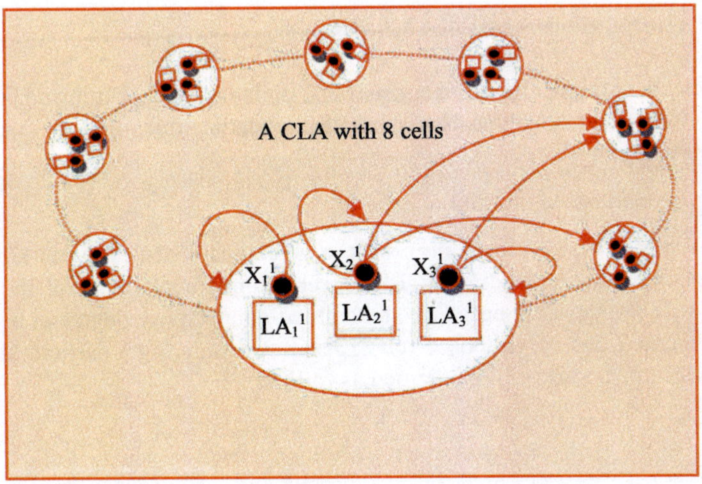

Fig. 4.2 A CLA with eight cells: each cell embeds a swarm of particles and a learning automaton for each particle

4.4.1.2 Enhancing Exploitation via Second Best Learning

Maintaining an archive of promising visited locations would be beneficial as the algorithm can possess more eligible historical information about the fitness landscape. Based on this idea, the second-best positions are employed in CLAMS. Each particle keeps track of the best-visited locations in its personal best and second-best positions. The second-best position of a particle is a position that has been experienced by the particle itself and did not become the particle's *pbest*; additionally, it is the fittest position satisfying this condition. The CLAMS uses the second-best position to enhance its exploration and exploitation abilities via two developed mechanisms: Second best mutation and Second best learning.

During the search process, which consists of ξ updating steps, particles move in the fitness landscape, and their positions are altered accordingly. Consequently, they may forget some of their promising visited locations and reside in inappropriate areas of the search space. When a particle is positioned in an unpleasant location, it might require much effort to find its way back to a suitable location. The search process will be enhanced if particles revisit some of their promising historical locations and exploit them further. The second-best learning is developed for this purpose. It allows the particles to revisit their favorable spotted locations and reexamine them furthermore. After every learning period, each particle is affected by the second-best learning operator. The operator simply sets the current position of the particle to its second-best position.

4.4.1.3 The Search Process of CLAMS

The search phase of the algorithm consists of ξ updating steps. During each updating step, the swarms are evolved based on PSO updating rules and the second-best mutation strategy.

Updating swarms based on PSO rules

During each step of the search process, the velocities of the particles are updated according to their adopted topologies. Several implementations for the velocity updating rule are possible based on the defined connectivity degree concept. Two velocity update rules are proposed for CLAMS. Considering a particle like X_i^k, its first velocity update rule will be defined as follows:

$$
\begin{aligned}
&v_i^k(t)\\
&= \begin{cases} W v_i^k(t-1) + \frac{\chi}{a_i^k(t)+1} \sum_{j=1}^{a_i^k(t)+1} r_j \odot \left(b_j - x_i^k(t-1)\right) & \text{if } a_i^k(t) > 1 \\ W v_i^k(t-1) + \frac{\chi}{2} r_1 \odot \left(p_i^k(t-1) - x_i^k(t-1)\right) + \frac{\chi}{2} r_2 \odot \left(p_g^k(t-1) - x_i^k(t)\right) & \text{if } a_i^k(t) = 1 \end{cases},\\
&b = randperm\left(p_i^k(t-1), p_g^k(t-1), p_g^{N(1,k)}(t-1), p_g^{N(2,k)}(t-1), \ldots, p_g^{N(\gamma,k)}(t-1)\right),\\
&r_j = \begin{cases} \left[r_{j,1}, r_{j,2}, \ldots, r_{j,D}\right]^D; r_{j,h} \sim u(0,1) & \text{with probability } 0.5 \\ r[1, 1, \ldots, 1]^D; r \sim u(0,1) & \text{otherwise} \end{cases}
\end{aligned} \tag{4.10}
$$

Here, $x_i^k(t)$ represents the position of the ith particle belonging to the kth swarm; $v_i^k(t)$ and $p_i^k(t)$ are its flying velocity and personal best position, respectively. $a_i^k(t)$ stands for the adopted connectivity degree of the particle (or the selected action of its associated learning automaton) during time t. It should be noted that the connectivity degree of particles remains unchanged for a period of ξ search steps between two consecutive action selection phases. $p_g^k(t)$ represents the global best position of the kth swarm. N(j, k) defines a function that returns the jth swarm in the neighborhood of swarm k, and randperm(A) is a function that arranges the elements of A in random order. W and χ are the inertia weight and the acceleration constant, respectively. Finally, $\odot$ stands for the Hadamard operator, which is the element by element multiplication of two matrices.

It is considering Eq. 4.10, there are some points worth noting. First, a particle only employs intra-swarm information when its connectivity degree is set to 2. Under this connectivity condition, the particle performs its search locally. When the connectivity degree of a particle becomes larger than one, it will be influenced by some attractors out of its parent swarm. In this case, it is also possible that the particle is merely guided by the outer-swarm elements, which helps it in escaping from poor local optima. The second point is about the random element in Eq. 4.10, i.e., r, which scales the summation terms in the update rule. When this element is a random scalar, the velocity update rule is called linear. On the other hand, the canonical velocity update rule typically uses a vector of random scalar numbers. Both updating rules have their benefits and drawbacks, which are studied in some literature like (Wilke et al. 2007). Equation (4.10) uses a combination of scalar and vector random numbers to utilize the benefits of both approaches.

The described velocity update rule can be modified in several ways in order to induce different behavioral dynamics in the particles. A variant of Eq. (4.10) can be obtained by replacing "randperm" with a function that returns a random combination of its arguments with repetitions allowed. In this case, particles can move with larger steps in specific directions, which results in high explorative behavior.

$$
\begin{aligned}
& v_i^k(t) \\
& = \begin{cases} W v_i^k(t-1) + \frac{\chi}{a_i^k(t)+1} \sum_{j=1}^{a_i^k(t)+1} r_j \odot \left(b_j - x_i^k(t-1)\right) & \text{if } a_i^k(t) > 1 \\ W v_i^k(t-1) + \frac{\chi}{2} r_1 \odot \left(p_i^k(t-1) - x_i^k(t-1)\right) + \frac{\chi}{2} r_2 \odot \left(p_g^k(t-1) - x_i^k(t)\right) & \text{if } a_i^k(t) = 1 \end{cases} \\
& b_j \sim u\left\{p_i^k(t-1), p_g^k(t-1), p_g^{N(1,k)}(t-1), p_g^{N(2,k)}(t-1), \ldots, p_g^{N(\gamma,k)}(t-1)\right\} \\
& r_j = \begin{cases} \left[r_{j,1}, r_{j,2}, \ldots, r_{j,D}\right]^D; r & \text{with probability}_{j,h} \\ r[1, 1, \ldots, 1]^D; r \upsilon u(0,1) & \text{otherwise} \end{cases}
\end{aligned} \tag{4.11}
$$

Here, b_j is obtained randomly according to a uniform distribution over the set of attractors.

Second best mutation (SBM)

PSO update rules affecting a particle in all dimensions would likely degrade its quality in some dimensions. As a result, by using canonical PSO update rules, it would require much effort to refine a particle when its personal best position has eligible information in most dimensions except a few ones. In order to alleviate this problem, The CLAMS utilizes a mutation mechanism which does not inflict changes in a particle. The second-best mutation is developed for this purpose. This mutation mechanism only affects the global best particle of each swarm.

In order to generate a mutant vector, SBM changes the personal best position of a particle in two random dimensions; if the newly generated mutant vector is evaluated to be fitter, it will replace the personal best position of its parent particle. The SBM operator changes the first selected dimension of a particle like X_i^k in the following way: first, a random particle is selected from a nearby swarm; then, the personal best position of X_i^k is replaced by the second-best position of the random particle in the corresponding mutation dimension. The second dimension of the particle to be affected by the SBM operator is perturbed using Cauchy distribution. This kind of perturbation enables the particle to escape from local optima and to explore areas beyond its swarm limits. A pseudo-code for the described operator is given in Fig. 4.3.

4.4.1.4 Learning the Effectiveness of the Selected Topologies

After a learning period, each learning automaton in a cell receives a reinforcement signal, which reflects the favorability of its selected action. Then, based on this signal,

Algorithm 4-1. Second best mutation (SBM operator)

Input:

K: swarms index.

1. Select the best particle of the swarm.
2. Let the personal best location of the selected particle be p_i^k.
3. Set $u = p_i^k$.
4. Select two random dimensions $s, t \in \{1, 2, \ldots, D\}$.
5. Select a random particle like X_j^l from a nearby swarm.
6. Modify u in its sth dimension based on a Cauchy distribution: $u_s = u_s + |p_{js}^l - u_s| C(1)$.
7. Modify u in its tth dimension based on the second-best position of X_j^l: $u_t = sb_{jt}^l$.
8. Replace p_i^k with u if it has a better fitness value.

Fig. 4.3 A pseudo-code for the second-best mutation

its initial probability vector will be updated using a learning algorithm. Similar to the velocity update rules, there are several ways to define reinforcement signals. The reinforcement signal for a taken action conjectures the effectiveness of the action. A suitable definition should exhibit this effectiveness as accurately as possible in order to steer eligible topology selection. To define the reinforcement signals formally, we first define two terms of average behavior and average improvement as follows:

Average behavior: the average behavior of a particle like X_i^k during a learning period from time t_1 to t_2 is the average fitness of its observed positions during the period, which will be denoted by $FM(X_i^k, t_1, t_2)$:

$$FM\left(X_i^k, t_1, t_2\right) = \frac{1}{\xi}\sum_{t=t_1}^{t=t_2} f\left(x_i^k(t)\right). \tag{4.12}$$

Average improvement: the average improvement of a particle like X_i^k during a learning period from time t_1 to t_2 is the amount of improvement in the particle's average behavior during the period, which will be denoted by $D(X_i^k, t_1, t_2)$:

$$D\left(X_i^k, t_1, t_2\right) = f\left(sb_i^k(t_1 - 1)\right) - FM_i^k(t_1, t_2). \tag{4.13}$$

Here, sb_i^k represents the second-best position of X_i^k. It should be noted that the second-best learning operator changes the position of each particle to its second-best location at the start of each learning period (Fig. 4.1). These two defined criteria can be used to measure the performance of a particle. In order to evaluate the effectiveness of the selected actions, the performance of their associated particles is compared with the performance of other particles in the closest nearby swarms; the set of these particles will be defined as follows:

$$NX^k = \left\{X_j^l | l \in \{N(1,k), N(2,k)\}\right\}. \tag{4.14}$$

Finally, in the CLA approach, each learning automaton learns from the experiences of other nearby entities. For this purpose, we wish to encourage the learning automata to select the most promising actions taken in their neighborhoods. A promising action will be defined as follows:

$$a_*^k(t_1, t_2) = a\left(\arg \max_{X_j^l \in NX^k} \left(D\left(X_j^l, t_1, t_2\right)\right), t_1\right), \tag{4.15}$$

where $a(X_i^j, t_1) = a_i^j(t_1)$ is the selected action of the ith learning automaton in the jth cell. Based on these definitions, the reinforcement signal for a particle like X_i^k will be defined as follows:

$$\beta_i^k = \begin{cases} 0\ if \begin{pmatrix} D\left(X_i^k, t_1, t_2\right) > median\left(\left\{D(X, t_1, t_2) | X \in NX^k\right\}\right) and \\ FM\left(X_i^k, t_1, t_2\right) > median\left(\left\{FM(X, t_1, t_2) | X \in NX^k\right\}\right) \end{pmatrix} \\ \qquad or\left(a\left(X_i^k, t_1\right) = a_*^k(t_1, t_2)\ and\ D\left(X_i^k, t_1, t_2\right) > 0\right) \\ 1\ otherwise \end{cases}. \tag{4.16}$$

Here, the 'median' is the median statistical operation. The reinforcement signals defined using Eq. 4.16 utilizes both personal experience of a learning automaton as well as the social experience of its nearby learning automata.

General pseudo-code for the described CLAMS algorithm consisting of all of its building blocks is given in Fig. 4.4.

4.4.2 Experimental Results

In order to evaluate the performance of CLAMS, it is examined on the CEC2013 benchmark set (Suganthan et al. 2014). CEC2013 presents a comprehensive set of optimization problems, which are useful in the verification of optimization approaches. Several nature-inspired optimization methods are chosen for comparison with CLAMS. A brief description of these methods is given in Table 4.2.

Like other stochastic optimization methods, the CLAMS has some parameters which require appropriate tuning. These parameters are listed in Table 4.3. Some of these parameters can be set intuitively, while some others can be chosen from the related literature. Also, the CLAMS has some specific parameters like ξ and χ which require suitable adjustment. Imperial analysis of different settings for these parameters is provided in (Vafashoar and Meybodi 2018).

In this section, the CLAMS method based on Eq. 4.10 is referred to as CLAMS-I, and the one which is based on Eq. 4.11 is referred to as CLAMS-II.

Algorithm 4-2. CLA based multi swarm algorithm (CLAMS)

Input:

γ: Maximum number of neighboring swarms.

M: number of swarms.

N: number of particles in each swarm.

D: problem dimension

Initialization:

- Initialize the position and velocity of each particle randomly in the allowable domains.
- Set the current location of each particle as its personal best.
- Select the particle with the highest fitness value in each swarm as the global best of that swarm.
- Initialize a CLA with M cells, each residing N learning automata. Each LA has $\gamma+1$ actions corresponding to $\gamma+1$ allowable topologies.
- Set $t = 0$.

Repeat until a termination condition is satisfied:

1. Action selection: Each learning automaton selects an action based on its internal probability vector. Assume $a_i^j(t+1)$ represents the selected action for the ith learning automaton in the jth cell.
2. Consider $a_i^j(t+2) = a_i^j(t+3) = ... = a_i^j(t+\xi) = a(X_i^j, t+1) = a_i^j(t+1)$.
3. Perform Second best learning: Set le current position of each particle to its second-best position.
4. Repeat for ξ iterations
    1. Set $t = t + 1$.
    2. Update the velocity of each particle using Eq. 4-10 (or 4-11).
    3. Update the position of each particle, according to Eq. 4-7.
    4. Evaluate each obtained position, and update the personal best and second-best positions of each particle
    5. Update the global best position of each swarm.
    6. For each swarm, k performs the second-best mutation, as illustrated in Fig. 4-3.
5. Using Eq. 4-16, generate a reinforcement signal for each learning automaton.
6. Using the obtained signals, update each learning automaton according to the learning algorithm.

Fig. 4.4 A pseudo-code for CLAMS

4.4.2.1 Comparison Results

As suggested in (Suganthan et al. 2014), each run of an algorithm on a benchmark is repeated 51 times, and its result is recorded after a maximum number of 1E4D fitness evaluations. Figures 4.5, 4.6, 4.7 and 4.8 depict the Friedman ranks of the compared methods on each of the tested problems. Also, the average Freidman rank of each method provided in Fig. 4.9.

Considering the obtained results, none of the compared methods are superior to others on all of the tested problems. However, the two CLAMS methods are very competitive with the best-performing ones on most of the benchmarks. Also, considering the average Freidman ranks in Fig. 4.9, CLAMS-II has the best performance, and CLAMS-I comes in the second place. According to the average ranks, we can sort the algorithms in the following order: CLAMS-II, CLAMS-I, CooA, NGDE/rand/1, MEABC, F_k-PSO, ABC-SPSO, CoFFWA, and IVbBoPSO.

On the unimodal benchmarks f_1–f_5, NGDE has the best performance and obtained the smallest average errors on four problems. On these benchmarks, the two CLAMS approaches are very competitive with NGDE. CLAMS-II obtained the best average results on two problems and the second-best averages on the reminder. Also,

Table 4.2 Compared stochastic optimization approaches for CEC2013

Algorithm	A brief description
NGDE/rand/1 (Cai et al. 2017)	In the neighborhood guided DE (NGDE), a neighborhood guided selection (NGS) is introduced to guide the mutation process by extracting the promising search directions
COOA (Feng et al. 2015)	A novel creativity-oriented optimization model (COOM) and algorithm (COOA) inspired by the creative thinking process
IVbBoPSO (Gunasundari et al. 2016)	Improved Velocity Bounded Boolean Particle Swarm Optimization (IVbBoPSO) that introduces a velocity min parameter
CoFFWA (Zheng et al. 2017)	A cooperative framework for fireworks algorithm (CoFFWA), which can greatly enhance the exploitation ability of non-core fireworks by using independent selection operator. It also increases the exploration capacity by crowdedness-avoiding cooperative strategy among the fireworks
MEABC (Wang et al. 2014)	A novel multi-strategy ensemble artificial bee colony (MEABC) algorithm in which a pool of distinct solution search strategies coexists throughout the search process and competes to produce offspring
ABC-SPSO (El-Abd 2013)	A hybrid PSO and ABC, the PSO algorithm is augmented with an ABC method to improve the personal bests of the particles
fk-PSO (Nepomuceno and Engelbrecht 2013)	A dynamic Heterogeneous particle swarm optimizer which allows particles to use different update equations referred to as behaviors. fk-PSO learns the favorability of each behavior to select the next behavior of a particle

Table 4.3 Parameter settings for CLAMS

Parameter	Description	Value
M	Number of swarms (Number of cells in the CLA)	14
N	Number of particles in each swarm (Number of learning automata in each cell)	3
$\lambda + 1$	Number of actions in a learning automaton	5
χ	Acceleration constant	3
ξ	Learning period	7
w	inertia weight	$w \in [0.4, 0.9]$
a	reward parameter for LA	0.2
b	Penalty parameter for LA	0.02

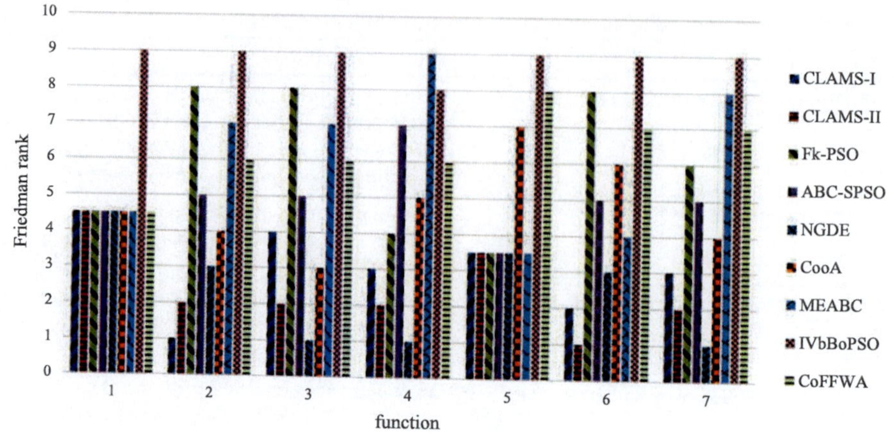

Fig. 4.5 Freidman ranks of the compared methods on the CEC2013 benchmark set, for benchmarks f1–f7

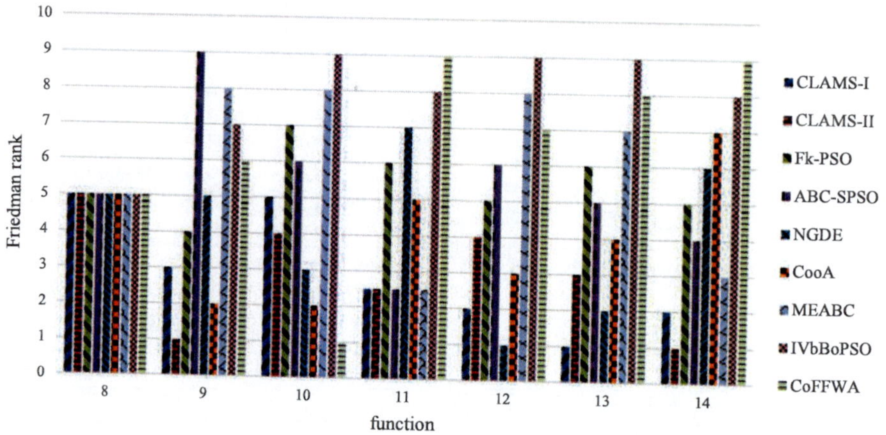

Fig. 4.6 Freidman ranks of the compared methods on the CEC2013 benchmark set, for benchmarks f_8–f_{14}

CLAMS-I possesses the best average on three problems of this group. The introduced CLAMS-II is the best performing method on the basic multimodal problems, f_6–f_{20}, while CLAMS-I comes in the second place. CLAMS-II has the smallest rank on seven problems of this category, while CLAMS-I achieved the smallest ranks in 3 cases. CooA obtained the third average rank on this group of problems. It obtained the lowest average errors on the composition problems, f_{21}–f_{28}, where it possesses the best ranks in 5 cases. However, the results of CLAMS-II are very close, and it achieved the second overall ranking on the composition problems.

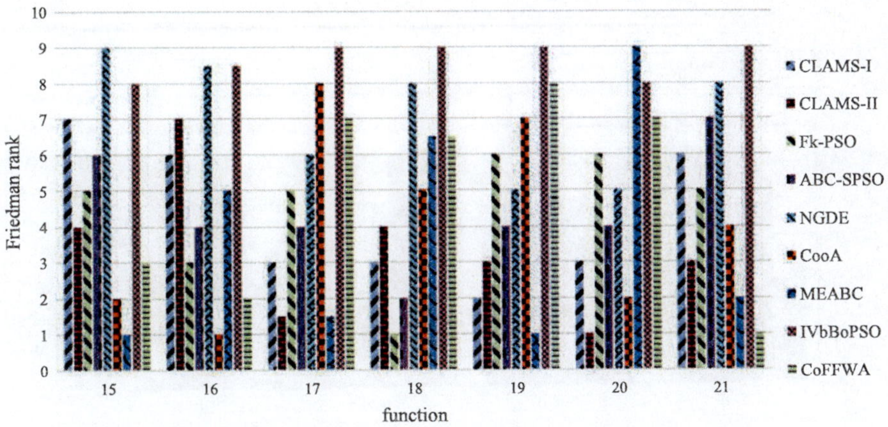

Fig. 4.7 Freidman ranks of the compared methods on the CEC2013 benchmark set, for benchmarks f_{15}–f_{21}

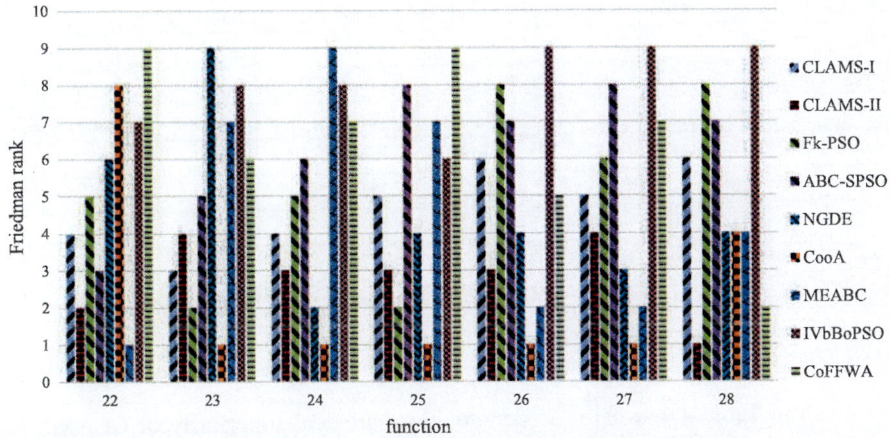

Fig. 4.8 Freidman ranks of the compared methods on the CEC2013 benchmark set, for benchmarks f_{15}–f_{21}

4.4.2.2 A Study on the Learning Capabilities of CLAMS

In order to exhibit the impact of the learning scheme on the performance of CLAMS, CLAMS is statistically compared with its pure chance version. In the pure chance version, the learning automata are disabled. Consequently, each particle chooses its topology in a completely random manner. The Wilcoxon rank-sum test for a confidence level of 95% is performed on each benchmark (Demšar 2006). Wilcoxon test is a nonparametric alternative to the paired t-test and is based on the null hypothesis that the two compared algorithms are statistically equivalent. The results of the test

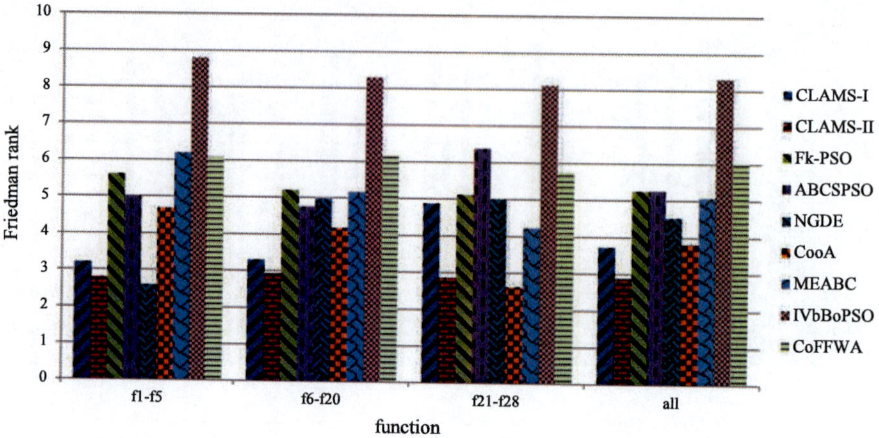

Fig. 4.9 Average Freidman ranks of the compared methods on the CEC2013 benchmark set

Table 4.4 Wilcoxon rank-sum test results on CLAMS-II and its pure chance variant

f1	f2	f3	f4	f5	f6	f7
(=)NAN	(+)7E − 12	(=)4E − 01	(+)4E − 17	(=)NAN	(+)5E − 16	(+)2E − 07
f8	f9	f10	f11	f12	f13	f14
(=)9E − 01	(+)3E − 02	(=)8E − 01	(=)NAN	(+)3E − 14	(+)8E − 11	(+)4E − 06
f15	f16	f17	f18	f19	f20	f21
(+)4E − 12	(=)2E − 01	(+)4E − 18	(+)4E − 17	(+)7E − 15	(+)8E − 10	(+)5E − 04
f22	f23	f24	f25	f26	f27	f28
(+)5E − 10	(+)2E − 10	(=)7E − 01	(+)5E − 02	(+)5E − 17	(+)4E − 04	(+)3E − 03

The p-values of the tests are provided for each benchmark in CEC2013

are given in Table 4.4, where '+' denotes the statistical superiority of CLAMS to its pure chance version, '−' denotes the opposite case, and '=' is used if the two versions of CLAMS are statistically indistinguishable.

From the obtained results, it is clear that the introduced CLAMS-II is statistically different from its pure chance version. Also, it can be seen that CLAMS-II is statistically superior to the pure chance CALMS on most of the tested benchmarks. There are only eight tested problems that CLAMS is statistically indistinguishable from its pure chance version. Additionally, considering Figs. 4.5, 4.6, 4.7 and 4.8, almost every ordinary optimization method performs well on some of these eight problems. These observations indicate the effectiveness of the CLA based learning method on different optimization problems.

4.5 Multi-swarm Bare Bones Particle Swarm Optimization with Distribution Adaption

The updating distributions of BBPSO have a significant impact on its performance. As a result, this section investigates the use of different techniques for determining these updating distributions. Four strategies are developed that are based on Gaussian or multivariate Gaussian distributions. Each strategy can aid the search process at different stages depending on the shape of the landscape where the particles reside. The choice of an appropriate updating strategy for the particles greatly depends on the characteristics of the fitness landscape that surrounds them. For adaptive learning and choosing suitable updating strategies for the particles, the cellular learning automata approach can be utilized. Through the interactions among its elements and the learning capabilities of its learning automata, cellular learning automata gradually learns to select the best updating rules for the particles based on their surrounding fitness landscape.

4.5.1 CLA Based Multi Swarm Bare-Bones Optimization Method (CLA-BBPSO)

After its initial development, several variants of BBPSO have been introduced in the literature. Various probability distributions or strategies are investigated for updating particle in BBPSO. Some strategies exhibit better exploration characteristics, while some others are suitable for exploitation. In order to obtain better results, like other swarm approaches, BBPSO requires an appropriate balance between its exploitation and exploration skills. However, achieving this issue is not an easy task, as it is highly dependent on the problem landscape and the state of the search procedure. Accordingly, an appropriate updating strategy depends on the characteristics of the fitness landscape and the positions of particles. In CLA-BBPSO, four updating strategies with different properties are introduced. Also, a CLA is utilized for adaptively choosing an appropriate strategy for each particle during the search procedure.

In the CLA-BBPSO, the entire population of particles is partitioned into several swarms. Each swarm resides in one cell of a CLA. These cells are connected in a ring topology (Fig. 4.10). Every cell of the CLA has a neighborhood radius of 1, and the immediate left and right cells (besides itself) in the ring constitute its neighbors. Each cell also contains a set of learning automata. Each LA has four actions, corresponding to four updating rules of CLA-BBPSO, and is associated with a particular particle in the cell. The learning automata govern the strategic application of their related particles. A general block diagram for CLA-BBPSO is given in Fig. 4.11, and its details are described in the following sections. Before continuing, Table 4.5 summarizes the notations that will be used in the description of CLA-BBPSO.

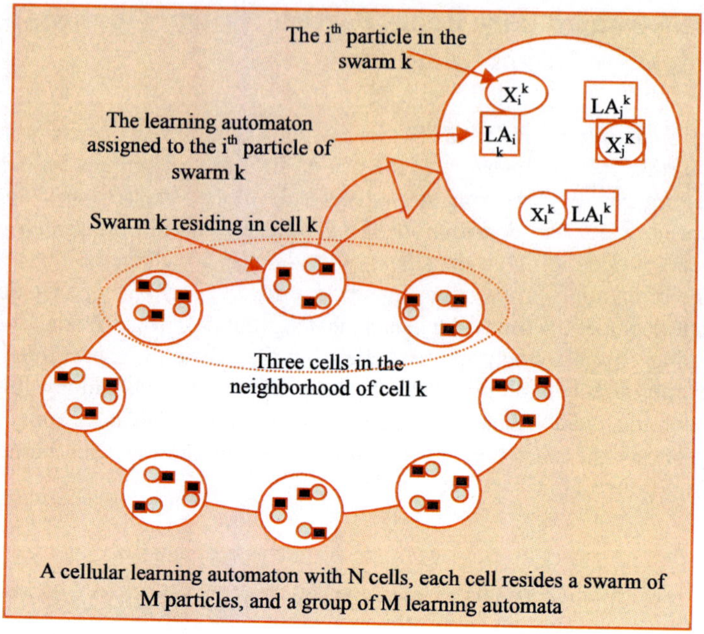

Fig. 4.10 Topology of CLA-BBPSO

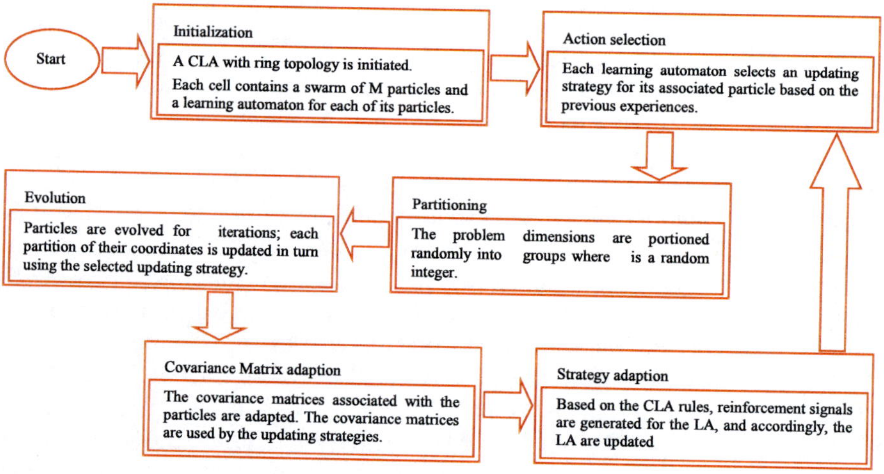

Fig. 4.11 A block diagram for CLA-BBPSO

4.5.1.1 Action Selection and Partitioning Phase

The evolutionary cycle of CLA-BBPSO starts with the action selection phase. During this phase, each learning automaton selects an action based on its internal probability

Table 4.5 A brief definition of notations used in the CLA-BBPSO

Notation	Description
X_i^k	The *i*th particle in the *k*th swarm
$p_i^k(t)$	Personal best position of X_i^k at time *t*
$p_{id}^k(t)$	The value of $p_i^k(t)$ in the *d*th dimension
$p_{iI}^k(t)$	The value of $p_i^k(t)$ in dimensions presented by *I*, where $I \subseteq \{1, 2, \ldots, D\}$
$g^k(t)$	the index of the best particle in the *k*th swarm
M	The size of each swarm
$x_i^k(t)$	The current position of X_i^k at time *t*
$p_g^k(t)$	The global best position of the *k*th swarm
$\theta(k, j)$	the *j*th neighbor of the *k*th cell (or swarm). $j = 0, 1, 2$ respectively refer to the *k*th cell (swarm), its left cell (swarm), and right cell (swarm)
$I_i^k(t)$	The partition of dimensions which is used by X_i^k at time *t*
D	Problem dimension
LA_i^k	The learning automaton associated with X_i^k

vector. Each selected action corresponds to one updating strategy (updating strategies of CLA-BBPSO will be described in Sect. 4.1.2) that will be applied to the particle associated with the corresponding learning automaton. In the beginning, all actions of a learning automaton have equal probabilities. As the search proceeds, the learning automata adjust their action probabilities in the strategy adaption phase of the algorithm, and accordingly, more appropriate strategies can be selected for the particles.

The next phase of the algorithm involves partitioning the coordinates of each particle. The coordinates associated with each partition will be evolved separately using the selected strategy for the particle. Accordingly, if the coordinates of a particle are divided into χ partitions, the particle will be updated for χ times during the next evolutionary phase. As we intend to evolve all of the particles for the same number of times during each evolutionary phase (to attain a fair comparison of their performances), the coordinates of each particle are partitioned in the same manner as other particles. To achieve this objective, first, *D* is partitioned into χ random summands (*D*, problem dimensionality, is an integer number):

$$D = \sum_{i=1}^{\chi} a_i,$$
$$\text{where } a_i = 2^k,\ k = \{1, 2, \ldots, MAXP\};\ \text{and } a_\chi \in \mathbb{Z}^+ - \{1\}. \tag{4.17}$$

Here *MAXP* is a parameter controlling the maximum size of each partition. As the summands are generated randomly, their number (χ) may vary in different partitioning phases. Using these ordered summands, the coordinates of each particle are partitioned randomly into χ groups where the size of the ith group equals to the ith summand (a_i):

$$\bigcup_{j=1}^{\chi} I_i^k(t+j) = \{1, 2, \ldots, D\}, \quad \text{where } \left|I_i^k(t+j)\right| = a_i. \tag{4.18}$$

Here $I_i^k(t+j)$ is the jth partition of coordinates for the ith particle of swarm k, which will be used during the $(t+j)$th iteration of the evolution phase.

4.5.1.2 Updating Rules of CLA-BBPSO

Four updating strategies are developed for CLA-BBPSO. The strategies have different characteristics that enhance various aspects of the search procedure. The first strategy is the classical update rule of the BBPSO algorithm. It uses a Gaussian distribution, which is based on the information present in the neighborhood of the particles.

Operator A, Gaussian operator:

If $d \in I_i^k(t)$:

$$x_{id}^k(t) = \begin{cases} \left(p_{id}^k(t-1) + p_{gd}^k(t-1)\right)/2 + N(0,1) \times \left|p_{gd}^k(t-1) - p_{id}^k(t-1)\right| & if \ (i,k) \neq \left(g^k(t-1), k\right) \\ \left(p_{id}^k(t-1) + p_{gd}^{\theta(k,r)}(t-1)\right)/2 + N(0,1) \times \left|p_{gd}^{\theta(k,r)}(t-1) - p_{id}^k(t-1)\right| & \text{otherwise} \end{cases} \tag{4.19}$$

If $d \notin I_i^k(t)$:

$$x_{id}^k(t) = p_{id}^k(t-1).$$

This operator alters the dth dimension of a particle using a random number drawn from a Gaussian distribution. In order to explore the promising areas out of the swarm, $(p_{id}^k(t-1) + p_{gd}^{\theta(k,r)}(t-1))/2$ and $|p_{id}^k(t-1) + p_{gd}^{\theta(k,r)}(t-1)|$ are used as the parameters of the Gaussian distribution for updating the global best particle of each swarm k. Here, $r \in \{1,2\}$ is a random number identifying a random neighboring swarm. For the other particles of the swarm, $(p_{id}^k(t-1) + p_{gd}^k(t-1))/2$ and $|p_{id}^k(t-1) - p_{gd}^k(t-1))|$ are used as the mean and standard deviation of the Gaussian distribution. Therefore, these particles use their intra-swarm information to conduct their movements. Accordingly, they can exploit the promising areas covered by their swarm in a better way. In order to prevent drastic changes and utilize previously learned information, a particle is only updated in limited dimensions. $I(t)$ identifies the dimensions of a particle to be updated during the iteration t.

The generated position of a particle in a dimension like d may fall out of the search range. In this case, CLA-BBPSO resamples the position of the particle in that dimension (by using operator A) until it sticks inside the search range.

Operator B, multivariate Gaussian operator with covariance matrix Σ:

$$x_i^k(t) = \begin{cases} N\left(\left(p_i^k(t-1) + p_g^k\right)/2, \sum_i^k\right) & \text{if}(i,k) \neq \left(g^k(t-1), k\right) \\ N\left(\left(p_i^k(t-1) + p_g^{\theta(k,r)}(t-1)\right)/2, \sum_i^k\right) & \text{otherwise} \end{cases}. \tag{4.20}$$

Operator A changes the value of each variable in the position vector independent from the values of other variables. However, it is useful to consider the existent interdependencies among the problem variables. Covariance is a measure that represents how much two random variables change together. Also, a multivariate Gaussian distribution utilizing a covariance matrix can represent the dependencies among the problem variables. Motivated by this idea, some evolutionary algorithms express the correlations among problem variables through covariance matrices (Hansen and Ostermeier 1996, 2001; Suttorp et al. 2009). Operator B changes the position of a particle based on the multivariate Gaussian distribution with mean μ and covariance matrix Σ. Whenever a particle is not the best particle of its swarm, $(p_i^k + p_g^k)/2$ is used as the mean of its associated distribution; otherwise, $(p_i^k + p_g^{\theta(k,r)})/2$ is used (similar to operator A). The covariance matrix associated with a particular particle is adapted during the search procedure (which will be discussed later).

Re-generating a vector for handling out of range search may be computationally inefficient; therefore, the method presented in (Li et al. 2012) will be used when operator B results in out of range positions:

$$x_{id}^k(t) = \begin{cases} x_{id}^k(t) & \text{if } x_{id}^k(t) \in [x_{d,\min}, x_{d,\max}] \\ u\left(p_{id}^k(t-1), x_{d,\max}\right) & \text{if } x_{id}^k(t) > x_{d,\max} \\ u\left(x_{d,\min}, p_{id}^k(t-1)\right) & \text{if } x_{id}^k(t) < x_{d,\min} \end{cases}, \tag{4.21}$$

where $u(a, b)$ generates a random number between a and b.

Operator C, multivariate Gaussian operator with partial covariance matrix:

Operator B resamples all of the coordinates of a personal best position to generate a new position. Therefore, some promising information may be lost. In order to relax this issue, Operator C is defined to affect only some coordinates. Let $I = I_i^k(t) \subseteq 1, 2, 3, \ldots, D\}$ represent these coordinates, and let Σ_{iI}^k be the projection of Σ_i^k into these coordinates; then, operator C defines a new position as:

If $d \in I_i^k(t)$:

$$x_{iI}^k(t) = \begin{cases} N\left(\left(p_{iI}^k(t-1) + p_{gI}^k\right)/2, \sum_{iI}^k\right) & \text{if } (i,k) \neq \left(g^k(t-1), k\right) \\ N\left(\left(p_{iI}^k(t-1) + p_{gI}^{\theta(k,r)}(t-1)\right)/2, \sum_{iI}^k\right) & \text{otherwise} \end{cases} \tag{4.22}$$

If $d \notin I_i^k(t)$:
$x_{id}^k(t) = p_{id}^k(t-1).$

Operator C handles the out of range search in the same way as Operator B.

Operator D, multivariate Gaussian operator with partial sample covariance matrix:

Estimation of distribution algorithms uses explicit distribution models like multivariate Gaussian distribution for sampling new points. Then, employing the promising newly generated points, they refine their probabilistic models. Inspiring from this idea, operator D uses the sample covariance matrix and the sample mean in the Multivariate Gaussian distribution (Let $I = I_i^k(t)$):

If $d \in I_i^k(t)$:

$$x_{iI}^k(t) = N\left(\mu_{iI}^k, \widehat{\Sigma}_{iI}^k\right) \tag{4.23}$$

If $d \notin I_i^k(t)$:
$x_{id}^k(t) = p_{id}k(t-1),$
where the sample means and the sample covariance matrices are defined as follows:

$$\mu_{iI}^k(t) = \begin{cases} \frac{1}{M}\sum_{j=1}^{M} p_{jI}^k(t-1) & \text{if } (i,k) \neq \left(g^k(t-1), k\right) \\ \frac{1}{3}\sum_{j=0}^{2} p_{gI}^{\theta(k,j)}(t-1) & \text{otherwise} \end{cases}, \tag{4.24}$$

$$\widehat{\Sigma}_{iI}^k = \begin{cases} \frac{1}{M}\sum_{j=1}^{M}\left(p_{jI}^k(t-1) - \mu_{iI}^k(t)\right)\left(p_{jI}^k(t-1) - \mu_{iI}^k(t)\right)^T & \text{if } (i,k) \neq \left(g^k(t-1), k\right) \\ \frac{1}{3}\sum_{j=0}^{2}\left(p_{gI}^{\theta(k,j)}(t-1) - \mu_{iI}^k(t)\right)\left(p_{gI}^{\theta(k,j)}(t-1) - \mu_{iI}^k(t)\right)^T & \text{otherwise} \end{cases} \tag{4.25}$$

4.5.1.3 Covariance Matrix Adaption

Each particle has a covariance matrix incorporated with it, which is evolved alongside the particle itself. Operators B and C use this covariance matrix in order to direct the particle in the landscape. Hansen and his colleagues proposed some covariance matrix adaptation mechanisms for evolution strategies (Hansen and Ostermeier 1996, 2001; Suttorp et al. 2009). They introduced rank-1 and rank-μ update rules to ensure a reliable estimator. At each generation, the information from the previous generations is used to refine the covariance matrix. Rank 1 uses the concept of an evolutionary path which is the path taken by the population over several generations, and is defined as follows:

$$p_c^{(g+1)} = (1 - c_c)p_c^{(g)} + \sqrt{c_c(2 - c_c)\mu_{eff}}\frac{m^{(g+1)} - m^{(g)}}{\sigma^{(g)}}, \tag{4.26}$$

where c_c is the backward time horizon, and $p_c^{(0)} = 0$; $m^{(g)}$ is the weighted mean of the best-selected samples at generation g. rank-μ update rule uses an estimation for the distribution of the selected steps along with the previous information about the covariance matrix:

$$C^{(g+1)} = (1 - c_{\text{cov}})C^{(g)} + c_{\text{cov}}\sum w_i y_i^{(g+1)} y_i^{(g+1)^T}, \tag{4.27}$$

where $y_i^{(g+1)} = (x_i^{(g+1)} - m^{(g)})/\sigma^{(g)}$, and w_i is the weight of the selected sample; σ represents the step size, which is also updated at each generation. Using these rules together, the overall update rule can be like the following:

$$\begin{aligned} C^{(g+1)} &= (1 - c_{\text{cov}})C^{(g)} + c_{\text{cov}}\left(1 - \frac{1}{\mu_{\text{cov}}}\right)\sum w_i y_i^{(g+1)} y_i^{(g+1)^T} \\ &+ \frac{c_{\text{cov}}}{\mu_{\text{cov}}} p_c^{(g+1)} p_c^{(g+1)^T}, \end{aligned} \tag{4.28}$$

BBPSO with scale matrix adaptation (SMA-BBPSO) uses multivariate t-distribution to update the position of particles (Campos et al. 2014). It employs a simple method for adapting the scale matrix of the t-distribution. At each iteration, the scale matrix related to the kth particle is updated using the following equation:

$$\sum_k = (1 - \beta)\sum_k + \beta n_k n_k^T, \tag{4.29}$$

where β is the learning rate controlling the portion of the scale matrix to be inherited from the experience, and n_k is the best position in the neighborhood of the kth particle. The idea is simple and is based on sampling the best-found positions into the scale matrix.

Employing the previous concepts, CLA-BBPSO uses two mechanisms to adapt the covariance matrix of each particle. First, it utilizes the evolution direction of the particle. This concept resembles the concept of evolution path. The evolution direction (κ) of particle X_i^k will be defined as follows:

$$\kappa_i^k(t) = \left(p_i^k(t) - p_i^k(t')\right)\left(p_i^k(t) - p_i^k(t')\right)^T, \tag{4.30}$$

where t and t', respectively, represent the current iteration number and the iteration number during the previous covariance matrix adaption phase. Accordingly, $p_i^k(t)$ is the current personal best location of the particle X_i^k, and $p_i^k(t')$ is its personal best position at the time of the previous covariance matrix adaption phase. Evolution direction samples the last improvement of a particle. Like Eq. (4.29), we also sample the best nearby position into the covariance matrices. However, considering the sample covariance matrix definition, which is given in Eq. (4.25), one can interpret the sample covariance as the average dependencies of the data samples concerning their mean. In this interpretation if we only consider the dependency of the best point, we can obtain:

$$\begin{aligned}&\left(n_i^k(t) - m\right)\left(n_i^k(t) - m\right)^T,\\&\text{where } n_i^k(t) = \arg\min(f(p)),\ p \in \left\{p_j^l(t) \| l \in \theta(k,s), s = 0, 1, 2\right\} - p_i^k(t).\end{aligned} \tag{4.31}$$

Here $n_i^k(t)$ is the best position in the neighborhood of the particle X_i^k. m is the mean of the distribution, and as it is unknown, it should somehow be estimated. We, simply, will use $p_i^k(t)$ as the mean of the distribution. In this way, the best nearby dependency concerning $p_i^k(t)$ will be sampled into the covariance matrix. Using this interpretation, if X_i^k happens to be the best particle in its neighborhood, then the above equation evaluates to zero. Therefore, we defined $n_i^k(t)$ as the best nearby position other than $p_i^k(t)$ itself. Equation (4.31) is very similar to Eq. (4.29); though, in Eq. (4.29), m is missing, which is similar to assuming a zero-value mean. As we shall see in the experimental section, assuming $m = 0$ may be troublesome on shifted test functions that have their optimum in a location other than the origin. Based on the ideas as mentioned above, the overall equation for covariance matrix adaption in CLA-BBPSO is:

$$\sum_i^k(t) = (1-\beta)\sum_i^k(t') + \beta S_i^k, \tag{4.32}$$

where

$$S_i^k = \frac{1}{2}\left(p_i^k(t) - p_i^k(t')\right)\left(p_i^k(t) - p_i^k(t')\right)^T + \frac{1}{2}\left(n_i^k(t) - p_i^k(t)\right)\left(n_i^k(t) - p_i^k(t)\right)^T. \tag{4.33}$$

4.5.2 Strategy Adaption

After the action selection phase, each particle is evolved for a period of several iterations. Afterward, the strategy adaption phase is applied to adjust the action probabilities in favor of the most promising ones. The favorability of a selected action for a learning automaton is measured based on the comparison of its outcome with the outcomes of the selected actions in the adjacent cells. The outcome of an action will be measured according to the amount of improvement in the fitness value of the associated particle, which is defined as follows:

$$\psi_i^k = \frac{1}{f\left(p_i^k(t)\right)}\left(f\left(p_i^k(t')\right) - f\left(p_i^k(t)\right)\right), \tag{4.34}$$

where t is the current iteration number, and t' is the iteration number in the previous strategy adaption phase. Accordingly, $f(p_i^k(t))$ and $f(p_i^k(t'))$ represent the fitness of the particle X_i^k, respectively, in the current and previous adaption phases. Based on the improvement values of the particles in its adjacent cells, the reinforce signal for a learning automaton like LA_i^k is obtained as follows:

$$\beta_i^k = \begin{cases} 0 \text{ if } \psi_i^k > median\left(\left\{\psi_j^l | l = \theta(k,h), h = 1,2\right\}\right) \\ 1 \text{ otherwise} \end{cases}. \tag{4.35}$$

Here *median* returns the statistical median of its argument set. Based on the received reinforcement signal, each learning automaton updates its internal action probabilities using a learning algorithm. This definition of reinforcement signals has some advantages. On average, the promising actions of about half of the learning automata are rewarded. Hence, these actions can be selected with higher probabilities next time. The selected actions of the other half are panelized; therefore, these learning automata will explore other actions (other operators) with higher probabilities than before. This way, we can ensure that each operator always would have a chance of selection.

After the strategy adaption phase, the algorithm continues a new cycle starting with the action selection phase. The general pseudocode for the proposed model is given in Fig. 4.12.

4.5.3 Experimental Studies

This section demonstrates the effectiveness of CLA-BBPSO through experimental study and comparison. The behavior of the method is studied on a set of benchmark functions (Table 4.6) that are taken from (Leung and Wang 2001; Liang et al. 2006). These benchmarks can be categorized into three groups (Liang et al. 2006): unimodal, unrotated multimodal, and rotated multimodal. Unimodal functions are rather easy to optimize and are mostly used to demonstrate the convergence speed of the algorithms. Unrotated multimodal functions mostly possess a large number of local optima. They provide an excellent means for investigating the ability of optimization algorithms in escaping from poor local optima. One issue with many test functions is their

Algorithm 4-3. CLA-BBPSO

I. Initialization:
   1. Initialize cellular learning with N cells; each cell is composed of a swarm of M randomly generated particles and a group of M learning automata with four actions of equal probability.

II. Set time to zero: $t = 0$.

III. Set: $t' = 0$.

IV. While the termination condition is not satisfied, repeat:
   1. Perform action selection: each LA selects an action based on its internal probability vector. Let Ac_i^k denote the selected action of LA_i^k.
   2. Partition D into random summands according to Eq. (4-17), and let χ denote the number of these summands.
   3. Using the generated summands, partition the coordinates of each particle according to Eq. (4-18). Let $I_i^k(t+j)$, $j=1,\ldots,\chi$ represent the generated partitions for particle X_i^k.
   4. For $t = t+1$ to $t+\chi$ do:
      a. Compute the new position of each particle X_i^k using $I_i^k(t)$ and the updating rule related to Ac_i^k.
      b. Evaluate the new positions and update the personal best position of each particle.
   5. Update the associated covariance matrix of each particle using Eq. (4-32).
   6. Generate reinforcement signals for each learning automata using Eq. (4-35).
   7. Based on the generated reinforcement signals, update each learning automata using the learning algorithm.
   8. Set $t' = t$.

Fig. 4.12 A general pseudocode for CLA-BBPSO

separability, meaning that they can be solved using D (the number of dimensions) independent searches.

To alleviate this problem, we can make a function inseparable through the rotation. To rotate a function $f(x)$, the original vector x is left multiplied with an orthogonal matrix M: $y = M \times x$; then, the newly generated vector y is used to calculate the fitness of the associated individual (Liang et al. 2006). Herein, Salomon's method will be used to generate the orthogonal matrices (Salomon 1996). Most of the benchmarks presented in Table 4.6 have their optimum point in the center of the search space. These benchmarks are also symmetric with the property that their local optima are around their global optimum. These characteristics result in a symmetric distribution of the population in the stochastic optimization algorithms. Therefore, when their individuals are attracted to each other using evolutionary operators, they may accidentally fall near the optimum global point. In order to deal with this issue, we employ the shifted version of the given functions. Consequently, $f(x - \alpha)$ is used instead of $f(x)$ as the fitness function. We set α to 1/10 of the search range. This shifting is only used for the functions with their optimum point at the origin.

In the experiments of this section, the 30-dimensional instances of the benchmark problems are utilized. Also, the parameter settings of CLA-BBPSO are summarized as follows: $MAXP = 1$, Number of swarms $= 10$, number of particles in a swarm $= 4$, reward parameter of a LAs $= 0.2$, penalty parameter of LAs $= 0.02$, $\beta = 0.4$ (controls the portion of the covariance matrix to be inherited from the past experiences Eq. 4.32).

Table 4.6 Classical benchmark problems along with their General attributes

Function	Range	Opt
$f_1 = 418.9829 \times D + \sum_{i=1}^{D}\left(-x_i \sin\left(\sqrt{\lvert x_i \rvert}\right)\right)$	$[-500,500]^D$	0
$f_2 = \sum_{i=1}^{D}\left(-x_i^2 - 10\cos(2\pi x_i) + 10\right)$	$[-5.12,5.12]^D$	0
$f_3(X) = f_2(Z), z_i = \begin{cases} x_i \lvert x_i \rvert < \frac{1}{2} \\ \frac{round(2x_i)}{2} \lvert x_i \rvert \geq \frac{1}{2} \end{cases}$	$[-5.12,5.12]^D$	0
$f_4 = -20\exp\left[-0.2\sqrt{\frac{1}{D}\sum_{i=1}^{D} x_i^2}\right] - \exp\left[\frac{1}{D}\sum_{i=1}^{D}\cos s(2\pi x_i)\right] + 20 + \exp(1)$	$[-32,32]^D$	0
$f_5 = \frac{1}{4000}\sum_{i=1}^{D} x_i^2 - \prod_{i=1}^{D}\cos\left(\frac{x_i}{\sqrt{i}}\right) + 1$	$[-600,600]^D$	0
$f_6 = \frac{\pi}{D}\left\{\begin{matrix} 10\sin^2(\pi y_1) + (y_D - 1)^2 + \\ \sum_{i=1}^{D-1}(y_i - 1)^2\left[1 + 10\sin^2(\pi y_{i+1})\right]\end{matrix}\right\} + \sum_{i=1}^{N} u(x_i, 10, 100, 4)$, $y_i = 1 + \frac{1}{4}(x_i + 1)$ $u(x_i, a, k, m) = \begin{cases} k(x_i - 1)^m & x_i > a \\ 0 & -a \leq x_i \leq a \\ k(-x_i - a)^m & x_i < -a \end{cases}$	$[-50,50]^D$	0
$f_7 = 0.1\left\{\begin{matrix} \sin^2(3\pi x_1) + (x_D - 1)^2\left[1 + \sin^2(2\pi x_D)\right] \\ + \sum_{i=1}^{D-1}(x_i - 1)^2\left[1 + \sin^2(3\pi x_{i+1})\right]\end{matrix}\right\}$ $+ \sum_{i=1}^{D} u(x_i, 10, 100, 4)$	$[-50,50]^D$	0
$f_8 = \sum_{i=1}^{D} x_i^2$	$[-100,100]^D$	0
$f_9 = \sum_{i=1}^{D}\lvert x_i \rvert + \prod_{i=1}^{D}\lvert x_i \rvert$	$[-10,10]^D$	0
$f_{10} = \sum_{i=1}^{D}\left[\sum_{j=1}^{i} x_j\right]^2$	$[-100,100]^D$	0
$f_{11} = \sum_{j=1}^{D-1}\left[100(x_{j+1} - x_j^2)^2 + (x_j - 1)^2\right]$	$[-100,100]^D$	0
$f_{12}: f_2(y), y = M \times x$	$[-5.12,5.12]^D$	0
$f_{13}: f_5(y), y = M \times x$	$[-600,600]^D$	0

(continued)

Table 4.6 (continued)

Function	Range	Opt
f_{14}: $f_4(y)$, $y = M \times x$	$[-32,32]^D$	0
f_{15}: $f_3(y)$, $y = M \times x$	$[-5.12,5.12]^D$	0
f_{16}: $f_1(y)$, $y = M \times x$	$[-500,500]^D$	0

4.5.4 Compared PSO Algorithms

CLA-BBPSO is compared with a group of PSO algorithms whose configurations are summarized in Table 4.7. The results of all algorithms are taken from 30 independent runs. Each run of an algorithm on a benchmark function is terminated after 10^5 fitness function evaluations on a 30-*D* version of the function.

PS^2OS_R resembles CLA-BBPSO in that both are utilizing multiple swarms of particles. In PS^2OS_R, a particle is attracted to three informers: historical best positions of itself and its swarm along with the best historical position in its neighboring swarms. Comprehensive learning particle swarm optimizer (CLPSO) is mainly developed for dealing with complex multimodal functions. Each coordinate of a particle learns comprehensively from the corresponding coordinate of another particle or the corresponding coordinate of the historical best position of itself until it fails to improve for a while. CLPSO updating procedure results in a diversified population, which ensures its high exploration capability; however, the diversified population results in a low convergence rate. In fully informed PSO (FIPS), all the neighbors contribute to the velocity adjustment. The sum of acceleration coefficients is considered to be a constant value, which is equally divided between the acceleration coefficients. Frankenstein PSO (FPSO) starts with a fully connected topology, and gradually decreases the connectivity of the particles. Adaptive PSO (APSO) uses the population distribution and fitness information to adjust the inertia weight and acceleration coefficients.

Table 4.7 Parameter settings for the compared PSO algorithms

Method	Parameter settings
PS^2OS_R (Chen et al. 2010)	$c = 1.3667$, $\chi = 0.729$
CLPSO (Liang et al. 2006)	$c = 1.49445$, $w \in [0.4, 0.9]$
FIPS (Mendes et al. 2004)	$\varphi = 4.1$, $\chi = 0.7298$
FPSO (Montes de Oca et al. 2009)	$\varphi = 4.0$, $w \in [0.4, 0.9]$
APSO (Zhan et al. 2009)	$1.5 < c_1, c_2 < 2.5$ and $c_1 + c_2 < 4$, $w \in [0.4, 0.9]$

4.5.4.1 Experimental Results

Each method is tested on each of the benchmark functions for 30 runs, and the statistical information of the achieved final results on the benchmark functions is used for comparison. Table 4.8 gives this statistical information in terms of average, median, best, and worst on each benchmark.

Observing Table 4.8, it is clear that the CLA-BBPSO has the highest accuracy among all of the compared methods on most of the tested problems. CLA-BBPSO has the best average on 12 benchmarks; while, its average results on the other four benchmarks are very near to the best ones. The obtained results of the purposed method are very distinctive from all of the other peer methods on the benchmarks: f_1, f_2, f_{10}, and f_{11}. Also, only three algorithms were capable of finding the global optimum of f_3, with CLA-BBPSO having the highest average. Although CLA-BBPSO does not possess the best average results on f_4 and f_{14}, its obtained results are very competitive to the best-obtained ones. Also, CLA-BBPSO has the second-best results on f_{12} and f_{16}, and its achieved results are very near to those of FPSO (the best method on these benchmarks).

We finish this section with a discussion on the behavior of CLA-BBPSO according to the obtained results. The adaptability of the algorithm is very beneficial to its performance on problems like f_{11}. The strategy adaption mechanism plays a significant role in the achieved results, as it can select suitable strategies for the particles based on the characteristics of the problems and the fitness information of the swarm. This idea will be evident if we consider the average selection probability of different operators during the search. For problems like f_2, f_3, and f_5, CLA-BBPSO decreases the selection probability of operators B and C. There exists no correlation among the coordinates of these problems, which can be learned into the covariance matrices of the particles. However, CLA-BBPSO increases the probability of operator B on problems like f_{10}, where its coordinates are fully correlated. Also, the algorithm balances the exploration and exploitation characteristics of the search through operator selection; as, depending on the fitness dynamics, some of the proposed operators tend to have more exploration or exploitation characteristics.

4.5.4.2 A Study on the Learning Capabilities of CLA-BBPSO

This section studies the effect of the proposed action selection mechanism on the performance of the algorithm and illustrates its functioning. Figure 4.13 shows how the operator probabilities vary throughout the algorithm's iterations on a group of benchmarks from Table 4.6. Each figure represents the average operator probability in the population over the spent number of fitness evaluations. As can be seen from the figures, each test problem requires operators with different characteristics. For separable problems, the probabilities of operators B and C reduce over time and get lower than the probabilities of operators A and D. This may be because no dependency exists among their coordinates, which can be learned over time. However, the type of dependencies among the coordinates of hyper-ellipsoid function completely fits

Table 4.8 Statistical information (in this order: average, median, best, and worst) of obtained results for the compared algorithms

		CLABBPSO	APSO	CLPSO	FIPS	FPSO	PS2OSR
f1	Average	1.81e−12	1722.92	4.27e−7	4718.4	7998.8	4809.3
	Median	1.81e−12	1598.96	3.24e−7	4730.5	7915.1	4709.7
	Best	1.81e−12	710.63	1.18e−7	3339.8	6619.8	3594.4
	Worst	1.81e−12	3396.83	1.34e−6	5531.8	9226.1	6021.3
f2	Average	0	4.04	2.09e−3	46.63	20.87	39.63
	Median	0	3.97	1.97e−3	43.48	20.68	38.3
	Best	0	0.99	2.92e−4	24.27	12.63	22.88
	Worst	0	7.95	4.97e−3	74.16	28.09	58.7
f3	Average	1.18e−13	6.30e−10	1.32e−2	42.903	25.58	37.4
	Median	3.64e−14	0	1.10e−2	42.95	25	37.5
	Best	0	0	2.70e−3	29.1	20.15	24
	Worst	1.34e−12	1.89e−8	5.46e−2	61.108	41	52
f4	Average	1.32e−14	4.43e−2	9.20e−5	8.23e−15	6.38e−2	7.40e−15
	Median	1.15e−14	1.50e−14	9.48e−5	7.99e−15	6.06e−2	7.99e−15
	Best	7.99e−15	7.99e−15	4.81e−5	7.99e−15	2.41e−2	4.44e−15
	Worst	2.22e−14	1.34	1.39e−4	1.50e−14	1.24e−1	7.99e−15
f5	Average	0	1.33e−2	5.64e−6	5.02e−4	9.64e−1	2.70e−3
	Median	0	7.31e−3	4.45e−6	5.53e−12	9.36e−2	0
	Best	0	0	7.27e−7	0	1.87e−2	0
	Worst	0	9.52e−2	3.95e−5	7.4128e−3	2.69e−1	2.94e−2
f6	Average	1.57e−32	1.03e−2	1.44e−9	7.49e−28	9.81e−5	3.10e−2
	Median	1.57e−32	2.01e−31	1.34e−9	1.57e−32	8.04e−5	1.57e−32
	Best	1.57e−32	3.63e−32	3.61e−10	1.57e−32	9.70e−6	1.57e−32
	Worst	1.57e−32	2.07e−1	2.55−9	2.24e−26	2.56e−4	6.21e−1
f7	Average	1.34e−32	3.29e−3	2.34e−8	1.34e−32	2.36e−3	1.34e−32
	Median	1.34e−32	1.11e−30	1.91e−8	1.34e−32	2.23e−3	1.34e−32
	Best	1.34e−32	3.29e−31	1.00e−8	1.34e−32	4.31e−4	1.34e−32
	Worst	1.34E-32	4.39e−2	5.42e−8	1.34e−32	6.81e−3	1.34e−32
f8	Average	0	6.22e−29	3.80e−8	0	3.88e−3	0
	Median	0	5.04e−29	3.79e−8	0	3.81e−3	0
	Best	0	0	1.49e−8	0	1.03e−2	0
	Worst	0	1.64e−28	8.46e−8	0	1.20e−3	0
f9	Average	0	4.57e−15	6.39e−6	0	7.99e−3	0
	Median	0	9.99e−16	6.65e−6	0	7.33e−3	0
	Best	0	0	3.96e−6	0	2.50e−3	0

(continued)

Table 4.8 (continued)

		CLABBPSO	APSO	CLPSO	FIPS	FPSO	PS2OSR
	Worst	0	8.85e−14	9.45e−6	0	1.77e−3	0
f10	Average	9.15e−7	9.501e−2	2842.7	9.99	154.35	1.67e−3
	Median	2.89e−7	7.1836e−2	2868.8	9.03	149.71	1.08e−3
	Best	3.71e−8	1.56e−2	2001.5	2.1	48.27	1.09e−4
	Worst	9.89e−6	4.48e−1	3917.3	26.285	249.4	5.02e−3
f11	Average	6.2	34.02	47.14	25.85	51.66	17.54
	Median	3.98	21.88	46.96	22.09	36.33	17.91
	Best	3.37e−6	7.21e−1	18.04	18.93	30.32	6.71
	Worst	21.89	175.01	81.61	76.58	350.59	22.73
f12	Average	29.16	50.42	49.18	85.2	23.71	40.89
	Median	29.35	48.75	51.42	87.23	22.65	38.83
	Best	16.92	29.84	27.42	35.88	12.35	19.89
	Worst	44.77	89.54	71.06	134.69	38.09	60.69
f13	Average	1.28e−13	8.94e−3	8.03e−4	5.82e−4	1.08e−1	2.13e−3
	Median	0	8.62e−3	7.54e−4	8.63e−10	8.85e−2	0
	Best	0	0	1.82e−5	1.11e−16	1.62e−2	0
	Worst	2.64e−12	2.46e−2	2.89e−3	9.12e−3	3.59e−1	1.47e−2
f14	Average	1.88e−14	1.53	1.01e−2	7.99e−15	6.05e−2	6.95e−2
	Median	1.86e−14	1.57	8.96e−3	7.99e−15	5.55e−2	7.99e−15
	Best	1.50e−14	7.99e−15	4.21e−3	7.99e−15	2.22e−2	4.44e−15
	Worst	2.93e−14	2.66	2.79e−2	7.99e−15	1.47e−1	1.15
f15	Average	28.64	47.68	44.38	72.77	26.21	46.51
	Median	28.76	46.52	42.16	70.94	26.65	46.59
	Best	19.65	23.39	32.43	45.23	16.12	30.27
	Worst	38	79.07	61.91	104.64	36.01	62.91
f16	Average	2370.52	3614.3	2583.4	7797.6	8543.5	5035.2
	Median	2422.65	3561.9	2606.4	7828.5	8585.5	5087.5
	Best	1578.12	2591.2	1238.1	6715.2	7505.2	3651.3
	Worst	3249.96	5367.1	3235.8	8698.7	9592.3	6789.6

with operator B. Therefore, the probability of operator B is increased over time. For the other two test functions, the probabilities change according to the state of the population at different stages of the search process.

To demonstrate that the algorithm has learning capabilities, it is statistically compared with its pure chance version (actions are selected with equal probability all the time). For this purpose, the Wilcoxon test is used to demonstrate the statistical difference between the two versions. The results of this test are provided in Table 4.9. From Table 4.9, CLA-BBPSO is statistically superior to the pure chance version on

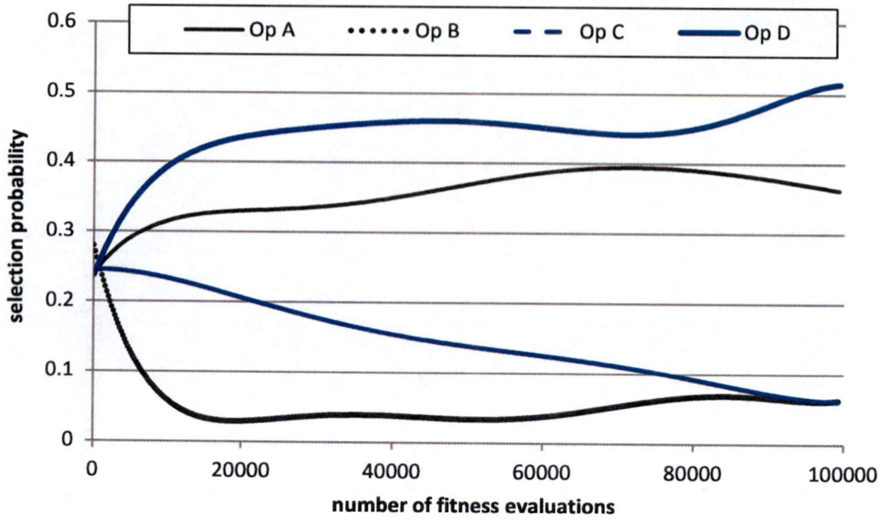

(a)

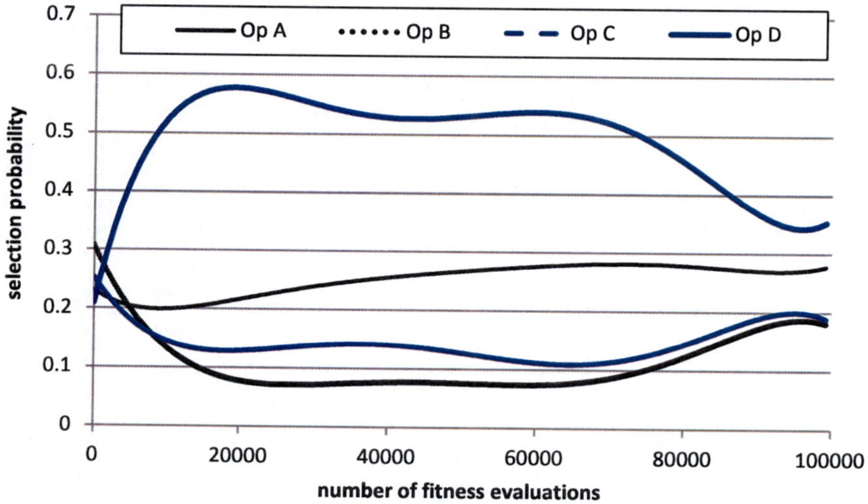

(b)

Fig. 4.13 The average probability of each operator in the population throughout the search process on: f_3 **(a)**, f_5 **(b)**, f_{10} **(c)**, f_{11} **(d)**, f_{12} **(e)**

11 benchmarks and is statistically indistinguishable on the reminder. Besides, in the latter instances, CLA-BBPSO was better than its pure chance version in terms of

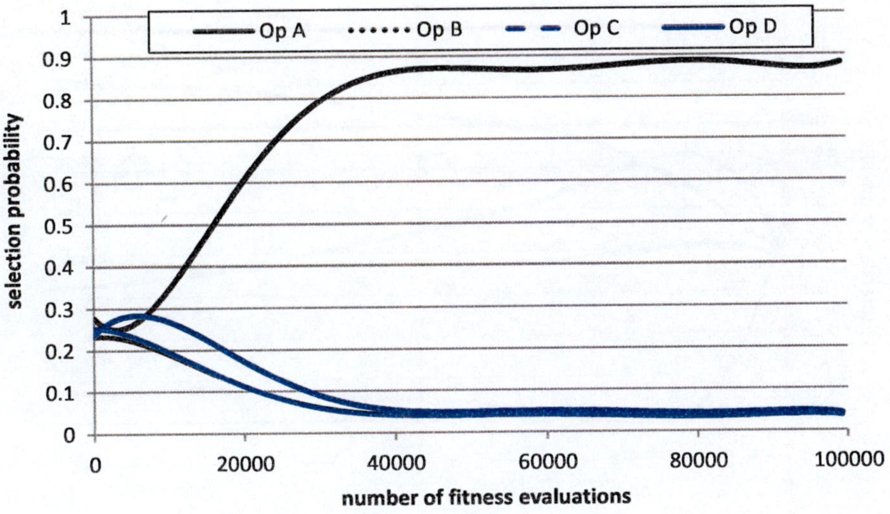

(c)

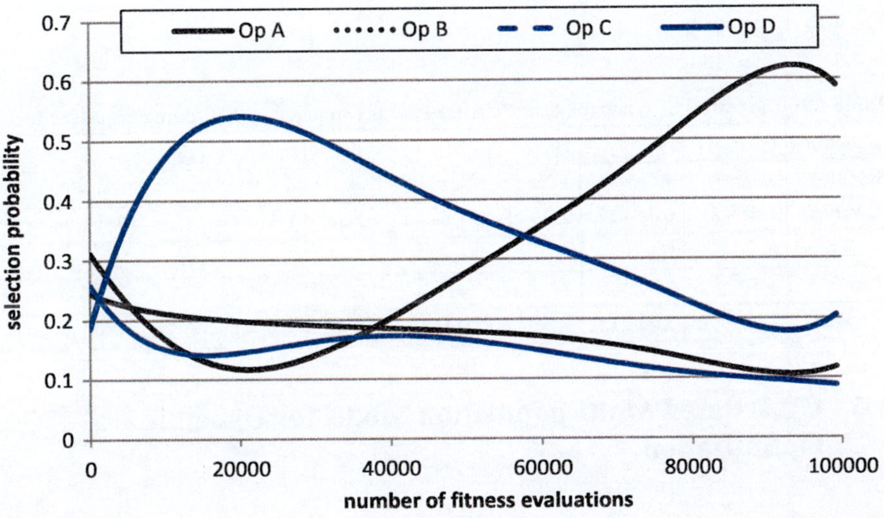

(d)

Fig. 4.13 (continued)

the convergence speed (which would have been clear if we compared them using convergence curves).

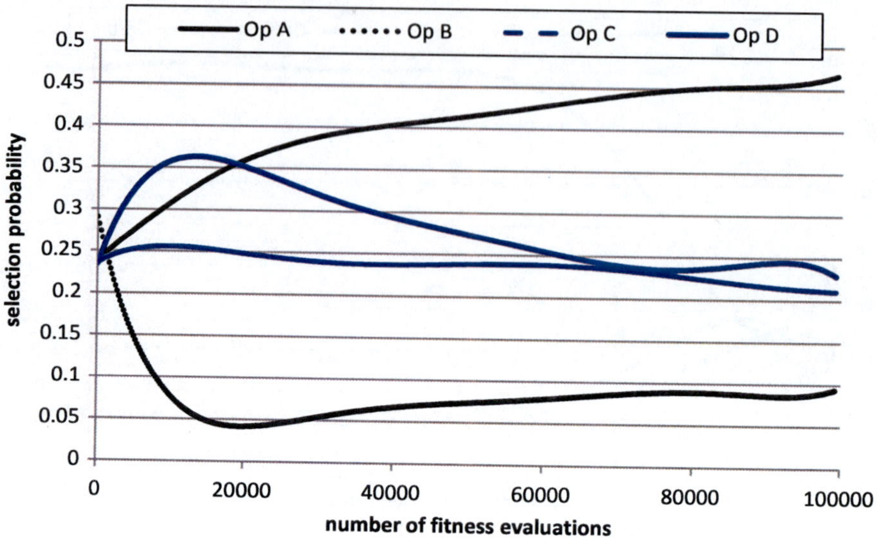

(e)

Fig. 4.13 (continued)

Table 4.9 The results of Wilcoxon test: CLA-BBPSO is compared with its pure chance version

Function	f_1	f_2	f_3	f_4	f_5	f_6	f_7	f_8
Result	+	+	+	+	=	+	+	+
P-value	1.5e−8	6.4e−12	2.9e−11	1.9e−11	NaN	1.5e−11	3.1e−12	1.2e−12
Function	f_9	f_{10}	f_{11}	f_{12}	f_{13}	f_{14}	f_{15}	f_{16}
Result	+	+	+	=	=	+	=	=
P-value	1.2e−12	3.0e−11	1.3e−6	0.66	0.52	2.2e−11	0.54	0.09

4.6 CLA-Based Multi-population Model for Dynamic Optimization

This section investigates the CLA-based method for dynamic optimization, which is presented in (Vafashoar and Meybodi 2020). Like previous sections, CLA is employed for embedding the population elements for adaptively controlling their behavior. Accordingly, the proposed method is based on the same concepts as the ones in previous sections. To demonstrate how these concepts can be employed on various nature-inspired methods, this section considers differential evolution as the based optimization method.

Some optimization problems have a dynamic nature, which means that the problem landscape is changing over time. This changing property in a problem landscape also affects its optimum solutions. Accordingly, an optimization method employed for solving these problems should be able to cope with this dynamicity of the optimal solutions. It should be able to track the optimum solutions of the problem and respond to the environmental changes. Classical stochastic optimization methods like particle swarm optimization and differential evolution algorithm, when employed on static optimization problems, require some degree of population convergence to achieve high precision results. However, total convergence in the case of dynamic optimization problems significantly declines the performance of an optimizer as it cannot respond to the environmental changes effectively under total convergence.

Additionally, an optimizer for dynamic problems requires some degree of local convergence to grantee solutions with appropriate accuracies. Accordingly, maintaining appropriate diversity for effectively responding to the environmental changes, while achieving high-quality solutions, is a major challenge in dynamic optimization problems. Mechanisms like multi-population, anti-convergence, and archival memories are often utilized in stochastic optimization methods when dealing with dynamic environments.

In the presented approach of this section, a cellular learning automaton adjusts the behavior of each subpopulation by adaptively controlling its updating rules. As the environment changes over time, an evolving population may go through a quite number of state transitions. Each state demands specific characteristics from an optimizer; hence, an adapted strategy for one state may be unsuitable for the upcoming ones. Additionally, a learning scheme may have limited time to adapt to a newly encountered state due to the frequentness of the environmental changes. Hence, it is infeasible for a dynamic optimizer to unlearn its existing beliefs to accommodate the practices required to embrace newly encountered states. In order to address this issue, a context-dependent learning scheme is introduced, which can adapt the behavior of each subpopulation according to the contexts of its different states. In each state, a subpopulation exploits its previously learned knowledge for the state and improves this knowledge until a state transition occurs. This scheme allows the adaptation mechanism to adjust each subpopulation to its residing states in a short time. The proposed DE uses an ensemble of three updating strategies, each inducing different behavior on a subpopulation. This strategy ensemble consists of rand/1/bin, which is commonly used in DE literature, and two strategies specifically designed to promote the exploitation and exploration characteristics of the subpopulations.

4.6.1 CLA Based Multi-population Method for Optimization in Dynamic Environments (CLA-MPD)

CLA-MPD is an ensemble approach, which incorporates a cellular learning automaton to adaptively control the behavior of its subpopulations according to their context information.

4.6.1.1 Motivation and Contribution

Several introduced approaches for adaptively controlling the evolutionary operators have demonstrated effective performances in static environments. However, such approaches cannot be employed directly in dynamic environments due to their time-varying characteristics. Here it is demonstrated that learning and choosing different operators according to the context of the state the environment can significantly improve the performance of a dynamic optimizer. It should be noted that most of the developed adaptive schemes for ensemble evolutionary optimization methods only utilize the fitness information of individuals. However, positional information of individuals is very important in dynamic optimizers because diversity preservation plays an important role in the current dynamic optimizers. Accordingly, the utilized context information in the proposed method includes both positional and fitness information.

4.6.1.2 Population Structure

In CLA-MPD, the entire population of individuals is partitioned into several subpopulations. These subpopulations are connected in a ring topology. Each subpopulation resides in one cell of a CLA (Fig. 4.14). The CLA has M cells, each embedding a particular subpopulation and a set of six learning automata. Each LA is associated with a specific context state (which will be introduced in the following sections) and is activated according to the context state of its related subpopulation.

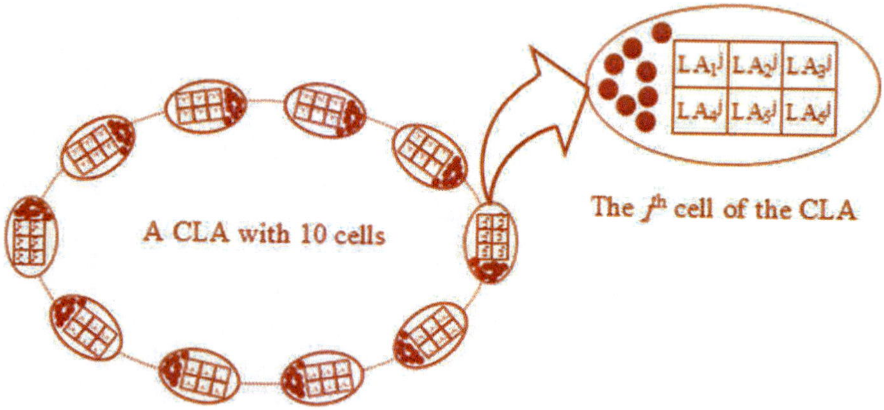

Fig. 4.14 Population structure in CLA-MPD

4.6.1.3 General Framework

Figure 4.15 shows the building blocks of the CLA-MPD. This section briefly introduces these building blocks, and their detailed descriptions are provided in the subsequent sections. The CLA-MPD uses an adaptive mechanism to evolve each subpopulation based on its context information. This context information is retained in a structure called context vector. The context vector of a subpopulation embodies its relative fitness along with some specific information about the distribution of its individuals. In the proposed method, the whole context vector space is partitioned into six different classes. The class of the context vector of a subpopulation will be called its context state. The CLA-MPD calculates the context vectors after population initialization and updates them during the evolutionary procedure.

The CLA-MPD utilizes a set of specially designed updating strategies, each inducing a different behavior on a subpopulation. Each subpopulation learns the effectiveness of each strategy in its different context states to make good use of the features provided by these strategies. The LAs associated with the subpopulation handle this learning task. Each LA of a subpopulation is associated with a particular

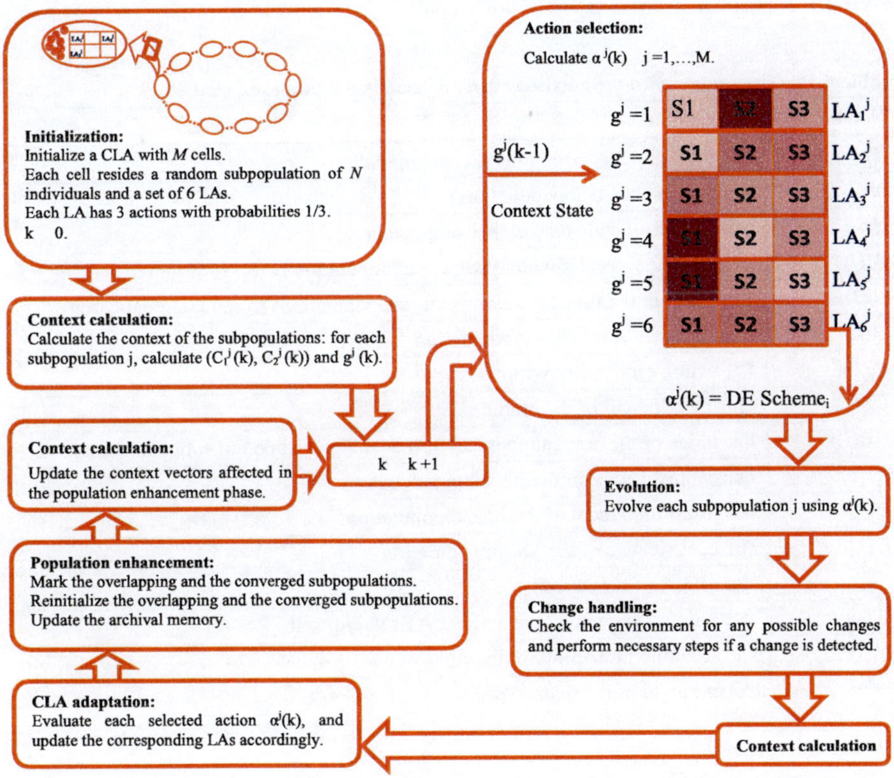

Fig. 4.15 The general framework of CLA-MPD

context state. During the action selection phase of a subpopulation, the LA associated with the current context state of the subpopulation is activated. The activated LA selects a strategy according to its estimated effectiveness in the current context state of the subpopulation.

After the action selection phase, using the chosen strategies, the subpopulations evolve in the evolution phase. Then, CLA-MPD performs change detection. Following environmental change detection, some necessary steps like re-initialization and reevaluation of individuals can take place. As the individuals change during the latter two phases, their corresponding context vectors and context states are recalculated.

As stated previously, each LA is responsible for selecting an appropriate updating strategy for its associated subpopulation in a specific context state. The LA achieves this through repeated cycles of strategy selection, strategy evaluation, and strategy adaption. During the CLA adaption phase, the effectiveness of the utilized strategies is evaluated, and the LAs are updated accordingly. After the CLA adaption phase, CLA-MPD reinitializes the overlapping and converged subpopulations to maintain diversity. If a subpopulation changes in this phase, the affected context vectors will be recalculated. Table 4.10 summarizes the notations that are used in the following sections.

Table 4.10 The summary of the notations that are used in the proposed method

Notation	Description
X_i^j	The ith individual within the jth subpopulation
M	Number of cells (subpopulations)
N	Number of individuals within a subpopulation
BI(j)	Index of the fittest individual within subpopulation j
NE(j, i)	The index of the ith neighboring cell (subpopulation) to cell (subpopulation) j
$g^j(k)$	Context state of the jth subpopulation
$P_i^j(k)$	The action probability vector of LA_i^j
$L^{\min}$	The lower bound of the search space
NB(j)	The index of the best subpopulation in the neighborhood of subpopulation j
(C_1^j, C_2^j)	The context vector of the jth subpopulation
X_B^j	The fittest individual of the jth subpopulation
$g^j(k)$	The context state of the jth subpopulation
LA_i^j	The ith LA of the jth cell
$\alpha^j(k)$	The action is chosen by the active LA of the jth cell
$P_{i,l}^j(k)$	The selection probability of the lth action for LA_i^j
$L^{\max}$	Upper bound of the search space

4.6.1.4 Context Calculation

The context vector of each subpopulation j during generation k is a tuple like $(C_1^j(k), C_2^j(k))$. C_1^j represents the number of fitter nearby subpopulations, where the fitness of a subpopulation is defined as the fitness of its best individual. Considering the employed neighborhood structure in CLA-MPD, $C_1^j \in \{0,1,2\}$. C_1^j will be called the rank of the jth subpopulation. C_2^j gives some information about the distribution of individuals in the subpopulation and is obtained as follows. First, the average distance of each individual in the subpopulation to all other individuals of the same subpopulation is obtained:

$$d_i^j(k) = \frac{1}{N-1}\sum_{l=1}^{N}\left|X_i^j(k) - X_l^j(k)\right|, \tag{4.36}$$

where |.| is the norm operator. After obtaining the average distances within the subpopulation, the average distance of the subpopulation to the nearest best individual is obtained according to the following equation:

$$\tilde{d}^j(k) = \begin{cases} d_{BI(j)}^j(k) & \text{if } C_1^j(k) = 0 \\ D^j(k) & \text{if } C_1^j(k) > 0 \end{cases}, \tag{4.37}$$

where

$$D^j(k) = \begin{cases} \frac{1}{N}\sum_{l=1}^{N}\left|X_B^{NB(j)}(k) - X_l^j(k)\right| & \text{if } C_1^j(k) = 1 \\ \min\left(\frac{1}{N}\sum_{l=1}^{N}\left|X_B^{NE(j,1)}(k) - X_l^j(k)\right|, \frac{1}{N}\sum_{l=1}^{N}\left|X_B^{NE(j,2)}(k) - X_l^j(k)\right|\right) & \text{if } C_1^j(k) = 2 \end{cases}$$

Based on the Eqs. 4.37 and 4.38, $C_2^j(k)$ is calculated as follows:

$$C_2^j(k) = \frac{\tilde{d}^j(k)}{\sqrt{median_{l=1\ldots N}\left(d_l^j(k)\right)}}. \tag{4.38}$$

To utilize the described context information in the proposed learning approach, we partition the context vector space into a manageable number of classes. The class of the context vector of a subpopulation will be called its context state. Based on our empirical studies, we use six different classes to represent the context state of each subpopulation j as follows:

Algorithm 4-4. Context Calculation

Input j: index of a subpopulation

1. If $f(X_B^j(k)) \leq f(X_B^{NE(j,1)}(k))$ And $f(X_B^j(k)) \leq f(X_B^{NE(j,2)}(k))$, then $C_1^j(k) \leftarrow 0$.
2. Else if $f(X_B^j(k)) > f(X_B^{NE(j,1)}(k))$ And $f(X_B^j(k)) > f(X_B^{NE(j,2)}(k))$, then $C_1^j(k) \leftarrow 2$.
3. Else $C_1^j(k) \leftarrow 1$.
4. For each individual X_i^j do: Calculated $d_i^j(k)$ according to Eq. 4-36.
5. Calculate $\bar{d}^j(k)$ according to Eq. 4-37.
6. Calculate $C_2^j(k)$ using Eq. 4-38.

Obtain the context state of the jth subpopulation, $g^j(k)$, based on Eq. 4-39.

Fig. 4.16 Context calculation procedure for the jth subpopulation

$$g^j(k) = \begin{cases} 1 \text{ if } C_1^j(k) > 0 \; And \; 0 \leq C_2^j(k) \leq 1 \\ 2 \text{ if } C_1^j(k) > 0 \; And \; 1 < C_2^j(k) \leq 2 \\ 3 \text{ if } C_1^j(k) > 0 \; And \; 2 < C_2^j(k) \\ 4 \text{ if } C_1^j(k) = 0 \; And \; 0 \leq C_2^j(k) \leq 1 \\ 5 \text{ if } C_1^j(k) = 0 \; And \; 1 < C_2^j(k) \leq 2 \\ 6 \text{ if } C_1^j(k) = 0 \; And \; 2 < C_2^j(k) \end{cases}, \tag{4.39}$$

where $g^j(k)$ is the context state of the jth subpopulation at generation k. Figure 4.16 represents the context calculation procedure for a typical subpopulation. For each subpopulation j, its context state and context vector values can be interpreted as follows:

Case $C_1^j(k) > 0$. After (re)initialization of a subpopulation, its individuals gradually cluster around a peak of the landscape as the algorithm proceeds. Whenever this cluster gets closer to a peak already covered by a fitter neighboring subpopulation, $C_2^j(k)$ becomes smaller ($g^j(k) = 1$). Whenever the cluster draws closer to an uncovered peak, $C_2^j(k)$ becomes larger ($g^j(k) = 3$). $g^j(k) = 2$ denotes a transition state for the jth subpopulation.

Case $C_1^j(k) = 0$. $C_2^j(k)$ merely shows the convergence level of the jth subpopulation: individuals tend to be distributed evenly around the best individual of their subpopulations in the convergence conditions.

4.6.1.5 Action Selection

A CLA controls the behavior of subpopulations in their contexts by adaptively controlling their updating rules. The CLA has M cells, and each cell resides a subpopulation and a group of six learning automata. Each learning automaton is associated with a particular context state of its related subpopulation and can choose an action

Algorithm 4-5. Action selection

1. For each cell (subpopulation) j do:
   a. $k \leftarrow g^j(k\text{-}1)$.
   b. $r \leftarrow$ rand(0,1). // *r is a uniformly distributed random number in (0, 1).*
   c. If $r < P_{i,1}^j$, then $\alpha^j(k) \leftarrow$ Strategy1.
   d. Else if $r < P_{i,2}^j$, then $\alpha^j(k) \leftarrow$ Strategy2.

Else $\alpha^j(k) \leftarrow$ Strategy3.

Fig. 4.17 Strategy (action) selection phase of the CLA-MPD

from the set $A = \{\text{Strategy1, Strategy2, Strategy3}\}$. These actions correspond to the updating rules employed by CLA-MPD and are described in the following section.

During generation k, each cell j activates the learning automaton associated with $g^j(k-1)$. Each activated learning automaton selects a strategy (an action) according to its action probability vector. The chosen strategy of the active LA within the jth cell during iteration k is denoted by $\alpha^j(k)$ (Fig. 4.17). Initially, the selection probability of each strategy for each LA is 1/3 (i.e. $P_{i,l}^j(0) = 1/3, \forall i \in \{1,\ldots,6\}, \forall j \in \{1, \ldots, M\}, \forall l \in \{1,2,3\}$). As the algorithm proceeds, the learning automata adaptively adjust their action probabilities according to their performances.

4.6.1.6 Evolution of Subpopulations

During the evolution phase, each subpopulation j is updated using its selected action, $\alpha^j(k)$ (Fig. 4.18). As stated previously, $\alpha^j(k) \in \{$ Strategy1, Strategy2, Strategy3$\}$, where each strategy defines a DE mutation scheme. Like jDE, the proposed method utilizes an adaptive scheme to control the scaling factor and crossover rate of

Algorithm 4-6. Evolution

1. For each subpopulation j do:
   a. For i=1 to N do:
      i. Obtain the control parameters as follows:
         - Calculate $\hat{F}_i^j(k)$ according to Eq. 4-40.
         - Calculate $\hat{Cr}_i^j(k)$ according to Eq. 4-41.
      ii. Obtain the mutant vector $V_i^j(k)$ as follows:
         - If $\alpha^j(k)$=1, use strategy 1 (Eq. 4-42).
         - If $\alpha^j(k)$=2, use strategy 2 (Eq. 4-43).
         - If $\alpha^j(k)$=3, use strategy 3 (Eq. 4-44).
      iii. Perform binomial crossover to generate a trial vector $U_i^j(k)$, according to Eq. 4-5.
      iv. Evaluate $U_i^j(k)$ and obtain $X_i^j(k)$:
      v. $X_i^j(k) = \begin{cases} X_i^j(k-1) & \text{if } f(X_i^j(k-1)) < f(U_i^j(k)) \\ U_i^j(k) & \text{otherwise} \end{cases}$

Update the control parameters using Eq. 4-45 and Eq. 4-46.

Fig. 4.18 The evolution procedure of subpopulations

target vectors. Each individual has its independent parameters for each strategy, and these parameters are adjusted employing evolution. During generation k, the control parameters are generated for the selected strategies according to the following rules:

$$\widehat{F}_i^j(k) = \begin{cases} F_l + r_1 \cdot F_u & \text{with probability } 0.1 \\ F_i^j\left(k-1, \alpha^j(k)\right) & \text{otherwise} \end{cases}, \tag{4.40}$$

$$\widehat{Cr}_i^j(k) = \begin{cases} U(0,1) & \text{with probability } 0.1 \\ Cr_i^j\left(k-1, \alpha^j(k)\right) & \text{otherwise} \end{cases}, \tag{4.41}$$

where F_l and F_u define, respectively, the lower and upper bounds for the scale factor (Their values are adopted from [100]). $F_i^j(k, s)$ is the archived scale factor of X_i^j for strategy s. Similarly, $Cr_i^j(k, s)$ is the archived crossover rate of X_i^j for strategy s. After calculating its control parameters, each individual X_i^j is mutated according to one of the following strategies.

Strategy 1, anti-convergence:

$$\begin{aligned} V_i^j(k) &= X_i^j(k-1) + \widehat{F}_i^j(k) \cdot \left(X_{br}^j(k-1) - X_i^j(k-1)\right) \\ &\quad + \widehat{F}_i^j(k) \cdot \left(X_{r1}^j(k-1) - X_{r2}^j(k-1)\right), \\ br &= \arg\min_l \left\{ f\left(X_l^j(k-1)\right) \middle| |X_i^j(k-1) - X_B^n(k-1) \right. \\ &\quad \left. | < \left|X_l^j(k-1) - X_B^n(k-1)\right| \right\}, \\ \text{with } n &= \begin{cases} NE\,(j,1) & \text{with probability } 0.5 \\ NE\,(j,2) & \text{otherwise} \end{cases}, \end{aligned} \tag{4.42}$$

where X_{br}^j is the fittest individual such that the distance between X_{br}^j and X_B^n is greater than the distance between X_i^j and X_B^n. Whenever all of the individuals of the jth subpopulation are closer than X_i^j to X_B^n, br is selected randomly from the set $\{1, \ldots, N\}$-$\{i\}$. r_1, r_2 are two mutually exclusive indices randomly chosen from the set $\{1, \ldots, N\}$ such that $r_1, r_2 \neq i, br$.

Strategy 2, local search:

$$\begin{aligned} V_i^j(k) &= X_i^j(k-1) + \widehat{F}_i^j(k) \cdot \left(X_{bl}^j(k-1) - X_i^j(k-1)\right) \\ &\quad + \widehat{F}_i^j(k) \cdot \left(X_{r1}^j(k-1) - X_{r2}^j(k-1)\right), \\ bl &= \arg\min_l \left\{ |X_l^j(k-1) - X_i^j(k-1)| \middle| f\left(X_l^j(k-1)\right) \right. \\ &\quad \left. < f\left(X_i^j(k-1)\right) \right\}, \end{aligned} \tag{4.43}$$

where X^j_{bl} is the closest individual of the jth subpopulation to X^j_i which is fitter than $X^j_i(k)$. Whenever X^j_i is the fittest individual of its subpopulation, bl is selected randomly from the set $\{1, \ldots, N\} - \{i\}$.

Strategy 3, rand/1/bin:

$$V^j_i(k) = X^j_{r1}(k) + \widehat{F}^j_i(k) \cdot \left(X^j_{r2}(k) - X^j_{r3}(k)\right). \tag{4.44}$$

After mutation, the crossover operation is performed between the target vector $X^j_i(k-1)$ and the mutant vector $V^j_i(k)$ to produce a trial vector according to Eq. 4.15. Once the trial vector is produced, the selection operator compares the target vector $X^j_i(k-1)$ and the trial vector $U^j_i(k)$ and selects the fitter one for the next generation. Additionally, promising control parameters are propagated to the next generation according to the following equations:

$$F^j_i\left(k, \alpha^j(k)\right) = \begin{cases} \widehat{F}^j_i(k) & \text{if } f\left(U^j_i(k)\right) < f\left(X^j_i(k-1)\right) \\ F^j_i\left(k-1, \alpha^j(k)\right) & \text{otherwise} \end{cases}, \tag{4.45}$$

$$Cr^j_i\left(k, \alpha^j(k)\right) = \begin{cases} \widehat{Cr}^j_i(k) & \text{if } f\left(U^j_i(k)\right) < f\left(X^j_i(k-1)\right) \\ Cr^j_i\left(k-1, \alpha^j(k)\right) & \text{otherwise} \end{cases}. \tag{4.46}$$

Mutation strategy 1 encourages an individual to search promising areas farther from its neighboring subpopulations. Accordingly, this strategy can improve the overall population diversity and can increase the chance of detecting multiple optima. By using the local search strategy, an individual is encouraged to search its nearby promising areas. Local search is a useful strategy for exploiting promising areas that are covered by a subpopulation. The third strategy is one of the most popular mutation strategies in the DE literature. Mutation strategies 1 and 2, although very helpful in dynamic environments, would produce limited and specific variations on a population. Hence, the use of a more general mutation strategy like rand/1/bin can help the algorithm to overcome this drawback.

4.6.1.7 Change Detection

Changes in the environment are detected by reevaluating the fitness of the best individual of each subpopulation. At the end of each change detection phase, we archive the best individual of each subpopulation along with its fitness value. Then, at the beginning of the next change detection phase, the archived individuals are reevaluated. Next, the newly obtained fitness values are compared with the archived ones. If there is a mismatch in the compared values, it can be inferred that an environmental change must have taken place.

Algorithm 4-7. Change detection phase

1. For each subpopulation j do:
   a. $Tem \leftarrow objective\_fun(X_{Arch}^{j})$.
   b. If $Tem \neq f_{Arch}(X_{Arch}^{j})$, then signal change.
2. If a change is detected
   a. For each subpopulation j do:
      i. $f_{mid}^{j} \leftarrow \text{median}\{ f(X_1^j(k)),\dots, f(X_N^j(k))\}$.
      ii. $A \leftarrow \{i \mid f(X_i^j(k)) = f_{mid}^{j}\}$.
      iii. If $A \neq \emptyset$, then $r \leftarrow$ a random element form A; else $r \leftarrow N+1$.
      iv. For $i=1$ to N do:
         - If $f(X_i^j(k)) \geq f_{mid}^{j}$ And $i \neq r$, then randomly initialize X_i^j within the search space.
      v. For each X_i^j do: $f(X_i^j(k)) \leftarrow objective\_fun(X_i^j(k))$.

For each subpopulation j do: $(X_{Arch}^{j}, f_{Arch}(X_{Arch}^{j})) = (X_B^j(k), f(X_B^j(k)))$

Fig. 4.19 Change detection procedure of the CLA-MPD

Upon the detection of an environmental change, CLA-MPD reinitializes each subpopulation j in the following manner. First, it finds the median of the stored objective function values of the subpopulation individuals. Then, it randomly reinitializes the individuals which have larger objective function values than the obtained median. Whenever more than one individual has an objective function value equal to the calculated median, CLA-MPD randomly retains one of them and reinitializes the rest. After the initialization step, all individuals are reevaluated. Finally, CLA-MPD archives the best individual of each subpopulation along with its fitness value for the next change detection phase. The pseudocode of this phase is given in Fig. 4.19.

4.6.1.8 CLA Adaptation

The learning automata are updated in the CLA adaptation phase (Fig. 4.20) to select effective strategies at future generations. To do this, CLA evaluates the effectiveness of each performed action; then, it generates a reinforcement signal to the LA, which emitted the action. Finally, using the received reinforcement, the LA updates its probability vector. We can measure the effectiveness of the utilized strategies from two perspectives: their impact on the population diversity and the fitness improvement they can provide. In order to evaluate the amount of fitness improvement, the following three metrics will be employed:

$$\Lambda^{j}(k) = f\left(X_B^j(k-1)\right) - f\left(X_B^j(k)\right), \tag{4.47}$$

$$\Gamma^{j}(k) = \frac{1}{N}\sum_{i=1}^{N}\left(f\left(X_i^j(k-1)\right) - f\left(X_i^j(k)\right)\right), \tag{4.48}$$

Algorithm 4-8. CLA adaptation phase

1. For each subpopulation *j* do:
   a. Calculate $\Lambda^j(k)$, $\Gamma^j(k)$, and $\Delta^j(k)$ according to Eq. 4-47, Eq. 4-48, and Eq. 4-49, respectively.
   b. If one of the conditions in Eq. 4-50 to 4-53 is satisfied, $\beta^j(k) \leftarrow 0$.
   c. Else $\beta^j(k) \leftarrow 1$.
   d. $h \leftarrow g^j(k\text{-}1)$. // *context state of the subpopulation = active LA*
   e. If $\beta^j(k)=0$ // *Reward the selected action*
   $$p^j_{h,l}(k+1) = \begin{cases} p^j_{h,l}(k) + a(1 - p^j_{h,l}(k)) & \text{if } l = \alpha^j(k) \\ p^j_{h,l}(k) + (1-a) & \text{if } l \neq \alpha^j(k) \end{cases}$$
   f. Else // *penalize the selected action*
   $$p^j_{h,l}(k+1) = \begin{cases} p^j_{h,l}(k)(1-b) & \text{if } l = \alpha^j(k) \\ \frac{b}{r-1} + p^j_{h,l}(k)(1-b) & \text{if } l \neq \alpha^j(k) \end{cases}$$

Fig. 4.20 The pseudo-code of the CLA adaptation phase

$$\Delta^j(k) = C^j_1(k-1) - C^j_1(k). \tag{4.49}$$

Here, Λ_j represents the amount of improvement in the fitness of the *j*th subpopulation, while Γ_j represents the average improvement of its individuals. Finally, Δ_j represents the amount of improvement in the rank of the *j*th subpopulation (first component of its context vector). Based on these metrics, the selected action of the *j*th cell is rewarded if one of the following conditions is satisfied; otherwise, it is penalized.

- Condition 1:

$$C^j_1(k) = 0 \text{ and } \left[\Lambda^j(k) > \min\left(\Lambda^{N(j,1)}(k), \Lambda^{N(j,2)}(k)\right) \text{ or } \Gamma^j(k) > \min\left(\Gamma^{N(j,1)}(k), \Gamma^{N(j,2)}(k)\right)\right] \tag{4.50}$$

- Condition 2:

$$C^j_1(k) > 0 \text{ and} C^j_2(k) > 2 \text{ and } \Delta^j(k) > 0 \tag{4.51}$$

- Condition 3:

$$C^j_2(k) > 2 \text{ and}\left[\Lambda^j(k) > \max\left(\Lambda^{N(j,1)}(k), \Lambda^{N(j,2)}(k)\right) \text{ or } \Gamma^j(k) > \max\left(\Gamma^{N(j,1)}(k), \Gamma^{N(j,2)}(k)\right)\right] \tag{4.52}$$

- Condition 4:

$$C_2^j(k) > C_2^j(k-1) \text{ and } \Delta^j(k) \geq 0 \text{ and}$$
$$\left[\Lambda^j(k) > \min\left(\Lambda^{N(j,1)}(k), \Lambda^{N(j,2)}(k)\right)\right.$$
$$\left.\text{or } \Gamma^j(k) > \min\left(\Gamma^{N(j,1)}(k), \Gamma^{N(j,2)}(k)\right)\right] \tag{4.53}$$

The defined conditions are the formulation of some general concepts, which can express the effectiveness of the employed strategies. Whenever a subpopulation with rank 0 experiences an acceptable amount of fitness improvement, its utilized strategy is rewarded (Condition 1). As there is not any clear way of determining this acceptable amount of improvement, we defined it relatively in comparison with the neighboring subpopulations. Similarly, a diversity improving strategy can be rewarded if it does not reduce the rank of its target subpopulation (Condition 4). We reward a utilized strategy that moves a subpopulation toward an undetected peak. If a subpopulation has a good distance from other fitter subpopulations and enjoys a great amount of fitness improvement, the subpopulation is likely moving toward an uncovered optimum (Condition 3). In the same way, we reward the strategies that cause an improvement in the rank of a subpopulation while maintaining its diversity (Condition 2).

4.6.1.9 Population Enhancement

An anti-convergence mechanism can improve an optimizer's ability to detect multiple optima. To this end, CLA-MPD is incorporated with an anti-convergence mechanism, that reinitializes a converged or an overlapping subpopulation. This re-initialization is performed with the aid of the external memory of promising positions. Figure 4.21 gives the pseudocode of the population enhancement phase.

Anti-convergence

Whenever the individuals of a subpopulation are converged well to a small Euclidean neighborhood, they cannot contribute much to the optimization process. In this case, the individuals are pretty close to each other. Accordingly, they cannot improve further through mutation and crossover. CLA-MPD reinitializes the converged subpopulations so that they can start to search for new optima. The proposed method considers each subpopulation j as a converged subpopulation if its radius becomes smaller than a threshold value:

$$converged^j = \min_{i=1:N}\left(\hat{d}_i^j(k)\right) < (0.1)^{2-C_1^j(k)}\varepsilon_1, \tag{4.54}$$

$$\text{where } \hat{d}_i^j(k) = median_{l=1:N}\left(\left|X_i^j - X_l^j\right|\right).$$

Here, $(0.1)^{2-C_1^j(k)}\varepsilon_1$ is the convergence threshold. The defined convergence threshold gives a more evolution budget to the promising peaks. Accordingly, CLA-MPD can achieve a high degree of accuracy on detected global optima.

Algorithm 4-9. Population Enhancement

1. $A \leftarrow \emptyset$.
2. For each subpopulation j do:
   a. For each individual, X_i^j do: calculate $\partial_i^j(k)$ (Eq. 4-54).
   b. If $min_{i=1:N}(\partial_i^j(k)) < (0.1)^{2-C_1^j(k)}\varepsilon_1$, then $converged^j \leftarrow$True.
   c. If $\exists l \neq j(\ f(X_B^j(k)) < f(X_B^l(k))$ And $|X_B^j(k) - X_B^l(k)| < \varepsilon_2$), then $overlapping^j \leftarrow$True.
   d. If $converged^j$=True And $C_1^j(k)=0$, then $A \leftarrow A \cup X_B^j(k)$.
3. For each subpopulation j do:
   a. If $converged^j$=True Or $overlapping^j$=True, then reinitialize the subpopulation:
      i. With probability 0.5:
         Re-initialize the individuals of the subpopulation randomly within the search space.
      ii. Else: Re-initialize the subpopulation using the external memory
4. For each $X \in A$ do:
   a. If $|M_E| < MN$, then $M_E \leftarrow M_E \cup X$.
   b. Else let Y be a memory element such that:
      $\exists Z \in (M_E \cup X)\ (Z \neq Y$ And $\forall W, U \in (M_E \cup X)\ (W \neq Y$ and $U \neq Y \Rightarrow |X-Z| \leq |U-W|)\ )$.

$M_E \leftarrow M_E - Y$. $M_E \leftarrow M_E \cup X$.

Fig. 4.21 Pseudocode of the population enhancement phase

CLA-MPD uses an exclusion rule to ensure that different subpopulations are located around different basins of attraction. The exclusion rule marks each pair of subpopulations as overlapping if the distance between their best individuals is less than a predefined threshold, ε_2, called exclusion radius. Once a pair of subpopulations is marked as overlapping, the population with the highest fitness is kept, and the other is reinitialized.

Memorization and re-initialization

Like many dynamic optimizers, CLA-MPD uses an external memory, M_E, to recall useful information from the past. Mainly, this idea would be beneficial when the changing optimum repeatedly returns to previous positions. The size of the external memory is identical to the population size (i.e., *MN*). Whenever, a subpopulation like j marked as a converged subpopulation has a better fitness in comparison to both its neighbors, its best individual is added to the external memory. If the memory is already full, CLA-MPD finds two closest elements from the set $M_E \cup X_B^j - (k)$ and replaces one of them with $X_B^j(k)$.

As stated previously, a subpopulation marked as a converged or overlapping subpopulation is reinitialized to discover new optima. Such a subpopulation is initialized either randomly within the search space (step 3.a.i in Fig. 4.21) or by using an external memory element (Fig. 4.22). In order to initialize a subpopulation j by using a memory position, an external memory element like X_m is selected randomly. Then, the position of each X_i^j is randomly generated within the search space or around the chosen memory element. In the latter case, each component $X_{i,l}^j$ of X_i^j is drawn from

Algorithm 4-10. Initialization using the external memory

Input: j // subpopulation index

1. Select a random element like X_m from M_e.
2. For each individual X_i^j do:
   a. With probability 0.5, redistribute the individual randomly within the search space.
   b. Otherwise:
      i. $R \leftarrow 0.5|L^{Min}-L^{Max}|$
      ii. Generate σ randomly within the search range: $\sigma_l \sim U[0, R_l]\ \forall k \in \{1,\ldots,D\}$.
      iii. Repeat:
         - Generate σ' randomly within the search range: $\sigma'_l \sim [0, R_l]\ \forall k \in \{1,\ldots,D\}$.
         - Set $\sigma_l \leftarrow \min(\sigma_l, \sigma'_l)\ \forall l$.
      iv. Until rand(0,1)<0.5.

$X_{il} \leftarrow \sigma_l N(0,1) + X_{ml}\ \forall l$.

Fig. 4.22 Subpopulation redistribution using a memory element

Gaussian distribution with mean $X_{m,l,}$ and standard deviation σ_l. $\sigma = [\sigma_1, \ldots, \sigma_D]$ is generated in the following manner. First, each σ_l is randomly sampled from a uniform distribution within the interval $[0, 0.5|L_l^{\text{Min}} - L_l^{\text{Max}}|]$. Then, σ is decreased randomly so that we can produce various distributions around X_m.

The idea behind this initialization procedure is as follows. Random initialization (step 3.a.i in Fig. 4.21) provides an excellent exploration of the landscape, while initialization around a memory element is useful for tracking the previously detected optima. The position of a peak may change drastically after a sequence of environmental changes. It is also possible that some landscape peaks will be diminished or eliminated after some time. To address these issues, we initialize almost half of a tracking subpopulation using a uniform distribution. The remaining individuals are sampled at different distances from the memory element using different Gaussian distributions.

4.6.2 Experimental Studies

In this section, we evaluate the performance of the CLA-MPD using the IEEE CEC 2009 benchmark problems for dynamic optimization (generated by using the GDBG system) (Li and others 2008). GDBG presents greater challenges to dynamic optimizers in comparison to the other commonly used benchmarks such as moving peaks problems due to its incorporated rotation methods, higher dimensionalities, and a larger number of local optima (Li and others 2008). There are seven change types in this testbed, which are small step (T_1), large step (T_2), random (T_3), chaotic (T_4), recurrent (T_5), recurrent with noise (T_6), and dimensional change (T_7). The benchmark set consists of six test functions, which are rotation peak function (F_1), the composition of sphere functions (F_2), the composition of Rastrigin's functions

Table 4.11 Parameter settings of the CLA-MPD

Parameter	Values
Number of subpopulations (M)	10
Number of individuals within each subpopulation (N)	10
Learning rate (a, b)	(0.1,0.01)
Convergence threshold (ε_1)	1E−3
Exclusion radius (ε_2)	5E−2

(F_3), composition of Griewank's functions (F_4), composition of Ackley's functions (F_5), and hybrid composition functions (F_6). F_1 is a maximization problem. For this problem, two different instances are considered, one using ten peaks and the other using 50 peaks. The remaining problems are to be minimized. We follow the experimental settings described in IEEE CEC 2009 (Li and others 2008). The summary of these settings is as follows. The number of dimensions is 10 (D = 10) in scenarios T_1–T_6 and is changing between 5 and 15 in scenario T_7. The number of changes per run is set to 60, and the change frequency is set to 10000D fitness evaluations. Simulation results are obtained over 20 independent runs on each benchmark. The average mean error criterion is used as a performance measure, which is defined as follows:

$$Avg\_mean = \frac{1}{runs \times num\_change} \sum_{i=1}^{run\_num} \sum_{j=1}^{num\_change} E_{i,j}^{last}(t), \qquad (4.55)$$

where E^{last} is the absolute function error recorded just before each change in the environment. For the rest of this section, the parameter settings that are summarized in Table 4.11 are used for CLA-MPD.

4.6.2.1 Evaluation Based on the Average Error

In this section, CLA-MPD is compared experimentally with seven dynamic optimization approaches given in Table 4.12. All of the peer methods in this section have already been tested on the GDBG benchmark set and exhibited effective performances. jDE, EDESAC, and mSQDE-i are adaptive multi-population DE algorithms. Accordingly, they have been chosen to evaluate the effectiveness of the proposed multi-population adaptive DE algorithm. Moreover, the building blocks of jDE are much similar to those of CLA-MPD. The main difference between the two algorithms lies in the context-aware learning scheme of CLA-MPD and its incorporated updating strategies. To compare CLA-MPD with approaches that are not based on DE, four algorithms, including ShrBA, PLBA, bCCEA, and DASA are selected.

The comparative results in terms of average mean errors and standard deviations are provided in Tables 4.13 and 4.14. A two-tailed t-test at a 0.05 significance level

Table 4.12 A brief description of the comparison methods

Reference	Method
ShrBA (Das et al. 2014)	Shrinking-based bees algorithm (BA) is an improved variant of basic BA. It includes an additional step called neighborhood shrinking, which is implemented in all of the patches simultaneously using global memory
PLBA (Das et al. 2014)	A BA in which the population initialization, local search, and global search stages are performed according to the patch concept and Levy motion
bCCEA (Hussein et al. 2017)	A cooperative coevolutionary approach for dynamic optimization which uses two types of individuals: random immigrants and elitist individuals
EDESAC (Au and Leung 2014)	A multi-population variant of DE that incorporates an ensemble of adaptive mutation strategies alongside different parameter settings. It also uses an archive of the past, promising solutions with an adaptive clearing scheme to enhance its population diversity
mSQDE-i (Sheldon Hui and Suganthan 2012)	A multi-population differential evolution algorithm with a self-adaptive strategy. Additionally, it uses an interaction mechanism between quantum and conventional individuals
DASA (Novoa-Hernández et al. 2013)	An ant colony based approach that uses pheromone as a means of communication between ants and uses the so-called differential graph representation of the search space
jDE (Brest et al. 2013)	Multi-population self-adaptive DE, which uses mechanisms such as aging, archival memory, and overlapping control to cope with the dynamicity of environments

is employed to compare the results of CLA-MPD against those of the other peer approaches. Also, Table 4.15 summarizes the overall comparative results. Each entry in Table 4.15 has the form "*w*/*t*/*l*," which means that CLA-MPD is better than, equal to and worse than the corresponding compared method on *w*, *t*, and *l* test instances, respectively.

Considering the obtained results, it is evident that CLA-MPD is superior to the other compared methods. It outperforms ShrBA and PLBA in a statistically significant fashion over all tested instances. Also, CLA-MPD obtains better results on most of the tested problems in comparison to EDESAC and DASA. CLA-MPD performs weaker than bCCEA on f_3; however, it outperforms bCCEA over the remaining problems. It should be noted that CLA-MPD is the second-best performer or is statistically indistinguishable from the second-best performers on f_3. Considering the poor performance of bCCEA on problems other than f_3, it can be observed that

Table 4.13 The average errors and the standard deviations of the peer methods on the functions f_1–f_3

		C-MPD	EDESAC	bCCEA	ShrBA	PLBA	mSQDE-i	DASA	jDE
F1-10	T1	**6.68E−03** **8.04E−02**	1.06E−01 7.89E−01[a]	1.30E+01 1.79E+01[a]	4.38E+00 8.22E+00[a]	8.07E+00 1.30E+01[a]	1.10E−01 7.80E−01[a]	1.86E−01 1.33E+00[a]	3.42E−02 4.25E−01[a]
	T2	2.78E+00 6.33E+00	3.55E+00 8.14E+00[a]	1.87E+01 2.10E+01[a]	1.20E+01 1.68E+01[a]	1.31E+01 1.52E+01[a]	**1.06E+00** **4.12E+00[b]**	4.20E+00 7.28E+00[a]	3.60E+00 7.92E+00[a]
	T3	3.77E+00 7.99E+00	6.86E+00 1.10E+01[a]	2.06E+01 1.91E+01[a]	1.13E+01 1.41E+01[a]	6.10E+00 1.07E+01[a]	**9.10E−01** **3.52E+00[b]**	6.41E+00 9.34E+00[a]	3.11E+00 8.26E+00[b]
	T4	3.49E−01 8.74E−01	1.77E+00 2.90E+00[a]	5.56E+00 5.80E+00[a]	1.10E+01 1.74E+01[a]	1.49E+01 1.69E+01[a]	1.10E−01 5.10E−01[b]	5.01E−01 1.74E+00[a]	**2.14E−02** **6.28E−01[b]**
	T5	**9.05E−04** **3.84E−04**	4.12E−01 2.04E+00[a]	2.93E+01 2.49E+01[a]	5.68E+00 6.57E+00[a]	2.99E+00 5.31E+00[a]	5.80E−01 2.07E+00[a]	1.22E+00 3.09E+00[a]	2.33E+00 5.38E+00[a]
	T6	**2.16E−02** **2.70E−01**	8.48E−01 4.45E+00[a]	3.49E+01 2.71E+01[a]	1.18E+01 1.83E+01[a]	1.45E+01 1.99E+01[a]	1.40E−01 6.30E−01[a]	2.86E+00 8.21E+00[a]	1.24E+00 5.54E+00[a]
F1−50	T1	2.02E−01 1.46E+00	4.57E−01 1.32E+00[a]	1.37E+01 1.77E+01[a]	9.24E+00 9.75E+00[a]	1.16E+01 1.15E+01[a]	2.78E+00 4.21E+00[a]	4.37E−01 1.86E+00[a]	**1.80E−01** **4.56E−01[b]**
	T2	3.93E+00 7.67E+00	5.47E+00 7.63E+00[a]	1.92E+01 1.97E+01[a]	1.10E+01 1.19E+01[a]	1.20E+01 1.35E+01[a]	**2.31E+00** **3.90E+00[b]**	5.03E+00 8.56E+00[a]	4.09E+00 6.46E+00
	T3	4.24E+00 6.78E+00	6.69E+00 8.16E+00[a]	2.61E+01 1.79E+01[a]	9.48E+00 1.06E+01[a]	8.92E+00 7.99E+00[a]	**2.03E+00** **3.65E+00[b]**	8.77E+00 1.07E+01[a]	4.29E+00 6.45E+00
	T4	5.64E−01 9.95E−01	1.47E+00 2.54E+00[a]	4.39E+00 4.92E+00[a]	8.22E+00 1.24E+01[a]	1.67E+01 1.89E+01[a]	3.39E+00 6.13E+00[a]	5.39E−01 1.68E+00	**8.77E−02** **2.46E−01[b]**
	T5	**4.43E−01** **1.22E+00**	6.25E−01 9.65E−01[a]	3.06E+01 2.45E+01[a]	2.29E+00 3.82E+00[a]	3.56E+00 4.43E+00[a]	6.10E−01 1.28E+00[a]	9.82E−01 3.69E+00[a]	9.48E−01 1.77E+00[a]
	T6	**7.29E−01** **1.52E+00**	1.71E+00 5.64E+00[a]	4.44E+01 2.72E+01[a]	1.94E+01 2.11E+01[a]	2.00E+01 2.43E+01[a]	1.27E+00 2.23E+00[a]	2.16E+00 4.86E+00[a]	1.77E+00 5.83E+00[a]

(continued)

Table 4.13 (continued)

		C-MPD	EDESAC	bCCEA	ShrBA	PLBA	mSQDE-i	DASA	jDE
F2	T1	**2.65E−01** **1.59E+00**	3.92E+00 6.71E+00[a]	4.32E+01 4.80E+01[a]	1.18E+01 1.48E+01[a]	1.37E+01 1.22E+01[a]	1.40E+00 4.64E+00[a]	2.15E+00 7.65E+00[a]	9.28E−01 2.75E+00[a]
	T2	**6.10E−01** **2.42E+00**	4.77E+00 8.15E+00[a]	5.36E+01 5.87E+01[a]	8.89E+01 1.64E+02[a]	7.71E+01 1.50E+02[a]	4.82E+00 1.52E+01[a]	2.43E+01 8.85E+01[a]	4.41E+01 1.13E+02[a]
	T3	**2.54E−01** **1.83E+00**	3.63E+00 6.36E+00[a]	4.10E+01 5.82E+01[a]	9.77E+01 1.66E+02[a]	7.27E+01 1.43E+02[a]	5.93E+00 2.19E+01[a]	1.89E+01 6.71E+01[a]	5.26E+01 1.21E+02[a]
	T4	2.05E+00 6.35E+00	5.04E+00 8.10E+00[a]	2.10E+02 1.37E+02[a]	3.16E+01 3.14E+01[a]	2.87E+01 2.57E+01[a]	**5.90E−01** **1.36E+00[b]**	1.36E+00 4.73E+00[b]	7.36E−01 3.07E+00[b]
	T5	**1.20E−02** **5.44E−02**	2.19E+00 5.43E+00[a]	5.44E+01 5.68E+01[a]	1.81E+02 1.90E+02[a]	1.24E+02 1.72E+02[a]	1.02E+01 3.01E+01[a]	4.69E+01 1.11E+02[a]	6.29E+01 1.36E+02[a]
	T6	**4.67E−03** **1.28E−03**	1.89E+00 5.03E+00[a]	6.59E+01 7.81E+01[a]	1.99E+01 3.48E+01[a]	1.29E+01 1.77E+01[a]	9.00E−01 1.74E+00[a]	2.76E+00 4.95E+00[a]	3.97E+00 1.47E+01[a]
F3	T1	1.38E+01 2.08E+01	3.08E+01 7.05E+01[a]	**7.10E+00** **9.34E+00[b]**	6.56E+02 1.50E+02[a]	5.97E+01 1.60E+02[a]	1.98E+01 1.12E+02	1.53E+01 6.60E+01	1.14E+01 6.18E+01
	T2	1.10E+01 2.08E+01	3.16E+01 6.79E+01[a]	**7.28E+00** **1.03E+01[b]**	9.13E+02 1.97E+02[a]	7.90E+02 2.48E+02[a]	9.86E+02 8.01E+01[a]	8.22E+02 1.97E+02[a]	5.63E+02 3.27E+02[a]
	T3	1.67E+01 5.80E+01	3.05E+01 7.32E+01[a]	**6.38E+00** **8.05E+00[b]**	8.91E+02 1.01E+02[a]	7.10E+02 2.90E+02[a]	9.79E+02 8.73E+01[a]	6.91E+02 3.13E+02[a]	5.87E+02 3.75E+02[a]
	T4	7.13E+01 1.32E+02	7.53E+01 1.70E+02	**2.08E+00** **4.22E+00[b]**	7.94E+02 3.05E+02[a]	4.77E+02 4.50E+02[a]	6.01E+02 5.36E+02[a]	4.37E+02 4.69E+02[a]	6.61E+01 2.11E+02
	T5	2.82E+01 7.13E+01	2.47E+01 5.35E+01	**6.01E+00** **7.80E+00[b]**	9.01E+02 6.07E+01[a]	7.24E+02 1.93E+02[a]	9.58E+02 9.48E+01[a]	6.99E+02 3.26E+02[a]	4.76E+02 3.27E+02[a]
	T6	1.38E+01 5.03E+01	2.29E+01 5.18E+01[a]	**5.81E+00** **7.89E+00[b]**	9.15E+02 2.99E+02[a]	5.34E+02 4.36E+02[a]	8.60E+02 4.72E+02[a]	6.33E+02 4.54E+02[a]	2.63E+02 3.93E+02[a]

[a]indicates that CLA-MPD performs better than the compared algorithm by t-test
[b]means that the corresponding algorithm is better than CLA-MPD
The best average errors are highlighted in boldface

Table 4.14 The average errors and the standard deviations of the peer methods on the functions f_4–f_6

		CL-MPD	EDESAC	bCCEA	ShrBA	PLBA	mSQDE-i	DASA	jDE
F4	T1	**5.06E−01** **2.19E+00**	6.55E+00 9.66E+00[a]	5.25E+01 7.83E+01[a]	8.99E+00 1.12E+01[a]	2.03E+01 1.91E+01[a]	9.40E−01 3.13E+00[a]	5.48E+00 2.87E+01[a]	1.57E+00 4.88E+00[a]
	T2	**5.37E−01** **2.53E+00**	7.24E+00 1.08E+01[a]	6.84E+01 9.29E+01[a]	1.19E+02 2.12E+02[a]	2.90E+02 2.67E+02[a]	6.83E+00 1.98E+01[a]	6.60E+01 1.46E+02[a]	4.98E+01 1.33E+02[a]
	T3	**6.96E−01** **3.00E+00**	5.01E+00 8.26E+00[a]	5.94E+01 8.35E+01[a]	1.50E+02 2.29E+02[a]	1.49E+02 2.30E+02[a]	6.84E+00 1.93E+01[a]	5.30E+01 1.48E+02[a]	5.21E+01 1.57E+02[a]
	T4	1.64E+00 4.14E+00	8.88E+00 1.55E+01[a]	3.18E+02 1.18E+02[a]	2.90E+01 3.03E+01[a]	1.64E+01 2.09E+01[a]	**6.00E−01** **1.30E+00[b]**	3.64E+00 6.68E+00[a]	1.67E+00 5.03E+00
	T5	**1.35E+00** **4.85E+00**	4.99E+00 8.48E+00[a]	9.12E+01 1.15E+02[a]	2.57E+02 2.30E+02[a]	2.50E+02 2.37E+02[a]	1.75E+01 5.42E+01[a]	1.21E+02 1.89E+02[a]	6.69E+01 1.49E+02[a]
	T6	**5.56E−03** **2.26E−03**	4.59E+00 7.86E+00[a]	7.75E+01 1.02E+02[a]	2.72E+01 8.76E+01[a]	2.47E+01 6.62E+01[a]	1.21E+00 3.20E+00[a]	2.35E+00 8.54E+00[a]	2.56E+00 5.85E+00[a]
F5	T1	**6.62E−02** **3.25E−01**	4.92E+00 7.95E+00[a]	7.55E+01 1.88E+02[a]	2.86E+01 2.09E+01[a]	1.04E+01 1.19E+01[a]	1.32E+00 2.56E+00[a]	9.75E−01 3.42E+00[a]	1.60E−01 1.24E+00[a]
	T2	**2.26E−02** **2.84E−02**	5.55E+00 8.74E+00[a]	7.76E+01 1.82E+02[a]	1.39E+01 1.84E+01[a]	1.49E+01 1.92E+01[a]	2.97E+00 5.30E+00[a]	2.11E+00 4.75E+00[a]	3.46E−01 1.39E+00[a]
	T3	7.01E−01 2.53E+00	5.56E+00 8.56E+00[a]	7.13E+01 1.65E+02[a]	7.43E+00 1.03E+01[a]	1.28E+01 1.50E+01[a]	2.76E+00 4.53E+00[a]	9.38E−01 3.27E+00[a]	**3.61E−01** **2.08E+00[b]**
	T4	**8.88E−02** **2.85E−01**	5.21E+00 8.71E+00[a]	3.28E+01 2.02E+01[a]	2.76E+01 2.75E+01[a]	2.43E+01 2.78E+01[a]	1.28E+00 2.25E+00[a]	4.10E−01 1.71E+00[a]	1.08E−01 9.18E−01
	T5	**1.88E−02** **6.71E−03**	4.52E+00 7.85E+00[a]	1.00E+02 2.31E+02[a]	1.52E+01 1.60E+01[a]	1.56E+01 1.71E+01[a]	2.13E+01 1.30E+02[a]	2.43E+00 5.65E+00[a]	4.34E−01 1.58E+00[a]
	T6	**1.80E−02** **6.90E−03**	3.63E+00 7.05E+00[a]	7.13E+01 1.81E+02[a]	1.13E+01 1.47E+01[a]	1.06E+01 1.51E+01[a]	4.75E+00 1.81E+01[a]	4.61E−01 1.70E+00[a]	3.75E−01 9.87E−01[a]

(continued)

Table 4.14 (continued)

		CL-MPD	EDESAC	bCCEA	ShrBA	PLBA	mSQDE-i	DASA	jDE
F6	T1	3.53E+00 5.12E+00	9.80E+00 1.29E+01[a]	1.70E+02 2.62E+02[a]	3.02E+01 1.76E+01[a]	9.34E+00 1.38E+01[a]	**3.13E+00** **5.46E+00**	6.33E+00 1.64E+01[a]	6.34E+00 1.07E+01[a]
	T2	**6.42E+00** **1.10E+01**	1.16E+01 1.38E+01[a]	1.87E+02 2.71E+02[a]	4.74E+01 9.83E+01[a]	3.46E+01 6.85E+01[a]	1.95E+01 6.39E+01[a]	2.62E+01 1.21E+02[a]	1.05E+01 1.38E+01[a]
	T3	**6.74E+00** **6.96E+00**	8.98E+00 1.07E+01[a]	1.78E+02 2.69E+02[a]	6.97E+01 1.78E+02[a]	4.83E+01 3.68E+02[a]	2.67E+01 8.90E+01[a]	1.86E+01 6.93E+01[a]	1.04E+01 2.58E+01[a]
	T4	**5.13E+00** **7.87E+00**	9.31E+00 1.36E+01[a]	1.38E+02 2.40E+02[a]	3.33E+01 6.28E+01[a]	1.84E+01 2.37E+01[a]	6.90E+00 9.65E+00[a]	7.46E+00 2.37E+01[a]	6.24E+00 1.33E+01[a]
	T5	**1.16E+01** **1.77E+01**	1.93E+01 2.81E+01[a]	2.17E+02 2.88E+02[a]	1.95E+02 3.06E+02[a]	1.10E+02 2.26E+02[a]	6.03E+01 1.64E+02[a]	3.68E+01 1.22E+02[a]	1.50E+01 4.43E+01[a]
	T6	1.23E+01 1.82E+01	2.09E+01 2.46E+01[a]	1.87E+02 2.70E+02[a]	2.87E+01 6.64E+01[a]	3.20E+01 7.95E+01[a]	1.30E+01 5.59E+01	1.37E+01 5.47E+01	**7.65E+00** **1.08E+01**[b]

[a]indicates that CLA-MPD performs better than the compared algorithm by t-test
[b]means that the corresponding algorithm is better than CLA-MPD
The best average errors are highlighted in boldface

Table 4.15 Summary of comparison results of CLA-MPD against other peer approaches

f	EDESAC	bCCEA	ShrBA	PLBA	mSQDE-i	DASA	jDE
1–10	6/0/0	6/0/0	6/0/0	6/0/0	3/0/3	6/0/0	4/0/2
1–50	6/0/0	6/0/0	6/0/0	6/0/0	4/0/2	6/0/0	4/0/2
2	6/0/0	6/0/0	6/0/0	6/0/0	5/0/1	5/0/1	5/0/1
3	4/2/0	0/0/6	6/0/0	6/0/0	5/1/0	5/1/0	4/2/0
4	6/0/0	6/0/0	6/0/0	6/0/0	5/0/1	6/0/0	5/1/0
5	6/0/0	6/0/0	6/0/0	6/0/0	6/0/0	6/0/0	4/1/1
6	6/0/0	6/0/0	6/0/0	6/0/0	4/2/0	5/1/0	5/0/1
All	40/2/0	36/0/6	42/0/0	42/0/0	32/3/7	39/2/1	31/4/7

CLA-MPD is more robust compared to bCCEA. When compared to jDE, CLA-MPD is statically superior on 31 test instances and obtains some distinguishable results on some benchmarks like f_2, f_3, f_4. Accordingly, it can be inferred that the proposed context-aware learning scheme and the adopted strategies are beneficial in dynamic environments. The comparison with the other multi-population adaptive DE method, mSQDE-i, also demonstrates the effectiveness of the presented adaptive DE approach. CLA-MPD obtains better results than mSQDE-i on 32 test instances, while its results are weaker on only 7 test cases. The consistently good performance over the majority of the benchmark instances indicates that the introduced schemes in CLA-MPD for tackling dynamically changing fitness landscapes do have the edge over the compared dynamic optimizers.

4.7 Conclusion

After an introduction to nature-inspired optimization methods, some of the developed hybrid optimizations approaches based on nature-inspired methods, and reinforcement learning was briefly reviewed. Next, we described in detail three of the recently developed cellular learning automata (CLA) based nature-inspired optimization algorithms. First, we investigated two recently developed CLA-based nature-inspired methods for static optimization. In the first algorithm, a CLA adaptively controls the topology connectivity of the particles. Through experimental studies, it was shown that the algorithm using an adaptive topology surpasses its variant, which uses a random topology.

Consequently, the employed topology adaption mechanism is profitable for PSO. The second CLA-based nature-inspired static optimization method discussed in this chapter is CLA-BBPSO. This method uses different probability distributions for directing the movements of a particle. These distributions possess different characteristics that can be helpful in various types of landscapes and different stages of the optimization process. The parameters of these distributions are learned using different methods. A CLA is responsible for the selection of an appropriate updating distribution for each particle. The experimental studies also confirmed the learning behavior of the method, where CLA chooses a suitable distribution based on the problem landscape and the search stage of the execution. Also, the learning ability of the algorithm is demonstrated through comparison with its pure chance version.

In the final part of the chapter, it was demonstrated that CLA could be utilized effectively for adaptively learning appropriate strategies for a nature-inspired method in a dynamic environment. The described method incorporates a set of updating strategies that are specifically designed for dynamic environments. A context-aware learning mechanism controls the utilization of each strategy. This mechanism enables each subpopulation to choose a proper strategy according to its context state. Accordingly, it can appropriately react to drastic changes such as re-initialization and environmental changes. The proposed learning approach is based on cellular learning automata. The algorithm maintains a CLA, and each cell of the CLA resides a subpopulation. The LAs of each cell control the evolutionary procedure of their associated subpopulation.

References

Abshouri, A.A., Meybodi, M.R., Bakhtiary, A.: New firefly algorithm based on multi swarm & learning automata in dynamic environments. In: IEEE Proceedings, pp. 989–993 (2011)

Ali, K.I., Brohi, K.: An adaptive learning automata for genetic operators allocation probabilities. In: 2013 11th International Conference on Frontiers of Information Technology, pp 55–59. IEEE (2013)

Alipour, M.M., Razavi, S.N., Feizi Derakhshi, M.R., Balafar, M.A.: A hybrid algorithm using a genetic algorithm and multiagent reinforcement learning heuristic to solve the traveling salesman problem. Neural Comput. Appl. **30**, 2935–2951 (2018). https://doi.org/10.1007/s00521-017-2880-4

Arora, S., Anand, P.: Learning automata-based butterfly optimization algorithm for engineering design problems. Int. J. Comput. Mater. Sci. Eng. **07**, 1850021 (2018). https://doi.org/10.1142/S2047684118500215

Au, C.-K., Leung, H.-F.: Cooperative coevolutionary algorithms for dynamic optimization: an experimental study. Evol. Intell. **7**, 201–218 (2014). https://doi.org/10.1007/s12065-014-0117-3

Brest, J., Korošec, P., Šilc, J., et al.: Differential evolution and differential ant-stigmergy on dynamic optimisation problems. Int. J. Syst. Sci. **44**, 663–679 (2013). https://doi.org/10.1080/00207721.2011.617899

Cai, Y., Zhao, M., Liao, J., et al.: Neighborhood guided differential evolution. Soft Comput. **21**, 4769–4812 (2017). https://doi.org/10.1007/s00500-016-2088-z

Campos, M., Krohling, R.A., Enriquez, I.: Bare bones particle swarm optimization with scale matrix adaptation. IEEE Trans. Cybern. **44**,1567–1578 (2014). https://doi.org/10.1109/TCYB.2013.2290223

Chen, H., Zhu, Y., Hu, K.: Discrete and continuous optimization based on multi-swarm coevolution. Nat. Comput. **9**, 659–682 (2010). https://doi.org/10.1007/s11047-009-9174-4

Dai, C., Wang, Y., Ye, M., et al.: An orthogonal evolutionary algorithm with learning automata for multiobjective optimization. IEEE Trans. Cybern. **46**, 3306–3319 (2016). https://doi.org/10.1109/TCYB.2015.2503433

Das, S., Mandal, A., Mukherjee, R.: An adaptive differential evolution algorithm for global optimization in dynamic environments. IEEE Trans. Cybern. **44**, 966–978 (2014). https://doi.org/10.1109/TCYB.2013.2278188

Demšar, Janez: Statistical comparisons of classifiers over multiple data Sets. J. Mach. Learn. Res. **7**, 1–30 (2006)

Eiben, A.E., Smith, J.E.: Introduction to Evolutionary Computing. Springer, Berlin Heidelberg (2015)

El-Abd, M.: Testing a particle swarm optimization and artificial bee colony hybrid algorithm on the CEC13 benchmarks. In: 2013 IEEE Congress on Evolutionary Computation, pp. 2215–2220 (2013)

El Hatri, C., Boumhidi, J.: Q-learning based intelligent multi-objective particle swarm optimization of light control for traffic urban congestion management. In: 2016 4th IEEE International Colloquium on Information Science and Technology (CiSt), pp. 794–799. IEEE (2016)

Enayatifar, R., Yousefi, M, Abdullah, A.H., Darus, A.N.: LAHS: A novel harmony search algorithm based on learning automata. Commun. Nonlinear Sci. Numer. Simul. **18**, 3481–3497 (2013). https://doi.org/10.1016/j.cnsns.2013.04.028

Feng, X., Zou, R., Yu, H.: A novel optimization algorithm inspired by the creative thinking process. Soft Comput. **19**, 2955–2972 (2015). https://doi.org/10.1007/s00500-014-1459-6

Geshlag, M.B.M., Sheykhzadeh, J.: A new particle swarm optimization model based on learning automata using deluge algorithm for dynamic environments. J. Basic Appl. Sci. Res. **3**, 394–404 (2012)

Gunasundari, S., Janakiraman, S., Meenambal, S.: Velocity Bounded Boolean Particle Swarm Optimization for improved feature selection in liver and kidney disease diagnosis. Expert Syst. Appl. **56**, 28–47 (2016). https://doi.org/10.1016/j.eswa.2016.02.042

Hansen, N., Ostermeier, A.: Adapting arbitrary normal mutation distributions in evolution strategies: the covariance matrix adaptation. In: Proceedings of IEEE International Conference on Evolutionary Computation, pp. 312–317. IEEE (1996)

Hansen, N., Ostermeier, A.: Completely derandomized self-adaptation in evolution strategies. Evol. Comput. **9**,159–195 (2001). https://doi.org/10.1162/106365601750190398

Hasanzadeh, M., Meybodi, M.R., Ebadzadeh, M.M.: A robust heuristic algorithm for cooperative particle swarm optimizer: a learning automata approach. In: ICEE 2012—20th Iranian Conference on Electrical Engineering, Tehran, Iran, pp. 656–661 (2012)

Hasanzadeh, M., Meybodi, M.R., Ebadzadeh, M.M.: Adaptive cooperative particle swarm optimizer. Appl. Intell. **39**, 397–420 (2013). https://doi.org/10.1007/s10489-012-0420-6

Hasanzadeh M, Meybodi MR, Ebadzadeh MM (2014) A learning automata approach to cooperative particle swarm optimizer. J. Inf. Syst. Telecommun. Tehran, Iran, 656–661

Hashemi, A.B., Meybodi, M.R.: A note on the learning automata based algorithms for adaptive parameter selection in PSO. Appl. Soft Comput. J. **11**, 689–705 (2011). https://doi.org/10.1016/j.asoc.2009.12.030

Howell, M.N., Gordon, T.J., Brandao, F.V.: Genetic learning automata for function optimization. IEEE Trans. Syst. Man Cybern. Part B **32**, 804–815 (2002). https://doi.org/10.1109/TSMCB.2002.1049614

Hui, S., Suganthan, P.N.: Ensemble Differential Evolution with dynamic subpopulations and adaptive clearing for solving dynamic optimization problems. In: 2012 IEEE Congress on Evolutionary Computation, pp. 1–8. IEEE (2012)

Hussein, W.A., Abdullah, S.N.H.S., Sahran, S.: The patch-levy-based bees algorithm applied to dynamic optimization problems. Discret. Dyn. Nat. Soc. **2017**, 1–27 (2017). https://doi.org/10.1155/2017/5678393

Iima, H., Kuroe, Y.: Swarm reinforcement learning algorithms based on Sarsa method. In: 2008 SICE Annual Conference, pp. 2045–2049. IEEE (2008)

Kennedy, J.: Bare bones particle swarms. In: Proceedings of the 2003 IEEE Swarm Intelligence Symposium. SIS'03 (Cat. No.03EX706), pp. 80–87. IEEE (2003)

Kordestani, J.K., Ahmadi, A., Meybodi, M.R.: An improved Differential Evolution algorithm using learning automata and population topologies. Appl. Intell. **41**, 1150–1169 (2014). https://doi.org/10.1007/s10489-014-0585-2

Kordestani, J.K., Firouzjaee, H.A,, Meybodi, M.R.: An adaptive bi-flight cuckoo search with variable nests for continuous dynamic optimization problems. Appl. Intell. **48**, 97–117 (2018). https://doi.org/10.1007/s10489-017-0963-7

Kordestani, J.K., Ranginkaman, A.E., Meybodi, M.R., Novoa-Hernández, P.: A novel framework for improving multi-population algorithms for dynamic optimization problems: a scheduling approach. Swarm Evol. Comput. **44**, 788–805 (2019). https://doi.org/10.1016/j.swevo.2018.09.002

Leung, Y.-W., Wang, Y.: An orthogonal genetic algorithm with quantization for global numerical optimization. IEEE Trans. Evol. Comput. **5**, 41–53 (2001). https://doi.org/10.1109/4235.910464

Li, C., others: Benchmark Generator for {CEC}'2009 Competition on Dynamic Optimization. Department of Computer Science, University of Leicester, Leicester, UK, Tech Rep (2008)

Liang, J.J., Qin, A.K., Suganthan, P.N., Baskar, S.: Comprehensive learning particle swarm optimizer for global optimization of multimodal functions. IEEE Trans. Evol. Comput. **10**, 281–295 (2006)

Lim W.H., Mat Isa, N.A.: Particle swarm optimization with increasing topology connectivity. Eng. Appl. Artif. Intell. **27**, 80–102 (2014). https://doi.org/10.1016/j.engappai.2013.09.011

Li, C., Yang, S.: An adaptive learning particle swarm optimizer for function optimization. In: 2009 IEEE Congress on Evolutionary Computation, pp. 381–388. IEEE (2009)

Li, C., Yang, S., Nguyen, T.T.: A self-learning particle swarm optimizer for global optimization problems. IEEE Trans. Syst. Man, Cybern. Part B **42**,627–646 (2012). https://doi.org/10.1109/TSMCB.2011.2171946

Li, W., Ozcan, E., John, R.: A learning automata based multiobjective hyper-heuristic. IEEE Trans. Evol. Comput., 1–1 (2018). https://doi.org/10.1109/TEVC.2017.2785346

Mahdaviani, M., Kordestani, J.K., Rezvanian, A., Meybodi, M.R.: LADE: learning automata based differential evolution. Int. J. Artif. Intell. Tools **24**, 1550023 (2015). https://doi.org/10.1142/S0218213015500232

Mendes, R., Kennedy, J., Neves, J.: The fully informed particle swarm: simpler, maybe better. IEEE Trans. Evol. Comput. **8**, 204–210 (2004). https://doi.org/10.1109/TEVC.2004.826074

Montes de Oca, M.A., Stutzle, T., Birattari, M, Dorigo M (2009) Frankenstein's PSO: a composite particle swarm optimization algorithm. IEEE Trans. Evol. Comput. **13**, 1120–1132. https://doi.org/10.1109/TEVC.2009.2021465

Mozafari, M., Shiri, M.E., Beigy, H.: A cooperative learning method based on cellular learning automata and its application in optimization problems. J. Comput. Sci. **11**, 279–288 (2015). https://doi.org/10.1016/j.jocs.2015.08.002

Nabizadeh, S., Rezvanian, A., Meybodi, M.R.: A multi-swarm cellular PSO based on clonal selection algorithm in dynamic environments. In: 2012 International Conference on Informatics, Electronics and Vision, ICIEV 2012. Dhaka, Bangladesh, pp. 482–486 (2012)

Nepomuceno, F.V., Engelbrecht, A.P.: A self-adaptive heterogeneous pso for real-parameter optimization. In: 2013 IEEE Congress on Evolutionary Computation. IEEE, pp. 361–368 (2013)

Novoa-Hernández, P., Corona, C.C., Pelta, D.A.: Self-adaptive, multipopulation differential evolution in dynamic environments. Soft Comput. **17**, 1861–1881 (2013). https://doi.org/10.1007/s00500-013-1022-x

Rastegar, R., Meybodi, M.R.: A new evolutionary computing model based on cellular learning automata. In: IEEE Conference on Cybernetics and Intelligent Systems, 2004, pp. 433–438. IEEE (2004)

Rezapoor Mirsaleh, M., Meybodi, M.R.: A learning automata-based memetic algorithm. Genet. Program. Evolvable Mach. **16**, 399–453 (2015). https://doi.org/10.1007/s10710-015-9241-9

Rezapoor Mirsaleh, M., Meybodi, M.R.: Balancing exploration and exploitation in memetic algorithms: a learning automata approach. Comput. Intell. **34**, 282–309 (2018). https://doi.org/10.1111/coin.12148

Rezvanian, A., Meybodi, M.R.: An adaptive mutation operator for artificial immune network using learning automata in dynamic environments. In: 2010 Second World Congress on Nature and Biologically Inspired Computing (NaBIC), pp. 479–483. IEEE (2010a)

Rezvanian, A., Meybodi, M.R.: Tracking extrema in dynamic environments using a learning automata-based immune algorithm. Communications in Computer and Information Science, pp. 216–225. Springer, Berlin Heidelberg (2010a)

Rezvanian, A., Meybodi, M.R.: LACAIS: Learning automata based cooperative artificial immune system for function optimization. Communications in Computer and Information Science, pp. 64–75. Springer, Berlin Heidelberg (2010b)

Salomon, R.: Re-evaluating genetic algorithm performance under coordinate rotation of benchmark functions. A survey of some theoretical and practical aspects of genetic algorithms. Biosystems **39**, 263–278 (1996). https://doi.org/10.1016/0303-2647(96)01621-8

Samma, H., Lim, C.P., Mohamad Saleh, J.: A new reinforcement learning-based memetic particle swarm optimizer. Appl. Soft Comput. **43**, 276–297 (2016). https://doi.org/10.1016/j.asoc.2016.01.006

Sengupta, A., Chakraborti, T., Konar, A., et al.: an adaptive memetic algorithm using a synergy of differential evolution and learning automata. In: 2012 IEEE Congress on Evolutionary Computation, pp. 1–8. IEEE (2012)

Sheng, X., Xu, W.: Solving the economic dispatch problem with Q-learning quantum-behaved particle swarm optimization method. In: 2015 14th International Symposium on Distributed Computing and Applications for Business Engineering and Science (DCABES), pp. 98–101. IEEE (2015)

Shen, X.-N., Minku, L.L., Marturi, N., et al.: A Q-learning-based memetic algorithm for multi-objective dynamic software project scheduling. Inf. Sci. (Ny) **428**, 1–29 (2018). https://doi.org/10.1016/j.ins.2017.10.041

Shi, Y., Eberhart, R.: A modified particle swarm optimizer. In: 1998 IEEE International Conference on Evolutionary Computation Proceedings. IEEE World Congress on Computational Intelligence (Cat. No. 98TH8360), pp. 69–73. IEEE (1998)

Suganthan, P.N., Hansen, N., Liang, J.J., et al.: Problem definitions and evaluation criteria for the CEC 2006 special session on constrained real-parameter optimization. KanGAL **2005005**, 251–256 (2014)

Suttorp, T., Hansen, N., Igel, C.: Efficient covariance matrix update for variable metric evolution strategies. Mach. Learn. **75**, 167–197 (2009). https://doi.org/10.1007/s10994-009-5102-1

Vafashoar, R., Meybodi, M.R., Momeni Azandaryani, A.H.: CLA-DE: a hybrid model based on cellular learning automata for numerical optimization. Appl. Intell. **36**, 735–748 (2012). https://doi.org/10.1007/s10489-011-0292-1

Vafashoar, R., Meybodi, M.R.: Multi swarm bare bones particle swarm optimization with distribution adaption. Appl. Soft Comput. J. **47**, 534–552 (2016). https://doi.org/10.1016/j.asoc.2016.06.028

Vafashoar, R., Meybodi, M.R.: Multi swarm optimization algorithm with adaptive connectivity degree. Appl. Intell. **48**, 909–941 (2018). https://doi.org/10.1007/s10489-017-1039-4

Vafashoar, R., Meybodi, M.R.: A multi-population differential evolution algorithm based on cellular learning automata and evolutionary context information for optimization in dynamic environments, Appl. Soft Comput. **88**, 106009 (2020).

Wang, H., Wu, Z., Rahnamayan, S., et al.: Multi-strategy ensemble artificial bee colony algorithm. Inf. Sci. (Ny) **279**, 587–603 (2014). https://doi.org/10.1016/j.ins.2014.04.013

Wilke, D.N., Kok, S., Groenwold, A.A.: Comparison of linear and classical velocity update rules in particle swarm optimization: notes on diversity. Int. J. Numer. Methods Eng. **70**, 962–984 (2007). https://doi.org/10.1002/nme.1867

Wu, G., Mallipeddi, R., Suganthan, P.N.: Ensemble strategies for population-based optimization algorithms—a survey. Swarm Evol. Comput. **44**, 695–711 (2019). https://doi.org/10.1016/j.swevo.2018.08.015

Yu, X., Gen, M.: Introduction to Evolutionary Algorithms. Springer, London, London (2010)

Zhang, J., Xu, L., Li, J., et al.: Integrating Particle Swarm Optimization with Learning Automata to solve optimization problems in noisy environment. In: 2014 IEEE International Conference on Systems, Man, and Cybernetics (SMC), pp. 1432–1437. IEEE (2014)
Zhang, J., Xu, L., Ma, J., Zhou, M.: A learning automata-based particle swarm optimization algorithm for noisy environment. In: 2015 IEEE Congress on Evolutionary Computation (CEC), pp. 141–147 (2015)
Zhang, J., Zhu, X., Zhou, M.: Learning automata-based particle swarm optimizer. In: 2018 IEEE Congress on Evolutionary Computation (CEC), pp. 1–6. IEEE (2018)
Zhan, Z.-H., Zhang, J., Li, Y., Chung H.S.-H.: Adaptive particle swarm optimization. IEEE Trans. Syst. Man, Cybern. Part B **39**, 1362–1381 (2009). https://doi.org/10.1109/TSMCB.2009.2015956
Zheng, S., Li, J., Janecek, A., Tan, Y.: A cooperative framework for fireworks algorithm. IEEE/ACM Trans. Comput. Biol. Bioinforma **14**, 27–41 (2017). https://doi.org/10.1109/TCBB.2015.2497227

Chapter 5
Applications of Multi-reinforcement Cellular Learning Automata in Channel Assignment

5.1 Introduction

Cellular learning automaton (CLA) (Beigy and Meybodi 2004; Rezvanian et al. 2018b) is a combination of cellular automata (Wolfram 1986) and learning automata (Kumpati and Narendra 1989; Rezvanian et al. 2018a). In this model, each cell of a cellular automaton (CA) contains one or more learning automata (LA). The LA or LAs residing in a particular cell define the state of the cell. Like CA, a CLA rule controls the behavior of each cell. At each time step, the local rule defines the reinforcement signal for a particular LA based on its selected action and the actions chosen by its neighboring LAs.

Consequently, the neighborhood of each LA is considered to be its local environment. A formal definition of CLA is provided by Beigy and Meybodi (2004). The authors also investigated the asymptotic behavior of the model and provided some theoretical analysis on its convergence. One of the first studies of CLA was the solution for channel assignment problem in cellular networks (Beigy and Meybodi 2009).

This chapter considers the channel assignment problem in wireless networks. Two different problems are considered in this section. First, we investigate the applicability of multi-reinforcement models in channel assignment in multiple collision domains. Next, we propose a method based on multi-reinforcement LA for distributed channel assignment in the mesh network.

The rest of the chapter is organized as follows: first, in Sect. 5.1, we solve the channel assignment in multiple collision domains problem using the multi-reinforcement models, which are designed explicitly for subset selection. Section 5.2 proposes a distributed method for distributed channel assignment in mesh networks. Finally, a summary of the chapter is provided in Sect. 5.3.

R. Vafashoar et al., *Cellular Learning Automata: Theory and Applications*, Studies in Systems, Decision and Control 307,
https://doi.org/10.1007/978-3-030-53141-6_5

5.2 Distributed Channel Assignment in Multiple Collision Domains

In this section, we investigate the application of the multi-reinforcement CLA models on a simple channel assignment problem described in (Felegyhazi et al. 2007; Vallam et al. 2011; Yang et al. 2012).

5.2.1 *System Model*

The network model in this section closely follows the models in (Felegyhazi et al. 2007; Vallam et al. 2011; Yang et al. 2012). We consider a static wireless network consisting of 2 N wireless devices/nodes. It is assumed that these devices are paired together. The set of all such pairs is denoted by $L = \{L_1, L_2, \ldots, L_N\}$. We assume that the two devices of a pair like L_i are communicating with each other. Additionally, each device only communicates with its paired device. As in (Felegyhazi et al. 2007; Vallam et al. 2011; Yang et al. 2012), we assume that the nodes are backlogged and always have packets to transmit. We model each communication as an undirected link between two nodes v_i and u_i, where v_i and u_i are two wireless devices communicating with each other. The use of the undirected link model is based on the fact that the IEEE 802.11 DCF requires the sender to be able to receive the acknowledgment message from the receiver for every transmitted packet. Each wireless device is equipped with multiple radio interfaces. Moreover, there are $C > 1$ orthogonal channels available in the network, e.g., 12 orthogonal channels in the IEEE 802.11a standard. The set of available channels is denoted by $C = \{c_1, c_2, \ldots, c_C\}$. To communicate, two devices of a pair have to tune at least one of their radios to the same channel(s). Parallel communications are allowed between two devices if they share multiple channels on radios. However, different radios on a node should be tuned to different channels to avoid the co-radios interference in a device. As two devices in a pair like L_i only communicate with each other, it is reasonable to assume that they both have the same number of radios which is denoted by k_i. Also, it is reasonable to assume that $k_i < C$ for all $L_i \in L$ as the channel assignment becomes trivial otherwise.

As the two nodes of a pair can possess several radios, they can communicate with each other at the same time over multiple channels. The communication between two nodes in a pair like L_i over a channel may be unsuccessful due to interference from other nodes. In this section, the potential interference model is employed. If u_i or v_i are in the interference range of the third node like v_j, they cannot communicate on a channel like c_l while v_j is sending or receiving data on c_l.

In the system model, each pair L_i tries to maximize its throughput. We also assume that different pairs do not cooperate, and each one tries selfishly to maximize its throughput. The throughput of each pair is maximized if it can select the maximum number of noninterfering channels. *Accordingly, the objective of each pair is to select the maximum number of noninterfering channels*. Moreover, some pairs may leave some radios idle if they cannot find any noninterfering channels for these radios. In the investigated channel assignment algorithm (Sect. 5.2.2), pairs are encouraged to leave some of their radios idle rather than assigning them with interfering channels by getting small rewards on their idle radios. Idle radios on some pairs would provide chances for some other pairs to tune their radios on more channels. An illustrative example of the described model is given in Fig. 5.1.

Here, we use a conflict graph to model possible interferences between different pairs/links. Each pair L_i in the system is represented by a node in the conflict graph, and there is an edge between two nodes in the conflict graph if and only if their associated pairs interfere with each other. Hence, communication in a pair is successful if none of the adjacent pairs in the conflict graph use that channel for their communications. An example conflict graph of a network with 25 nodes is depicted in Fig. 5.2.

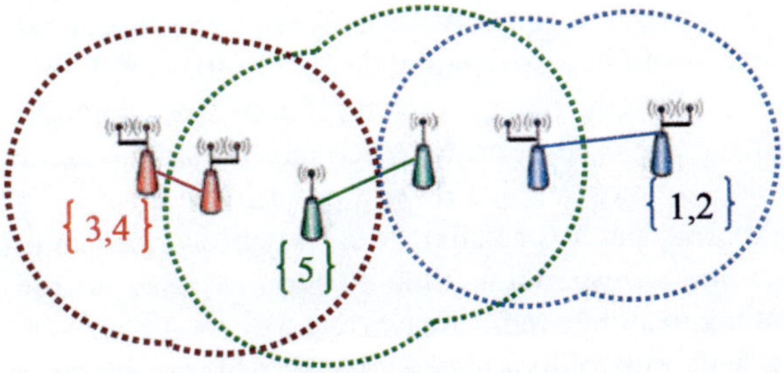

Fig. 5.1 A network with three pair of nodes and C = {1, 2, 3, 4, 5}

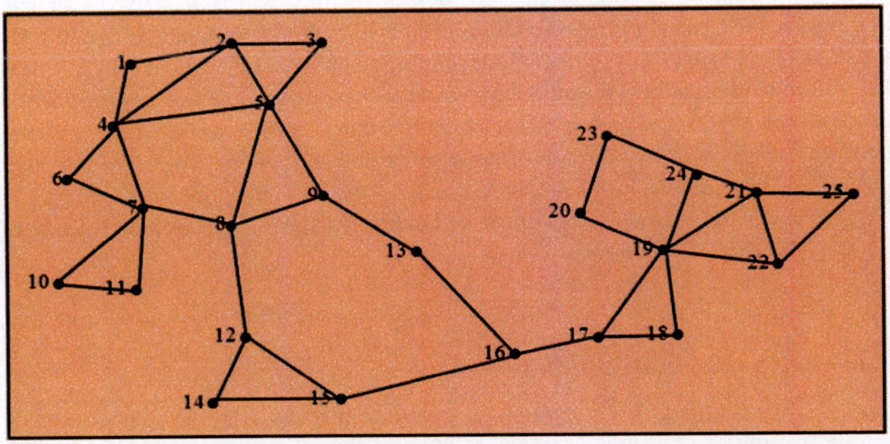

Fig. 5.2 A sample conflict graph of a wireless network with 25 pairs of users. Each pair of users is modeled as a node in the conflict graph, and the possible interferences between nodes are represented by the edges of the graph

5.2.2 Channel Assignment Algorithm Based on CLA

The approach presented in (Vafashoar and Meybodi 2019) is used for solving the introduced channel assignment problem by the CLA approach. A CLA with N cells is considered where N represents the number of nodes in the conflict graph (or pair of devices/nodes in the network). The CLA is isomorphic with the considered conflict graph, i.e., cells i and j are connected if and only if nodes i and j are adjacent in the conflict graph. Each cell of the CLA contains a LA which controls the channel assignment strategy of its associated pair of devices.

Each LA, like LA_i, has an action set of size $|C| + k_i$, where k_i represents the number of equipped radios on each node/device in the ith pair, and C represents the available channels of the node. Here, the action set size is $|C| + k_i$ rather than $|C|$ since each pair can choose to leave some of its radios in an idle state. During each iteration of the assignment procedure, each LA, like LA_i, selects a set of k_i simple-actions according to its internal state. Then, these chosen simple-actions are assigned to the radios of the associated nodes. After the action selection step, each LA receives a reinforcement signal, which represents the favorability of its selected simple-actions. In the proposed algorithm, each selected simple-action corresponding to an available channel receives one reward point if the channel is not selected in any of the neighboring pairs (or cells). If a chosen channel is also selected in a neighboring cell, the corresponding simple-action receives no reward. A simple-action corresponding to an idle radio always receives $\varepsilon < 1$ reward point (or receives one reward point with probability ε). Accordingly, pairs are encouraged not to select interfering channels so that these channels can be utilized by the nearby nodes. The CLA of the proposed channel assignment algorithm can be implemented based on any one of the LA models that are described in Sect. 3.1 in Chap. 3 or can be implemented based on the model described in Sect. 3.2 in Chap. 3. Figure 5.3 illustrates the channel assignment algorithm based on the CLA, which uses LA type III. In order to illustrate the operation of the channel assignment algorithm, the procedure in Fig. 5.3 is presented in sequential form. However, each pair L_i of the network runs its code synchronously with other nodes of the network, as illustrated in Fig. 5.4. The CLA in Fig. 5.4 is based on LA type II. The algorithm based on LA type I is much similar to the ones illustrated in Fig. 5.4; hence, it is not provided here for the sake of space economy.

5.2.3 Experimental Studies

In this section, we examine the performance of the CLA models on the channel assignment problem described in Sect. 5.2.1. Similar to (Felegyhazi et al. 2007; Vallam et al. 2011; Yang et al. 2012), we consider random networks for performance evaluations. In this regard, two networks are generated by randomly placing 20 and 25 nodes in $1000 \times 1000\ \text{m}^2$ of area. The associated conflict graphs of these networks

Algorithm 5-1. Channel assignment procedure based on CLA with multi-signal LA type III

Input:

$G = (V, E)$: conflict graph of the network

$C = \{c_1, c_2, \ldots, c_N\}$: The set of available Channels in the network.

$K = (K_1, K_2, \ldots, K_N)$. K_i: number of equipped radios in pair L_i.

λ: learning rate of each LA.

Initialization:

Consider an irregular CLA isomorphic with the network.

LA_i in the CLA has the following characteristics:

- It is a multi-signal LA with $r_i = |C| + K_i$ simple-actions and super-action size of K_i.
- $A_i = \{a_{i1}, \ldots, a_{ir_i}\}$.
- The energy level of the LA is represented by a r_i-dimensional vector x_i, where $x_{ij} \in [0,1]$ represents the energy level of the jth simple-action
- $x_i(0) = [K_i/r_i, K_i/r_i, \ldots, K_i/r_i]^T$.

Set $t \leftarrow 0$.

Repeat

1. For each cell like cell i do:
   a. Select a set of simple-actions according to their energy levels; name it $S'_i(t)$.
   b. Randomly select one of the closest super-actions to $S'_i(t)$; denote it by $\alpha_i(t)$.
2. Generate a reinforcement vector for each LA like LA_i:
   a. For each action, $a_{ij} \in \alpha_i(t)$ do:
      i. If $a_{ij} \in C$:
      $$\beta_i^j(t) = \begin{cases} 1 \text{ if } a_{ij} \text{ is not selected in any neighboring pair} \\ 0 \text{ otherwise} \end{cases}$$
      ii. If $a_{ij} \notin C$:
      $$\beta_i^j(t) = \begin{cases} 1 \text{ with probability } \varepsilon \\ 0 \text{ otherwise} \end{cases}$$
3. Update each LA like LA_i as follows:
   a. Define $J = \{j | a_{ij} \in \alpha_i(t)\}$.
   b. Define $I = \{j | a_{ij} \in A_i - \alpha_i(t)\}$.
   c. Update the energy level of the LA as follows:
   $$x_{ij}(t+1) = x_{ij}(t) + \lambda \beta_i^j(t)\left(1 - x_{ij}(t)\right) \forall j \in J$$
   $$\chi = \frac{\lambda \sum_{l \in J} \beta_i^l (1 - x_{il}(t))}{\sum_{l \in I} x_{il}(t)}$$
   $$x_{ij}(t+1) = x_{ij}(t) - \chi x_{ij}(t) \forall j \notin J$$
4. Set $t \leftarrow t+1$

until some stopping condition is satisfied

Fig. 5.3 Channel assignment procedure based on CLA with multi-signal LA type III

are provided in Figs. 5.1, and 5.5. Moreover, a bidirectional single-hop flow between each node pairs is considered, which has a constant bit rate.

In order to evaluate the performance of the algorithms, we use the total number of none conflicting channels that are assigned to the different nodes in a network. Using this number, one can also obtain the total throughput of the network. Five instances of each network are considered in this section. These instances are summarized in Table 5.1. In this section, the network depicted in Fig. 5.5 will be denoted by PA, and its instances will be identified by PA-instance_number. Similarly, the network in Fig. 5.2 and its instances will be denoted by PB and PB-instance-number, respectively.

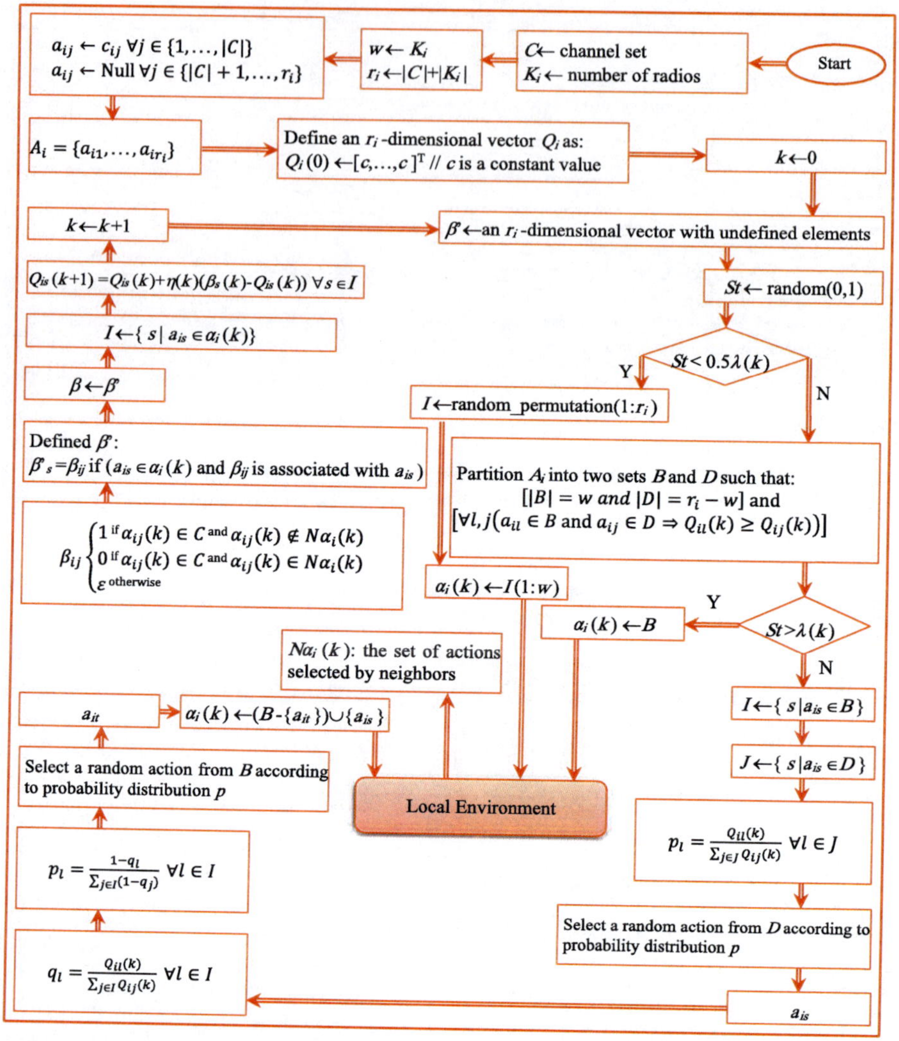

Fig. 5.4 Block diagram for the operation of the *i*th cell of the CLA based on LA type II. The operation of the *i*th cell is performed in node *i* of the network. All nodes synchronously run similar codes

5.2.3.1 The Sensitivity of the Proposed CLA Models to the Learning Rate

This section investigates the effects of the learning rate on the CLA model, which is based on multi-reinforcement LA type III. As the results of the other developed models are much similar, they are excluded for the sake of space economy. The results of this section are obtained over 25 independent runs on the different instances of PA, and each run of the algorithm on a particular instance is terminated after 1E+4

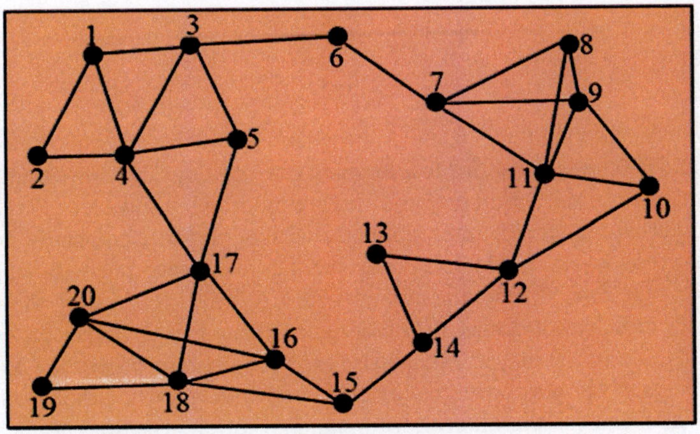

Fig. 5.5 A sample conflict graph of a wireless network with 20 pair of users

Table 5.1 Different instances of the channel assignment problems

	Instance 1	Instance 2	Instance 3	Instance 4	Instance 5
Channel num	3	12	12	12	12
Radio num	2	2	4	6	8

Each instance determines the number of radios and the number of channels for each node in the considered networks

iterations. The total number of the channels assigned without interference (which is called system performance (Yang et al. 2012)) is used as a comparison metric. In order to define system performance metric, we define the following terms.

The channel vector of each pair L_i:

$$s_{ij} = \begin{cases} 1 \text{ if } c_i \text{ is assigned to the radios of } L_i \\ 0 \text{ otherwise.} \end{cases} \tag{1}$$

Interference set of L_i as follows:

$$I(L_i) = \{\text{The set of pairs that are adjacent to } L_i \text{ in the conflict graph}\} \tag{2}$$

The number of noninterfering channels of L_i:

$$p_i = \sum_{j=1}^{C} s_{ij} - \sum_{j \in I(L_i)}^{C} s_i . s_j, \tag{3}$$

where '.' Stands for dot product.

Finally, system performance is defined as:

$$P = \sum_{i=1}^{N} p_i. \tag{4}$$

They are considering Fig. 5.6, the proposed model obtains inferior results when its learning rate is 0.001. Using this low learning rate prolongs the convergence time of the algorithm since the action probabilities are modified in tiny steps. Additionally, the proposed model demonstrates a slightly weaker performance under $\lambda = 0.1$ in some instances like PA-5, which is due to premature convergence without sufficient exploration of actions. Besides the settings with $\lambda \in \{0.001, 0.025, 0.1\}$, the algorithm obtains very close results under the other experimented settings. For the rest of the experiments, a learning rate of 0.01 will be used for all of the proposed CLA models as it exhibits appropriate convergence behavior. Moreover, using the small learning rate $\lambda = 0.01$ enables the algorithm to examine different actions for a sufficient number of times before convergence.

5.2.3.2 Comparison Results of Different CLA Models

The results of different CLA methods on the channel assignment problems are given in Table 5.2. The statistical results of each method on each one of the considered problem instances are obtained over 25 independent runs, and each run is terminated after 10,000 iterations. As it is evident from Table 5.2, all CLA models have very close results. However, the CLA models based on multi-reinforcement LA type III and maximum expected reward exhibit slightly better performances. LA type III is mainly based on the L_{RI} learning model. The convergence properties of CLA with the L_{RI} learning scheme are investigated in several works, where some appropriate convergence behaviors are proven for this model (Beigy and Meybodi 2004; Rezvanian et al. 2018b).

Problem instance 1 uses 3 orthogonal channels with two radios for each user. Accordingly, at most 40 and 50 nonconflicting assignments are possible for problems PA and PB, respectively. However, as it can be observed from the related conflict graphs of the problems, many nodes have neighborhoods larger than 2. Accordingly, none conflicting assignments may be impossible for some radios. Scenario 2 is a very easy problem, and all methods can achieve an optimal channel assignment as there are 12 different channels with only two radios in each node. This problem instance gets harder in scenarios 3, 4, and 5 as we increase the number of radios for each user.

Although in terms of convergence rate, a single LA III demonstrated rather weaker performances in comparison to the other developed LAs (see Chap. 3), a group of LA III learning agents exhibits better performances in a CLA. CLA-LA-III has obtained slightly better results on eight problem instances in comparison to CLA-LA-I. It also obtained better results on seven problems in comparison to CLA-LA-II. Accordingly, LA I seems to be a preferable choice for the CLA model. Moreover, CLA-LA-II is

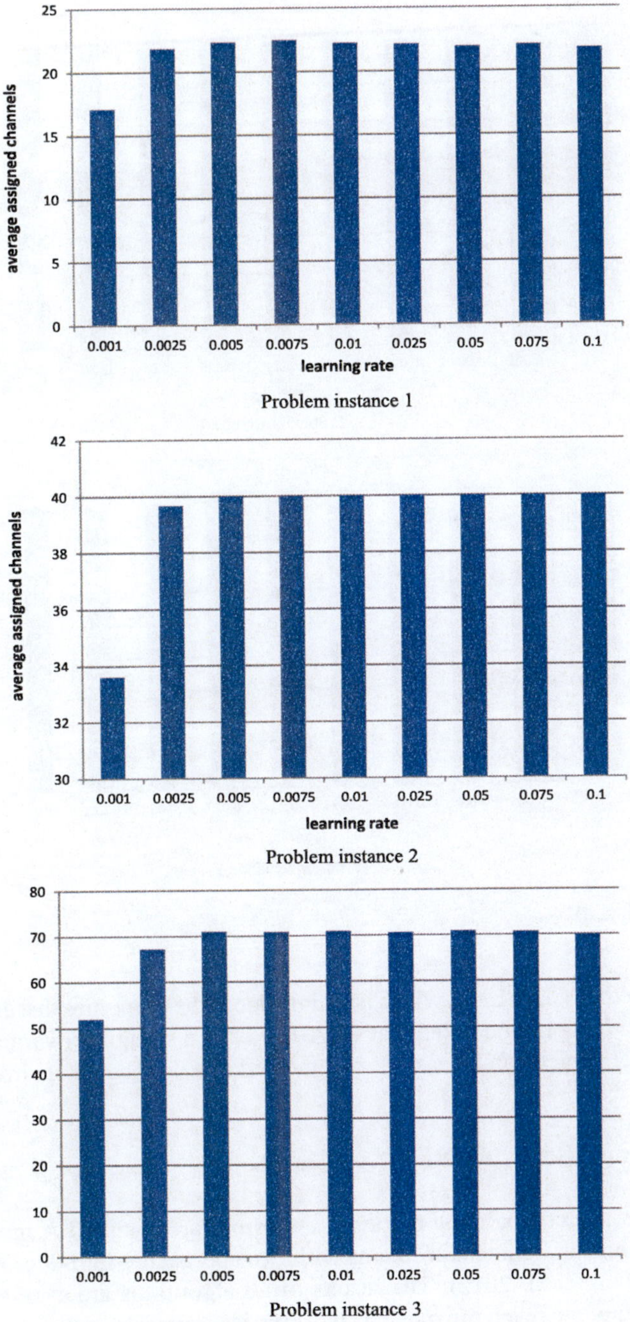

Fig. 5.6 The sensitivity of CLA with multi-reinforcement LA type III under different settings of learning rate on PA. The results indicate the average number of non-interfering channel assignments obtained after 1E+4 iterations

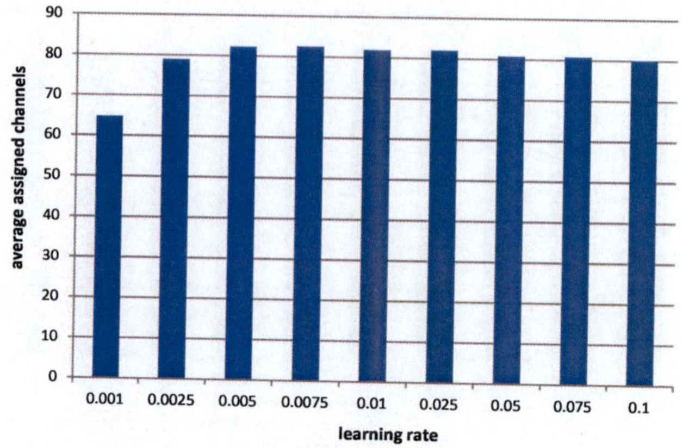

Problem instance 4

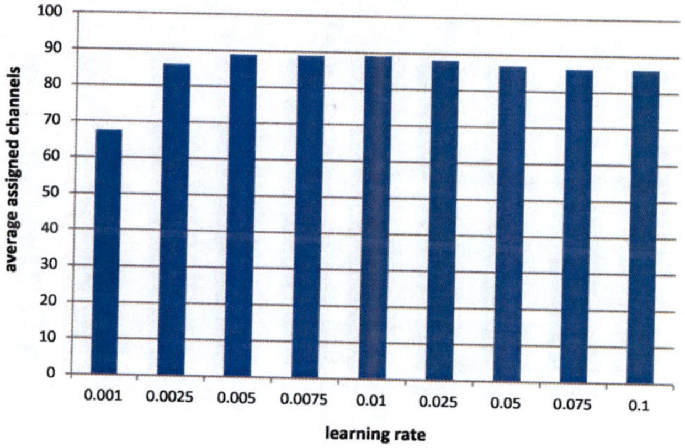

Problem instance 5

Fig. 5.6 (continued)

slightly better than CLA-LA-I, which is mainly due to the properties that are discussed in Sect. 3.1.2. It should be noted that CLA-LA-III is a straightforward and straight forward model, which can be easily implemented in a distributed environment.

5.2.3.3 Comparison with Other Approaches

In this section, we compare the proposed multi-reinforcement CLA models on the described channel assignment problems with two approaches introduced in (Vallam et al. 2011; Yang et al. 2012). The results of all algorithms are averaged over 25 independent runs, and each run of each algorithm is terminated after 1E+4 iterations. The "total payoff" criterion, as described in (Felegyhazi et al. 2007), will be employed as a performance metric in comparisons. This criterion is also closely related to the throughput of the system. Let Y_{ic} denotes the number of radios in the neighborhood

Table 5.2 Comparison results of the different proposed CLA models on the channels assignment problems in terms of the mean, median, and std of the obtained solutions

	PA-1	PA-2	PA-3	PA-4	PA-5	PB-1	PB-2	PB-3	PB-4	PB-5
CLA-LA-I	2.13E+01	4.00E+01	6.98E+01	8.02E+01	8.60E+01	2.91E+01	5.00E+01	9.46E+01	1.10E+02	1.16E+02
	2.10E+01	4.00E+01	7.00E+01	8.00E+01	8.60E+01	2.90E+01	5.00E+01	9.50E+01	1.10E+02	1.16E+02
	8.91E−01	0.00E+00	6.24E−01	1.11E+00	1.14E+00	5.72E−01	0.00E+00	1.08E+00	1.21E+00	1.77E+00
CLA-LA-II	2.22E+01	4.00E+01	7.04E+01	8.12E+01	8.79E+01	2.92E+01	5.00E+01	9.53E+01	1.12E+02	1.17E+02
	2.20E+01	4.00E+01	7.00E+01	8.10E+01	8.80E+01	2.90E+01	5.00E+01	9.50E+01	1.13E+02	1.17E+02
	7.07E−01	0.00E+00	6.45E−01	7.64E−01	1.22E+00	6.88E−01	0.00E+00	1.07E+00	8.79E−01	1.01E+00
CLA-LA-III	2.22E+01	4.00E+01	7.12E+01	8.21E+01	8.96E+01	2.98E+01	5.00E+01	9.70E+01	1.13E+02	1.18E+02
	2.20E+01	4.00E+01	7.10E+01	8.20E+01	8.90E+01	3.00E+01	5.00E+01	9.70E+01	1.13E+02	1.18E+02
	5.97E−01	0.00E+00	6.88E−01	7.81E−01	7.68E−01	4.08E−01	0.00E+00	7.90E−01	9.80E−01	1.16E+00

of node i which are tuned on channel c; it should be noted that node i is also counted in Y_{ic} if it has a radio on channel c. The maximum data rate which can be achieved in the collision domain of node i on channel c can be denoted by $\tau_{\max}(Y_{ic})$. In practice, $\tau_{\max}(Y_{ic})$ is a non-increasing function of Y_{ic}. However for this section, $\tau_{\max}(Y_{ic}) = \tau$ will be considered independent of Y_{ic} and c. Additionally, it will be assumed that $\tau_{\max}(Y_{ic})$ is shared equally among the interfering radios which are assigned to channel c (Yang et al. 2012). Based on these assumptions, the utility or payoff of a pair of nodes like L_i can be obtained according to the following equation (Yang et al. 2012):

$$u_i = \sum_{c \in C} \tau_{ic} \text{ with: } \tau_{ic} = \begin{cases} \frac{1}{Y_{ic}} \tau & \text{if node pair } L_i \text{ has a radio on channel } c \\ 0 & \text{otherwise} \end{cases} \tag{5}$$

According to the defined utility for the node pairs, the total payoff or the total utility of the network can be defined as:

$$U = \sum_{i=1:N} u_i \tag{6}$$

Herein, the compared methods corresponding to Algorithm 1 presented in (Yang et al. 2012) and algorithm 2 presented in (Vallam et al. 2011) will be denoted by CAL1 and CAL2, respectively. Figure 5.7 gives the comparison results based on the total payoff criteria. It should be noted that we only considered the non-conflicting assigned channels in the computation of the total utility in the proposed CLA models. In the results, τ is set to 54Mbps.

As it can be noticed from the obtained results, the proposed CLA models can be effectively utilized for the given channel assignment problem. One advantage of the CLA models over the two other compared approaches is the synchronous nature of their updating procedures. In the CLA models at each instant of time, all nodes chose their channels and updated their variables simultaneously. Accordingly, all nodes can be updated several times in a short period, which significantly reduces the convergence time of the algorithms. However, although the two compared methods are distributed, they update the network nodes asynchronously and mostly one node at each time step. Accordingly, for a network consisting of N nodes, CLA approaches update each node $o(N)$ times more in the same period when they are compared with CLA1 and CAL2. Besides, the comparisons based on the total payoff criterion demonstrates the superiority of the CLA approaches over CLA1 and CLA2. CAL1 and CLA2 achieved identical or close solutions on the second and third instances of the two considered networks. However, they attained a lesser average total payoffs on the other hard instances of the described problems. Considering PA-1, PA-4, PA-5, PB-1, BP-4, and PB-5, there are significant differences between the achieved results of the proposed CLA based methods and CAL1 and CAL2.

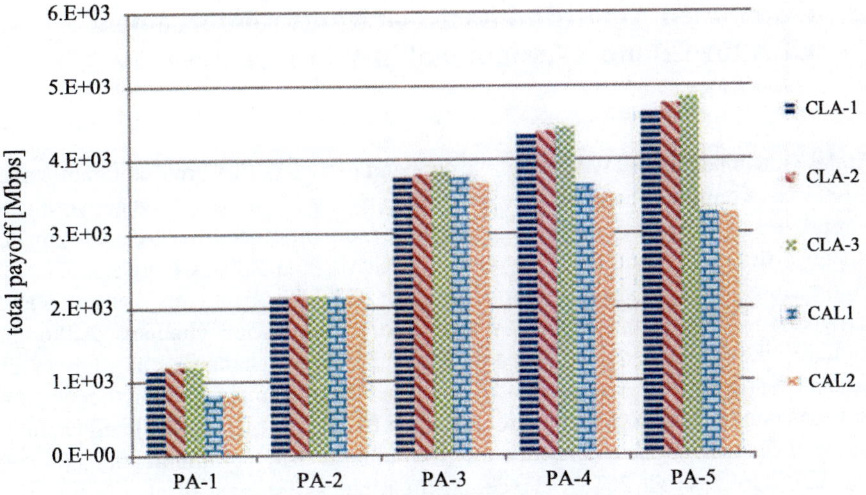

(a) Problem A

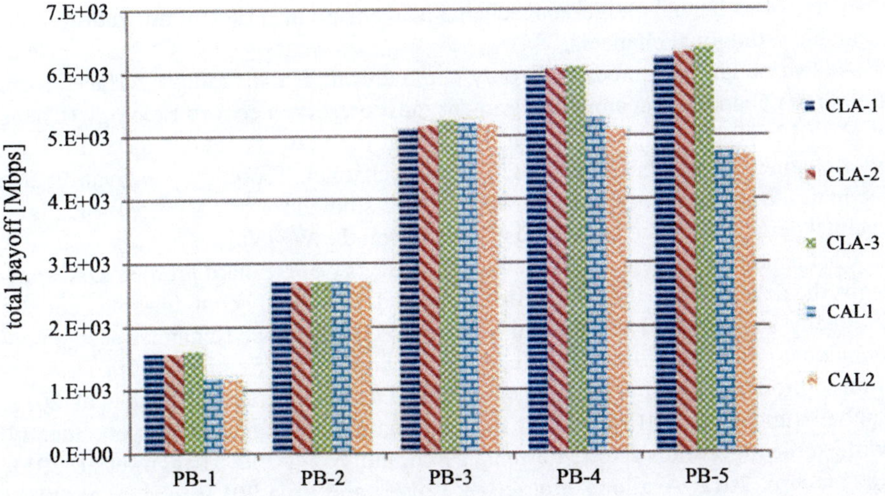

(b) Problem B

Fig. 5.7 Total payoff of the compared methods on the different instances of the channel allocation problems

5.3 Distributed Algorithm Based on Multi-reinforcement CLA for Channel Assignment in Wireless Mesh Networks

Wireless mesh networks (WMNs) are broadband networks that provide flexible and cost-effective communication structures. The capacity of the backbone network can be improved significantly by equipping multiple radio interfaces that are operating on multiple orthogonal (noninterfering) channels (Skalli et al. 2007; Cheng et al. 2012). Some current wireless LAN standards like IEEE 802.11 motivate this idea and allow each node to transmit simultaneously over multiple orthogonal channels. Although standards like 802.11a support up to 12 noninterfering channels, the number of channels is still limited. Consequently, some nearby transmissions may occur on a common channel. The communication between two nodes over an assigned channel may be unsuccessful due to interference from other nearby communications (Wang et al. 2015; Iqbal et al. 2018). Hence, an improper assignment of channels to the radios can lead to the under-utilization of the network. Several channel assignment approaches have been proposed in the literature to alleviate this issue. The main objective of a channel assignment scheme is to ensure the efficient utilization of the available orthogonal channels.

As two neighboring nodes can only communicate if their radio interfaces share a common channel, a channel assignment must preserve network topology (Cheng et al. 2012). Topology preservation means that the network retains its entire links after channel assignment as if it uses a single channel. Topology preservation also facilitates the routing mechanisms, and the well-studied routing protocols of single-channel networks can be employed effectively on the WMNs.

Channel assignment generally falls in the category of NP-hard problems. Accordingly, the developed channel assignment methods commonly contain some heuristic or greedy steps to find suboptimal solutions. The developed channel assignment techniques can be classified as static, dynamic, and hybrid (Skalli et al. 2007). Algorithms introduced in (Subramanian et al. 2008; Duarte et al. 2012; Cheng et al. 2012; Subbaiah and Naidu 2015) are static meaning that they assign channels permanently, while some algorithms like (Mohsenian Rad and Wong 2009; Riggio et al. 2011; Rezgui et al. 2012; Avallone et al. 2013; Gálvez and Ruiz 2013; Iqbal et al. 2018) update their assignments in a short-term or long-term basis. In a hybrid channel assignment, some radio interfaces may be assigned permanent channels, while the channels on the other radios may change with time (Li and Feng 2010; Deng et al. 2019). From another point of view, channel assignment methods can be categorized as distributed or central. In the distributed approaches, each node performs its assignments by employing local information (Mohsenian Rad and Wong 2009; Duarte et al. 2012; Deng et al. 2019); however, in centralized approaches, a controller unit collects the global topology information and assigns a channel to each link to maximize the whole network performance (Ali and Ngadi 2016; Zhao et al. 2018; Jayaraman et al. 2018; Park et al. 2018; Zeng et al. 2019). In centralized methods, a single controller unit acquires topology and traffic information from the network nodes and performs the channel allocation. Consequently, centralized methods may be unsuitable for dynamic channel allocation purposes as they require strong coordination, which results in high computational complexity and significant signaling overhead. On the other hand, a distributed method would have less optimization power in comparison to central approaches since channels are assigned based on local and partial information.

Most of the developed channel assignment methods are based on some heuristic and greedy schemes (Marina et al. 2010; Vallati and Mingozzi 2015; Chaudhry et al. 2016a, b; Islam et al. 2018). Approaches based on particle swarm optimization (Cheng et al. 2012, 2013), genetic algorithm (Selvakumar and Revathy 2018, 2019), simulated annealing (Lin et al. 2012), gravitational search (Ding et al. 2012), and Tabu search (Subramanian et al. 2008; Cheng and Yang 2011; Rezgui et al. 2012; Doraghinejad et al. 2014; Chen and Chen 2015) have been previously investigated for channel assignment. Subramanian et al. used the Tabu search and a greedy approximation algorithm related to the Max K-cut problem for centralized and distributed channel allocation (Subramanian et al. 2008). The CLICA is based on a graph-theoretic formulation of channel assignment and tries to minimize the maximum interference in the network (Marina et al. 2010). It greedily finds low interference topologies while preserving network connectivity. Linear array beam-forming based channel assignment (LAB-CA) involves a repeated greedy procedure which starts with finding a weighted maximal independent set in the conflict graph. Then, it assigns a common channel to the members of this set and removes them from the conflict graph (Chaudhry et al. 2016b).

Due to the limitations of the central channel assignment methods (Alim Al Islam et al. 2016), distributed schemes are favorable in many scenarios. However, as discussed in (Alim Al Islam et al. 2016), distributed channel assignment schemes have their drawbacks. These approaches are not amenable to a high degree of optimization as each node makes its own decisions based on local and incomplete information. A distributed channel assignment can be modeled as a multi-player game. Several methods have been proposed which can guarantee convergence to Nash equilibria in multi-player games under certain conditions. However, it can be shown that some equilibrium channel assignments may not preserve all links when utility functions are based on local information. To alleviate this issue, the proposed distributed channel assignment schemes operate asynchronously. During each iteration of these algorithms, only one node is updated, and it only considers the assignments which preserve its connections. However, these asynchronous algorithms may have less optimization power as they can follow limited trajectories. Consequently, some solutions may be unreachable from initial assignments. The second issue with the asynchronous mechanisms is their slow searching process: only one node (or few nodes) can be updated during each iteration.

As stated previously, due to the limited number of available channels and equipped radio interfaces in a node, efficient utilization of multi-channel architecture requires an intelligent channel assignment mechanism. This section introduces a distributed algorithm for intelligent channel assignment in wireless mesh networks. Here, we want to illustrate the application of LA models provided in Sect. 3.4. For this purpose, the network is modeled using a CLA, which is based on the MNLA model. Each LA is responsible for the selection of an appropriate set of channels for its corresponding node.

5.3.1 Network Model and Problem Definition

This section formulates the network model to be considered, as well as the definition of the channel assignment problem.

5.3.1.1 Network Model

A wireless mesh network with stationary wireless nodes (i.e., wireless mesh routers) is considered. Each mesh node is equipped with several radio interfaces. The number of radio interfaces on node u is denoted by $R(u)$. It is assumed that all radios in the network operate in half-duplex mode with Omni-directional antennas. There are a certain number of non-interfering channels available in the network. This set of available channels is represented by $\Omega = \{\varsigma_1, \ldots, \varsigma_\Gamma\}$. Each radio interface can be turned on one of the available channels. Transmission can occur between two nodes u and v if their radios share a common channel, and the receiver radio lies within the transition range of the sender. For the ease of exposition, it is assumed that all radios have identical transmission ranges. Also, we consider bidirectional (DATA-ACK) communication as in 802.11.

Consequently, the network topology can be modeled as an undirected graph $G = (V, E)$ called potential communication graph (Avallone and Akyildiz 2008). Here, V and E denote the set of nodes and links, respectively. A communication link like $(v, u) \in E$ signifies that the nodes u and v can communicate with each other as long as they have radio interfaces on a common channel.

The broadcasting nature of the wireless medium and accordingly, so-called co-channel interference are the major factors affecting a WMN's performance. There are two categories of interference models commonly used in the related references protocol model and physical model. As the presented approach in this chapter is a traffic-independent approach, the protocol model is utilized here. In the protocol model, an interference range is considered for each transmitter, which identifies the nearby transmissions that the transmitter interferes with. We assume that all transmitters have an identical interference range. Based on the interference model, the pairs of communication links that can interfere with each other are represented using a conflict graph. The set of communication links constitute the vertex set of the conflict graph. There is an edge between two vertices $e_i = (s, t)$ and $e_j = (v, u)$ in the conflict graph if communication on e_i can interfere with communication on e_j (assuming that both of the links are tuned on an identical channel). Formally, the conflict graph for G can be defined as:

Table 5.3 Summary of the used notations in CLACA

Notation	Description
$G(V, E)$	Potential communication graph
$v_j \in V$	A network node
(v_i, v_j)	The link between nodes v_i and v_j
n_j	Number of adjacent nodes to node v_j
$\mathrm{Ne}(v_i, s)$	The sth adjacent node to node v_i
$cell^j$	The jth cell of the CLA
LA^j	The MLA associated with node v_j
M^j	The probability matrix of LA^j
$\chi(k)$	Candidate assignment
$\chi(k)(v_l, v_j)$	Candidate channel for (v_l, v_j)
$\chi^M(k)$	A special assignment, which assigns 1 to each edge
$\alpha^j(k)$	Chosen channels by LA^j during iteration k
$\beta^j(k)$	Reinforcement signal for LA^j in iteration k
$\mathrm{Int}(\chi(k), (v_l, v_j))$	Number of links interfering with (v_l, v_j) in the assignment $\chi(k)$

$$G = (V_c, E_c) \text{ with } V_c = E \text{ and } E_c = \{(e_i, e_j)\\ = ((s, t), (u, v))|e_i, e_j \in V_c \text{ and } \min(d(s, u), d(s, v), d(t, u), d(t, v)). \quad (7)$$

Here, ξ is the interference range and $d(v_i, v_j)$ returns the Euclidean distance between the transmitters in nodes v_i and v_j.

A channel assignment χ assigns a channel $\chi(e)$ to each edge $e \in E$. Since we assume uniform traffic on the links incident on each node, the minimization of the total number of interfering links is considered as the channel allocation objective. It should be noted that the number of the different channels assigned to the links incident on a particular node cannot exceed its number of equipped radios. The next statement, usually termed as the interface constraint of the channel assignment problem.

5.3.2 *The Proposed CLA-Based Channel Assignment Method (CLACA)*

This section introduces the proposed channel assignment approach. The potential communication graph $G(V, E)$ is utilized to represent the network topology. Also, we use n_j to represent the number of adjacent nodes to node v_j and $\mathrm{Ne}(v_j, s)$ to represent its sth neighbor. A brief description of the employed notations in CLACA is given in Table 5.3. The candidate solution to the channel assignment problem is obtained using a CLA, which is isomorphic to the network. It has $|V|$ cells, and each pair of cells $cell^j$, and $cell^k$ are adjacent if and only if $(v_k, v_j) \in E$. Each cell $cell^j$ contains an MNLA, which is denoted by LA^j. LA^j has an action set $A = \{\varsigma_1, \ldots, \varsigma_\Gamma\}$ and restriction R_j on the number of distinct actions that it can choose. The CLA represents a stochastic solution to the channel assignment problem. A deterministic solution associated with this solution will be represented by $\chi(k)$. The assignment $\chi(k)$: $E \rightarrow A \cup \emptyset$ defines a channel for each link of the network.

In CLACA, each LA^j tries to learn the most suitable channels for its associated node (v_j). In doing so, it must consider the chosen channels in its neighborhood. Learning suitable channels is an iterative procedure. During each iteration k, LA^j chooses a set of channels for its associated links. Then, it evaluates its decision according to the chosen channels in its neighborhood. Based on this evaluation, M^j is updated so that LA^j would be able to make better decisions in future generations.

We can view each generation of CLACA as three phases: channel selection, channel negotiation, and population update. During the channel selection phase, each MNLA selects a set of channels for its associated links. Then, channel negation happens. During channel negotiation, adjacent nodes decide on common channels for their incident links. The chosen channels are evaluated, and the MNLAs and fitness values are updated in the population update phase. These three phases are discussed in detail in the following subsections. In what follows, the terms MNLA and cell may be used interchangeably when there is no ambiguity.

5.3.2.1 Channel Selection

As stated previously, each cell of a CLA contains an MNLA and is associated with a specific node of the network. During the channel selection phase, each MNLA selects a set of channels for the associated links of its node. An MLA selects an n-tuple of channels according to its probability matrix. Each n-tuple defines a set of suggested channels, which will be denoted by $\alpha^j(k)$. Each component $\alpha_t^j(k)$ of $\alpha^j(k)$ defines a suggested channel for link $(v_j, \mathrm{Ne}(v_j, t))$.

5.3.2.2 Channel Negotiation

During the channel selection phase, two adjacent cells like $cell^j$ and $cell^l$ may choose distinct channels for their incident link (v_j, v_l). Accordingly, a series of negotiations happen so that such adjacent cells can decide on a common channel. First, each cell $cell^j$ computes the interference level of its sth selected channel based on $\chi(k)$. To do this, it hypostatically sets $\chi(k)(v_j, \mathrm{Ne}(v_j, s))$ to $\alpha_s^j(k)$ and then computes $\mathrm{Int}(\chi(k), (v_j, \mathrm{Ne}(v_j, s)))$. This obtained interference level will be denoted by $\mathrm{Int}\left(\chi(k), \alpha_s^j(k)\right)$. After computing the interference levels, the negotiations occur in three steps as follows.

During the first phase of negotiations, each cell accepts the chosen channel by a neighboring cell on their incident link if their chosen channels are identical. Assume $\mathrm{Ne}(v_j, s) = v_l$ and $\mathrm{Ne}(v_l, t) = v_j$. If $\alpha_s^j(k) = \alpha_t^l(k)$, then set $Accept\left(\alpha_t^l(k)\right) = Accept\left(\left(\alpha_s^j(k)\right) = 1\right.$.

During the second phase of negotiations, CLACA tries to define a channel for each undefined link (v_j, v_l) as follows. Let the chosen channels for this link be $\alpha_t^l(k)$ and $\alpha_s^j(k)$. CLACA accepts $\alpha_t^l(k)$ if $\alpha_t^l(k) \in \alpha^j(k)$ and $\mathrm{Int}\left(\chi(k), \alpha_t^j(k)\right) < \mathrm{Int}\left(\chi(k), \alpha_s^j(k)\right)$. After accepting $\alpha_t^l(k)$, $\alpha_s^j(k)$ is considered as rejected. (In the case of tie, i.e. $\mathrm{Int}\left(\chi(k), \alpha_t^j(k)\right) = \mathrm{Int}\left(\chi(k), \alpha_s^j(k)\right)$, either $\alpha_s^j(k)$ or $\alpha_t^j(k)$ is accepted randomly).

During the third phase of negotiations, each cell considers its undefined links: the links that their channels have not been defined in the previous two phases. For each undefined link like (v_j, v_l), assuming $\mathrm{Ne}(v_j, s) = v_l$ and $\mathrm{Ne}(v_l, t) = v_j$, $cell^j$ can use either $\alpha_t^l(k)$ or $\alpha_s^j(k)$. Among all possible choices for its corresponding undefined links, $cell^j$ chooses the one which has the total lowest interference level and satisfies the radio constraint of the cell. We denote the chosen channels of $cell^j$ in this phase with $\breve{\alpha}^j(k)$. Next, each cell presents its chosen channels for its undefined links to its adjacent cells. Each cell $cell^j$ accepts $\alpha_t^l(k)$ if $\alpha_t^l(k) = \breve{\alpha}_s^j(k)$ and rejects its own action $\alpha_s^l(k)$. It should be noted that, according to our simulations, the number of possible allowable choices for the links of a node in this stage is small. However, this number may be large in very dense networks. To avoid computational complexities, we only consider 20 random choices whenever the number of possible choices is greater than 20.

After the above three negotiation phases, CLACA assigns Null to each remaining undefined link.

5.3.2.3 CLA Update

After channel negotiation, the CLA is updated. In this phase of the algorithm, a set of reinforcement signals is generated for each MNLA, and the probability matrices are updated accordingly. The candidate solution is also obtained based on the accepted channels. Furthermore, CLACA updates the local fitness of each cell in this block of the algorithm.

First, the candidate solution is obtained based on its associated accepted channels. Assume that $\mathrm{Ne}(v_j, s) = v_l$ and $\mathrm{Ne}(v_l, t) = v_s$. The channel associated with the link (v_l, v_j) in the ith candidate solution is obtained according to the following rule:

$$\chi(k+1)(v_i, v_l) = \begin{cases} \alpha_s^j(k) & \text{if } \alpha_s^j(k) \text{ is accepted} \\ \alpha_t^l(k) & \text{if } \alpha_t^l(k) \text{ is accepted} \\ \phi & \text{otherwise} \end{cases} \quad (8)$$

Based on the obtained candidate solution $\chi(k)$, the fitness of each link is obtained according to the following equitation.

$$fl(v_l, v_j) = \frac{Int(\chi(k+1), (v_l, v_j))}{Int(\chi^M, (v_l, v_j))}, \tag{9}$$

where $fl(v_l, v_j)$ represents the number of links interfering with (v_l, v_j), which is normalized by the maximum possible interference on (v_l, v_j).

According to the obtained link fitness values, the reinforcement signal β_s^j for the *s*th chosen channel in the selected n-tuple of $cell^j$ is calculated as follows:

$$\beta_s^j(k)\begin{cases} 1 & \text{if } \alpha_s^{ij}(k) \text{ is rejected} \\ fl(v_l, v_j) & \text{otherwise} \end{cases} \tag{10}$$

Here, it is assumed that $\mathrm{Ne}(v_j, s) = v_l$. Based on its received reinforcement, each MNLA updates its probability matrix (Eqs. 3.56–3.60 from Chapt. 3). The reinforcement signal for each chosen channel is generated based on the effectiveness of the channel for its link. This effectiveness can be measured using the interference level of the channel for the link. Complete pseudocode for the proposed method is given in Fig. 5.8. It should be noted that for better understanding, the algorithm is presented in central from.

5.3.2.4 Experimental Studies

In this section, the performance of the proposed approach is evaluated through simulation studies. The simulations are based on two network topologies. These topologies are generated by randomly placing 70 and 100 nodes on a 1000 m × 1000 m area. These topologies are denoted by N_{70} and N_{100}, and their characteristics are summarized in Table 5.4. Two instances for each network topology are considered in this work, one with 12 orthogonal channels and the other with three orthogonal channels. These two instances are identified by the suffixes Ch_{12} and Ch_3 in this section.

Additionally, we assume that all nodes are equipped with the same number of radio interfaces. The number of radio interfaces is chosen from the set {2, 4, 6, 8, 10, 12} when there are 12 orthogonal channels available. It is chosen from the set {2, 3} when there are three channels available for the network. To distinguish between different network instances with different radio interfaces, the suffix $R_{\#num\_radio}$ is employed.

To evaluate the performance of the proposed approach, it is compared with three evolutionary approaches based on particle swarm optimization (Cheng et al. 2012, 2013), and Tabu search (Subramanian et al. 2008). For conciseness, we use PSOI, PSOII, and Tabu to represent the approaches introduced in Cheng et al. (2012, 2013) and Subramanian et al. (2008), respectively. The parameter settings of these

Algorithm 5-2. CLACA

a. Initialization:
   1. $k \leftarrow 0$.
   2. Set $Z^j_{st}(1)$ to some initial value like 1. $\forall j \in \{1,\ldots,|V|\}, \forall s \in \{1,\ldots,n_j\}, \forall t \in \{1,\ldots,r\}$.
   3. Set $\mu^j_{st}(1)$ to some initial value like 1. $\forall j \in \{1,\ldots,|V|\}, \forall s \in \{1,\ldots,n_j\}, \forall t \in \{1,\ldots,r\}$.
   4. Set $M^j_{st}(1)$ to $1/r$. $\forall j \in \{1,\ldots,|V|\}, \forall s \in \{1,\ldots,n_j\}, \forall t \in \{1,\ldots,r\}$.
   5. For $j=1$ to $|V|$ do:
      i. For $s=1:n_j$ do:
         - $v_l \leftarrow \mathrm{Ne}(v_j,s)$ and obtain t such that $\mathrm{Ne}(v_l,t)=v_j$.
         - Set $\chi(1)(v_j,v_l)$ to some value like 0.

b. While stopping condition is not satisfied, do:
   1. $k \leftarrow k+1$.

   Action selection:

   2. for $j=1$ to $|V|$ do:
      i. Using M^j, select n_j channels, $\alpha^j(k)$, such that the number of distinct selected channels is at most R_j (according to the approach described in chapter 3).

   Channel negotiation:

   3. For $j=1$ to $|V|$ do:// *first phase of negotiations*
      i. For $s=1$ to n_j do:
         - $v_l \leftarrow \mathrm{Ne}(v_j,s)$ and assume that $\mathrm{Ne}(v_l,t)=v_j$.
         - Accept $\alpha^l_t(k)$ if $\alpha^j_s(k)=\alpha^l_t(k)$.
   4. For $j=1$ to $|V|$ do:// *second phase of negotiations*
      i. For $s=1$ to n_j do:
         - $v_l \leftarrow \mathrm{Ne}(v_j,s)$ and assume that $\mathrm{Ne}(v_l,t)=v_j$.
         - Accept $\alpha^l_t(k)$ if $\alpha^l_t(k) \in \alpha^j(k)$ and $\mathrm{Int}(\chi(k), \alpha^l_t(k)) < \mathrm{Int}(\chi(k), \alpha^j_s(k))$.
         - Reject $\alpha^l_t(k)$ if $\alpha^j_s(k) \in \alpha^l(k)$ and $\mathrm{Int}(\chi(k), \alpha^j(k)) > \mathrm{Int}(\chi(k), \alpha^j_s(k))$.
   5. For $j=1$ to $|V|$ do:// *third phase of negotiations*
      i. For $s=1$ to n_j do:
         - $v_l \leftarrow \mathrm{Ne}(v_j,s)$ and assume that $\mathrm{Ne}(v_l,t)=v_j$.
         - Find $\breve{\alpha}^j(k)$ which minimizes $\sum_{s=1}^{n_j} \mathrm{Int}\left(\chi(k), \breve{\alpha}^j_s(k)\right)$

           Such that:

$$\breve{\alpha}^j_s(k) \in \begin{cases} \{\alpha^j_s(k)\} \text{ if } \alpha^j_s(k) \text{ is accepted} \\ \{\alpha^l_t(k)\} \text{ if } \alpha^l_t(k) \text{ is accepted} \\ \{\alpha^j_s(k), \alpha^l_t(k)\} \quad \text{otherwise} \end{cases}.$$

   6. For $j=1$ to $|V|$ do:// the *third phase of negotiations cont.*
      i. For $s=1$ to n_j do:
      ii. $v_l \leftarrow \mathrm{Ne}(v_j,s)$ and assume that $\mathrm{Ne}(v_l,t)=v_j$.
      iii. Accept $\alpha^l_t(k)$ if $\breve{\alpha}^j_s(k)=\breve{\alpha}^l_t(k)$ and $\alpha^l_t(k)=\breve{\alpha}^j_s(k)$.
      iv. Reject $\alpha^l_t(k)$ otherwise.

   CLA update:

   7. For $j=1$ to $|V|$ do:
      i. For $s=1$ to n_j do:
         - $v_l \leftarrow \mathrm{Ne}(v_j,s)$ and assume that $\mathrm{Ne}(v_l,t)=v_j$.
         - Compute $\chi(k+1)(v_j,v_l)$ according to Eq. 5-8.
         - Compute $f(v_l,v_j)$ according to Eq. 5-9.
   8. For $j=1$ to $|V|$ do:
      i. For $s=1$ to n_j do:
         - Calculate β^{ij}_s
      ii. Update LA^j

Set the channel of each link (v_i,v_j) to $\chi(k)(v_i,v_j)$.

Fig. 5.8 Pseudocode for CLACA

Table 5.4 The characteristics of the employed random network topologies

Network	Number of nodes	Number of links	Min node degree	Max node degree
N_{70}	70	350	4	17
N_{100}	100	779	3	26

approaches are adopted from their related references. Also, whenever the corresponding setting of a parameter is not given in the related literature, it is obtained based on a limited number of trials and errors. The results of each algorithm on each test instance are obtained over 20 independent runs, and the average results are reported in the corresponding sections. Additionally, all runs of the compared methods, except for Tabu, are terminated after 5E+4 candidate solution updates for a fair comparison. As Tabu is the simplest method among all peer methods, its maximum number of allowable updates per run is extended to 2E+5 in the comparisons.

Two performance metrics are commonly used in the related literature: fractional network interference and network throughput (Cheng et al. 2013). Fractional network interference is defined as the ratio of the total network interference to the potential network interference number. For approaches that are based on assigning a single channel to each link, like CLACA and PSOII, the value of interference ratio is smaller than one. However, as PSOI is based on assigning channels to radio interfaces, each link may be assigned with more than one channel. Hence, the fractional network interference of PSOI may grow larger than one (see the simulation results of PSOI given in (Cheng et al. 2013)). Therefore, fractional network interference cannot be used as a fair performance metric here. The network throughput is defined as the sum of the capacity of all links in the network. Note that there might be several links on different channels between two adjacent nodes in PSOI. The following equation is used to calculate the capacity of each link *e*.

$$Capacity(e) = \frac{1}{1 + \text{number of links interfering with } e}. \quad (11)$$

$$\text{Network Throughput} = \sum_{e} \text{capacity}(e). \quad (12)$$

5.3.2.5 Learning Rate

The learning rate is one of the most important parameters in the LA-based systems. Consequently, this section investigates the behavior of the proposed channel assignment approach under different settings of this parameter. To this end, CLACA is tested on $N_{70}CH_{12}$ using different settings of λ from the set {0.001, 0.0025, 0.005, 0.0075, 0.01, 0.02, 0.04, 0.06}. The obtained results of the test are given in Fig. 5.9.

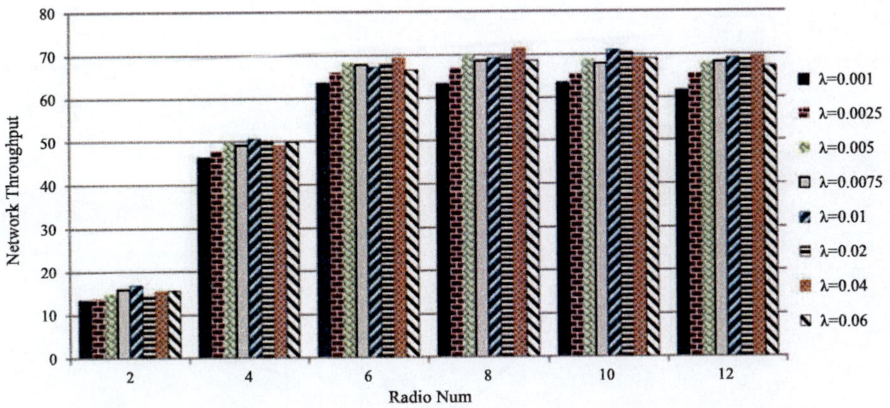

Fig. 5.9 Average network throughput of CLACA on the sample network of size 70 under different settings of λ

The most obvious point about the obtained results is related to the effect of the learning rate on the convergence speed of CLACA. With lower learning rates such as 0.001, the learning automata exhibit more explorative and random behavior. Accordingly, it takes a relatively long time for the algorithm to reach an appropriate solution. In contrast, the algorithm behaves more greedily when it uses a higher learning rate, such as 0.06. The lack of proper exploration, in this case, may result in premature convergence. As a result, the algorithm obtained relatively poor results on some tested instances when the learning rate is set to the values 0.001, 0.0025, and 0.06. CLACA shows very close performance for different settings of λ in the range [0.005, 0.04]. However, $\lambda = 0.01$ seems to be the most appropriate choice on several occasions. Consequently, this setting is adopted for the following experiments.

5.3.2.6 Topology Preservation

In the proposed method, all links associated with a candidate assignment are updated synchronously in each iteration. Accordingly, it is impossible to grantee that all links are preserved in the proposed method. However, as each cell strives to improve the connectivity of its associated node (according to the definition of local fitness and reinforcement signals), it can be assumed that most links are preserved in the obtained assignment. In this section, we experimentally investigate the ratio of the links that are preserved after the final assignment. To this end, the average ratio of the preserved links is recorded during the execution of the CLACA. This experiment is repeated over 100 independent runs, and the average results on $N_{70}CH_{12}$ are given in Fig. 5.10.

As can be observed from Fig. 5.10, some links may be left unassigned when each radio is equipped with only two radios. The probability that a link is left unassigned in $N_{70}R_2CH_{12}$ is about 0.023. This probability drops to less than 0.009 in $N_{70}R_4CH_{12}$.

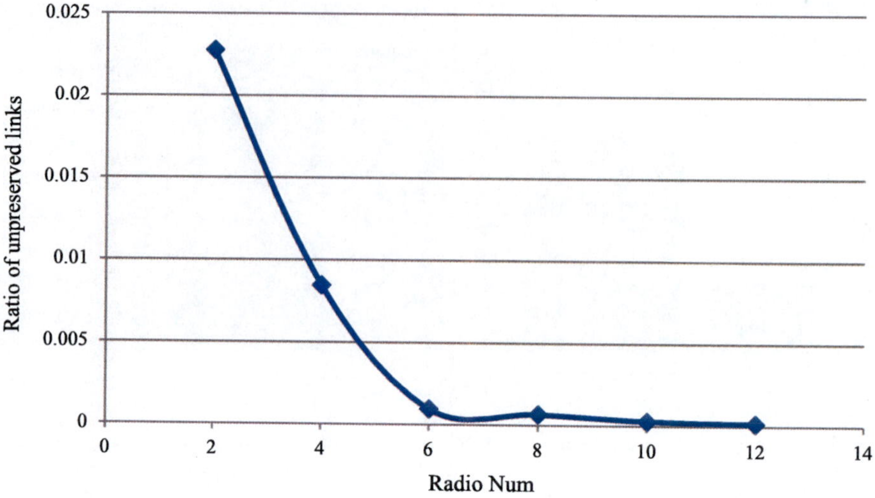

Fig. 5.10 The average ratio of unassigned links in $N_{70}CH_{12}$

These two test instances are rather hard for CLACA as they put stronger constraints on the channel selection operation of each node. For test instances with larger than four radios, the probability of an unassigned link in the final solution is neglectable (about 0.00016 in $N_{70}R_{12}CH_{12}$). Moreover, considering the results of the previous sections, it can be seen that the obtained throughput of the network is very close to instances using more than four radios. Consequently, one can simply reserve one radio for a default channel in these test instances to grantee the preservation of the network topology.

5.3.2.7 Comparison with Different Approaches

This section experimentally compares the performance of CLACA with those of other approaches. Figure 5.11 gives the comparison results in terms of average throughput over 20 independent runs. Considering the obtained results, it can be observed that the proposed method can achieve the best average throughput on most of the tested instances. The first point about the comparison results is about the type of the compared algorithms. CLACA is a distributed method that relies only on local information. However, the other three compared algorithms are central and are using global network information.

It should be noted that local search can only guarantee the convergence to a local optimum, which means that a distributed method that only utilizes local information cannot generally outperform well designed global optimization methods. The answer to the higher throughput of CLACA in comparison to the other approaches in cases like $N_{70}CH_{12}R_{10}$ can be described as follows. First, most of the developed channel central assignment methods like PSOII are based on just local operations.

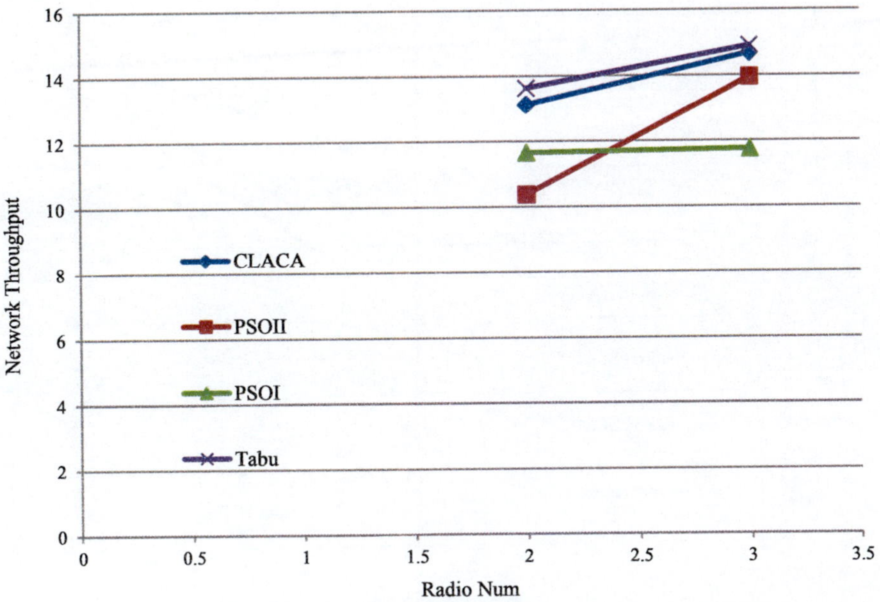

(a) $N_{70}CH_3$

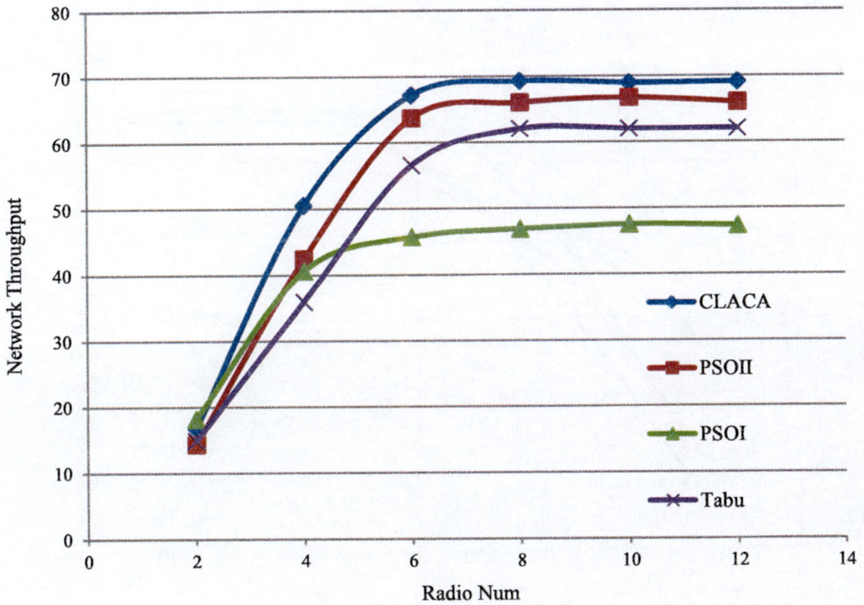

(b) $N_{70}CH_{12}$

Fig. 5.11 The comparison results of the peer methods in terms of network throughput

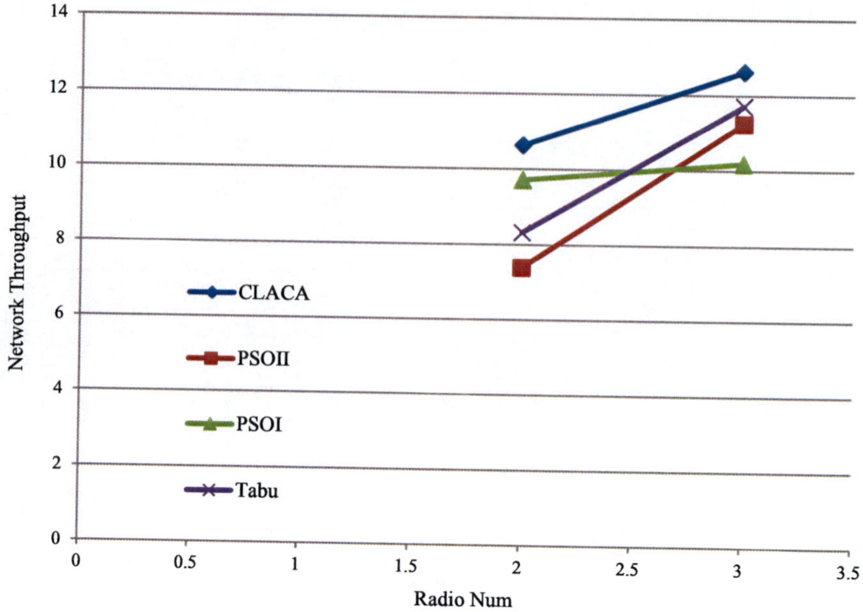

(c) $N_{100}CH_3$

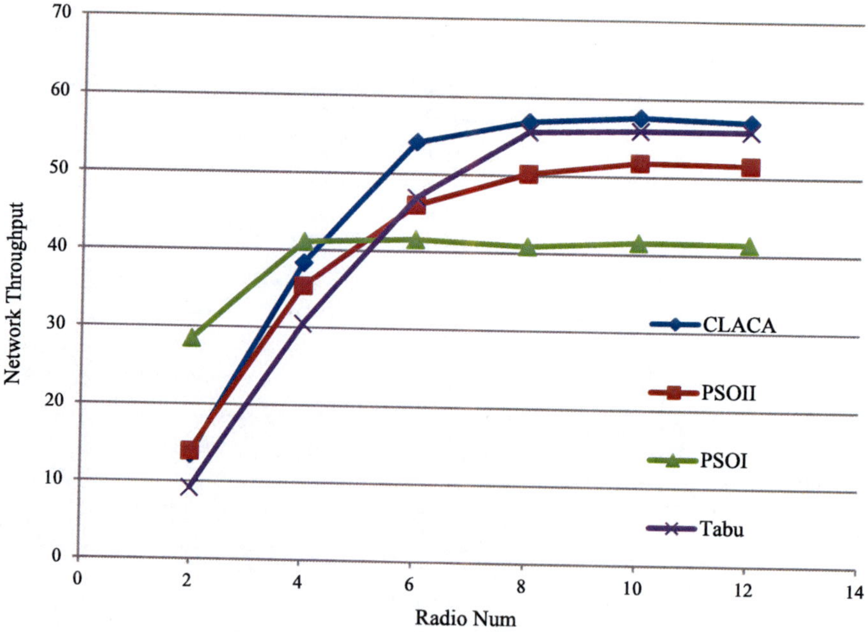

(d) $N_{100}CH_{12}$

Fig. 5.11 (continued)

The Update rule of PSOII only modifies one or two links in each iteration, and worst this modification is performed only if it preserves the topology of the current topology of the network. Accordingly, PSOII is just based on local operations. Also, as this algorithm can only perform limited modifications on a particle, many solutions may be theoretically unreachable for the algorithm from an initial population.

Moreover, during each iteration of CLACA, all LAs are updated synchronously at the same time so that they can learn from their neighbors' choices and select better channels in the next iterations. However, in other algorithms, only a few links are changed in each time step. Accordingly, their searching and learning procedure is much slower than CLACA. This issue is apparent when we compare the results of algorithms on $N_{70}CH_{12}$ and $N_{100}CH_{12}$. While PSOII is superior to Tabu on $N_{70}CH_{12}$, its results are inferior to $N_{100}CH_{12}$. This is because the dimensionality of $N_{100}CH_{12}$ is more than twice of the dimensionality of $N_{70}CH_{12}$ (considering the number of links in the networks). Accordingly, 5E + 4 number of fitness evaluations seems to be inadequate for PSOII to outperform Tabu on $N_{100}CH_{12}$. However, as Tabu is granted with a larger fitness evaluation number (2E + 5), it can explore the ample search space of $N_{100}CH_{12}$ better than PSOII.

PSOI outperforms other approaches on $N_{100}CH_{12}R_2$ and $N_{100}CH_{12}R_4$. This is mainly since, in PSOI, data transmission between two nodes can occur over multiple links. Higher dimensionality of N_{100} and a limited number of radios (2 and 4) dramatically degrades the performance of other approaches. However, PSOI can cope with this limitation and exploit the 12 utilized channels in a better way. Finally, the proposed CLACA method obtains some weaker results when the number of radios is limited. These relatively weaker results are mainly due to the channel negation phase of the algorithm. As discussed in the previous sections, it may be difficult for adjacent nodes to come up with a common channel when the number of radios is limited.

5.4 Conclusion

This chapter investigated the application of CLA on two-channel assignment problems. It was demonstrated that multi-reinforcement CLA models, which are designed for subset selection, can effectively solve the non-cooperative channel assignment problem in multiple collision domains. The comparison results with two effective peer methods demonstrate the superiority of the proposed CLA based channel assignment approach on this problem.

In the next part of the chapter, it is demonstrated that a multi-reinforcement CLA can be employed for solving the channel assignment problem in wireless mesh networks. The channel assignment problem can be modeled as an optimization problem, where the channel assignments for the network links are the problem variables. Due to the high correlation between these variables, obtaining optimum channel assignment is very hard for typical networks. To capture the existent dependencies among variables in the discrete search space of the channel assignment problem, we modeled the candidate solution of the problem by a cellular learning automaton. Based on this modeling, Sect. 5.3 introduced a distributed channel assignment approach for wireless mesh networks called CLACA. Unlike most of the proposed distributed channel assignment methods, there is no need for any asynchronization mechanism, and each node is capable of updating its associated channels in the candidate solutions along with other nodes. Because of the high explorative nature of the proposed method, it can effectively explore the search space and achieve appropriate solutions. We cannot guarantee topology preservation in a synchronous channel assignment approach unless we put some limitations on the search process. However, it was experimentally demonstrated that CLACA could preserve its links with a very high probability, especially when there are more than four radios equipped on each node. Additionally, we can always use a reserved radio to preserve topology in CLACA. Experimental studies also demonstrated the effectiveness of the proposed method in terms of the achieved throughput.

References

Ali, S., Ngadi, M.A.: Optimized interference aware joint channel assignment model for wireless mesh network. Telecommun. Syst. **62**, 215–230 (2016). https://doi.org/10.1007/s11235-015-0076-8

Alim Al Islam, A.B.M., Islam, M.J., Nurain, N., Raghunathan, V.: Channel assignment techniques for multi-radio wireless mesh networks: a survey. IEEE Commun. Surv. Tutor. **18**, 988–1017 (2016). https://doi.org/10.1109/COMST.2015.2510164

Avallone, S., Akyildiz, I.F.: A channel assignment algorithm for multi-radio wireless mesh networks. Comput. Commun. **31**, 1343–1353 (2008). https://doi.org/10.1016/j.comcom.2008.01.031

Avallone, S., Di, Stasi G., Kassler, A.: A traffic-aware channel and rate reassignment algorithm for wireless mesh networks. IEEE Trans. Mob. Comput. **12**, 1335–1348 (2013). https://doi.org/10.1109/TMC.2012.107

Beigy, H., Meybodi, M.R.: Cellular learning automata based dynamic channel assignment algorithms. Int. J. Comput. Intell. Appl. **8**, 287–314 (2009)

Beigy, H., Meybodi, M.R.: A mathematical framework for cellular learning automata. Adv. Complex Syst. **07**, 295–319 (2004). https://doi.org/10.1142/S0219525904000202

Chaudhry, A.U., Chinneck, J.W., Hafez, R.H.M.: Fast heuristics for the frequency channel assignment problem in multi-hop wireless networks. Eur. J. Oper. Res. **251**, 771–782 (2016a). https://doi.org/10.1016/j.ejor.2015.12.016

Chaudhry, A.U., Hafez, R.H.M., Chinneck, J.W.: Realistic interference-free channel assignment for dynamic wireless mesh networks using beamforming. Ad Hoc Netw. **51**, 21–35 (2016b). https://doi.org/10.1016/j.adhoc.2016.08.001

Chen, Y.-Y., Chen, C.: Simulated annealing for interface-constrained channel assignment in wireless mesh networks. Ad Hoc Netw. **29**, 32–44 (2015). https://doi.org/10.1016/j.adhoc.2015.01.019

Cheng, H., Xiong, N., Vasilakos, A.V., et al.: Nodes organization for channel assignment with topology preservation in multi-radio wireless mesh networks. Ad Hoc Netw. **10**, 760–773 (2012). https://doi.org/10.1016/j.adhoc.2011.02.004

Cheng, H., Xiong, N., Yang, L.T., et al.: Links organization for channel assignment in multi-radio wireless mesh networks. Multim. Tools Appl. **65**, 239–258 (2013). https://doi.org/10.1007/s11042-011-0800-7

Cheng, H., Yang, S.: Joint QoS multicast routing and channel assignment in multiradio multichannel wireless mesh networks using intelligent computational methods. Appl. Soft Comput. J. **11**, 1953–1964 (2011). https://doi.org/10.1016/j.asoc.2010.06.011

Deng, X., Luo, J., He, L., et al.: Cooperative channel allocation and scheduling in multi-interface wireless mesh networks. Peer-to-Peer Netw. Appl. **12**, 1–12 (2019). https://doi.org/10.1007/s12083-017-0619-8

Ding, Y., Huang, Y., Zeng, G., Xiao, L.: Using partially overlapping channels to improve throughput in wireless mesh networks. IEEE Trans. Mob. Comput. **11**, 1720–1733 (2012). https://doi.org/10.1109/TMC.2011.215

Doraghinejad, M., Nezamabadi-pour, H., Mahani, A.: Channel assignment in multi-radio wireless mesh networks using an improved gravitational search algorithm. J. Netw. Comput. Appl. **38**, 163–171 (2014). https://doi.org/10.1016/j.jnca.2013.04.007

Duarte, P.B.F., Fadlullah, Z.M., Vasilakos, A.V., Kato, N.: On the partially overlapped channel assignment on wireless mesh network backbone: a game theoretic approach. IEEE J. Sel. Areas Commun. **30**, 119–127 (2012). https://doi.org/10.1109/JSAC.2012.120111

Felegyhazi, M., Cagalj, M., Bidokhti, S.S., Hubaux, J.-P.: Non-cooperative multi-radio channel allocation in wireless networks. In: IEEE INFOCOM 2007—26th IEEE International Conference on Computer Communications. IEEE, pp 1442–1450 (2007)

Gálvez, J.J., Ruiz, P.M.: Efficient rate allocation, routing and channel assignment in wireless mesh networks supporting dynamic traffic flows. Ad Hoc Netw. **11**, 1765–1781 (2013). https://doi.org/10.1016/j.adhoc.2013.04.002

Iqbal, S., Abdullah, A.H., Ahsan, F., Qureshi, K.N.: Critical link identification and prioritization using Bayesian theorem for dynamic channel assignment in wireless mesh networks. Wirel. Netw. **24**, 2685–2697 (2018). https://doi.org/10.1007/s11276-017-1471-8

Islam, M., Abdur Razzaque, M., Mamun-Or-Rashid, M., et al.: Traffic engineering in cognitive mesh networks: Joint link-channel selection and power allocation. Comput. Commun. **116**, 212–224 (2018). https://doi.org/10.1016/j.comcom.2017.12.002

Jayaraman, R., Raja, G., Bashir, A.K., et al.: Interference mitigation based on radio aware channel assignment for wireless mesh networks. Wirel. Pers. Commun. **101**, 1539–1557 (2018). https://doi.org/10.1007/s11277-018-5776-4

Kumpati, S., Narendra, M.A.L.T.: Learning Automata: An Introduction. Prentice-Hall (1989)

Li, M., Feng, Y.: Design and implementation of a hybrid channel-assignment protocol for a multi-interface wireless mesh network. IEEE Trans. Veh. Technol. **59**, 2986–2997 (2010). https://doi.org/10.1109/TVT.2010.2047032

Marina, M.K., Das, S.R., Subramanian, A.P.: A topology control approach for utilizing multiple channels in multi-radio wireless mesh networks. Comput. Netw. **54**, 241–256 (2010). https://doi.org/10.1016/j.comnet.2009.05.015

Mohsenian Rad, A.H., Wong, V.W.S.: Congestion-aware channel assignment for multi-channel wireless mesh networks. Comput. Netw. **53**, 2502–2516 (2009). https://doi.org/10.1016/j.comnet.2009.05.006

Park, P., Jung, B., Lee, H., Jung, D.-J.: Robust channel allocation with heterogeneous requirements for wireless mesh backbone networks. Sensors **18**, 2687 (2018). https://doi.org/10.3390/s18082687

Rezgui, J., Hafid, A., Ben Ali, R., Gendreau, M.: Optimization model for handoff-aware channel assignment problem for multi-radio wireless mesh networks. Comput. Netw. **56**, 1826–1846 (2012). https://doi.org/10.1016/j.comnet.2012.02.002

Rezvanian, A., Saghiri, A.M., Vahidipour, S.M., et al.: Recent Advances in Learning Automata. Springer (2018a)

Rezvanian, A., Saghiri, A.M., Vahidipour, S.M., et al.: Cellular Learning Automata. pp. 21–88 (2018b)

Riggio, R., Rasheed, T., Testi, S., et al.: Interference and traffic aware channel assignment in WiFi-based wireless mesh networks. Ad Hoc Netw. **9**, 864–875 (2011). https://doi.org/10.1016/j.adhoc.2010.09.012

Selvakumar, K., Revathy, G.: Escalating quality of services with channel assignment and traffic scheduling in wireless mesh networks. Cluster Comput. **22**, 11949–11955 (2019). https://doi.org/10.1007/s10586-017-1528-6

Selvakumar, K., Revathy, G.: Channel assinment using Tabu search in wireless mesh networks. Wirel. Pers. Commun. **100**, 1633–1644 (2018). https://doi.org/10.1007/s11277-018-5658-9

Skalli, H., Ghosh, S., Das, S., Lenzini, L.: Channel assignment strategies for multiradio wireless mesh networks: issues and solutions. IEEE Commun. Mag. **45**, 86–95 (2007). https://doi.org/10.1109/MCOM.2007.4378326

Subbaiah, K.V., Naidu, M.M.: An efficient interference aware channel allocation algorithm for wireless mesh networks. In: 2015 International Conference on Signal Processing and Communication Engineering Systems. IEEE, pp. 416–420 (2015)

Subramanian, A.P., Gupta, H., Das, S.R., Cao, Jing: Minimum interference channel assignment in multiradio wireless mesh networks. IEEE Trans. Mob. Comput. **7**, 1459–1473 (2008). https://doi.org/10.1109/TMC.2008.70

Lin, T.-Y., Hsieh, K.-C., Huang, H.-C.: Applying genetic algorithms for multiradio wireless mesh network planning. IEEE Trans. Veh. Technol. **61**, 2256–2270 (2012). https://doi.org/10.1109/TVT.2012.2191166

Vafashoar, R., Meybodi, M.R.: Reinforcement learning in learning automata and cellular learning automata via multiple reinforcement signals. Knowl.-Based Syst. **169**, 1–27 (2019). https://doi.org/10.1016/j.knosys.2019.01.021

Vallam, R.D., Kanagasabapathy, A.A., Murthy, C.S.R.: A non-cooperative game-theoretic approach to channel assignment in multi-channel multi-radio wireless networks. Wirel. Netw. **17**, 411–435 (2011). https://doi.org/10.1007/s11276-010-0288-5

Vallati, C., Mingozzi, E.: Efficient design of wireless mesh networks with robust dynamic frequency selection capability. Comput. Netw. **83**, 15–29 (2015). https://doi.org/10.1016/j.comnet.2015.01.009

Wang, J., Shi, W., Cui, K., et al.: Partially overlapped channel assignment for multi-channel multi-radio wireless mesh networks. EURASIP J. Wirel. Commun. Netw. **2015**, 25 (2015). https://doi.org/10.1186/s13638-015-0259-8

Wolfram, S.: Theory and Applications of Cellular Automata. World Scientific Publication (1986)

Yang, D., Fang, X., Xue, G.: Channel allocation in non-cooperative multi-radio multi-channel wireless networks. In: 2012 Proceedings IEEE INFOCOM. IEEE, pp. 882–890 (2012)

Zeng, F., Zhao, N., Li, W.: Joint interference optimization and user satisfaction improvement for multicast routing and channel assignment in wireless mesh networks. Cluster Comput. **22**, 15059–15072 (2019). https://doi.org/10.1007/s10586-018-2497-0

Zhao, X., Zhang, S., Li, L., et al.: A multi-radio multi-channel assignment algorithm based on topology control and link interference weight for a power distribution wireless mesh network. Wirel. Pers. Commun. **99**, 555–566 (2018). https://doi.org/10.1007/s11277-017-5132-0

Chapter 6
Cellular Learning Automata for Collaborative Loss Sharing

6.1 Introduction

In real-world situations, we may face situations where offering insurance coverage to all users is not possible. The percentage of users who can benefit insurance coverage depends on the finances of a service provider. To increase the financial capability of service providers, we propose an irregular cellular learning automaton (ICLA) based loss-sharing approach among the service providers in this chapter. ICLA is a powerful mathematical model for decentralized applications that can operate in environments with high uncertainty and randomness. Due to the mentioned capabilities, we employed ICLA for modeling service providers and relation among them to present a reinsurance approach under collaborative conditions. In the proposed approach in this chapter, the service providers are mapped to cells, and relations among them are modeled using neighboring relations in ICLA. The located learning automata in cells decide about loss-sharing parameters. The goal of service providers is to reach a Pareto optimal loss-sharing agreement (LSA). With a Pareto optimal LSA, it is impossible to find another LSA to make the utility of a service provider better off without making at least the utility of one of the service provider's neighbors worse off. Because we mapped service providers to a cell of an ICLA, therefore all the service providers that collaborate in loss sharing with a service provider are called the service provider's neighbors.

Along with increasing the financial capabilities of service providers, the proposed loss-sharing approach decreases the variance of unpredictable losses, which is desirable for risk-averse service providers. A risk-averse service provider prefers more certain conditions when it is confronted with two choices with the same expected utility. Thus, if loss sharing or collaboration with other service providers can decreases uncertainty, a risk-averse service provider will be interested in collaborating. Such collaboration can be useful in environments such as federated clouds.

R. Vafashoar et al., *Cellular Learning Automata: Theory and Applications*, Studies in Systems, Decision and Control 307,
https://doi.org/10.1007/978-3-030-53141-6_6

In federated clouds, two or more independent geographically distinct Clouds share either authentication, files, computing resources, command, and control or access to storage resources (Javadi et al. 2012; Rajkumar Buyya 2013). The proposed approach in this chapter can be used to extend the sharing ideas to the losses as well. We show that by using ICLA, finding Pareto optimal LSAs do not need detailed information about distributions of the losses and even utility functions of the neighbor service providers. Due to the randomness of the losses and their unknown distributions currently establishing a theoretic method for finding an appropriate loss-sharing agreement is very difficult and complex. The results of the conducted experiments illustrate that the proposed approach can improve the utility of both the service provider and user. The rest of this chapter is organized as follows: Sect. 6.2 contains some preliminaries and notations which are used in the next sections. Section 6.3 review the related works and explains the proposed ICLA based approach for loss sharing. The results of the conducted experiments are presented in Sect. 6.4.

6.2 Preliminaries and Notations

This section offers a brief introduction to preliminaries and concepts which are needed in the following sections.

6.2.1 ICLA

Cellular learning automaton (CLA) as a new reinforcement learning approach is a combination of cellular automata (CA) (Wolfram 1986) and learning automata (LA) (Rezvanian et al. 2018a, 2019a). It is formed by a group of interconnected cells, which are arranged in some regular forms such as a grid or ring, in which each cell contains one or more LAs. A cellular learning automaton can be considered as a distributed model, which inherits both the computational power of cellular automata and the learning ability of learning automata. Accordingly, it is more powerful than a single learning automaton due to the ability to produce more complex patterns of behavior. Also, the learning abilities of cells enable cellular learning automata to produce these complex patterns by understandable behavioral rules rather than complicated mathematical functions commonly used in cellular automata (Rezvanian et al. 2018b). In basic cellular learning automaton (CLA), cells are arranged in some regular forms like a gird or ring. However, there exist some applications which require irregular arrangements of the cells. Esnaashari and Meybodi have proposed an irregular CLA (Esnaashari and Meybodi 2018) for clustering the nodes in a wireless sensor network. Irregular CLA is modeled as an undirected graph in which each vertex represents a cell of the CLA, and its adjacent vertices determine its neighborhood. Up to now, various applications of Irregular CLA are reported in the literature. The application areas of this model include, for instance, (but not limited to) wireless sensor networks (Rezvanian et al. 2018c), graph problems (Rezapoor Mirsaleh and Meybodi 2016; Vahidipour et al. 2019), cloud computing (Morshedlou and Meybodi 2017), complex networks (Rezvanian et al. 2018a; Khomami et al. 2018), and social network analysis (Rezvanian et al. 2019b).

6.2.2 Utility

Utility (or usefulness) is the perceived ability of things to satisfy needs. The utility is an essential concept in game theory, economics, and business because it represents satisfaction experienced by an individual (Adelson 1971). Frequently in the economics literature (Zweifel and Eisen 2012), utility is defined as a function of income or wealth. In this chapter, when a service provider provisions its users according to their agreed SLA, the income of service provider increases to I, and its utility will be u(I). When the service provider violates SLAs, it pays penalties to users, and its income decreases to (I-L). Therefore, its utility will be u(I-L).

6.2.3 Benefit of Loss Sharing

In this section, we provide a short discussion on the benefit of loss sharing for risk-averse service providers. We explain why risk-averse service providers are interested in sharing their losses. As said before, the utility function of a risk-averse service provider has two main features: $u' > 0$ and $u'' < 0$. Figure 6.1 shows a utility function for a risk-averse service provider. Let x to be a random variable and assume its value indicates the magnitude of the loss. When all requests are provided

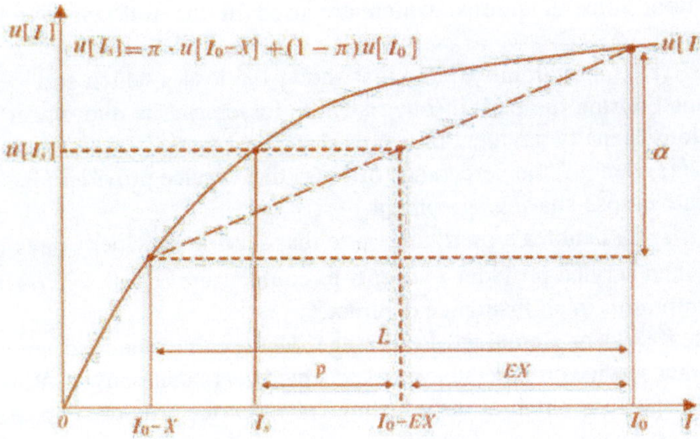

Fig. 6.1 The utility function of a risk-averse service provider

according to the agreed SLAs, x is zero, and the service provider's utility is $u(I_0)$. In case of SLA violation, X is L and utility will be decreased to $u(I_0 - X)$. Therefore, when there is a likelihood of SLA violation with probability π, the expected utility of the service provider is $\pi \times u(I_0 - X) + (1 - \pi) \times u(I_0)$. Now let to have another service provider precisely with the same utility function and loss distribution. If both service providers accept to share their losses according to an LSA such as $(((0.5, 0), (0.5, pr)), ((0.5, pr), (0.5, 0)))$, then the expected utility of both the service providers are:

$$\pi^2 \times u(I_0 - X) + 2 \times \pi \times (1 - \pi) \times u(I_0 - X/2) + (1 - \pi)^2 \times u(I_0) \quad (6.1)$$

Since $u' > 0$ and $u'' < 0$ as a result we have:

$$u(I_0) - u(I_0 - X/2) < u(I_0 - X/2) - u(I_0 - X) \quad (6.2)$$

Equation (6.2) means $u(I_0)+u(I_0 - X)$ is smaller than $2\times u(I_0 - X/2)$; therefore, we have:

$$\begin{aligned}&\pi^2 u(I_0 - X) + 2\pi(1 - \pi)u(I_0 - X/2) + (1 - \pi)^2 u(I_0)\\&> \pi u(I_0 - X) + (1 - \pi)u(I_0)\end{aligned} \quad (6.3)$$

Equation (6.3) means both the service providers can reach higher utility in case of loss sharing. So rational service providers are interested in collaborating and sharing their losses.

6.2.4 *Notations*

In this section, some notations, which are used in the following sections, are introduced:

$X_i(t)$: $X_i(t)$ is a random variable that shows the losses which service provider i has sustained during the period *(t-1, t)*. These losses can be due to, for example, SLA violations (penalty payment or ...).

$\alpha_i^j(t)$: $\alpha_i^j(t)$ denotes the percentage of $X_i(t)$ that service provider j has accepted to compensate in loss-sharing agreement.

$pr_i^j(t)$: $pr_i^j(t)$ denotes a premium value that service provider i pays to service provider j when service provider j accepts to compensate $\alpha_i^j(t) \times X_i(t).pr_i^j(t)$. is similar to premiums in an insurance contract.

$RP_i(t)$: $RP_i(t)$ or Remained Percentage denotes the remained percentage of $X_i(t)$ that other service providers have not accepted yet to compensate. When service provider j accepts to compensate $\alpha_i^j(t)$ from $X_i(t)$, service provider i updates $RP_i(t)$ and puts $RP_i(t) = RP_i(t) - \alpha_i^j(t)$.

LA_{ij}: LA_{ij} denotes a learning automaton in cell i that decides about $\alpha_j^i(t)$.

N_i: N_i denotes to neighbors of cell i(The service providers that collaborate with SP_i in loss sharing strategy). we assume that for each i, $i \in N_i$.

$I_i(t)$: $I_i(t)$ denotes income of service provider i from its business in iteration t.

$PI_i(t)$: $PI_i(t)$ denotes net income of service provider i in round t. $PI_i(t)$ is calculated as:

$$PI_i(t) = I_i(t) - \sum_{j \in N_i} \left(\alpha_i^j(t) \times X_j(t) \right) + \sum_{j \in N_i} pr_i^j(t) \tag{6.4}$$

Loss Sharing Agreement: A loss-sharing agreement (LSA) is denoted by a tuple as:

$$\begin{pmatrix} ((\alpha_1^1(t), pr_1^1(t))), (\alpha_2^1(t), pr_2^1(t)), \ldots, (\alpha_n^1(t), pr_n^1(t)), \ldots, \\ ((\alpha_1^i(t), pr_1^i(t)), (\alpha_2^i(t), pr_2^i(t)), \ldots, (\alpha_n^i(t), pr_n^i(t))), \ldots, \\ ((\alpha_1^n(t), pr_1^n(t)), (\alpha_2^n(t), pr_2^n(t)), \ldots, (\alpha_n^n(t), pr_n^n(t))) \end{pmatrix} \tag{6.5}$$

where $\sum_{j=1}^{n} \alpha_i^j(t) = 1$ and $\forall i, tpr_i^i(t) = 0$.

6.3 The Proposed Approach for Loss Sharing

Nowadays, sharing ideas to increase utility is common in many distributed computing infrastructures like Peer-to-Peer networks, grids, and clouds. For example, in federated clouds (Rochwerger et al. 2011; Goiri et al. 2012; Toosi et al. 2012; Samaan 2014), providers share their unused capacities to increase their profits. There are plenty of other works that focus on the idea of resource sharing in federated clouds. For example, (Buyya et al. 2009) present a market-oriented model among clouds for VM exchange and capacity sharing. Reference (Samaan 2014) offers a formulation for resource sharing among cloud providers that leads to optimal revenue. To the best knowledge of the authors, there is no notable work for loss sharing in distributed computing literature, and this work is the first one in this scope. However, the idea of loss sharing is not new, and there are plenty of works in economic and insurance literature that focused on the topic of risk and loss sharing (Aase 1995; Deelstra and Plantin 2014; Cheung et al. 2014; Chateauneuf et al. 2015). For example, reinsurance is one of the loss-sharing approaches in the insurance industry. The ultimate goal of reinsurance is to reduce insurance companies' exposure to loss by passing part of the risk of loss to a reinsurer or a group of reinsurers. By choosing a particular reinsurance method, the insurance company may be able to create a more balanced and homogeneous portfolio of insured risks. This would lend greater predictability to the portfolio results on a net basis and would be reflected in income smoothing. Because the loss-sharing process is similar to reinsurance approaches, some of the existing works on finding equilibriums of reinsurance markets (Aase 1992, 1995) can inspire designing an appropriate approach for loss sharing in distributed computing environments. Loss sharing approaches can reduce exposure to random losses that threaten profitability and utility. This work is the first one that tries to employ a loss-sharing approach to increase the utility of service providers in a distributed computing environment.

Now assume there are n risk-averse service providers. Let to show these service providers using a graph in which each vertex represents a service provider. Each edge in this graph represents the tendency between two service providers for collaboration (e.g., sharing losses). We call this graph a tendency graph. Since the service providers are risk-averse, so their utility functions satisfy two conditions: $u_i'(I_i) > 0$ *and* $u_i''(I_i) < 0$. Utility of each service provider is not known by the other service providers and X_i is announced just to the service providers that are interested in collaborating with service provider i. We also assume that service providers announce their satisfaction level from an LSA to each other using a satisfaction signal. Values of satisfaction signals are in the range [0, 1], and a bigger value means higher satisfaction level is obtained. Distribution of X_i is not known. Now we want to find an appropriate LSA such that collaboration of the service providers under this LSA increases their utility as possible.

To check does an LSA increases the utility of service providers, we need detailed information about the distribution of losses and utility functions of service providers. Generally, finding an appropriate LSA that increases the utility of all service providers is difficult, especially when there is no complete information about utility function forms and loss distributions. This will be much more complicated when there are too many service providers with different utility functions and different loss distributions. In addition to the mentioned difficulties, $X_i(t)$ and $X_j(t)(\forall j \in N_i)$ are random variables thus $PI_i(t)$ and $u_i(PI_i(t))$ (see Eq. 6.4) will be as well. This means that even with the same LSA in two different iterations, service providers may experience different utilities. This means a theoretic analysis is not possible here. It seems that non-theoretic methods must be considered as alternatives for finding an appropriate LSA. Due to the capabilities of learning automata and ICLA in environments with high uncertainty, in this section, we present an ICLA-based method for this problem.

In the proposed method, each cell of ICLA represents a service provider (a vertex of tendency graph), and neighborhood relation is defined based on the tendency for collaboration between service providers (edges of tendency graph). Per each neighbor j there is a learning automaton LA_{ij} in cell i. In an iterative procedure, this learning automaton selects one of its actions that this action determines $\alpha_j^i(t)$. Neighbor j (service provider j) pays $pr_j^i(t)$ to service provider i for $\alpha_j^i(t) \times X_j(t)$. Amount of $pr_j^i(t)$ can be zero or be determined according to a premium calculation method in insurance literature (Zweifel and Eisen 2012). Then all service providers calculate $u(PI(t))$ and generate satisfaction signals. Figure 6.2 illustrates the proposed method in detail. There is a local rule in ICLA that determines the reinforcement signal to any particular learning automaton. In our method, this rule uses *GenRSignal* to generate a reinforcement signal. Figure 6.3 shows the pseudo-code of *GenRSignal*.

Algorithm 6-1. loss-sharing algorithm

Each round t Do {
Simultaneously for each service provider i {
Simultaneously for each $SP_j \in N_i$ {
SP_j waits for a random time and then asks SP_i about $RP_i(t)$.
SP_i announces $RP_i(t)$ to SP_j.
SP_j rescales actions of LA_{ji} according to the response of SP_i.
Using the rescaled actions, LA_{ji} selects one action to specify $\alpha_i^j(t)$.
SP_j announces $\alpha_i^j(t)$ to SP_i.
SP_i updates $RP_i(t)$.
}
SP_i calculates $X_i(t)$ in round t and announces it to all the SP_j that $\alpha_i^j(t) \neq 0$.
SP_i calculates $PI_i(t) = I_i(t) - \sum_{j \in N_i} (\alpha_i^j(t) \times X_j(t)) + \sum_{j \in N_i} pr_i^j(t)$ using all the announced $X_j(t)$ s.
SP_i calculates satisfaction signal $(\beta_i(t))$ and announces it to all the SP_j that $\alpha_i^j(t) \neq 0$.
SP_i receives satisfaction signals $(\beta_j(t))$ of the other SP_j that $\alpha_i^j(t) \neq 0$.
SP_i generates reinforcement signals for LA_{ij} using GenRSignal($PI_i(t)$,$\beta_j(t)$,$\alpha_i^j(t)$, $\alpha_j^i(t)$).
Probability vector of LA_{ij} is updated using the generate reinforcement signal.
}
} While (is_Threshold_Satisfied())

Fig. 6.2 Pseudo code of the proposed method for loss sharing

Algorithm 6-2. GenRSignal

GenRSignal($PI_i(t)$,$\beta_j(t)$,$\alpha_i^j(t)$, $\alpha_j^i(t)$). { Rand = Generate_Random_Number(0,1); // 0 < Rand < 1
If (t==1) {
$u_i^{max_{ii}}$;
If (Rand > 0.5) {
Reinforcement_Signal =1;
} Else {
Reinforcement_Signal =0;
}
} else {
If ($u_i(PI_i(t)) > u_i^{max}$) { $u_i^{max_{ii}}$; }
$\beta_i(t) = f_1(-\frac{u''}{u'}) \times f_2(\alpha_i^j) \times f_3(\alpha_j^i) \times \frac{u_i(PI_i(t))}{u_i^{max}}$;
$\beta(t) = w_{ij}\beta_i(t) + (1 - w_{ij})\beta_j(t)$; // $0 \leq w_{ij} \leq 1$
If (Rand > $\beta(t)$) {
Reinforcement_Signal =1;
} Else {
Reinforcement_Signal =0;
}// end of If (Rand > $\beta(t)$)... Else ...
}// end of If (t==1) ... Else ...
Return Reinforcement_Signal;
}

Fig. 6.3 Pseudo code of GenRSignal

The mentioned procedure is repeated every iteration by every service provider until a predefined threshold to be met. Figure 6.4 illustrates pseudo-code to check the threshold satisfaction.

Algorithm 6-3. checking threshold satisfaction

```
n: Let n denotes the number of cells (or service providers).
ni: Let ni denotes the number of neighbors of cell i.
mij: Let mij denotes the number of actions of LAij.
pij^k: Let pij^k denotes the probability of k^th action in LAij.
is_Threshold_Satisfied() {
    for i=1 to i=n concurrently do:
        for j=1 to j =ni concurrently do {
            boolVar = false;
            for (k=1, k ≤ mij,k++ ){
                    if (1 − pij^k < ε)
                        boolVar = true;
            }
            If (boolVar)    STOP(LAij);
        }
    If (All LAij are stopped)    return true;
    Else                         return false;
    }
```

Fig. 6.4 Pseudocode for checking threshold satisfaction

6.4 Experiments

In this section, first, the evaluation approach is described and then to evaluate the applicability and performance of the proposed ICLA-based method, numerical experiments are conducted, and results are reported in Sect. 5.4.2.

6.4.1 Evaluation Approach

Due to the lack of similar work for comparison, the evaluation is based on comparing utility by adopting a loss-sharing strategy versus the utility without any loss sharing. Also, we show that the obtained LSAs in the conducted experiments reach results similar to the results of a Pareto optimal LSA. An LSA is Pareto optimal if it is impossible to make any service provider better off (according to utility measure) by changing its share ($\alpha_i^j(t)$) without making at least one neighbor service provider worse off. We focused on the Pareto optimality concept because it is assumed that loss sharing among service providers is a collaborative process.

However, the proposed method can operate under any conditions with unknown utility function forms and loss distributions. However, to prove the Pareto optimality of an LSA, it is needed to have an exact form of utility functions and loss distributions. In conducted experiments, we use an exponential form function similar to Eq. (6.6) to describe the utility function of service providers. This function form satisfies the conditions needed for the utility function of a risk-averse service provider.

$$u_i(PI_i) = c_i e^{-a_i PI_i} - c_i \ (c_i \langle 0, a_i \rangle 0) \tag{6.6}$$

Because it is common in the risk management literature (Haimes 2015) to model random losses with exponential distributions, we assume that random variable X_i has an exponential distribution with parameter θ_i. Using the utility function of Eq. (6.6) and an exponential distribution of losses, choosing an $\alpha_i^j(t)$ appropriate to inverse of risk aversion of service providers causes to reach a pareto optimal LSA. We prove this claim in following but first we need a definition and theorem.

Definition 6.1 Indifferent Level of Income (PI_{ILI}) : PI_{ILI} is certain amount of income for which, a risk averse service provider is indifferent between having a risky income PI_i(See Eq. 6.4) and having this certain amount of income (PI_{ILI}^i). In other word we have $u_i\left(PI_{ILI}^i\right) = E(u_i(PI_i))$

$$PI_{ILI}^i = u_i^{-1}(E(u_i(PI_i))) \tag{6.7}$$

Note If we substitute the utility function of Eq. (6.6) in Eq. (6.7) we obtain Eq. (6.8):

$$PI_{ILI}^i = u_i^{-1}(E(u_i(PI_i))) = \frac{\ln(E(e^{-a_i PI_i}))^{-1}}{a_i} \tag{6.8}$$

Theorem 6.1 *ALSA will be Pareto optimal if and only if for each* i *and* j, $\sum_{i \in N_j} PI_{ILI}^i$ *reaches its maximum value.*

Proof Assume that a non-Pareto optimal LSA such as *C*, whose income for provider *i* is PI_i, gets maximum value for $\sum_{i \in N_j} PI_{ILI}^i$ for each *j*. According to (6.8), we have (6.9).

$$\sum_{i \in N_j} PI_{ILI}^i = \sum_{i \in N_j} \frac{\ln(E(e^{-a_i PI_i}))^{-1}}{a_i} = \max_{\overline{PI'}} \left(\sum_{i \in N_j} \frac{\ln\left(E\left(e^{-a_i PI_i'}\right)\right)^{-1}}{a_i} \right) \tag{6.9}$$

$\overline{PI'}$ is a vector and its elements being net income of service provider *j* and its neighbors. Since *C* is not Pareto optimal so there is an LSA such as C'' whose income is PI_i'' for service provider *i* ($i \in N_j$) and for each $i \in N_j$ we have:

$$\begin{aligned} E\left(u_i\left(PI_i''\right)\right) > E(u_i(PI_i)) \Rightarrow E\left(c_i e^{-a_i PI_i''} - c_i\right) \\ > E\left(c_i e^{-a_i PI_i} - c_i\right) \Rightarrow \left(E\left(e^{-a_i PI_i''}\right)\right)^{-1} > \left(E\left(e^{-a_i PI_i}\right)\right)^{-1} \end{aligned} \tag{6.10}$$

For a reason that $a_i > 0$ and ln is an increasing function, so for each $i \in N_j$

$$\frac{\ln(E(e^{-a_i PI_i''}))^{-1}}{a_i} > \frac{\ln(E(e^{-a_i PI_i}))^{-1}}{a_i} \tag{6.11}$$

Therefore we have (6.10):

$$\sum_{i \in N_j} \frac{\ln(E(e^{-a_i PI_i''}))^{-1}}{a_i} > \sum_{i \in N_j} \frac{\ln(E(e^{-a_i PI_i}))^{-1}}{a_i} \tag{6.12}$$

According to (6.9), RHS of (6.12) is equal to a maximum value, and this means LHS is bigger than maximum. This is a contradiction and means **C is Pareto optimal LSA.**

Now assume a Pareto optimal LSA such as C whose income is PI_i for service provider *i,* but it does not satisfy (6.9). As a result, there is an LSA such as C'' whose income is PI_i'' for service provider i $(i \in N_j)$ and satisfy (6.13).

$$\sum_{i \in N_j} \frac{\ln(E(e^{-a_i PI_i''}))^{-1}}{a_i} > \sum_{i \in N_j} \frac{\ln(E(e^{-a_i PI_i}))^{-1}}{a_i} \tag{6.13}$$

Now consider an LSA like C' which has shares (α_i^j) similar to shares of $C''\left(\alpha_i^j(C') = \alpha_i^j(C'')\right)$ but different $pr_i^j\,(pr_i^j(C') \neq pr_i^j(C''))$. Let $pr_i^j(C')$ to be defined as (6.14).

$$pr_i^j(C') = pr_i^j(C'') - \frac{\ln(E(e^{-a_i PI_i''}))^{-1}}{a_i} + \frac{\ln(E(e^{-a_i PI_i}))^{-1}}{a_i} + \varphi_i \tag{6.14}$$

where φ_i is a positive value as (6.15):

$$\varphi_i = \frac{1}{|N_j|} \times \left(\sum_{i \in N_j} \frac{\ln(E(e^{-a_i PI_i''}))^{-1}}{a_i} - \sum_{i \in N_j} \frac{\ln(E(e^{-a_i PI_i}))^{-1}}{a_i} \right) \tag{6.15}$$

Because $\alpha_i^j(C') = \alpha_i^j(C'')$ thus

$$\begin{aligned} PI_i' &= I_i + \sum_{j \in N_i} pr_j^i(C') - \sum_{i \in N_j} \alpha_j^i(C')X_j = I_i \\ &+ \sum_{j \in N_i} \left(pr_j^i(C'') - \frac{\ln(E(e^{-a_i PI_i''}))^{-1}}{a_i} + \frac{\ln(E(e^{-a_i PI_i}))^{-1}}{a_i} + \varphi_i \right) \\ &- \sum_{i \in N_j} \alpha_j^i(C'')X_j \end{aligned} \tag{6.16}$$

Let $\gamma_i = |N_i| \times \left(\varphi_i - \frac{\ln(E(e^{-a_i PI_i''}))^{-1}}{a_i} + \frac{\ln(E(e^{-a_i PI_i}))^{-1}}{a_i}\right)$ so consequently we have $PI_i' = PI_i'' + \gamma_i$.

Since $E(\gamma_i) = \gamma_i$ as a result:

$$\frac{\ln(E(e^{-a_i PI_i'}))^{-1}}{a_i} = \frac{\ln(E(e^{-a_i PI_i''}))^{-1}}{a_i} + \gamma_i \tag{6.17}$$

By substitution of value of γ_i, we have (6.18).

$$\frac{\ln(E(e^{-a_i PI_i'}))^{-1}}{a_i} = \frac{\ln(E(e^{-a_i PI_i}))^{-1}}{a_i} + |N_i| \times \varphi_i \tag{6.18}$$

Since φ_i is positive, so we have (6.19).

$$\frac{\ln(E(e^{-a_i PI_i'}))^{-1}}{a_i} > \frac{\ln(E(e^{-a_i PI_i}))^{-1}}{a_i} \tag{6.19}$$

Equation (6.19) is valid for each $i \in N_j$ and this means LSA C is not Pareto optimal. This is a contradiction and means that **a Pareto optimal LSA maximizes** $\sum_{i \in N_i} PI^i_{ILI}$. □

Now using Theorem 6.1, we can prove Theorem 6.2.

Theorem 6.2 *Using the utility function of Eq. (6.6) and an exponential distribution of losses, choosing an* α_i^j *appropriate to the inverse of risk aversion of service providers cause to reach a Pareto optimal LSA.*

Proof From Theorem 6.1 we know that an LSA is Pareto optimal if and only if $\sum_{i \in N_j} PI^i_{ILI}$ reach its maximum value for each j. So, if we show that the shares appropriate to the inverse of risk aversion maximize $\sum_{i \in N_j} PI^i_{ILI}$, the proof will be complete. Using Eq. (6.4) we can write:

$$PI^i_{ILI}(t) = u_i^{-1}(E(u_i(PI_i(t)))) \tag{6.20}$$

For simplicity purpose, in the following we skip iteration index t and also, we use pr_i instead of $\sum_{j \in N_i} pr_i^j(t)$. According to (6.8), we obtain (6.21).

$$PI^i_{ILI} = -\frac{1}{a_i} \ln(E(e^{-a_i PI_i})) \tag{6.21}$$

At each iteration pr_i has a fixed value while $\sum_{j\in N_i} \alpha^i_j X_j$ depends on random losses. Thus we have (6.22).

$$PI^i_{ILI} = -\frac{1}{a_i}\ln(E(e^{-a_i PI_i})) = -\frac{1}{a_i}\ln(e^{-a_i pr_j} E\left(e^{a_i \sum_{j\in N_i} \alpha^i_j X_j}\right)) \quad (6.22)$$

Because losses are independent random variables thus

$$PI^i_{ILI} = -\frac{1}{a_i}(\ln(e^{-a_i pr_j}) + \sum_{j\in N_i} \ln(E\left(e^{a_i \alpha^i_j X_J}\right))) \quad (6.23)$$

(6.23) can be written in form of (6.24):

$$PI^i_{ILI} = pr_i - \frac{1}{a_i}\sum_{j\in N_i} \ln(E\left(e^{a_i \alpha^i_j X_j}\right)) \quad (6.24)$$

Since for each random loss X_j have an exponential distribution with the parameter of θ_j, so we have (6.25).

$$E\left(e^{a_i \alpha^i_j X_j}\right) = \frac{1}{1-\theta_j a_i X_j} \quad (6.25)$$

By substituting (6.25) in (6.24) we have:

$$PI^i_{ILI} = pr_i - \frac{1}{a_i}\sum_{j\in N_i} \ln(\frac{1}{1-\theta_j a_i \alpha^i_j}) = pr_i + \frac{1}{a_i}\sum_{j\in N_i} \ln(1-\theta_j a_i \alpha^i_j) \quad (6.26)$$

So for each z, we have:

$$\sum_{i\in N_z} PI^i_{ILI} = \sum_{i\in N_z} pr_i + \sum_{i\in N_z}\sum_{j\in N_i} \frac{1}{a_i} \ln(1-\theta_j a_i \alpha^i_j) \quad (6.27)$$

Since $\sum_{i\in N_z} pr_i$ are fixed so if we show that the shares equal to the inverse of risk aversions maximize the second term of RHS of (6.27) for each z such that $\sum_{j\in N_i} \alpha^j_i = 1$, the proof will be completed. This is the case using Karush-Kuhn-Tucker conditions, and it is straight forward that $\alpha^i_j = \frac{1}{\sum_{j\in N_i} \frac{a_i}{a_j}}$ maximizes $\sum_{i\in N_z} PI^i_{ILI}$. □

Now we want to conduct experiments to compare the LSA obtained empirically by our proposed method with the Pareto optimal LSA. Note that in our method, there is no need to know the form of utility functions and loss distributions.

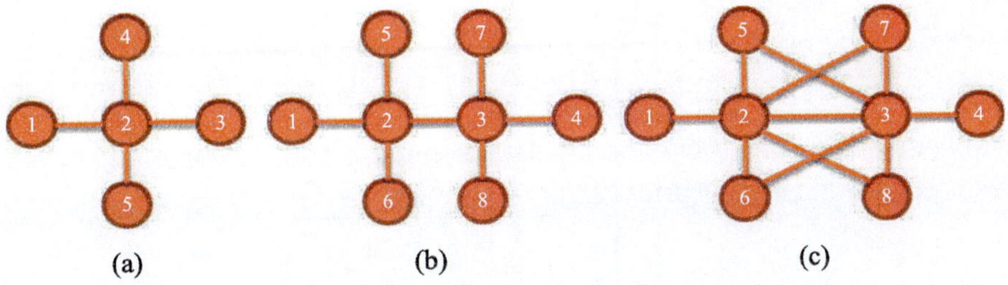

Fig. 6.5 Tendency graphs for **a** case A, **b** case B, **c** case C

6.4.2 Experiment Results

We conducted the experiments using 5 and 8 service providers in three different cases: cases A, B, and C. Figure 6.5 shows the tendency graphs for each case. Each learning automaton has 11 actions in our experiments, which are labeled with action 0, action 1,..., and action 10. Converging of learning automata to i^{th} action (action i) means that learning automaton has decided to put $\alpha_i^j(t)$ equal to $(0.1 \times i)$. The learning algorithm of all learning automata are L_{R-I}. We discuss the results of the experiment from different viewpoints: capability of ICLA in finding Pareto optimal LSA and usefulness of a Pareto optimal LSA due to the increasing utility of service providers. Finally, we discuss the applicability of the proposed method for a large number of service providers according to the results of experiments.

6.4.2.1 Capability of ICLA in Finding Pareto Optimal LSA

In this experiment, we compare the obtained LSA by ICLA with Pareto optimal LSA under four different settings, as shown in Table 6.1. In all settings and for all service providers, c_i in Eq. (6.10) and I_i are equal to -5 and 2000, respectively. I_i is used to calculate PI_i via Eq. (6.4). In each setting a_i in Eq. (6.6), which is the risk aversion of the service provider, has been set according to Table 6.1.

Table 6.2 illustrates the shares (α_i^j) in the obtained LSA by the proposed method (ICLA-based) and Pareto optimal LSA (Theoretic) for each service providers in case A. In this table value of α_i^j is located in column *ij*. The shares in cases B and C are illustrated in Tables 6.3 and 6.4, respectively. As shown in these tables, the proposed

Table 6.1 Risk aversion of service providers

Risk aversion	SP1	SP2	SP3	SP4	SP5	SP6	SP7	SP8
Setting 1	0.0005	0.0005	0.0005	0.0005	0.0005	0.0005	0.0005	0.0005
Setting 2	0.0001	0.0002	0.0003	0.0004	0.0005	0.0006	0.0007	0.0008
Setting 3	0.0008	0.0007	0.0006	0.0005	0.0004	0.0003	0.0002	0.0001
Setting 4	0.0002	0.0002	0.0004	0.0004	0.0005	0.0005	0.0008	0.0008

Table 6.2 The proposed shares by a Pareto optimal LSA (Theoretic) and the proposed method (ICLA-based) in case A

Case A		SP1 (α)		SP2 (α)					SP3 (α)		SP4 (α)		SP5 (α)	
		11	12	21	22	23	24	25	32	33	42	44	52	55
Setting 1	Theoretic	0.5	0.5	0.2	0.2	0.2	0.2	0.2	0.5	0.5	0.5	0.5	0.5	0.5
	ICLA-Based	0.5	0.5	0.2	0.2	0.2	0.2	0.2	0.5	0.5	0.5	0.5	0.5	0.5
Setting 2	Theoretic	0.666	0.334	0.438	0.219	0.146	0.109	0.088	0.6	0.4	0.667	0.333	0.714	0.286
	ICLA-Based	0.7	0.3	0.4	0.2	0.2	0.1	0.1	0.6	0.4	0.7	0.3	0.7	0.3
Setting 3	Theoretic	0.466	0.534	0.141	0.162	0.189	0.226	0.282	0.462	0.538	0.416	0.584	0.364	0.636
	ICLA-Based	0.4	0.6	0.2	0.1	0.2	0.2	0.3	0.5	0.5	0.4	0.6	0.4	0.7
Setting 4	Theoretic	0.5	0.5	0.294	0.294	0.147	0.147	0.118	0.66	0.34	0.66	0.34	0.71	0.29
	ICLA-Based	0.5	0.5	0.3	0.3	0.2	0.1	0.1	0.7	0.3	0.7	0.3	0.7	0.3

Table 6.3 The proposed shares by a Pareto optimal LSA (Theoretic) and the proposed method (ICLA-based) in case B

Case B		SP1 (α)		SP2 (α)					SP5 (α)		SP6 (α)	
		11	12	21	22	23	25	26	52	55	62	66
Setting 1	Theoretic	0.5	0.5	0.2	0.2	0.2	0.2	0.2	0.5	0.5	0.5	0.5
	ICLA-Based	0.5	0.5	0.2	0.2	0.2	0.2	0.2	0.5	0.5	0.5	0.5
Setting 2	Theoretic	0.666	0.334	0.454	0.227	0.151	0.091	0.077	0.714	0.286	0.75	0.25
	ICLA-Based	0.7	0.3	0.4	0.2	0.2	0.1	0.1	0.7	0.3	0.7	0.3
Setting 3	Theoretic	0.466	0.534	0.123	0.14	0.163	0.246	0.328	0.364	0.636	0.3	0.7
	ICLA-Based	0.5	0.5	0.1	0.2	0.2	0.2	0.3	0.4	0.6	0.3	0.7
Setting 4	Theoretic	0.5	0.5	0.3	0.3	0.15	0.125	0.125	0.71	0.29	0.71	0.29
	ICLA-Based	0.5	0.5	0.3	0.3	0.2	0.1	0.1	0.7	0.3	0.7	0.3
		SP4 (α)		SP3 (α)					SP7 (α)		SP8 (α)	
		43	44	32	33	34	37	38	73	77	83	88
Setting 1	Theoretic	0.5	0.5	0.2	0.2	0.2	0.2	0.2	0.5	0.5	0.5	0.5
	ICLA-Based	0.5	0.5	0.2	0.2	0.2	0.2	0.2	0.5	0.5	0.5	0.5
Setting 2	Theoretic	0.571	0.429	0.37	0.247	0.185	0.106	0.092	0.7	0.3	0.73	0.27
	ICLA-Based	0.6	0.4	0.4	0.2	0.2	0.1	0.1	0.7	0.3	0.7	0.3
Setting 3	Theoretic	0.455	0.545	0.073	0.084	0.099	0.249	0.499	0.25	0.75	0.143	0.857
	ICLA-Based	0.4	0.6	0	0.1	0.1	0.3	0.5	0.2	0.8	0.2	0.8
Setting 4	Theoretic	0.5	0.5	0.4	0.2	0.2	0.1	0.1	0.67	0.33	0.67	0.33
	ICLA-Based	0.5	0.5	0.4	0.2	0.2	0.1	0.1	0.7	0.3	0.7	0.3

Table 6.4 The proposed shares by a Pareto optimal LSA (Theoretic) and the proposed method (ICLA-based) in case C

Case C		SP1 (α)		SP2 (α)							SP5 (α)			SP6 (α)		
		11	12	21	22	23	25	26	27	28	52	53	55	62	63	66
Setting 1	Theoretic	0.5	0.5	0.143	0.143	0.143	0.143	0.143	0.143	0.142	0.333	0.333	0.334	0.333	0.333	0.334
	ICLA-Based	0.5	0.5	0.1	0.2	0.1	0.2	0.1	0.2	0.1	0.3	0.3	0.4	0.3	0.4	0.3
Setting 2	Theoretic	0.666	0.334	0.405	0.202	0.135	0.081	0.068	0.058	0.051	0.484	0.322	0.194	0.5	0.334	0.166
	ICLA-Based	0.7	0.3	0.4	0.2	0.1	0.1	0.1	0.1	0	0.5	0.3	0.2	0.5	0.3	0.2
Setting 3	Theoretic	0.466	0.534	0.05	0.056	0.066	0.099	0.132	0.199	0.398	0.255	0.298	0.447	0.222	0.259	0.519
	ICLA-Based	0.5	0.5	0	0	0.1	0.1	0.1	0.2	0.5	0.3	0.3	0.4	0.2	0.3	0.5
Setting 4	Theoretic	0.5	0.5	0.263	0.263	0.134	0.105	0.105	0.065	0.065	0.527	0.263	0.21	0.527	0.263	0.21
	ICLA-Based	0.5	0.5	0.3	0.3	0.1	0.1	0.1	0	0.1	0.5	0.3	0.2	0.5	0.3	0.2
		SP4 (α)		SP3 (α)							SP7 (α)			SP8 (α)		
		43	44	32	33	34	35	36	37	38	72	73	77	82	83	88
Setting 1	Theoretic	0.5	0.5	0.143	0.143	0.143	0.143	0.143	0.143	0.142	0.333	0.333	0.334	0.333	0.333	0.334
	ICLA-Based	0.5	0.5	0.2	0.1	0.2	0.1	0.1	0.2	0.1	0.4	0.3	0.3	0.3	0.3	0.4
Setting 2	Theoretic	0.571	0.429	0.291	0.194	0.145	0.116	0.097	0.083	0.074	0.512	0.341	0.147	0.52	0.348	0.132
	ICLA-Based	0.6	0.4	0.3	0.2	0.1	0.1	0.1	0.1	0.1	0.5	0.3	0.2	0.5	0.4	0.1
Setting 3	Theoretic	0.455	0.545	0.055	0.064	0.077	0.096	0.129	0.193	0.386	0.176	0.206	0.618	0.109	0.127	0.764
	ICLA-Based	0.5	0.5	0	0	0.1	0.1	0.2	0.2	0.4	0.2	0.2	0.6	0.1	0.1	0.8
Setting 4	Theoretic	0.5	0.5	0.306	0.151	0.151	0.121	0.121	0.075	0.075	0.572	0.285	0.143	0.572	0.285	0.143
	ICLA-Based	0.5	0.5	0.3	0.2	0.1	0.1	0.1	0.1	0.1	0.6	0.3	0.1	0.6	0.3	0.1

method has the capability of reaching an LSA near Pareto optimal LSA without even knowledge about loss distributions or utility function forms. For the visual comparison of the results, Figs. 6.6, 6.7, 6.8, 6.9, 6.10, 6.11, 6.12 and 6.13 illustrates the shares proposed by the obtained LSA using the proposed method (ICLA-based) and Pareto optimal LSA (Theoretic). *Aij-k* in the horizontal axis represents α_i^j under

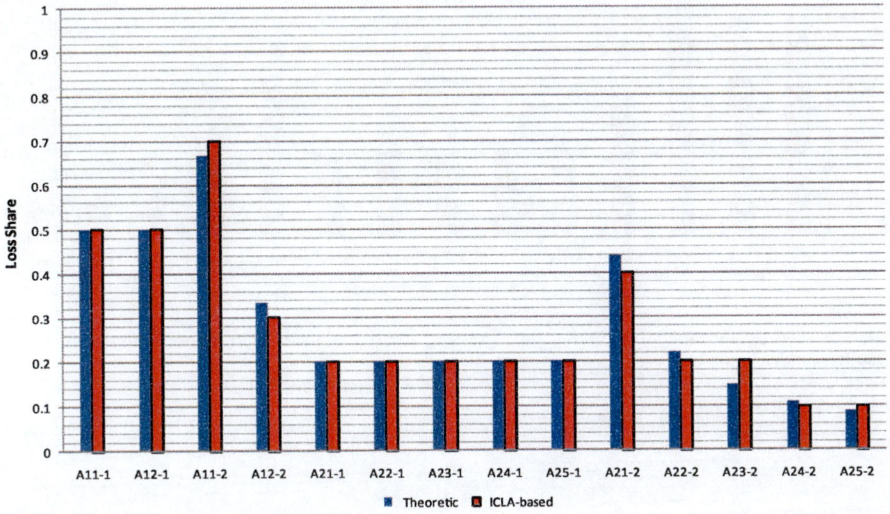

Fig. 6.6 The proposed shares for X_1 and X_2 using the ICLA-based and theoretic methods under case A

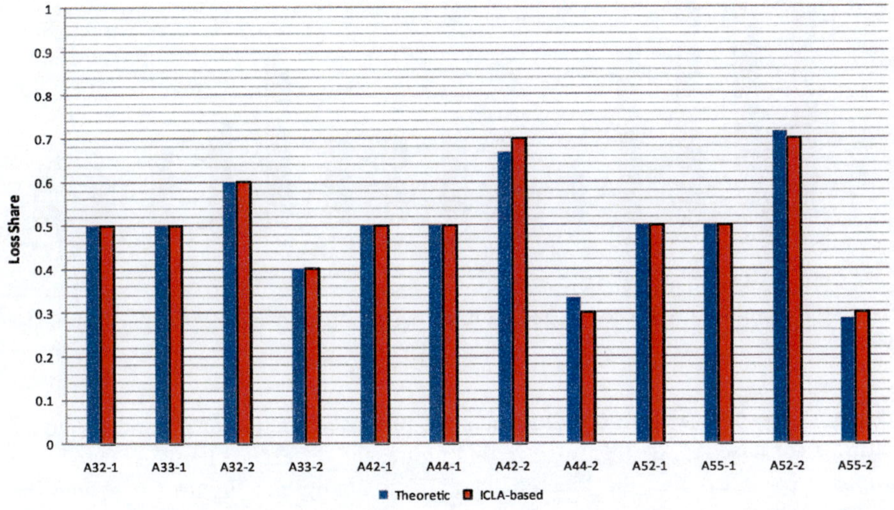

Fig. 6.7 The proposed shares for X_3, X_4 and X_5 using the ICLA-based and theoretic methods under case A

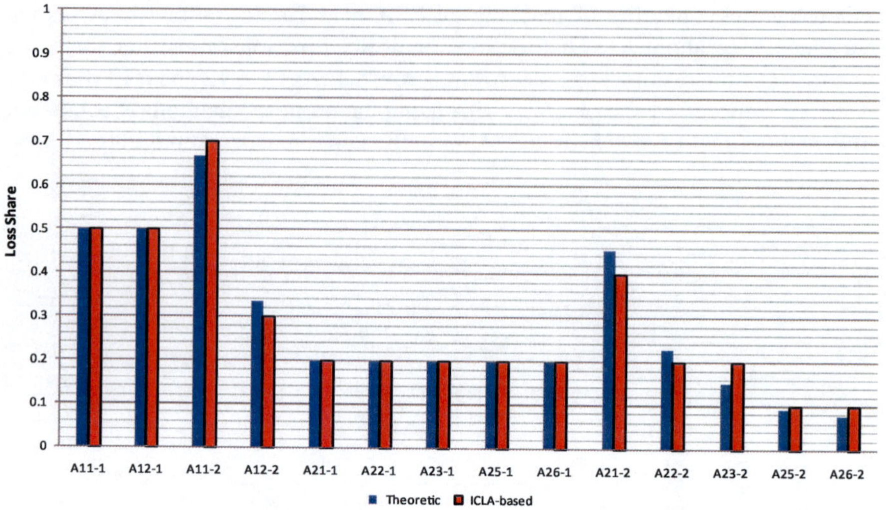

Fig. 6.8 The proposed shares for X_1 and X_2 using the ICLA-based and theoretic methods under case B

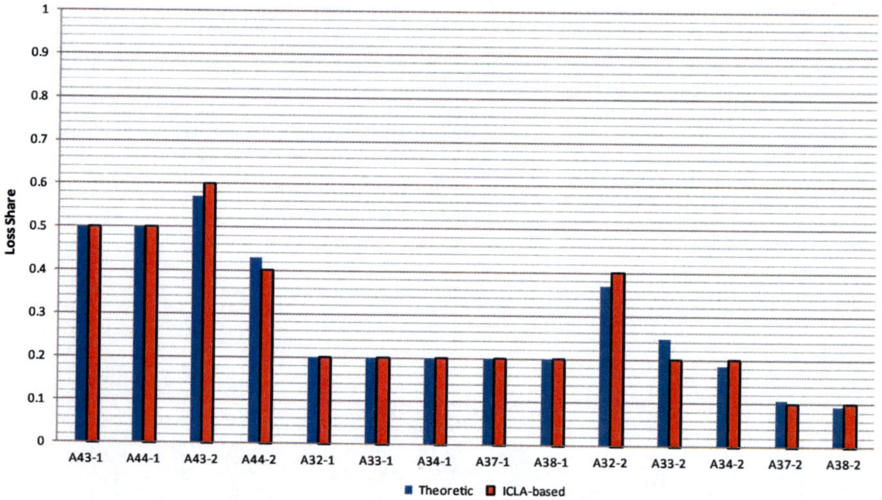

Fig. 6.9 The proposed shares for X_3 and X_4 using the ICLA-based and theoretic methods under case B

setting k ($k = 1{,}2$). Figure 6.6 illustrates the proposed shares for X_1 and X_2 under setting 1 for case A. As illustrated in this figure, the difference between the shares proposed by ICLA and the theoretic method is negligible. Figure 6.7 shows the proposed shares for X_3, X_4 and X_5 have minor differences as well. Diagrams of Figs. 6.8, 6.9, 6.10, 6.11, 6.12 and 6.13 illustrate that even for more complex tendency

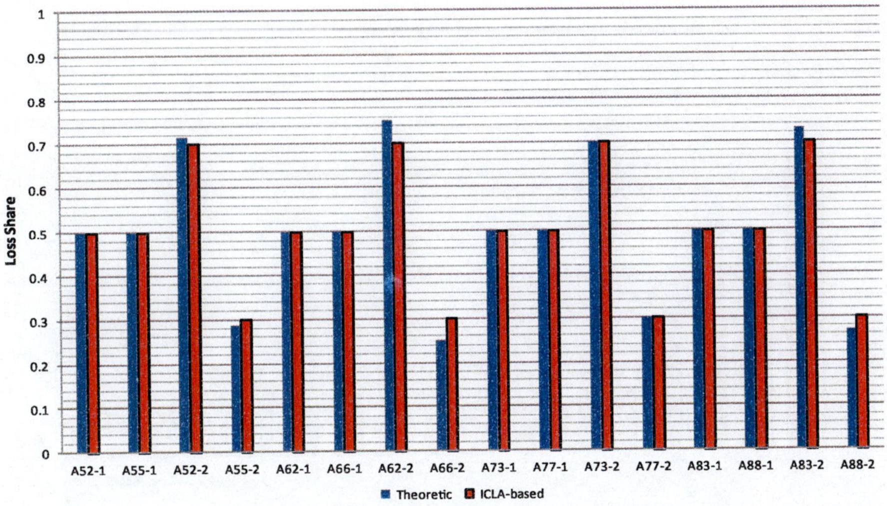

Fig. 6.10 The proposed shares for X_5, X_6, X_7 and X_8 using the ICLA-based and theoretic methods under case B

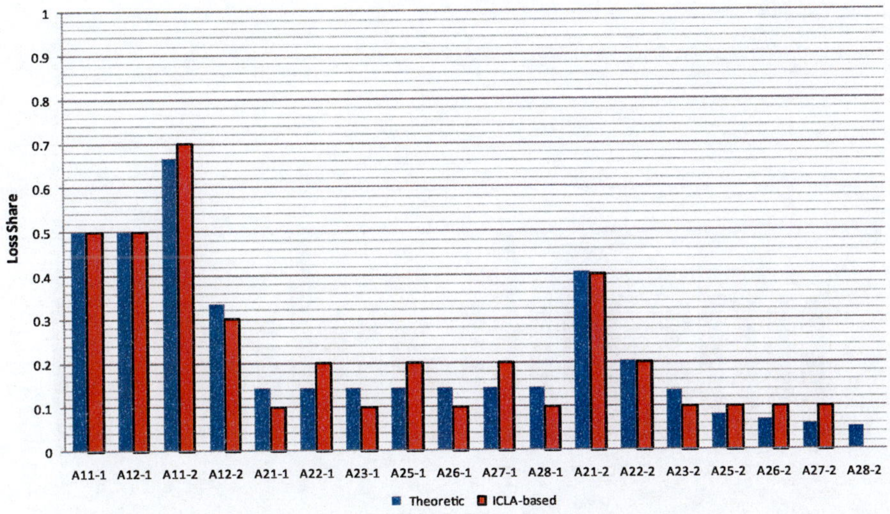

Fig. 6.11 The proposed shares for X_1 and X_2 using the ICLA-based and theoretic methods under case C

graphs, such as case B and C, the shares proposed by the ICLA-based method are very close to shares proposed by the theoretic approach. By comparing the diagrams for case A (Figs. 6.6 and 6.7) with diagrams of case B (Figs. 6.8, 6.9 and 6.10) and C (Figs. 6.11, 6.12 and 6.13), it can be seen that number of service providers (nodes in tendency graphs) have no considerable impact on the accuracy of the proposed

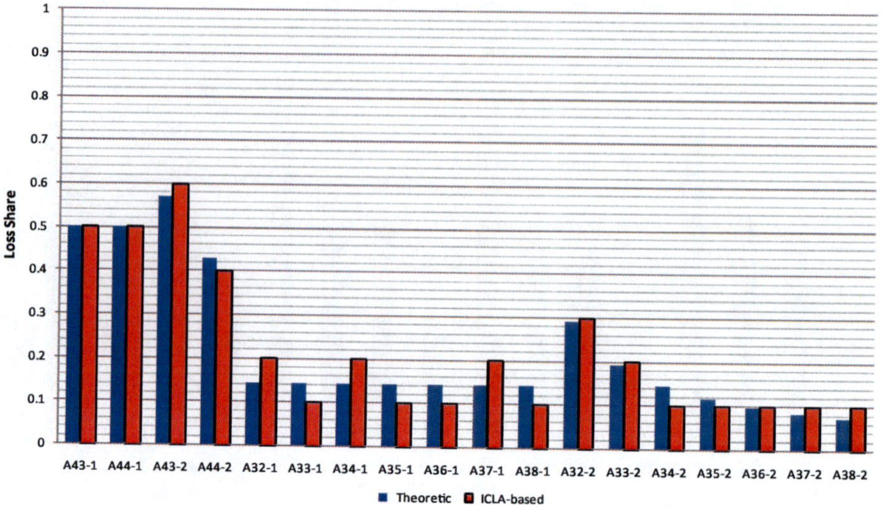

Fig. 6.12 The proposed shares for X_3 and X_4 using the ICLA-based and theoretic methods under case C

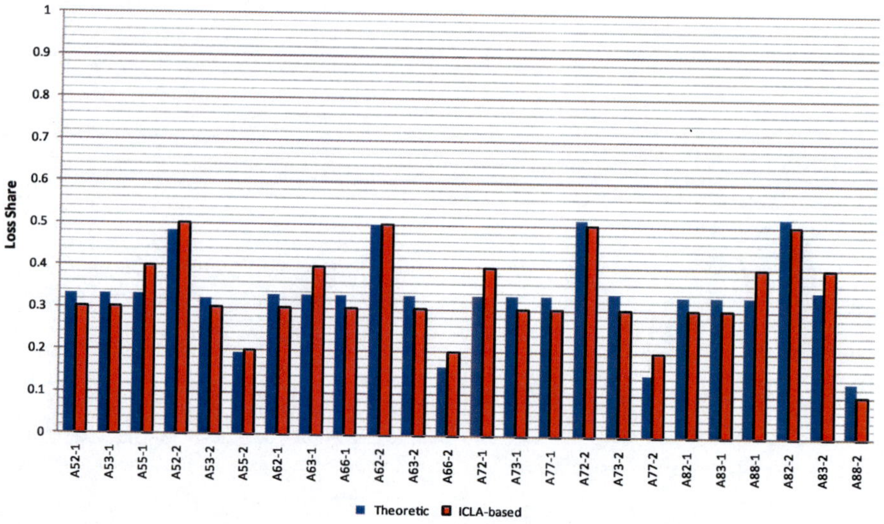

Fig. 6.13 The proposed shares for X_5, X_6, X_7 and X_8 using the ICLA-based and theoretic methods under case C

method. This is a valuable feature for the proposed method and shows the capability of ICLA to reach an appropriate LSA near to a Pareto optimal LSA without even knowledge about loss distributions or utility function forms. However, for more complex tendency graphs, ICLA needs more iteration for finding an appropriate LSA.

6.4.2.2 Usefulness of a Pareto Optimal LSA for Risk-Averse Service Providers

In this section, we compare the utility of service providers under three different situations: ICLA-sharing situation where the service providers share their losses according to the LSA which is obtained by the proposed method, Opt-Sharing situation where the loss-sharing is done according to the Pareto optimal LSA which is calculated theoretically and Non-Sharing situation which is a situation without any loss sharing.

The reported results in this section belong to a conducted experiment using setting 4. Table 6.5 compares the average utility of the service providers over 200 different iterations in case A, B, and C. As shown in this table by participating in loss sharing, all service providers can reach better utility. It is expected the Opt-Sharing to reach better utility than ICLA-Sharing, but there are some exceptions in Table 6.5. These are due to the randomness of the environment, and in a long time, opt-sharing always reaches better average utility. As illustrated in Table 6.5, the difference in average utility between ICLA-Sharing and Opt-Sharing is trivial. This shows that the proposed method has an excellent performance without even knowing about the form of utility functions and loss distributions. Let define utility difference as $\Delta u = u_{ICLA-Sharing} - u_{Non-Sharing}$. Figure 6.14a shows the utility difference of SP1 in case A over 200 different iterations. As illustrated in this figure, most of the time, the utility difference is positive. Figure 6.14b shows the average of the utility difference ($\overline{\Delta u}(t)$) from beginning up to iteration t. Figures 6.15 and 6.16 show the same diagrams for SP2 and SP5, respectively. Comparing the average utility difference of SP1 and SP5, it can be concluded that provided a similar condition for two service providers, $\overline{\Delta u}(t)$ will be greater for the more risk-averse one.

The utility difference and average utility difference in cases B and C also support the idea that most of the time, utility difference is positive and $\overline{\Delta u}(t)$ will be higher for more risk-averse service providers. Figure 6.17 illustrates the variances of utilities

Table 6.5 The average utility of the service providers over 200 different iterations

		SP1	SP2	SP3	SP4	SP5	SP6	SP7	SP8
Case A	ICLA-Sharing	**1.574**	1.597	**2.693**	**2.752**	3.104	–	–	–
	Opt-Sharing	**1.574**	**1.598**	2.686	**2.752**	**3.106**	–	–	–
	Non-Sharing	1.5	1.501	2.541	2.571	2.964	–	–	–
Case B	ICLA-Sharing	**1.578**	**1.589**	2.676	**2.652**	3.102	3.095	**3.932**	**3.94**
	Opt-Sharing	**1.578**	**1.589**	**2.678**	**2.652**	**3.104**	**3.097**	3.926	3.935
	Non-Sharing	1.507	1.505	2.57	2.539	2.951	2.925	3.764	3.799
Case C	ICLA-Sharing	**1.586**	1.611	2.72	2.668	**3.127**	**3.124**	**3.974**	**3.978**
	Opt-Sharing	1.585	**1.616**	**2.729**	**2.67**	3.126	3.123	3.966	3.971
	Non-Sharing	1.518	1.514	2.585	2.571	2.963	2.948	3.776	3.825

The best value is highlighted in boldface

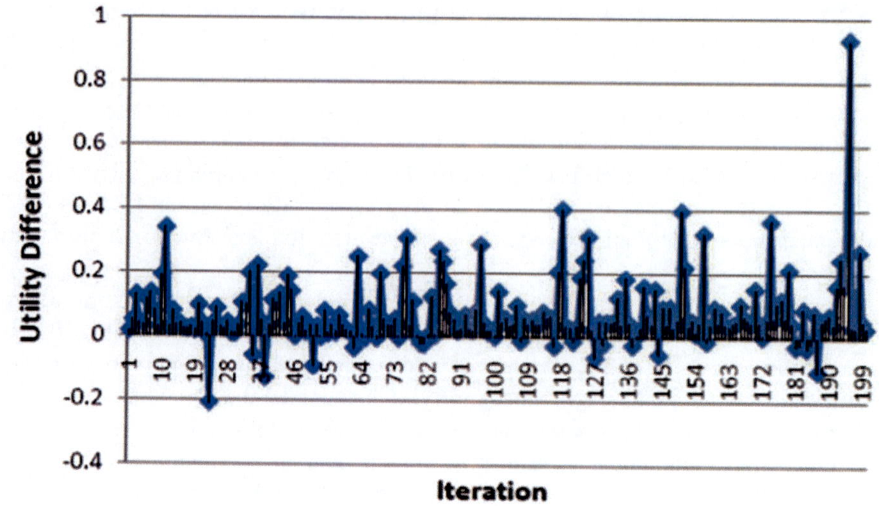

(a) The utility difference in case A over 200 sampled iterations

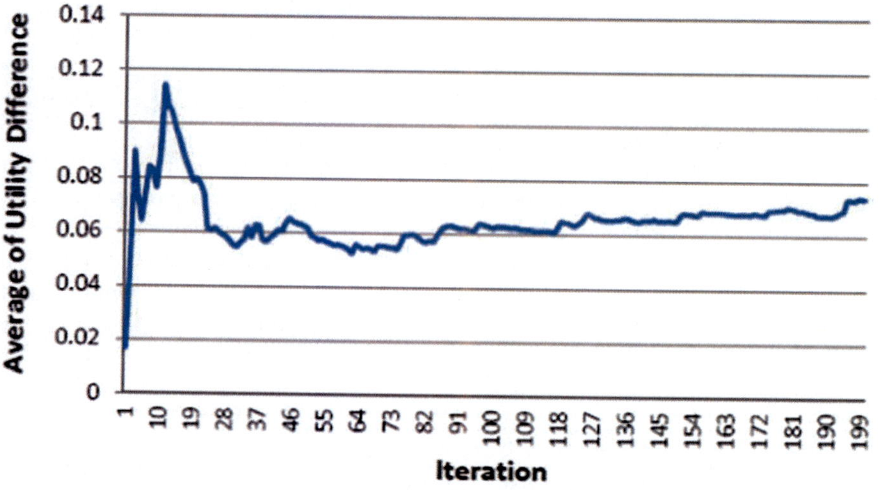

(b) The average utility difference for SP1

Fig. 6.14 The utility difference in case A over 200 sampled iterations and average for SP1

of SP1, SP2 and SP5 under Non-Sharing and ICLA-Sharing situations in Cases A, B, and C. Because the variance of utility in Non-Sharing situation is approximately identical for the three cases, so it is plotted once. As shown in the figure, under the ICLA-Sharing situation, the variance of utility drops considerably for all the three service providers. Because the service providers are risk-averse, so they prefer a more stable utility, and this drop is desirable for them. As illustrated in Fig. 6.17, differences

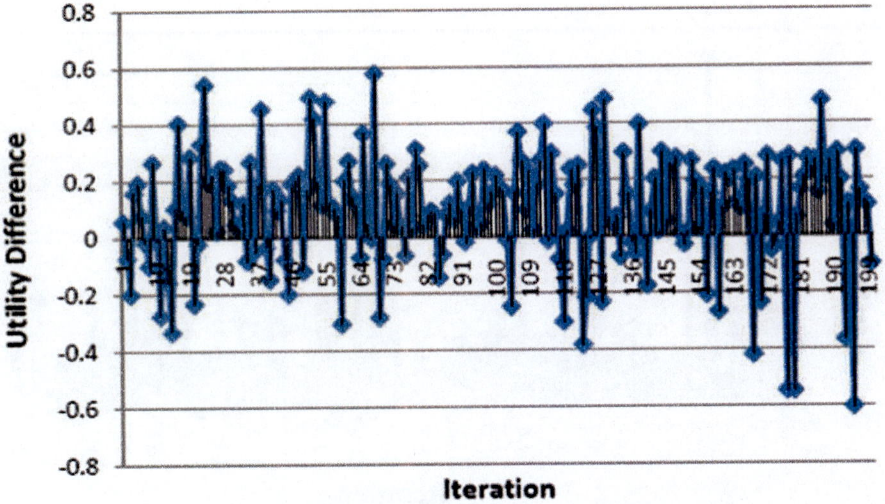

(a) The utility difference in case A over 200 sampled iterations

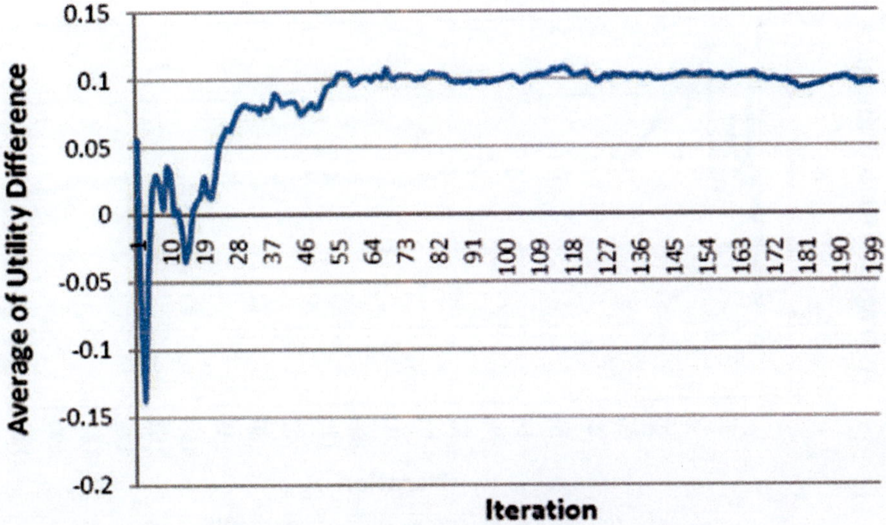

(b) The average utility difference for SP2

Fig. 6.15 The utility difference in case A over 200 sampled iterations and average for SP2

of variances in cases B and C are considerable for SP2 and SP5 while this is not the case for SP1. This is for the reason that SP1 has the same collaboration tendency in both cases B and C, while SP2 and SP5 extend their collaboration tendency in case C. Existence of a greater number of service providers in loss sharing causes more drops in the variance of utility and service providers to experience more stable utility.

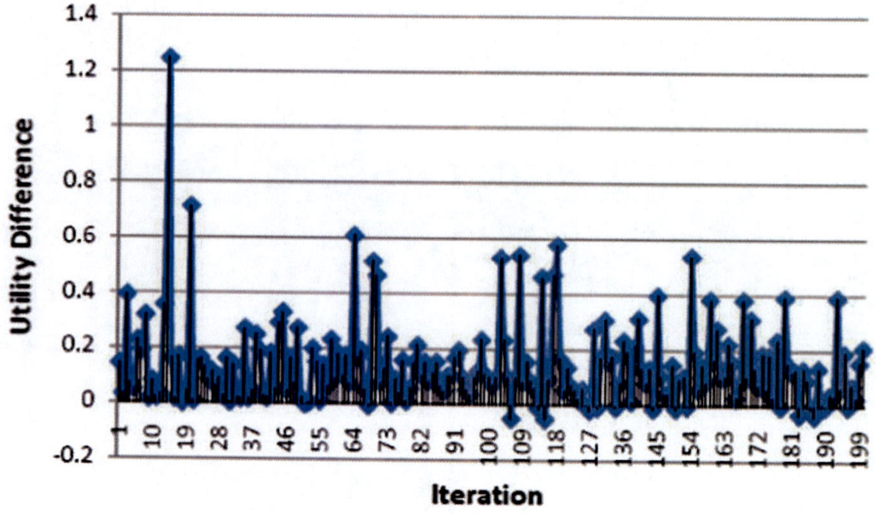

(a) The utility difference in case A over 200 sampled iterations

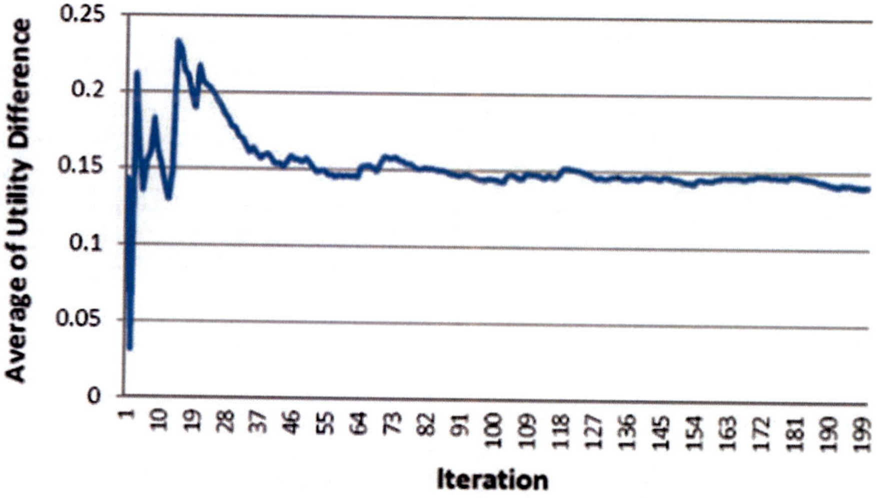

(b) The average utility difference for SP5

Fig. 6.16 The utility difference in case A over 200 sampled iterations and average for SP5

6.4.2.3 Scalability of the Proposed Method

In this section, we discuss the scalability of the proposed method. Figure 6.18a plots the evolution of the probability vector of LA_{21} which decides about X_1 in case A. Figures 6.18b and c show the same diagrams for LA_{22} and LA_{25} in case A. As illustrated in Fig. 6.5a, two, five and two service providers are involved in the sharing of X_1, X_2 and X_5 respectively. The needed iterations for convergence of

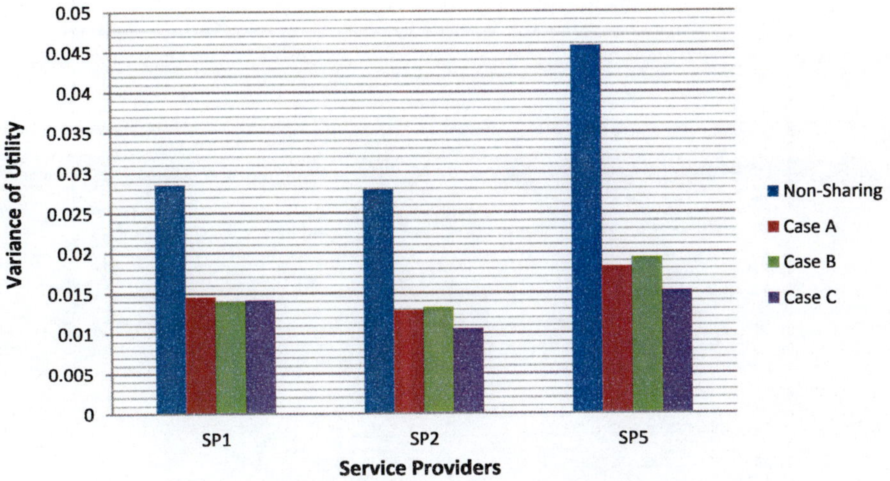

Fig. 6.17 Variances of utilities of SP1, SP2 and SP5 under Non-Sharing and ICLA-sharing situations in Cases A, B and C

LA_{21}, LA_{22} and LA_{25} to one of their actions is about 700, 1600, and 650 iterations. This shows that the needed iterations for convergence of a LA depend on the number of neighbors. Comparing the needed iterations for convergence of LA_{21}, LA_{22} and LA_{25} in case B with case A and C (see Figs. 6.19 and 6.20) illustrate that the needed iterations for convergence are almost invariant of the number of service providers. At the same time, it depends on the number of neighbors. So, using the proposed method, the needed iterations for finding an LSA depends on the maximum degree of tendency graph, not to the size of the graph. This result illustrates that the proposed method is usable even for large graphs with a proper maximum degree.

6.5 Conclusion

In real-world situations, we may face situations where offering insurance coverage to all users is not possible. As discussed, the percentage of users who can benefit insurance coverage depends on the financial capabilities of service providers. In this chapter, an approach is presented for loss sharing among collaborative service providers to increase their financial capabilities and utilities. Because service providers are risk-averse, they prefer more certain conditions when they are confronted with two choices with the same expected utility. This means that if collaboration with other service providers can decreases uncertainty, a risk-averse service provider is interested in collaborating. Due to the capabilities of learning automata and ICLA in environments with high uncertainty, we employed ICLA to present the proposed approach. In the proposed approach, learning automata decide about the

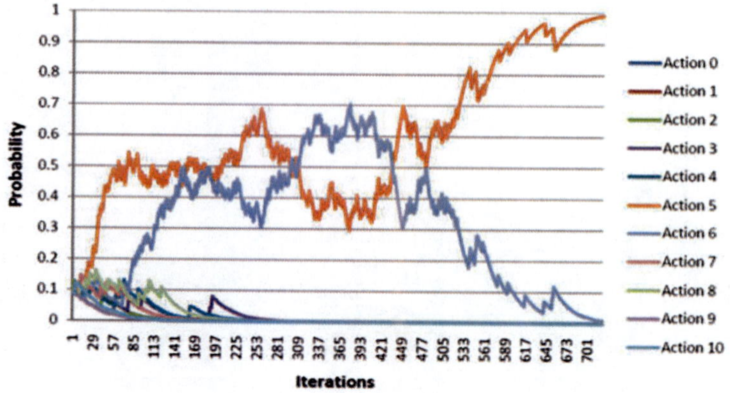

(a) LA_{21}

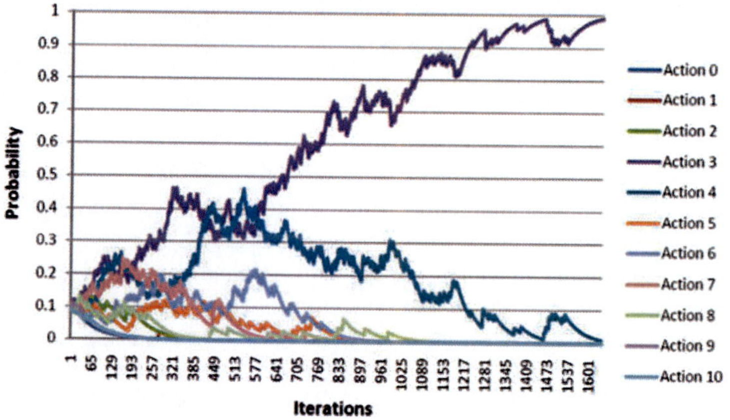

(b) LA_{22}

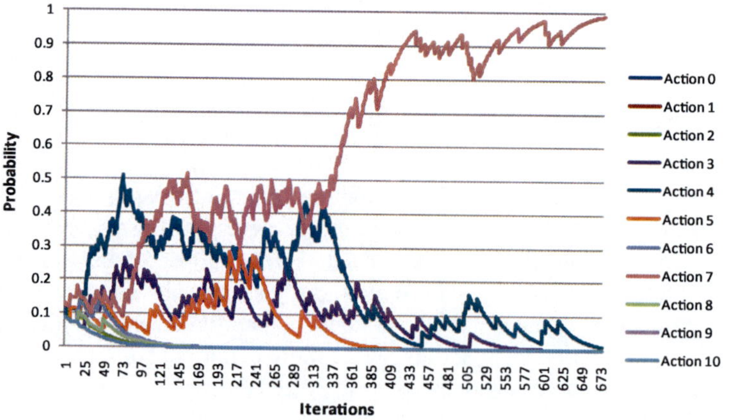

(c) LA_{25}

Fig. 6.18 Evolution of probability vector of **a** LA_{21} **b** LA_{22} **c** LA_{25} in case A

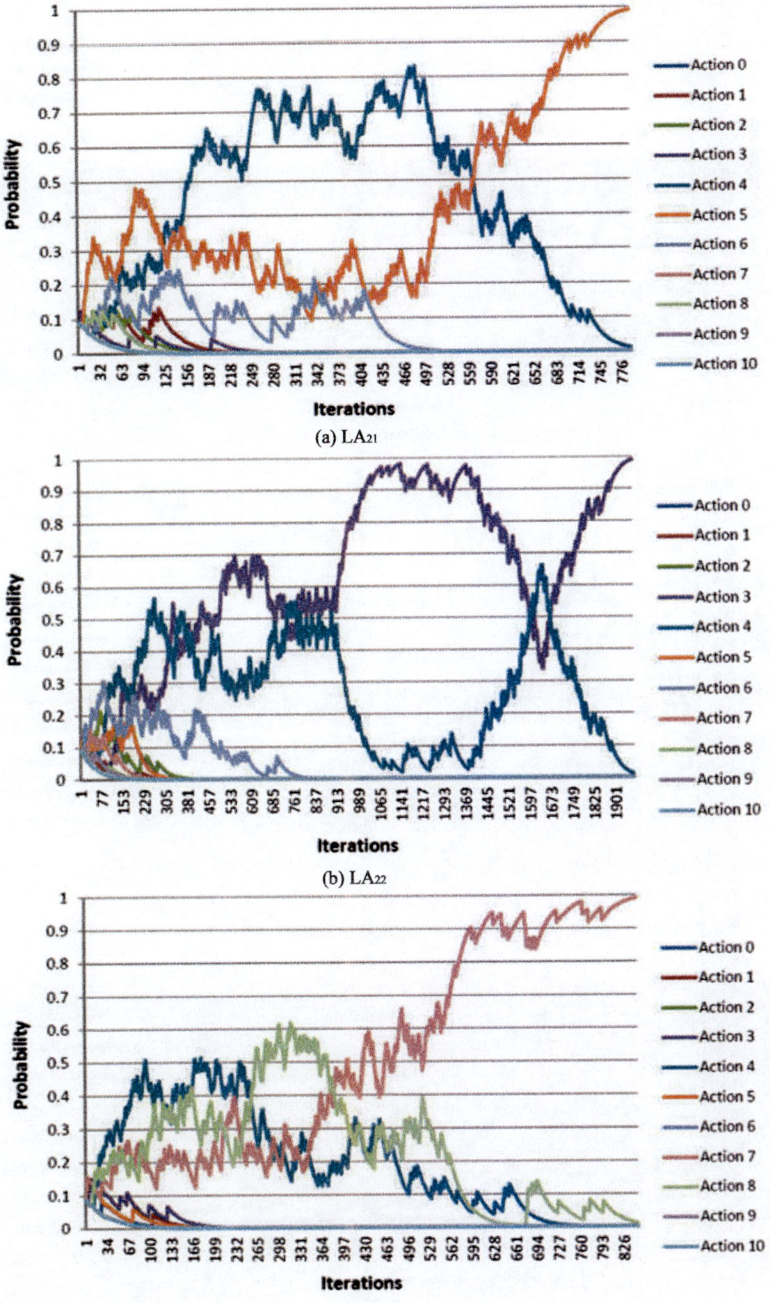

(a) LA_{21}

(b) LA_{22}

(c) LA_{25}

Fig. 6.19 Evolution of probability vector of **a** LA_{21} **b** LA_{22} **c** LA_{25} in case B

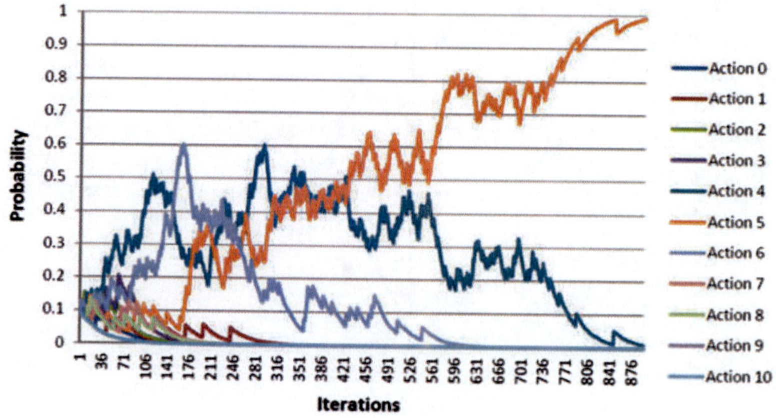

(a) LA_{21}

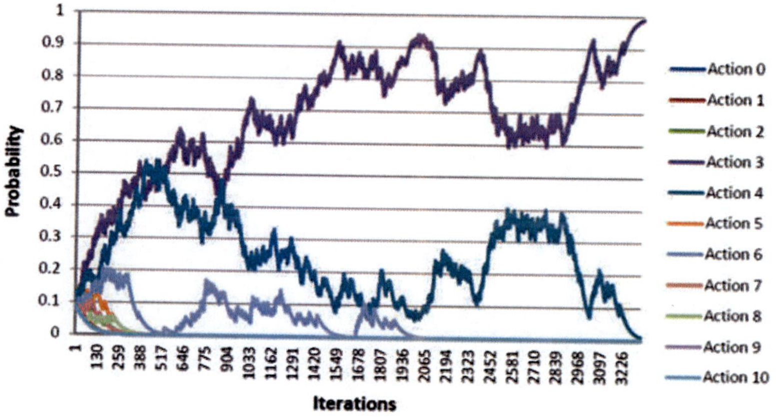

(b) LA_{22}

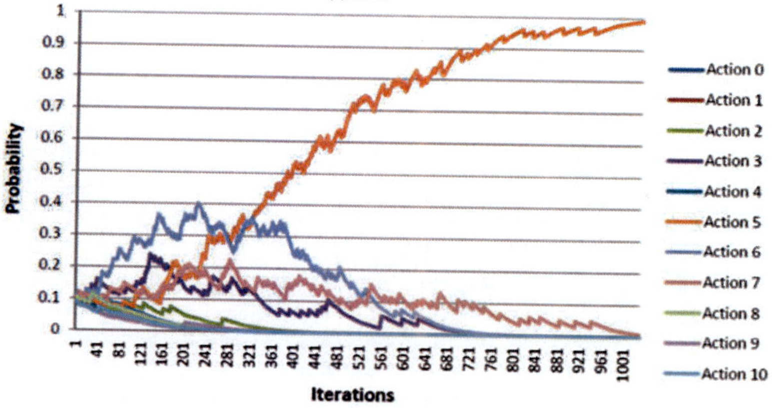

(c) LA_{25}

Fig. 6.20 Evolution of probability vector of **a** LA_{21} **b** LA_{22} **c** LA_{25} in case C level of loss sharing. The results of the conducted experiments illustrated that the proposed method is capable of finding an appropriate LSA such that it increases the utilities of all service providers. Besides, the utilities obtained by the proposed method have trivial differences with the utilities earned by a Pareto optimal LSA. Finding a Pareto optimal LSA needs detailed information about utility functions and loss distributions. At the same time, it is notable that there is no need for such information by the proposed ICLA-based approach. Also, the results illustrated that the proposed approach could be employed for a large number of service providers, and the time needed for finding an LSA is almost invariant of the number of service providers.

References

Aase, K.K.: Dynamic Equilibrium and the Structure of Premiums in a Reinsurance Market. Geneva. Pap. Risk Insur. Theory **17**, 93–136 (1992). https://doi.org/10.1007/BF00962709

Aase, K.K.: Equilibrium in a reinsurance syndicate; existence, uniqueness and characterization. Insur. Math. Econ. **283** (1995)

Adelson, R.M.: Utility Theory for Decision Making. J. Oper. Res. Soc. **22**, 308–309 (1971). https://doi.org/10.1057/jors.1971.67

Buyya, R., Yeo, C.S., Venugopal, S., et al.: Cloud computing and emerging IT platforms: Vision, hype, and reality for delivering computing as the 5th utility. Futur. Gener. Comput. Syst. **25**, 599–616 (2009). https://doi.org/10.1016/j.future.2008.12.001

Chateauneuf, A., Mostoufi, M., Vyncke, D.: Multivariate risk sharing and the derivation of individually rational Pareto optima. Math. Soc. Sci. **74**, 73–78 (2015). https://doi.org/10.1016/j.mathsocsci.2014.12.004

Cheung, K.C., Sung, K.C.J., Yam, S.C.P., Yung, S.P.: Optimal reinsurance under general law-invariant risk measures. Scand. Actuar. J. **2014**, 72–91 (2014). https://doi.org/10.1080/03461238.2011.636880

Deelstra, G., Plantin, G.: Risk Theory and Reinsurance. Springer, London, London (2014)

Esnaashari, M., Meybodi, M.R.: Irregular cellular learning automata. IEEE Trans. Cybern. **45**, 1622–1632 (2018). https://doi.org/10.1016/j.jocs.2017.08.012

Goiri, Í., Guitart, J., Torres, J.: Economic model of a Cloud provider operating in a federated Cloud. Inf. Syst. Front. **14**, 827–843 (2012). https://doi.org/10.1007/s10796-011-9325-x

Haimes, Y.Y.: Risk Modeling, Assessment, and Management. Wiley Publishing (2015)

Javadi, B., Abawajy, J., Buyya, R.: Failure-aware resource provisioning for hybrid cloud infrastructure. J. Parallel Distrib. Comput. **72**, 1318–1331 (2012). https://doi.org/10.1016/j.jpdc.2012.06.012

Khomami, M.M.D., Rezvanian, A., Meybodi, M.R.: A new cellular learning automata-based algorithm for community detection in complex social networks. J. Comput. Sci. **24**, 413–426 (2018). https://doi.org/10.1016/j.jocs.2017.10.009

Morshedlou, H., Meybodi, M.R.: A new local rule for convergence of ICLA to a compatible point. IEEE Trans. Syst. Man, Cybern. Syst. **47**, 3233–3244 (2017). https://doi.org/10.1109/TSMC.2016.2569464

Rajkumar Buyya, C.V.: Mastering Cloud Computing: Foundations and Applications Programming. Morgan Kaufmann Publishers Inc (2013)

Rezapoor Mirsaleh, M., Meybodi, M.R.: A new memetic algorithm based on cellular learning automata for solving the vertex coloring problem. Memetic. Comput. **8**, 211–222 (2016). https://doi.org/10.1007/s12293-016-0183-4

Rezvanian, A., Saghiri, A.M., Vahidipour, S.M., et al.: Learning automata for complex social networks. In: Recent Advances in Learning Automata, pp 279–334 (2018a)
Rezvanian, A., Saghiri, A.M., Vahidipour, S.M., et al.: Cellular learning automata. pp 21–88 (2018b)
Rezvanian, A., Saghiri, A.M., Vahidipour, S.M., et al.: Learning automata for wireless sensor networks. In: Recent Advances in Learning Automata, pp 91–219 (2018c)
Rezvanian, A., Moradabadi, B., Ghavipour, M., et al.: Learning Automata Approach for Social Networks. Springer International Publishing (2019a)
Rezvanian, A., Moradabadi, B., Ghavipour, M., et al.: Wavefront cellular learning automata: a new learning paradigm. In: Learning Automata Approach for Social Networks, pp. 51–74. Springer (2019b)
Rochwerger, B., Breitgand, D., Epstein, A., et al.: Reservoir—when one cloud is not enough. Comput. (Long Beach Calif) **44**, 44–51 (2011). https://doi.org/10.1109/MC.2011.64
Samaan, N.: A novel economic sharing model in a federation of selfish cloud providers. IEEE Trans. Parallel Distrib. Syst. **25**, 12–21 (2014). https://doi.org/10.1109/TPDS.2013.23
Toosi, A.N., Thulasiram, R.K., Buyya, R.: Financial option market model for federated cloud environments. In: 2012 IEEE Fifth International Conference on Utility and Cloud Computing, pp. 3–12. IEEE (2012)
Vahidipour, S.M., Esnaashari, M., Rezvanian, A., Meybodi, M.R.: GAPN-LA: a framework for solving graph problems using Petri nets and learning automata. Eng. Appl. Artif. Intell. **77**, 255–267 (2019). https://doi.org/10.1016/j.engappai.2018.10.013
Wolfram, S.: Theory and Applications of Cellular Automata. World Scientific Publication (1986)
Zweifel, P., Eisen, R.: Insurance Economics. Springer, Berlin Heidelberg, Berlin, Heidelberg (2012)

Chapter 7
Cellular Learning Automata for Competitive Loss Sharing

7.1 Introduction

In this chapter, the problem of loss sharing with competitive service providers is considered. Similar to the collaborative case in Chap. 6, we use irregular cellular learning automaton (ICLA) for modeling service providers and relation among them to present a loss-sharing approach under competitive conditions. In competitive conditions, service providers try to reach a compatible loss-sharing agreement (LSA). Under a compatible LSA, no service provider can improve its utility by changing strategies unilaterally (e.g., the shares from the LSA) while the neighbor service providers keep their strategies unchanged. To offer an ICLA based loss-sharing approach, we first extend the theory of ICLA by presenting a new local rule for convergence of ICLA to a compatible. This rule uses the response of the environment and the actions selected by the neighboring LAs to generate the reinforcement signal. Formal proofs for the convergence are provided, and results of the conducted experiments support the theoretical findings. The convergence of ICLA to the compatible point is of great importance because it can conduce to efficient solutions for the problems and applications. As an application of the presented local rule, we present an ICLA-based loss-sharing approach for competitive service providers. Similar to the collaborative condition in Chap. 6, by loss sharing along with increasing the financial capabilities of service providers, the proposed loss-sharing approach decreases the variance of random losses, which is attractive for risk-averse service providers. We demonstrate that by using ICLA, finding compatible LSAs do not need detailed information about distributions of the losses and even utility functions of the other service providers. Due to the randomness of the losses and their unknown distributions currently establishing a theoretic method for finding an appropriate loss-sharing agreement is very difficult and complex. The results of the conducted experiments illustrate that the proposed approach can improve the utility of both the service provider and user.

R. Vafashoar et al., *Cellular Learning Automata: Theory and Applications*, Studies in Systems, Decision and Control 307,
https://doi.org/10.1007/978-3-030-53141-6_7

The rest of this chapter is organized as follows: Sect. 7.2 contains some preliminaries and notations which are used in the next sections. In Sect. 7.3, we present a new local rule for convergence of ICLA to the compatible point. We also provide formal proofs for this convergence. Section 7.4 explains the proposed ICLA based approach for loss sharing. The results of the conducted experiments are presented in Sect. 7.5.

7.2 Preliminary Concepts

This section contains some preliminary concepts and definition which are needed for the understanding of the next sections. For any definition or concept which is not presented here, please refer to Chap. 6.

7.2.1 ICLA

Cellular learning automaton (CLA) (Beigy and Meybodi 2004; Rezvanian et al. 2018b) is a combination of cellular automata (Wolfram 1986) and learning automata (Kumpati Narendra 1989; Rezvanian et al. 2018a). In this model, each cell of a cellular automaton (CA) contains one or more learning automata (LA). The LA or LAs residing in a particular cell define the state of the cell. Like CA, a CLA rule controls the behavior of each cell. At each time step, the local rule defines the reinforcement signal for a particular LA based on its selected action and the actions chosen by its neighboring LAs. Consequently, the neighborhood of each LA is considered to be its local environment. A formal definition of CLA is provided by Beigy and Meybodi (Beigy and Meybodi 2004). The authors also investigated the asymptotic behavior of the model and provided some theoretical analysis on its convergence. A CLA starts from some initial state (it is the internal state of every cell); at each stage, each automaton selects an action according to its probability vector and performs it in its local environment. Next, the rule of the CLA determines the reinforcement signals (the responses) to the LAs residing in its cells and based on the received signals, each LA updates its internal probability vector. This procedure continues until a termination condition is satisfied

In the basic cellular learning automaton (CLA), cells are arranged in some regular forms like a gird or ring. However, there exist some applications which require irregular arrangements of the cells. Esnaashari and Meybodi have proposed an irregular CLA (Esnaashari and Meybodi 2018) for solving wireless sensor network problems. Irregular CLA is modeled as an undirected graph in which each vertex represents a cell of the CLA, and its adjacent vertices determine its neighborhood. Up to now, various applications of Irregular CLA are reported in the literature. The application areas of this model include, for instance, (but not limited to) wireless sensor networks (Rezvanian et al. 2018c), graph problems (Mousavian et al. 2014), peer-to-peer networks (Rezvanian et al. 2018d), cloud computing (Morshedlou and Meybodi 2017), complex networks (Khomami et al. 2018), network sampling (Rezvanian et al. 2019a), link prediction (Rezvanian et al. 2019b), and social network analysis (Ghavipour and Meybodi 2017).

7.2.2 ICLA Game

ICLA game is an extended form of a stochastic game of learning automata (Vrancx et al. 2008b). Consider a game with N players. Every iteration, the players select an action from their action sets. Then they receive stochastic payoffs from the environment. The payoff each player receives from the environment depends on the joint actions of all players. These payoffs can be dissimilar for different players. The stationary probability distributions which establish payoffs of joint actions are unknown for all players. Now assume a new game in which payoffs of player i and player j are independent of the selected actions of each other. Assume there exists a sequence of players i, i_1, …, i_m, j for $m \geq 1$ such that every two successive players in the sequence care about each other's strategy. Note that the payoffs of players i and j do not affect the selected action of each other instantly. However, in the long term, they would be influenced by each other's strategy indirectly and through other players who are in the sequence. Now, if in the former game and the latter game we represent each player by a learning automaton (LA), then we have a game of learning automata in the former case and an ICLA game in the latter case. Actions of a player constitute the action set of LA. Hence at any given instant, the probability vector of a LA is equivalent to the mixed strategy of a player. Since all parts of Definition 7.1 (except part 1) are valid for ICLA as well, so we can use them also for describing the ICLA game. The action set of each LA determines its state set (see Definition 7.1(2)). All LAs that LA i care about their strategy are in LA i's neighborhood vector. Local rule of ICLA game establishes the stochastic payoff of each learning automaton.

7.2.3 Compatible Loss Sharing Agreement

A loss-sharing agreement is compatible if and only if the unilateral deviation of service provider i (for any i) about any $\alpha_i^j(t)$ or $\alpha_j^i(t)$ (for any j) is not profitable. In other words, if service provider i unilaterally revise its decision about any $\alpha_i^j(t)$ or $\alpha_j^i(t)$ (for any j) then its expected utility $(Eu_i(PI_i(t)))$ will decrease.

7.3 A New Local Rule for Convergence of ICLA to Compatible Point

Potentials of a single LA to operate in a stochastic environment, made researchers extend its idea to a team of LAs as a group operating in a game setting. Common payoff (Kumpati and Narendra 1989; Tilak et al. 2011) and zero-sum (Lakshmivarahan 1981; Kumpati and Narendra 1989; Tilak et al. 2011) games are studied in the literature. A significant result is that learning automata reach a game-theoretic equilibrium point as the penalty parameter goes to zero (Lakshmivarahan 1981; Sastry et al. 1994). In (Billard 1998), a cooperative game-theoretic approach to model a multi-level decision-making process is presented. Players of this game are learning automata, which decide about the profitability of the group constitution. The reported results illustrate that delayed information, both types of penalties (asymmetric and symmetric), lead to chaos but with different Lyapunov exponents. These results are confirmed by Billard (1994), (1997) as well. The games of LAs have also been used in many applications such as multiple access channel selection (Zhong et al. 2010), congestion avoidance in wireless networks (Nicopolitidis et al. 2003; Misra et al. 2009), channel selection in radio networks (Tuan et al. 2010), clustering in wireless ad hoc networks (Akbari Torkestani and Meybodi 2010), power system stabilizers (Kashki et al. 2010), backbone formation in ad hoc wireless networks (Torkestani and Meybodi 2009), spectrum allocation in cognitive networks (Liu et al. 2010), reinforcement learning problems (Nowé et al. 2006; Rodríguez et al. 2012), and solving various NP-complete problems (Thierens 2005; Tilak et al. 2010). These applications show capabilities of LA when multiple learning automata interact with each other. Another class of games, which uses capabilities of learning automata, is Markov games category. The idea of using LAs in Markov games originates from the ability of LAs to manage and control a Markov Chain (Wheeler and Narendra 1986; Vrancx et al. 2008a) without knowledge about transition probabilities in an independent and non-centralized way. In (Vrancx et al. 2008b) it is illustrated that a group of independent LAs are able to converge to equilibrium of a limiting game, even with limited observations. Furthermore in (Masoumi and Meybodi 2012), some learning automata-based methods for finding optimal policies in Markov games are suggested.

In all the mentioned works, all LAs affect each other directly, but this may not be true for some applications. There are many applications that players do not directly influence each other. The mathematical framework presented in (Beigy and Meybodi 2004) is appropriate for modeling such situations. Also, the introduced concept of compatible point in (Beigy and Meybodi 2004) can be used as an equivalent equilibrium point to Nash. However, the capabilities and applications of CLA (Rezvanian et al. 2018b) are not restricted to stochastic games with transitive influences. CLA has been found to perform well in many application areas such as image processing (Gharehchopogh and Ebrahimi 2012; Nejad et al. 2014; Hasanzadeh Mofrad et al. 2015), modeling of commerce networks (Meybodi and Khojasteh 2001), channel assignment in cellular networks (Beigy and Meybodi 2003, 2009), clustering the wireless sensor networks (Esnaashari and Meybodi 2008; Ahmadinia et al. 2013), mobile wireless sensor network (Esnaashari and Meybodi 2013), vertex cover problem (Mousavian et al. 2014), community detection (Khomami et al. 2018), vertex coloring (Vahidipour et al. 2017a, b), sampling social networks (Ghavipour and Meybodi 2017), peer-to-peer networks (Saghiri and Meybodi 2018) and solving NP-hard problems (Eraghi et al. 2009; Akbari Torkestani and Meybodi 2011; Rezapoor Mirsaleh and Meybodi 2016), to name just a few. In this section, we present a new local rule for convergence of ICLA to a compatible point. Before presenting the new local rule first, we show two preliminary lemmas which are required to prove convergence of ICLA to a compatible point.

7.3.1 Preliminary Lemmas

In this section, first, we prove the existence of a compatible point in ICLA, and then a preliminary lemma will be presented. Definition 7.4 defines K in Lemma 7.1.

Lemma 7.1 *In the ICLA game, there is at least one compatible point.*

Proof The proof for ICLA is somewhat similar to Lemma 7.2 in (Beigy and Meybodi 2004). Let' us define a mapping $T : KK$ as Eq. (7.1).

$$\forall i, \forall r \in \{1, \ldots, m_i\} \quad \bar{p}_{ir} = \frac{p_{ir} + \varphi_{ir}}{1 + \sum_{j=1}^{m_i} \varphi_{ij}}, \tag{7.1}$$

where $\psi_{ir} = d_{ir}\left(\underline{p}\right) - D_i\left(\underline{p}\right)$ and $\varphi_{ir}\left(\underline{p}\right) = \max\{\psi_{ir}\left(\underline{p}\right), 0\}$.

Because T is continuous and K is closed, bounded and convex so based on Brouwer's fixed point theorem, T has at least one fixed point. We claim that a point will be a fixed point if and only if it be a compatible point of ICLA. To show this, assume that $\underline{p}$ is a fixed point of T. So, for all r and i: $\bar{p}_{ir} = p_{ir}$. Concerning this fact and Eq. (7.1), we have:

$$\forall i, \forall r \in \{1, \ldots, m_i\} \quad p_{ir} = \frac{p_{ir} + \varphi_{ir}}{1 + \sum_{j=1}^{m_i} \varphi_{ij}} \tag{7.2}$$

From Eq. (7.2) we can conclude Eq. (7.3).

$$\sum_{j=1}^{m_i} \varphi_{ij} = \frac{\varphi_{ir}}{p_{ir}} \tag{7.3}$$

Equation (7.3) results in Eq. (7.4):

$$\frac{\varphi_{i1}}{p_{i1}} = \frac{\varphi_{i2}}{p_{i2}} = \cdots = \frac{\varphi_{ir}}{p_{ir}} \tag{7.4}$$

Being $\varphi_{ir} > 0$ for all φ_{ir} is not possible, so the only case that satisfies Eq. (7.4) is the case that

$$\forall i, m \quad \varphi_{ir} = 0 \tag{7.5}$$

Now assume that $y_i(1 \leq i \leq m_i)$ is the action that reaches a maximum value for $d_{iy_i}\left(\underline{p}\right)$ among the others. Consider another action $y_j(j \neq i)$ that it is $d_{iy_j}\left(\underline{p}\right)$ be smaller than $d_{iy_i}\left(\underline{p}\right)$. Assume that probability distribution $\underline{p}$ puts a positive probability value on y_j , so we will have $D_i\left(\underline{p}\right) < d_{iy_i}\left(\underline{p}\right)$ that means $\varphi_{iy_i}\left(\underline{p}\right) > 0$. This is in contradiction with Eq. (7.5) so the assumption is not valid and probability distribution $\underline{p}$ puts zero probability values on all action that their $d_{iy_j}\left(\underline{p}\right)$ is smaller than $d_{iy_i}\left(\underline{p}\right)$. According to this fact, we have Eq. (7.6):

$$\forall \underline{q} \in \mathrm{K} \quad \sum_r d_{ir}\left(\underline{p}\right) \times p_{ir} \geq \sum_r d_{ir}\left(\underline{p}\right) \times q_{ir} \tag{7.6}$$

So $\underline{p}$ is a compatible point. Conversely, if $\underline{p} \in \mathrm{K}$ is a compatible point, then Eq. (7.6) will be valid for all i. configuration $\underline{q}$ also contains $\underline{q} = \left(\underline{p}_1, \ldots, \underline{e}_{r_i}, \ldots, \underline{p}_n\right)$ for fixed i. Since $d_{ir_i}\left(\underline{p}\right)$ is independent of $\underline{p}_i$, we have $\psi_{ir_i}\left(\underline{p}\right) \leq 0$. Therefore $\varphi_{ir_i}\left(\underline{p}\right) = 0$ for all i and all $r_i = 1\ldots, m_i$. So $\underline{p}$ is a fixed point of T. □

Lemma 7.2 *In algorithm 1 (Fig. 7.1), If* t *approaches infinity then* $\underline{\hat{p}}_i^t$ converges to $\underline{p}_i^t$.

Algorithm 7-1. Algorithm 1 for estimation of neighbor's strategy

$n_i^t(a_j)$ shows how many times, LA_i selected action a_j during epoch t.
L = length of epoch (number of times LA_i chooses an action)

```
t = 1;
p̂_ij^1 = n_i^1(a_j) / L
while(True) {
    p̂_ij^{t+1} = ( p̂_ij^t + n_i^t(a_j)/L ) / 2
    t = t + 1;
}
```

Fig. 7.1 Pseudo-code of Algorithm 1 for estimation of neighbor's strategy

Notations $\underline{p}_i^t$ is the probability vector of LA_i at iteration t and is not known to the other learning automata

$\hat{\underline{p}}_i^t$ is the estimation of $\underline{p}_i^t$ by observing the chosen actions of LA_i during L.

Proof When ICLA reaches a stable point, in that point there exist a $\underline{p}^*$ such that for every $\varepsilon > 0$ we have:

$$\lim_{t\to\infty} P(\left|\underline{p}^t - \underline{p}^*\right| > \varepsilon) = 0 \tag{7.7}$$

It's obvious that

$$\lim_{t\to\infty} P(\left|\underline{p}_i^t - \underline{p}_i^*\right| > \varepsilon) = 0 \tag{7.8}$$

Now, if we can prove (7.9), the proof will be completed. Please note that $\underline{p}^*$ is the stable configuration and $\underline{p}^t$ is the probability vectors of learning automata. $\hat{\underline{p}}^t$ is an estimation of $\underline{p}^t$ and the relation between $\hat{\underline{p}}^t$ and $\underline{p}^t$ is illustrated by (7.12).

$$\lim_{t\to\infty} P(|\hat{\underline{p}}_i^t - \underline{p}_i^*| > \varepsilon) = 0 \tag{7.9}$$

In algorithm 1, $n_i^t(a_j)$ can be replaced by summation of X^t random variables where X^t defined as:

$$X^t = \begin{cases} 1 \text{ if action } a_j \text{ was chosen during epoch t} \\ 0 \qquad\qquad\qquad \text{otherwise} \end{cases}$$

Because changes of strategies in each epoch are negligible, so the probability to choose action a_j during the epoch, t is a fixed value like as $\underline{p}_{ij}^t$. Then we have:

$$E(X^t) = 1 \times \underline{p}_{ij}^t + 0 \times \left(1 - \underline{p}_{ij}^t\right) = \underline{p}_{ij}^t \tag{7.10}$$

So

$$\begin{aligned} E\left(n_i^t(a_j)\right) =& E\left(\frac{X_1^t + \cdots + X_L^t}{L}\right) = \frac{1}{L} \times \left(E\left(X_1^t\right) + \cdots + E\left(X_L^t\right)\right) \\ =& \frac{1}{L} \times L \times \underline{p}_{ij}^t = \underline{p}_{ij}^t \end{aligned} \tag{7.11}$$

Now using algorithm 1, we can see that $\hat{\underline{p}}_{ij}^{t+1} = \frac{1}{2} \times \hat{\underline{p}}_{ij}^t + \frac{1}{2} \times \frac{n_i^t(a_j)}{L}$.

So $\hat{\underline{p}}_{ij}^{t+1} = \frac{1}{2^t} \times \hat{\underline{p}}_{ij}^1 + \frac{1}{2^{t-1}} \times \frac{n_i^2(a_j)}{L} + \frac{1}{2^{t-2}} \times \frac{n_i^3(a_j)}{L} + \cdots + \frac{1}{2} \times \frac{n_i^t(a_j)}{L}$

$$E\left(\underline{p}_{ij}^{t+1}\right) = E\left(\frac{1}{2^t} \times \frac{n_i^1(a_j)}{L} + \cdots + \frac{1}{2} \times \frac{n_i^t(a_j)}{L}\right) \Rightarrow E\left(\hat{\underline{p}}_{ij}^{t+1}\right)$$
$$= \frac{1}{2^t} \times \underline{p}_{ij}^1 + \cdots + \frac{1}{2} \times \underline{p}_{ij}^t$$

$$\lim_{t\to\infty} E\left(\hat{\underline{p}}_{ij}^{t+1}\right) = \lim_{t\to\infty}\left(\cdot\frac{1}{2^t} \times \underline{p}_{ij}^1 + \cdots + \frac{1}{2} \times \underline{p}_{ij}^t\right) = \cdots + 0 + 0 + \cdots + \frac{1}{2^k}$$
$$\times \lim_{t\to\infty}\left(\underline{p}_{ij}^{t-k+1}\right) + \cdots + \frac{1}{2} \times \lim_{t\to\infty}\left(\underline{p}_{ij}^t\right) \tag{7.12}$$

$$\lim_{t\to\infty} E\left(\hat{\underline{p}}_{ij}^{t+1}\right) = 0 + 0 + 0 + \cdots + \frac{1}{2^2} \times \underline{p}_{ij}^* + \frac{1}{2} \times \underline{p}_{ij}^* = \underline{p}_{ij}^*\left(\sum_{t=1}^{\infty} \frac{1}{2^t}\right) = \underline{p}_{ij}^* \tag{7.13}$$

So, in the general case, we have

$$\lim_{t\to\infty} E\left(\hat{\underline{p}}_i^t\right) = \underline{p}_i^* \tag{7.14}$$

Since $\underline{p}_i^*$ is the expected value of $\hat{\underline{p}}_i^t$ from the central limit theorem we get:

$$\lim_{t\to\infty} P(\left|\hat{\underline{p}}_i^t - \underline{p}_i^*\right| > \varepsilon) = 0 \tag{7.15}$$

So, the proof is completed. □

7.3.2 The Proposed Local Rule

Let $\underline{p}_i^c$ denotes the current probability vector of LA_i and the best response probability vector of LA_i (concerning the probability vectors of its neighbors) is indicated by $\underline{p}_i^{br}$. Algorithm 2 (see Fig. 7.2) illustrates our proposed local rule to generate a reinforcement signal (β). According to definition 7.7, for each LA unilaterally deviation is not profitable in compatible points. As a result, when ICLA is converged to a compatible point, the probability vector of every learning automaton is the best response probability vector for the probability vectors of its neighbors. Considering this fact, the concept behind the proposed local rule is as below.

In an iterative process, if each learning automaton selects its action according to the best response probability vector (respect to the observed and estimated probability vectors of its neighbors), then with accurate estimations, a non-compatible point cannot be a stable point. This is for the reason that in a non-compatible point, there is at least one learning automaton that unilateral deviation will be profitable for it. To estimate the best response probability vector for learning automaton *i,* we use the

Algorithm 7-2. proposed local rule

1 $\underline{p}_i^c$ = current probability vector of LA_i
2 $\underline{p}_i^{br}$ = best response probability vector acquired from Eq. (7-17)
3 d_i = vector of reward probabilities
4 *Begin*
5 $\underline{p}_i^{br}(\hat{\underline{p}}^t, Q_t^i) = \arg\max_{\underline{p}_i} \sum_{a^1} \sum_{a^2} \cdots \sum_{a^{\overline{m}_i}} \underline{p}_i(\hat{\underline{p}}^t, Q_t^i)(a^i) \prod_{\substack{j=1 \\ j\neq i}}^{\overline{m}_i} \hat{\underline{p}}_j^t(a^j).Q_t^i(a^1,\ldots,a^{\overline{m}_i});$
6 $\Delta\underline{p}_i^{RS} = \underline{p}_i^{br} - \underline{p}_i^c;$
7 $j_{\max} = \arg\max_j \Delta\underline{p}_i^{RS}(j);$
8 $j_{\min} = \arg\min_j \Delta\underline{p}_i^{RS}(j);$
9 $d_i = \left(\frac{\Delta\underline{p}_i^{RS}(1) - \Delta\underline{p}_i^{RS}(j_{\min})}{\Delta\underline{p}_i^{RS}(j_{\max}) - \Delta\underline{p}_i^{RS}(j_{\min})}, \ldots, \frac{\Delta\underline{p}_i^{RS}(\overline{m}_i) - \Delta\underline{p}_i^{RS}(j_{\min})}{\Delta\underline{p}_i^{RS}(j_{\max}) - \Delta\underline{p}_i^{RS}(j_{\min})} \right)$
10 if (selected action $== a^k \in \{a^1,\ldots,a^{\overline{m}_i}\}$) then
11   if ($\Delta\underline{p}_i^{RS}(k) \geq 0$) then
12   set $\beta = 1$ with probability $d_i(k)$ and $\beta = 0$ with probability $(1 - d_i(k))$;
13   else if ($\Delta\underline{p}_i^{RS}(k) < 0$) then
14   set $\beta = 0$ with probability $d_i(k)$ and $\beta = 1$ with probability $(1 - d_i(k))$;
15 update Q - values using Eq. (7.18);
16 return β;
17 End

Fig. 7.2 Algorithm 2 shows the proposed local rule

expression in line 5 in the proposed local rule. Please note that in line 5, except for the real probability vector of LA_i, the estimated probability vectors of neighbor learning automata are used. The proposed local rule tries to reduce difference ($\Delta \underline{R}_i^{RS}$) between the current probability vector ($\underline{p}_i^c$) and the best response probability vector ($\underline{p}_i^{br}$). The average reward for learning automaton i (d_i) is determined such that action j to be rewarded proportional to $\Delta \underline{R}_i^{RS}(j)$ to reduce the difference. For computation of $\underline{p}_i^{br}$ we need $\hat{\underline{p}}^t$ and Q_t^i. $\hat{\underline{p}}^t$ can be estimated using algorithm 1, but for Q_t^i, we should maintain a table of Q-values $Q^i(a^1, \ldots, a^{\overline{m}_i})$ for each joint action of $(a^1, \ldots, a^{\overline{m}_i})$. By inspiration from Q-learning, we define the optimal Q-values for LA_i with respect to its neighbors as Eq. (7.16).

$$Q_*^i(a^1, \ldots, a^{\overline{m}_i}) = Er_i(a^1, \ldots, a^{\overline{m}_i}) + \beta.v^i\left(\underline{p}_1^*, \underline{p}_2^*, \ldots, \underline{p}_i^{br}\left(\underline{p}^*, Q_*^i\right), \ldots, \underline{p}_{\overline{m}_i}^*\right) \tag{7.16}$$

Where $v^i\left(\underline{p}_1^*, \underline{p}_2^*, \ldots, \underline{p}_i^{br}\left(\underline{p}^*, Q_*^i\right), \ldots, \underline{p}_{\overline{m}_i}^*\right)$ is the total discounted reward of LA_i over infinite iterations when probability vectors of the neighbors are $\left(\underline{p}_1^*, \underline{p}_2^*, \ldots, \underline{p}_{i-1}^*, \underline{p}_{i+1}^*, \ldots, \underline{p}_{\overline{m}_i}^*\right)$ and $\underline{p}_i^{br}\left(\underline{p}^*, Q_*^i\right)$ is the best response probability vector concerning the probability vectors of the neighbors. Er_i is the expected value of the responses of the environment that learning automaton i has received until iteration t. The ordinary local rule uses these responses to reward or penalize the selected action of learning automaton simply. Since the environment is stochastic and information is incomplete, we have no idea about Q-values, and we should find a way to estimate them. Initially, we propose to assign a fixed value (e.g., zero) to Q-values and then update them using the Eq. (7.17).

$$Q_{t+1}^i\left(a^1,\ldots,a^{\overline{m}_i}\right) = (1-\alpha)\times Q_t^i\left(a^1,\ldots,a^{\overline{m}_i}\right) + \alpha \times \left[Er_i^t\left(a^1,\ldots,a^{\overline{m}_i}\right) + \beta.\sum_{a^1}\sum_{a^2}\cdots\sum_{a^{\overline{m}_i}} \underline{p}_i^{br}\left(\hat{\underline{p}}^t, Q_t^i\right)\left(a^i\right) \prod_{\substack{j=1 \\ j\neq i}}^{\overline{m}_i} \hat{\underline{p}}_j^t\left(a^j\right).Q_t^i\left(a^1,\ldots,a^{\overline{m}_i}\right)\right], \tag{7.17}$$

where $\underline{p}_i^{br}\left(\hat{\underline{p}}^t, Q_t^i\right)$ is the best response probability vector against $\left(\hat{\underline{p}}_1^t,\ldots,\hat{\underline{p}}_{i-1}^t,\hat{\underline{p}}_{i+1}^t,\ldots,\hat{\underline{p}}_{\overline{m}_i}^t\right)$. Now we point out a corollary from (Szepesvári and Littman 1999) that is the basis of our proof, and then we show that using Eq. (7.17) Q_t^i will converge to Q_*^i when $t\to\infty$.

Corollary 7.1 (Szepesvári and Littman 1999): *Consider the process generated by iteration of Eq.* (7.18), *where* $0\leq f_t(x)\leq 1$.

$$V_{t+1}(x) = (1-f_t(x))\times V_t(x) + f_t(x)\times[P_tV_t](x) \tag{7.18}$$

If the process defined by Eq. (7.19) converges to v^* w.p.1.

$$U_{t+1}(x) = (1-f_t(x))\times U_t(x) + f_t(x)\times\left[P_tv^*\right](x) \tag{7.19}$$

Moreover, the following conditions hold:

- There exist number $0<\gamma<1$ and a sequence $\lambda_t\geq 0$ converging to zero w.p.1 such that $||P_tV - P_tv^*||\leq\gamma||V-v^*||+\lambda_t$ holds for all $V\in\{V\}$.
- $0\leq f_t(x)\leq 1, t\geq 0$ and $\sum_{t=1}^{n} f_t(x)$ converges to infinity uniformly in x as $n\to\infty$.

Then, the defined iteration converges to v^* w.p.1.

Before presenting the convergence proof for the proposed local rule, we first explain some notations. We put $f_t(x)=\alpha$ and use Q instead of V. thus $\{V\}$ is replaced by $\{Q\}$ which is a set of all Q-functions ($Q: A_1\times\ldots\times A_{\overline{m}}\to R$) and P_t is a mapping from $\{Q\}$ to $\{Q\}$. Also $\|Q\|$ is equal to $\max_{(a^1,\ldots,a^{\overline{m}})}\left|Q\left(a^1,\ldots,a^{\overline{m}}\right)\right|$ which is a finite value.

Lemma 7.3 Using Eq. (7.17) to update Q-values, the Q-values converge to optimal Q-values of Eq. (7.16) w.p.1.

Proof Let' us define a mapping $P_t : \{Q\} \to \{Q\}$ as Eq. (7.20).

$$P_t^i Q_t^i(a^1, \ldots, a^{\overline{m}}) = Er_i^t(a^1, \ldots, a^{\overline{m}}) + \beta . \sum_{a^1} \sum_{a^2} \cdots \sum_{a^{\overline{m}}} \underline{p}_i^{br}\left(\hat{\underline{p}}^t, Q_t^i\right)(a^i) \prod_{\substack{j=1 \\ j \neq i}}^{\overline{m}} \hat{\underline{p}}_j^t(a^j).Q_t^i(a^1, \ldots, a^{\overline{m}}) \quad (7.20)$$

We also have Eq. (7.21).

$$P_t^i Q_*^i(a^1, \ldots, a^{\overline{m}}) = Er_i^t(a^1, \ldots, a^{\overline{m}}) + \beta . \sum_{a^1} \sum_{a^2} \cdots \sum_{a^{\overline{m}}} \underline{p}_i^{br}\left(\underline{p}^*, Q_*^i\right)(a^i) \prod_{\substack{j=1 \\ j \neq i}}^{\overline{m}} \underline{p}_j^*(a^j).Q_*^i\left(a^1, \ldots, a^{\overline{m}}\right) \quad (7.21)$$

It is straight forward that

$$\begin{aligned}||P_t^i Q_t^i(a^1, \ldots, a^{\overline{m}}) - P_t^i Q_*^i(a^1, \ldots, a^{\overline{m}})|| = ||Er_i^t(a^1, \ldots, a^{\overline{m}}) \\ + \beta . \sum_{a^1} \sum_{a^2} \cdots \sum_{a^{\overline{m}}} \underline{p}_i^{br}\left(\hat{\underline{p}}^t, Q_t^i\right)(a^i) \prod_{\substack{j=1 \\ j \neq i}}^{\overline{m}} \hat{\underline{p}}_j^t(a^j).Q_t^i(a^1, \ldots, a^{\overline{m}}) \\ - Er_i^t(a^1, \ldots, a^{\overline{m}}) - \beta . \sum_{a^1} \sum_{a^2} \cdots \sum_{a^{\overline{m}}} \underline{p}_i^{br}\left(\underline{p}^*, Q_*^i\right)(a^i) \\ \prod_{\substack{j=1 \\ j \neq i}}^{\overline{m}} \underline{p}_j^*(a^j).Q_*^i(a^1, \ldots, a^{\overline{m}})||\end{aligned} \quad (7.22)$$

Since the result of (7.22) is independent of $Er_i^t(a^1, \ldots, a^{\overline{m}})$, so we have Eq. (7.23)

$$||P_t^i Q_t^i(a^1, \ldots, a^{\overline{m}}) - P_t^i Q_*^i(a^1, \ldots, a^{\overline{m}})|| = \beta .| \sum_{a^1} \sum_{a^2} \cdots \sum_{a^{\overline{m}}} \underline{p}_i^{br}\left(\hat{\underline{p}}^t, Q_t^i\right)(a^i) \prod_{\substack{j=1 \\ j \neq i}}^{\overline{m}} \hat{\underline{p}}_j^t(a^j).Q_t^i(a^1, \ldots, a^{\overline{m}})$$

$$-\sum_{a^1}\sum_{a^2}\cdots\sum_{a^{\overline{m}}}\underline{p}_i^{br}\left(\underline{p}^*, Q_*^i\right)(a^i)\prod_{\substack{j=1\\ j\neq i}}^{\overline{m}}\underline{p}_j^*(a^j).Q_*^i(a^1,\ldots,a^{\overline{m}})| \tag{7.23}$$

Let to put $\hat{\underline{p}}_i^t(a_j)=\underline{p}_i^*(a_j)+\varepsilon_i^t(a_j)$, so we have Eq. (7.24):

$$\Bigg|\beta.\sum_{a^{1\cdots}}\sum_{a^m}\underline{p}_i^{br}(\hat{p}^t, Q_t^i)(a^i)\prod_{\substack{j=1\\ j\neq i}}^{\overline{m}}\hat{p}_j^t(a^j)\cdot Q_t^i(a^1,\ldots,a^{\overline{m}})$$

$$-\sum_{a^{1\cdots}}\sum_{a^{\overline{m}}}\underline{p}_i^{br}\left(\underline{p}^*, Q^i\right)(a^i)\prod_{\substack{j=1\\ j\neq i}}^{\overline{m}}\underline{p}_j^*(a^j)\cdot \mathbb{Q}_*^i(a^1,\ldots,a^{\overline{m}})\Bigg|$$

$$\leq\beta.\Bigg|\sum_{a^{1\cdots}}\sum_{a^{\overline{m}}}\underline{p}_i^{br}\left(\underline{p}^*, Q_t^i\right)(a^i)\prod_{\substack{j=1\\ j\neq i}}^{\overline{m}}\underline{p}_{j'}^*(a^j)\cdot Q_t^i(a^1,\ldots,a^{\overline{m}})$$

$$-\sum_{a^{1\cdots}}\sum_{a^{\overline{m}}}\underline{p}_i^{br}\left(\underline{p}^*, Q^i\right)(a^i)\prod_{\substack{j=1\\ j\neq i}}^{\overline{m}}\underline{p}_j^*(a^j)\cdot \underline{\rho}'(a^1,\ldots,a^{\overline{m}})\Bigg|$$

$$+\beta.\Bigg|\sum_{a^{1\cdots}}\sum_{a^{\overline{m}}}\underline{p}_i^{br}\left(\underline{p}^*, Q_t^i\right)(a^i)\prod_{\substack{j=1\\ j\neq i}}^{\overline{m}}\underline{\varepsilon}_j^t(a^j)\cdot Q_t^i(a^1,\ldots,a^{\overline{m}})\Bigg| \tag{7.24}$$

Using Eq. (7.15), we have:

$$\lim_{t\to\infty}\left|\sum_{a^1}\cdots\sum_{a^{\overline{m}}}\underline{p}_i^{br}\left(\underline{p}^*, Q_t^i\right)(a^i)\prod_{\substack{j=1\\ j\neq i}}^{\overline{m}}\underline{\varepsilon}_j^t(a^j)\cdot Q_t^i(a^1,\ldots,a^{\overline{m}})\right|=0 \quad (7.25)$$

Now let the following notations:

$$EQ_t=\sum_{a^1}\cdots\sum_{a^{\overline{m}}}\underline{p}_i^{br}\left(\underline{p}^*, Q_t^i\right)(a^i)\prod_{\substack{j=1\\ j\neq i}}^{\overline{m}}\underline{p}_j^*(a^j)\cdot Q_t^i(a^1,\ldots,a^{\overline{m}}) \quad (7.26)$$

$$EQ_*=\sum_{a^1}\cdots\sum_{a^{\overline{m}}}\underline{p}_i^{br}\left(\underline{p}^*, Q_*^i\right)(a^i)\prod_{\substack{j=1\\ j\neq i}}^{\overline{m}}\underline{p}_j^*(a^j)\cdot Q_*^i(a^1,\ldots,a^{\overline{m}}) \quad (7.27)$$

$$\lambda_t=\beta\times\left|\sum_{a^1}\cdots\sum_{a^{\overline{m}}}\underline{p}_i^{br}\left(\underline{p}^*, Q_t^i\right)(a^i)\prod_{\substack{j=1\\ j\neq i}}^{\overline{m}}\underline{\varepsilon}_j^t(a^j)\cdot Q_t^i(a^1,\ldots,a^{\overline{m}})\right| \quad (7.28)$$

Using Eqs. (7.23) and (7.24) we have Eq. (7.29).

$$||P_t^iQ_t^i(a^1,\ldots,a^{\overline{m}})-P_t^iQ_*^i(a^1,\ldots,a^{\overline{m}})||\leq\beta\cdot|EQ_t-EQ_*|+\lambda_t \quad (7.29)$$

From Eq. (7.15), we have $\underline{p}_i^{br}\left(\underline{p}^*, Q_t^i\right)\to\underline{p}_i^*(a^i)$ and $\underline{p}_i^{br}\left(\underline{p}^*, Q_*^i\right)\to\underline{p}_i^*(a^i)$ when $t\to\infty$. Now if $EQ_t\geq EQ_*$ then we have:

$$|EQ_t-EQ_*|=EQ_t-EQ_*=\sum_{a^1}\cdots\sum_{a^{\overline{m}}}\underline{p}_i^*(a^i)\prod_{\substack{j=1\\ j\neq i}}^{\overline{m}}\underline{p}_j^*(a^j)\cdot Q_t^i(a^1,\ldots,a^{\overline{m}})$$

$$-\sum_{a^1}\cdots\sum_{a^{\overline{m}}}\underline{p}_i^*(a^i)\prod_{\substack{j=1\\ j\neq i}}^{\overline{m}}\underline{p}_j^*(a^j)\cdot Q_*^i(a^1,\ldots,a^{\overline{m}})$$

$$= \sum_{a^1} \cdots \sum_{a^{\overline{m}}} \prod_{j=1}^{\overline{m}} \underline{p}_j^*(a^j)\left(Q_t^i\left(a^1, \ldots, a^{\overline{m}}\right) - Q_*^i\left(a^1, \ldots, a^{\overline{m}}\right)\right)$$

$$\leq \sum_{a^1} \cdots \sum_{a^{\overline{m}}} \prod_{j=1}^{\overline{m}} \underline{p}_j^*(a^j) \times \max_{(a^1,\ldots,a^{\overline{m}})} \left(Q_t^i\left(a^1, \ldots, a^{\overline{m}}\right) - Q_*^i\left(a^1, \ldots, a^{\overline{m}}\right)\right)$$

$$= 1 \times \max_{(a^1,\ldots,a^{\overline{m}})} \left(Q_t^i\left(a^1, \ldots, a^{\overline{m}}\right) - Q_*^i\left(a^1, \ldots, a^{\overline{m}}\right)\right)$$

$$\leq \max_{(a^1,\ldots,a^{\overline{m}})} \left|Q_t^i\left(a^1, \ldots, a^{\overline{m}}\right) - Q_*^i\left(a^1, \ldots, a^{\overline{m}}\right)\right| = \left\|Q_t^i\left(a^1, \ldots, a^{\overline{m}}\right) - Q_*^i\left(a^1, \ldots, a^{\overline{m}}\right)\right\| \quad (7.30)$$

So using Eqs. (7.29) and (7.30) we have:

$$\left\|P_t^i Q_t^i\left(a^1, \ldots, a^{\overline{m}}\right) - P_t^i Q_*^i\left(a^1, \ldots, a^{\overline{m}}\right)\right\| \leq \beta \cdot \left\|Q_t^i\left(a^1, \ldots, a^{\overline{m}}\right) - Q_*^i\left(a^1, \ldots, a^{\overline{m}}\right)\right\| + \lambda_t \quad (7.31)$$

If we have the other case ($EQ_t < EQ_*$) then:

$$|EQ_t - EQ_*| = EQ_* - EQ_t = \sum_{a^1} \cdots \sum_{a^{\overline{m}}} \prod_{j=1}^{\overline{m}} \underline{p}_j^*(a^j)(Q_*^i(a^1, \ldots, a^{\overline{m}})$$

$$-Q_t^i(a^1, \ldots, a^{\overline{m}}))$$

$$\leq \sum_{a^1} \cdots \sum_{a^{\overline{m}}} \prod_{j=1}^{\overline{m}} \underline{p}_j^*(a^j) \times \max_{(a^1,\ldots,a^{\overline{m}})} (Q_*^i(a^1, \ldots, a^{\overline{m}}) - Q_t^i(a^1, \ldots, a^{\overline{m}}))$$

$$\leq \max_{(a^1,\ldots,a^{\overline{m}})} \left|(Q_*^i(a^1, \ldots, a^{\overline{m}}) - Q_t^i(a^1, \ldots, a^{\overline{m}}))\right| = \max_{(a^1,\ldots,a^{\overline{m}})} \left|Q_t^i(a^1, \ldots, a^{\overline{m}})\right.$$

$$\left.-Q_*^i(a^1, \ldots, a^{\overline{m}})\right|$$

$$= \left\|Q_t^i(a^1, \ldots, a^{\overline{m}}) - Q_*^i(a^1, \ldots, a^{\overline{m}})\right\| \quad (7.32)$$

So Eq. (7.31) is valid for the case $EQ_t < EQ_*$ as well. This results in (According to Corollary 7.1, Eqs. (7.15) and (7.31)) that updating of Q-values using Eq. (7.17) will cause convergence to optimal Q-values of Eq. (7.16) w.p.1 and proof are completed. □

Theorem 7.1 *Using the local rule of Algorithm 2, ICLA converges to a compatible point.*

Proof In Lemma 7.3, we proved that updating of Q-values using Eq. (7.17) causes convergence to optimal Q-values of Eq. (7.16) w.p.1. So when $t \to \infty$, Eq. (7.17) will approach to (7.33)

$$Q_*^i\left(a^1, \ldots, a^{\overline{m}}\right) = (1 - \alpha) \times Q_*^i\left(a^1, \ldots, a^{\overline{m}}\right) + \alpha \times [Er_i^t\left(a^1, \ldots, a^{\overline{m}}\right)$$

$$+\beta.\sum_{a^1}\sum_{a^2}\cdots\sum_{a^{\overline{m}}}\underline{p}_i^{br}\left(\underline{p}^*,Q_*^i\right)(a^i)\prod_{\substack{j=1\\j\neq i}}^{\overline{m}}\underline{p}_j^*(a^j).Q_*^i(a^1,\ldots,a^{\overline{m}})] \quad (7.33)$$

It is evident that:

$$Q_*^i(a^1,\ldots,a^{\overline{m}})=Er_i^t(a^1,\ldots,a^{\overline{m}})$$
$$+\beta.\sum_{a^1}\sum_{a^2}\cdots\sum_{a^{\overline{m}}}\underline{p}_i^{br}\left(\underline{p}^*,Q_*^i\right)(a^i)\prod_{\substack{j=1\\j\neq i}}^{\overline{m}}\underline{p}_j^*(a^j).Q_*^i(a^1,\ldots,a^{\overline{m}}) \quad (7.34)$$

Comparing Eqs. (7.16) and (7.34) results in that:

$$v^i\left(\underline{p}_1^*,\underline{p}_2^*,\ldots,\underline{p}_i^{br}\left(\underline{p}^*,Q_*^i\right),\ldots,\underline{p}_{\overline{m}}^*\right)$$
$$=\sum_{a^1}\sum_{a^2}\cdots\sum_{a^{\overline{m}}}\underline{p}_i^{br}\left(\underline{p}^*,Q_*^i\right)(a^i)\prod_{\substack{j=1\\j\neq i}}^{\overline{m}}\underline{p}_j^*(a^j).Q_*^i(a^1,\ldots,a^{\overline{m}}) \quad (7.35)$$

In Algorithm 2, we specify the best response strategy using the RHS of Eq. (7.35) which is equal to $\underline{p}_i^{br}\left(\underline{p}^*,Q_*^i\right)$ in $v^i\left(\underline{p}_1^*,\underline{p}_2^*,\ldots,\underline{p}_i^{br}\left(\underline{p}^*,Q_*^i\right),\ldots,\underline{p}_{\overline{m}}^*\right)$. So, algorithm 2 specifies the best response strategy to $\left(\underline{p}_1^*,\ldots,\underline{p}_{i-1}^*,\underline{p}_{i+1}^*,\ldots,\underline{p}_{\overline{m}_i}^*\right)$. Based on concepts in Chap. 4 of (Kumpati and Narendra 1989), this local rule leads the probability vector of LA to the best response configuration. Since every *LA* in ICLA applies this rule, therefore probability vectors of all of them are the best response configuration. So, according to best response strategy playing concepts (Nisan et al. 2011), ICLA converges to a compatible point. □

7.3.3 Complexity Analysis of the Proposed Local Rule

As described before, each learning automaton in ICLA needs to maintain Q-values for each combination of joint actions of itself and its neighbors. These Q-values are maintained internally by the learning automaton, assuming that it can observe the actions of its neighbors. The learning automaton i updates $Q^i\left(a^1,\ldots,a^{\overline{m}_i}\right)$, where a^j ($j=1,\ldots,\overline{m}_i$) is the selected action of neighbor jth (assume the neighbors are labeled by index j). $\overline{m}_i$ is a number of the neighbors of learning automaton i in ICLA. Let $|A^i|$ be the size of action space of learning automaton i assuming $\overline{m}=\max_i m_i$

and $|A| = \max_i |A^i|$, $n \times |A|^{\overline{m}}$ is an upper bound for total space requirement for Q-values where n is the total number of learning automata in the ICLA. Besides, there is a similar space requirement to maintain $Er_i(a^1, \ldots, a^{\overline{m}_i})$ values. Therefore, the proposed local rule in terms of space complexity is linear in the number of learning automata (size of ICLA), polynomial in the number of actions of learning automata, but exponential in the number of neighbors of learning automata. Since reinforcement learning algorithms, like Q-learning, compute values for each joint-action or state-action pair, this leads to exponential computational complexity. Running time of algorithm 2 is dominated by the computation of $\underline{p}_i^{br}\left(\hat{\underline{p}}, Q_t^i\right)$. The computational complexity of this computation is $|A|^{\overline{m}}$. It is polynomial in the number of actions of learning automata but exponential in the number of neighbors of learning automata.

Note If we have multiple learning automata in each cell of ICLA (for example, see (Beigy and Meybodi 2010), then each learning automaton in a cell will be a neighbor for every learning automaton which is located in that cell or its neighboring cells.

7.3.4 Experiments and Results

This section contains the results of the conducted experiments. First, we present two numerical experiments, and then an example of the application of ICLA in the channel assignment problem is given.

7.3.4.1 Numerical Experiments

The numerical experiments are given to illustrate the superiority and effectiveness of the proposed local rule. These results are compared with the results of the ordinary local rule. In the experiments $ICLA_{m,n}$ refers to an ICLA with m cells and n actions for each learning automaton located in a cell. Because the notation is destitute of neighboring details, so a neighboring matrix is also needed to specify a unique ICLA. The learning algorithm of all learning automata is L_{R-I} .

Numerical Experiment 1

This experiment aims to compare the convergence rate of ICLA using the proposed local rule versus the ordinary local rule. For this propose, we investigate convergence rate in $ICLA_{3,2}$,$ICLA_{4,2}$ and $ICLA_{5,2}$ which is illustrated in Fig. 7.3. The edges in the graphs show neighborhood relations. Here response of the environment is generated using a response matrix. This matrix contains the probability of rewarding the selected actions of learning automata an ICLA. For example, when selected joint actions by LAs is (*Low, High, High*) and the corresponding entry in the response

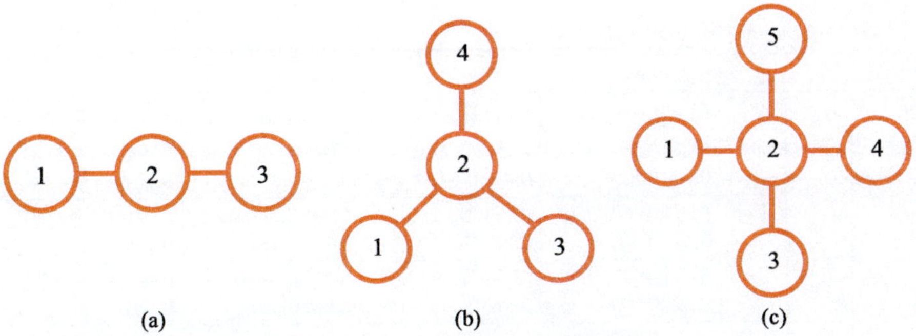

Fig. 7.3 ICLA structures for **a** $ICLA_{3,2}$, **b** $ICLA_{4,2}$ and **c** $ICLA_{5,2}$

Table 7.1 Response matrix for $ICLA_{3,2}$

a_1	a_3	$a_2 =$ Low	$a_2 =$ High
Low	Low	(0.54, 0.59, 0.72)	(0.91, 0.68, 0.81)
	High	(0.54, 0.82, 0.66)	**(0.91,0.88, 0.91)**
High	Low	(0.48, 0.75, 0.72)	(0.84, 0.70, 0.81)
	High	(0.48, 0.59, 0.66)	(0.84, 0.77, 0.91)

Table 7.2 Response matrix for $ICLA_{4,2}$

		$a_2 =$ Low		$a_2 =$ High	
a_1	a_3	$a_4 =$ Low	$a_4 =$ High	$a_4 =$ Low	$a_4 =$ High
Low	Low	(0.54, 0.59, 0.72, 0.62)	(0.54, 0.71, 0.72, 0.48)	(0.91, 0.68, 0.81, 0.95)	(0.91, 0.68, 0.81, 0.78)
	High	(0.54, 0.82, 0.66, 0.62)	(0.54, 0.69, 0.66, 0.48)	**(0.91,0.88, 0.91, 0.95)**	(0.91,0.76, 0.91, 0.78)
High	Low	(0.48, 0.75, 0.72, 0.62)	(0.48, 0.61, 0.72, 0.48)	(0.84, 0.70, 0.81, 0.95)	(0.84, 0.70, 0.81, 0.78)
	High	(0.48, 0.59, 0.66, 0.62)	(0.48, 0.79, 0.66, 0.48)	(0.84, 0.77, 0.91, 0.95)	(0.84, 0.77, 0.91, 0.78)

The best value is highlighted in boldface

matrix is (0.91, 0.88, 0.91) (see, e.g., Table 7.1) then the reinforcement signals to LA1, LA2 and LA3 is reward with probability 0.91, 0.88 and 0.91 respectively. Tables 7.1, 7.2 and 7.3 show the reinforcement matrices for $ICLA_{3,2}$,$ICLA_{4,2}$ and $ICLA_{5,2}$.

Figure 7.4 shows the convergence rate with different amounts of reward parameter a (see Eq. 7.1) in L_{R-I} learning algorithms using the proposed local rule and the ordinary local rule. The results is obtained from 1000 times running of learning process by starting from preliminary configuration of [(0.5, 0.5), (0.5, 0.5), (0.5, 0.5)]

Table 7.3 Response matrix for $ICLA_{5,2}$

			a_2 = Low		a_2 = High	
a_1	a_3	a_5	a_4 = Low	a_4 = High	a_4 = Low	a_4 = High
Low	Low	Low	(0.54, 0.59, 0.72, 0.62, 0.57)	(0.54, 0.71, 0.72, 0.48, 0.57)	(0.91, 0.68, 0.81, 0.95, 0.88)	(0.91, 0.68, 0.81, 0.78)
		High	(0.54, 0.67, 0.72, 0.62, 0.68)	(0.54, 0.81, 0.72, 0.48, 0.68)	(0.91, 0.77, 0.81, 0.95, 0.75)	(0.91, 0.78, 0.81, 0.78)
	High	Low	(0.54, 0.82, 0.66, 0.62, 0.57)	(0.54, 0.69, 0.66, 0.48, 0.57)	**(0.91,0.88, 0.91, 0.95, 0.88)**	(0.91,0.76, 0.91, 0.78)
		High	(0.54, 0.65, 0.66, 0.62, 0.68)	(0.54, 0.77, 0.66, 0.48, 0.68)	(0.91,0.72, 0.91, 0.95, 0.75)	(0.91,0.69, 0.91, 0.78)
High	Low	Low	(0.48, 0.75, 0.72, 0.62, 0.57)	(0.48, 0.61, 0.72, 0.48, 0.57)	(0.84, 0.70, 0.81, 0.95, 0.88)	(0.84, 0.70, 0.81, 0.78)
		High	(0.48, 0.83, 0.72, 0.62, 0.68)	(0.48, 0.70, 0.72, 0.48, 0.68)	(0.84, 0.69, 0.81, 0.95, 0.75)	(0.84, 0.80, 0.81, 0.78)
	High	Low	(0.48, 0.59, 0.66, 0.62, 0.57)	(0.48, 0.79, 0.66, 0.48, 0.57)	(0.84, 0.77, 0.91, 0.95, 0.88)	(0.84, 0.77, 0.91, 0.78)
		High	(0.48, 0.68, 0.66, 0.62, 0.68)	(0.48, 0.81, 0.66, 0.48, 0.68)	(0.84, 0.35, 0.91, 0.95, 0.75)	(0.84, 0.75, 0.91, 0.78)

The best value is highlighted in boldface

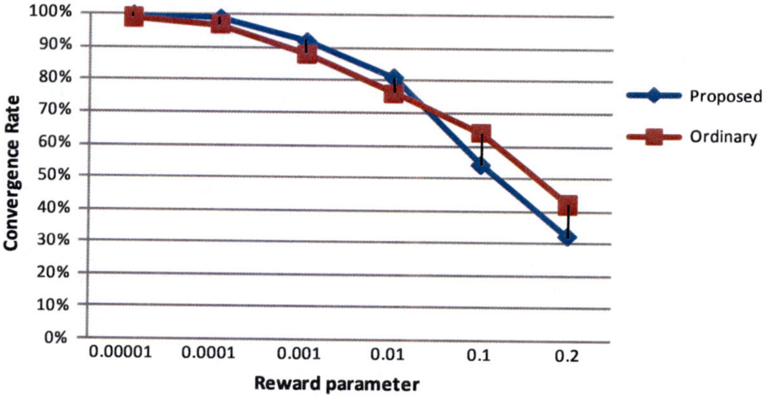

Fig. 7.4 Convergence rate using different reward parameters in $ICLA_{3,2}$

for (LA_1, LA_2, LA_3) in $ICLA_{3,2}$. Figures 7.5 and 7.6 show the same diagrams for $ICLA_{4,2}$ and $ICLA_{5,2}$.

As illustrated in Figs. 7.4, 7.5 and 7.6, by bigger reward parameters, we have a lower convergence rate by both the local rules. The problem of catching in absorbing states is a major reason for the lower convergence rate here. Except for big reward parameter values in $ICLA_{3,2}$. The proposed local rule always reaches a better convergence rate than the ordinary local rule. The difference in convergence rate is because

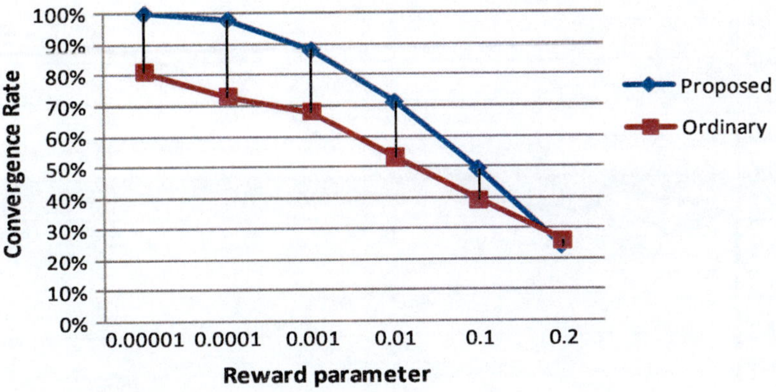

Fig. 7.5 Convergence rate using different reward parameters in $ICLA_{4,2}$

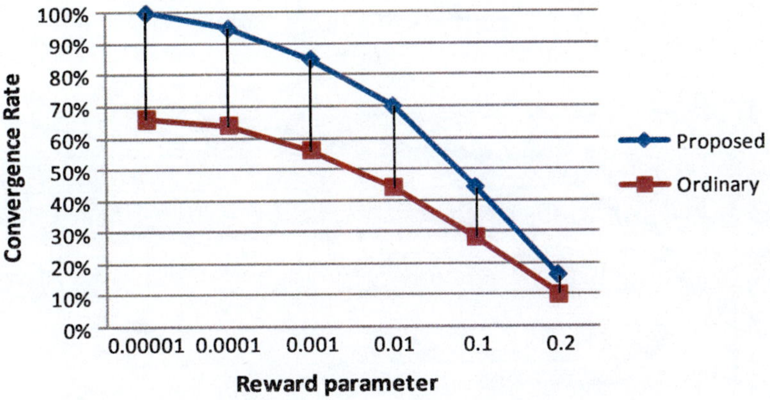

Fig. 7.6 Convergence rate using different reward parameters in $ICLA_{5,2}$

of the reason that the ordinary local rule uses just the response of the environment in the current round to generate a reinforcement sign. In contrast, the proposed local rule uses a history of the responses (using Q-values). Although the better convergence rate is obtained using the proposed local rule by increasing the reward parameter, a drop-in convergence rate using the proposed local rule is more sensible than drop using the ordinary local rule. This is for the reason that changes in ICLA configuration will not be negligible in two successive rounds using high reward parameter values. This causes drop in convergence rate because Algorithm 2 assumes the best response in iteration t (see $\underline{p}_i^{br}$ in Algorithm 2) is too close to the best response in iteration $t + 1$.

Using Table 7.1 as the response matrix to generate environment response, $ICLA_{3,2}$ converges to (*Low, High, High*) which is a compatible point. Figures 7.7 plots the evolution of the probability vector of the learning automaton of cell 2 in

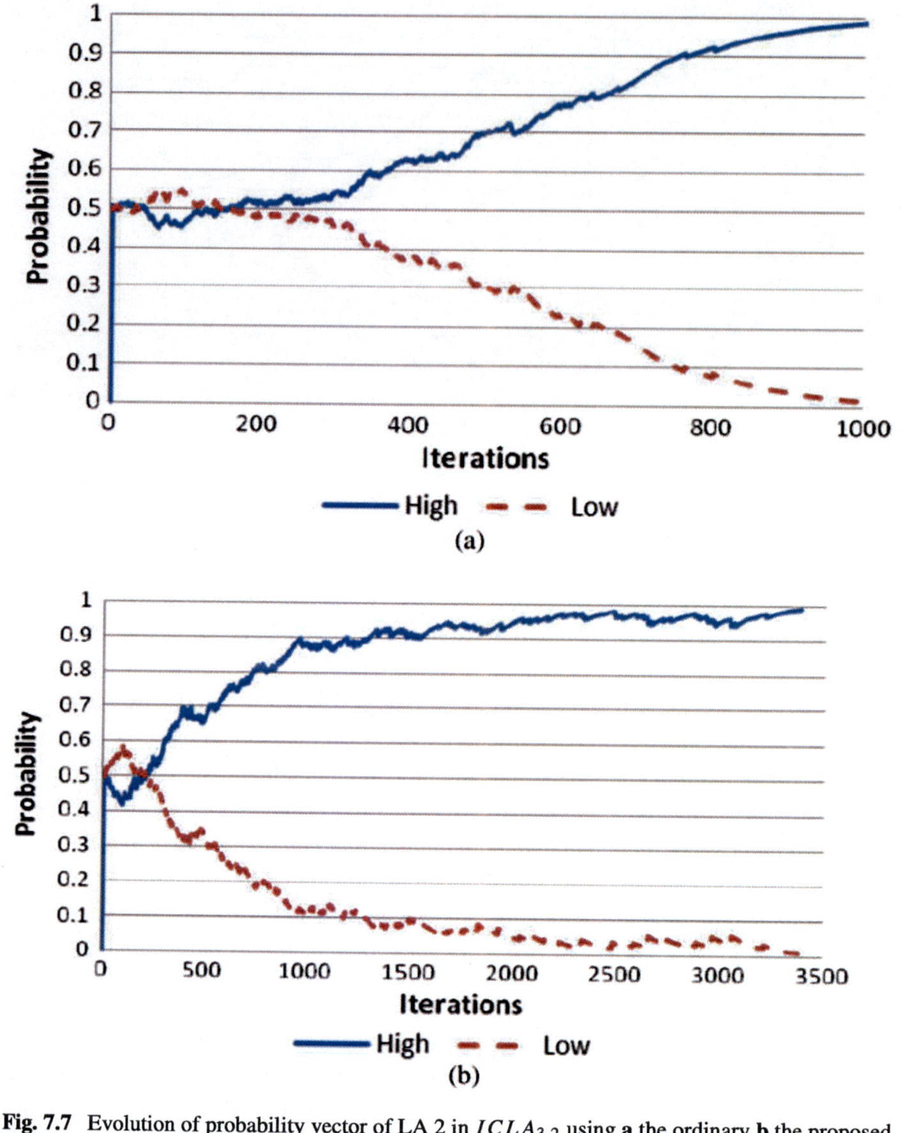

Fig. 7.7 Evolution of probability vector of LA 2 in $ICLA_{3,2}$ using **a** the ordinary **b** the proposed local rule

$ICLA_{3,2}$ using the ordinary (a) and the proposed local rule (b). As illustrated in this figure, by using the proposed local rule, ICLA needs further iterations to converge.

Figures 7.8 and 7.9 show that similar to Fig. 7.7 the proposed local rule needs more iterations to convergence in $ICLA_{4,2}$ and $ICLA_{5,2}$. Now let $L_{nm} = \frac{I_{nm}(\text{Proposed})}{I_{nm}(\text{Ordinary})}$, where $I_{nm}(R)$ denotes the number of iterations $ICLA_{n,m}$ needs for convergence using the local rule R. According to the results illustrated in Fig. 7.10, we have

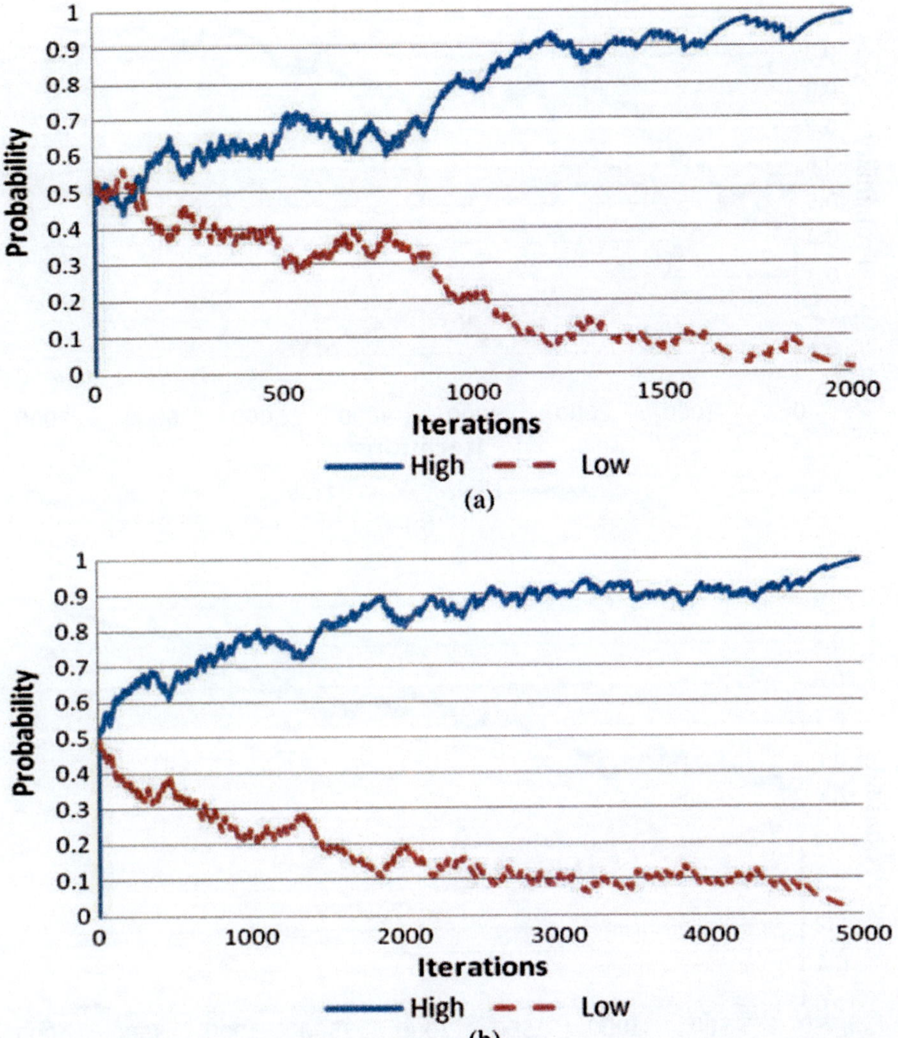

(b)

Fig. 7.8 Evolution of probability vector of LA 2 in $ICLA_{4,2}$ using **a** the ordinary **b** the proposed local rule

$L_{32} = 3.6$, $L_{42} = 2.4$ and $L_{52} = 1.7$. For $n \geq 9$, L_{n2} is less than 1. So, by increasing the number of cells in $ICLA_{n,2}$, it is expected that fewer iterations to be needed for convergence by the proposed local rule in comparison with the ordinary local rule. Figure 7.10b demonstrates that by using the ordinary local rule the necessary iterations for convergence in $ICLA_{3,2}$, $ICLA_{4,2}$ and $ICLA_{5,2}$ are 23%, 29% and 36% of (I_{n2}(Proposed) + I_{n2}(Ordinary)) for n = 3, 4 and 5 respectively.

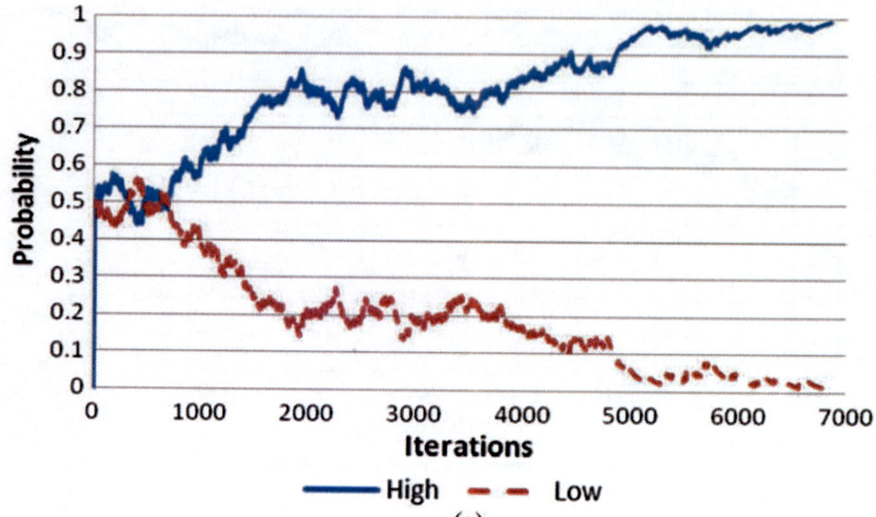

(a)

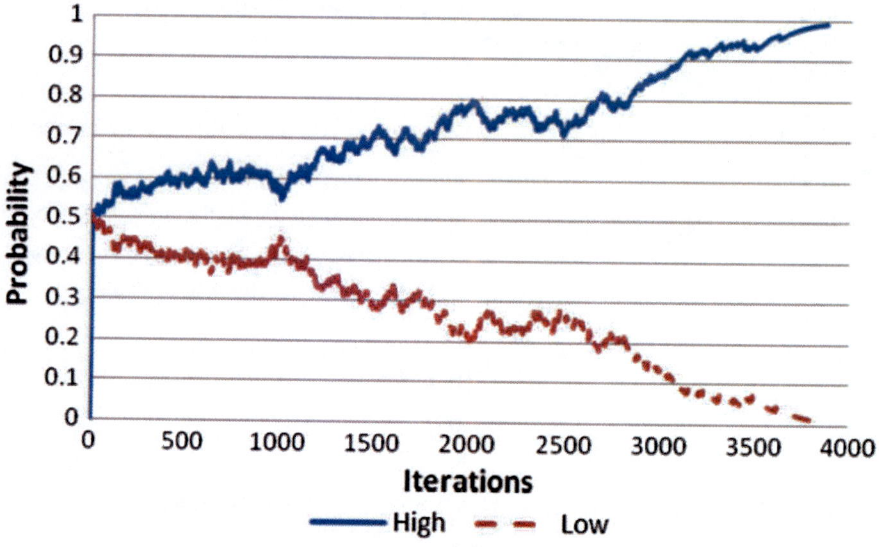

(b)

Fig. 7.9 Evolution of probability vector of LA 2 in $ICLA_{5,2}$ using **a** the ordinary **b** the proposed local rule

Numerical Experiment 2

This numerical experiment aims to study the convergence of ICLA when it starts learning from different initial configurations. For this experiment, we use an $ICLA_{3,3}$ with L_{R-I} learning algorithm. Structure of $ICLA_{3,3}$ is similar to $ICLA_{3,2}$ in

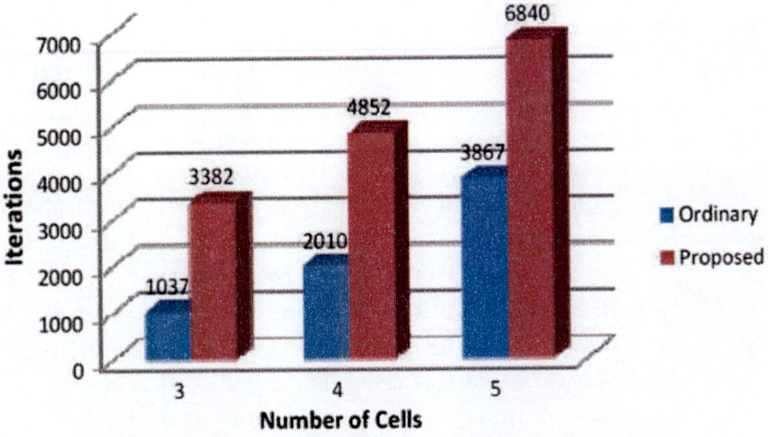

(a)

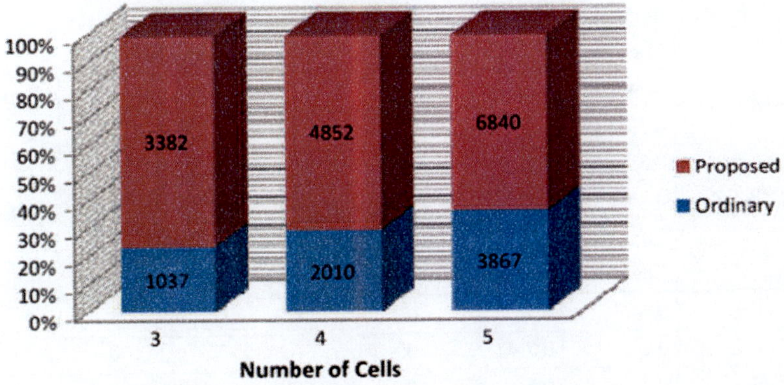

(b)

Fig. 7.10 Comparing the number of necessary iterations for convergence of ICLA using the proposed and ordinary local rule

Fig. 7.3a. Responses of the environment are generated using Table 7.4. Table 7.5 contains the initial configurations for this ICLA. For example, configuration 3 shows that the initial probability vector of the learning automaton in cell 2 is [0.2, 0.7, 0.1]. Using Table 7.4 to generate responses of environment, [(0,1,0), (0,0,1), (0,0,1)] is a compatible configuration for $ICLA_{3,3}$. Figures 7.11, 7.12, 7.13 and 7.14 show convergence rate of ICLA using different values of *a*. As illustrated in these figures, by having different initial configurations, the proposed local rule still has a better convergence rate in all the cases. Moreover, using the proposed local rule, the difference between convergence rates by starting from different initial configurations are smaller than the difference in case of using the ordinary local rule. This means the convergence rate is less dependent on initial configuration using the proposed local rule. Figures 7.15 and 7.16 compare this dependency. The best and worst convergence

Table 7.4 Response matrix for $ICLA_{3,3}$

Reward probability for (a_1, a_2, a_3)		a_2 = Low	a_2 = Normal	a_2 = High
a_1 = Low	a_3 = Low	(0.51, 0.81, 0.38)	(0.36, 0.25, 0.84)	(0.53, 0.44, 0.32)
	a_3 = Normal	(0.51, 0.76, 0.52)	(0.36, 0.47, 0.39)	(0.53, 0.25, 0.57)
	a_3 = High	(0.51, 0.45, 0.77)	(0.36, 0.66, 0.71)	(0.53, 0.36, 0.90)
a_1 = Normal	a_3 = Low	(0.87, 0.32, 0.38)	(0.65, 0.90, 0.84)	(0.92, 0.51, 0.32)
	a_3 = Normal	(0.87, 0.61, 0.52)	(0.65, 0.54, 0.39)	(0.92, 0.75, 0.57)
	a_3 = High	(0.87, 0.71, 0.77)	(0.65, 0.82, 0.71)	**(0.92, 0.95, 0.90)**
a_1 = High	a_3 = Low	(0.44, 0.37, 0.38)	(0.73, 0.61, 0.84)	(0.71, 0.65, 0.32)
	a_3 = Normal	(0.44, 0.49, 0.52)	(0.73, 0.37, 0.39)	(0.71, 0.51, 0.57)
	a_3 = High	(0.44, 0.56, 0.77)	(0.73, 0.80, 0.71)	(0.71, 0.73, 0.90)

The best value is highlighted in boldface

Table 7.5 Initial configurations

	Configuration 1			Configuration 2		
	Low	Normal	High	Low	Normal	High
LA in Cell 1	0.33	0.33	0.34	0.1	0.7	0.2
LA in Cell 2	0.33	0.33	0.34	0.2	0.2	0.6
LA in Cell 3	0.33	0.33	0.34	0.05	0.05	0.9
	Configuration 3			Configuration 4		
	Low	Normal	High	Low	Normal	High
LA in Cell 1	0.6	0.2	0.1	0.3	0.5	0.2
LA in Cell 2	0.2	0.7	0.1	0.6	0.2	0.2
LA in Cell 3	0.35	0.5	0.15	0.2	0.1	0.7

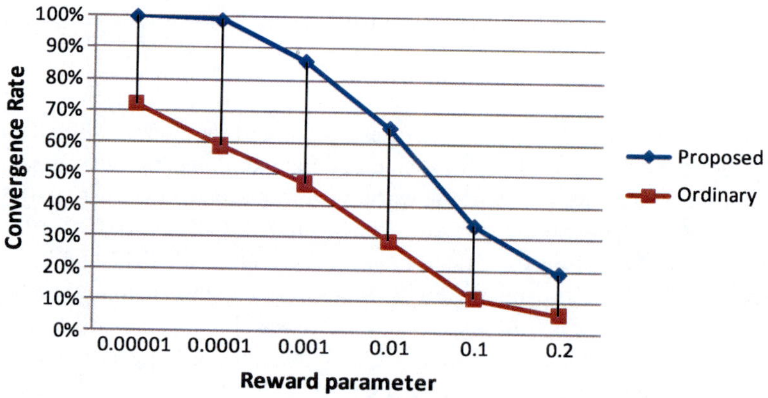

Fig. 7.11 Convergence rate using the proposed and ordinary local rule by starting from configuration 1

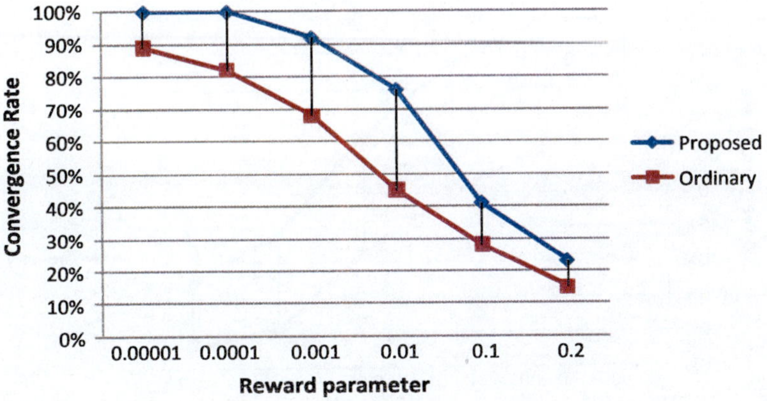

Fig. 7.12 Convergence rate using the proposed and ordinary local rule by starting from configuration 2

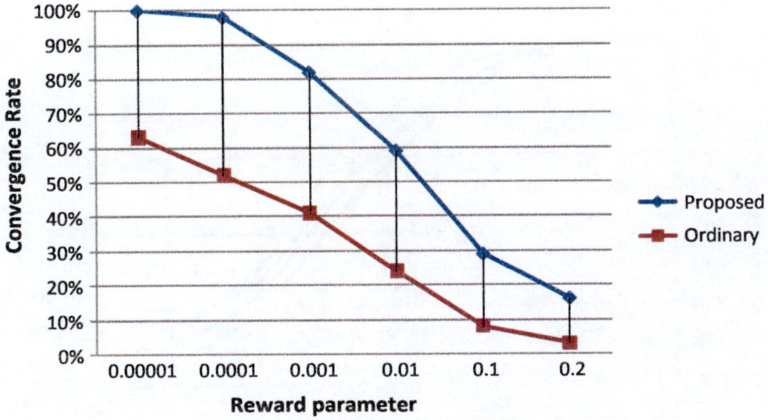

Fig. 7.13 Convergence rate using the proposed and ordinary local rule by starting from configuration 3

rates are obtained by starting from configuration 2 and configuration 3, respectively. This is for the reason that configuration 2 spatially is nearer to a compatible point than other configurations, and configuration 3 is the farthest configuration among the others.

Figure 7.17 plots the evolution of the probability vector of learning automaton in cell 2 of $ICLA_{3,3}$ by starting from (a) Configuration 1, (b) Configuration 2, (c) Configuration 3 and (d) Configuration 4. Figures 7.18, 7.19, 7.20 and 7.21 plot the evolution of the action probability for selected actions of learning automata using

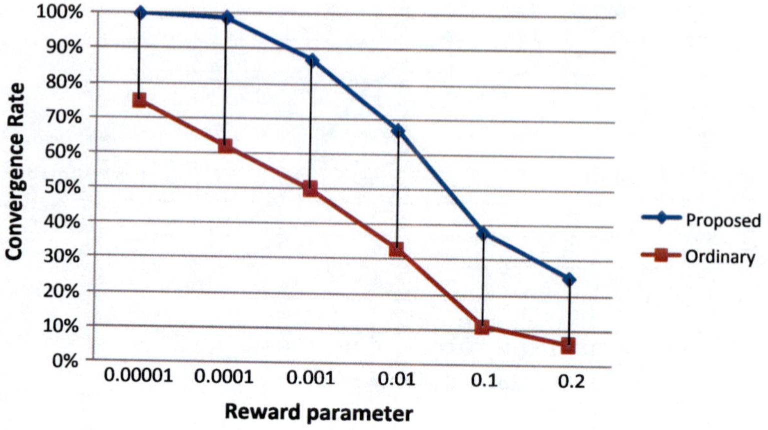

Fig. 7.14 Convergence rate using the proposed and ordinary local rule by starting from configuration 4

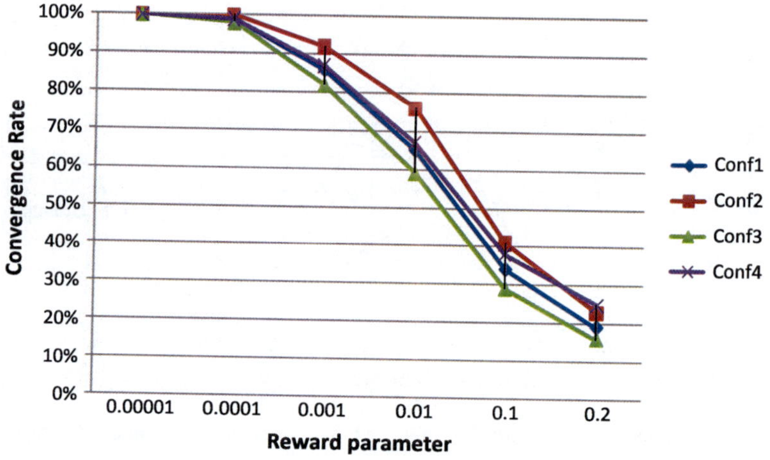

Fig. 7.15 Comparing convergence rate by starting from different initial configurations using the proposed local rule

the four mentioned initial configurations of Table 7.5. As can be seen from these figures, regardless of what the initial configuration is, ICLA converges to the same configuration. These figures show that the configuration, ICLA is converged to it, is [(0, 1, 0), (0, 0, 1), (0, 0, 1)] which is compatible point of Table 7.4.

ICLA is a powerful mathematical model for decentralized applications. The convergence of ICLA to a compatible point has great importance in studying the behavior of ICLA. With a simple local rule which rewards or punishes LAs just based on the response of the environment and the selected actions of neighbors, the convergence of ICLA to a compatible point is not guaranteed. In this section, we

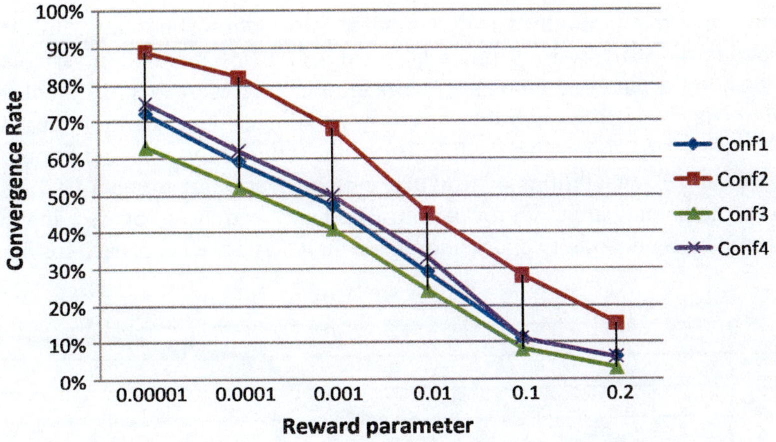

Fig. 7.16 Comparing convergence rate by starting from different initial configurations using the ordinary local rule

provided a new local rule that guarantees convergence of ICLA to a compatible point. Formal proofs for the existence of a compatible point and convergence are provided as well. We provided different experiments to show the usefulness and superiority of the proposed local rule. The obtained results from experiments support our idea about the superiority of the proposed local rule against the ordinary local rule. Now in the following sections, we present an ICLA-based loss-sharing approach that employs the proposed local rule in this section for reaching a compatible LSA.

7.4 An ICLA-Based Competitive Loss Sharing Approach for Cloud

This section describes a new approach for finding a compatible LSA for loss sharing problem. Assume that there are n risk-averse cloud service providers and their utility functions satisfy two conditions of risk aversion: $u_i'(I_i) > 0 \quad and \quad u_i''(I_i) < 0$. Service providers are rational competitors that may cooperate under a loss-sharing agreement in order to increase their utility and capability for loss compensation. Utility of a service provider is not known by the other service providers. Distribution of X_i is not known as well and service provider i announces X_i just to the service providers who are interested in cooperating. It is also assumed that service providers announce their satisfaction level from an LSA to each other. Bigger value of satisfaction signal means higher satisfaction level is obtained. Now the problem is finding an LSA such that cooperation of the service providers under the LSA maximizes their utility as much as possible.

Please note that here, there is no complete information about, e.g., utility functions, and loss distributions. $X_i(t)$, $PI_i(t)$ and $u_i(PI_i(t))$ are random variables (see Eq. 7.4), and the environment is uncertain and random. Moreover, different service providers may have different utility functions and different loss distributions. Therefore, it is very hard, indeed impossible, to use theoretic analysis to find a compatible LSA. Due to the capabilities of irregular cellular learning automata (ICLA), as a powerful mathematical model for decentralized applications, to operate in environments with high uncertainty and randomness, an ICLA-based approach for finding a

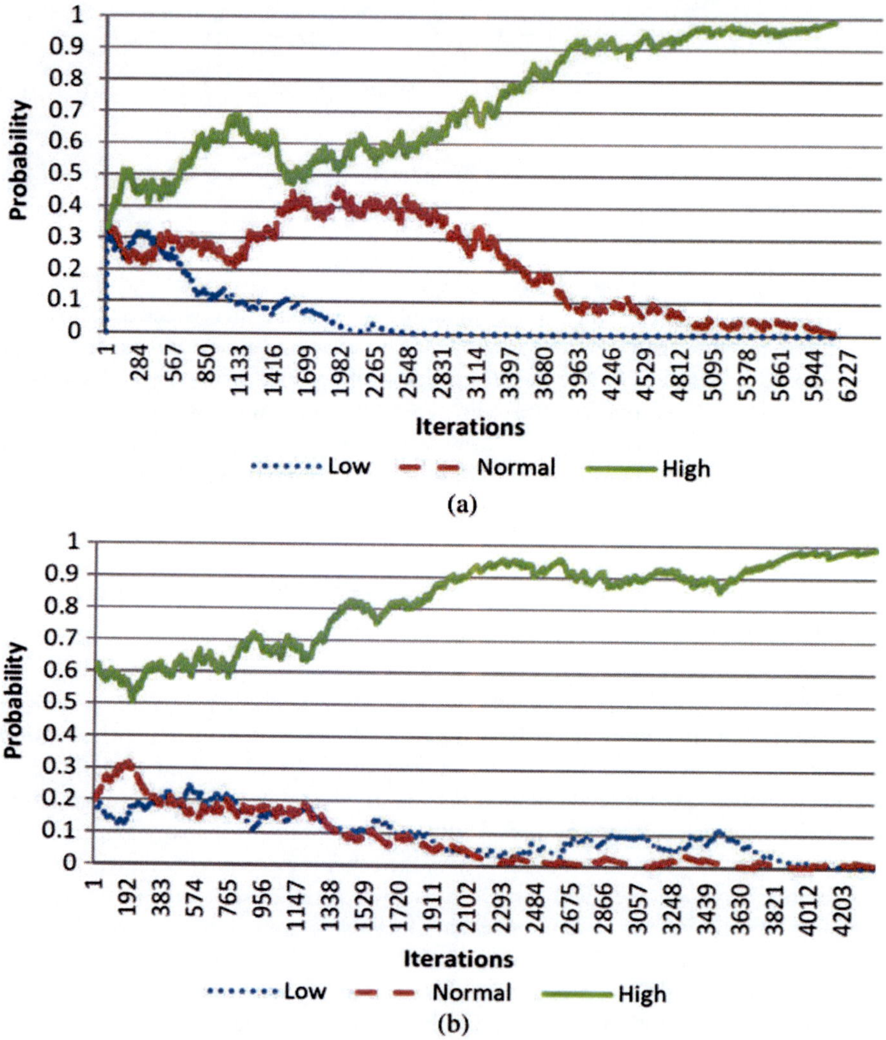

Fig. 7.17 Evolution of probability vector of LA 2 in $ICLA_{3,3}$ by starting from **a** Configuration 1 **b** Configuration 2 **c** Configuration 3 **d** Configuration 4

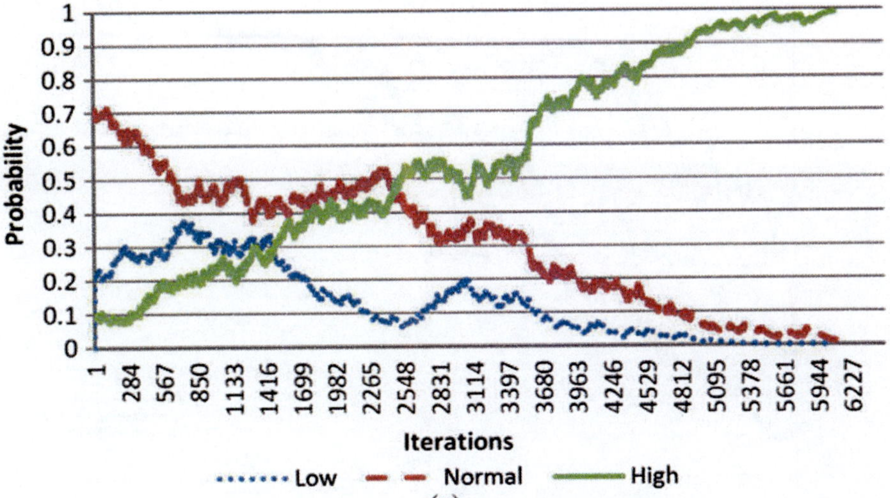

(c)

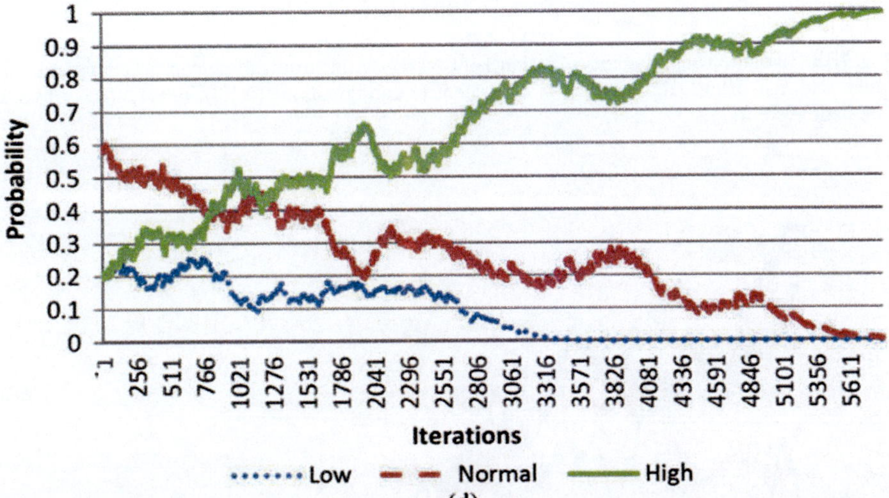

(d)

Fig. 7.17 (continued)

compatible LSA is proposed. In the proposed approach, each cell of ICLA represents a service provider, and neighborhood relation is defined based on the tendency for cooperation between two service providers. For each neighbor j, there is a learning automaton LA_{ij} in cell *i*. Over an iterative process, this learning automaton selects one of its actions. This action determines $\alpha_j^i(t)$. Neighbor *j* (service provider *j*) should pay $pr_j^i(t)$ to service provider *i* for $\alpha_j^i(t) \times X_j(t)$. Amount of $pr_j^i(t)$ can be zero or be determined according to a premium calculation method (Zweifel and Eisen 2012). Then all service providers calculate $u(PI(t))$ and generate satisfaction signals.

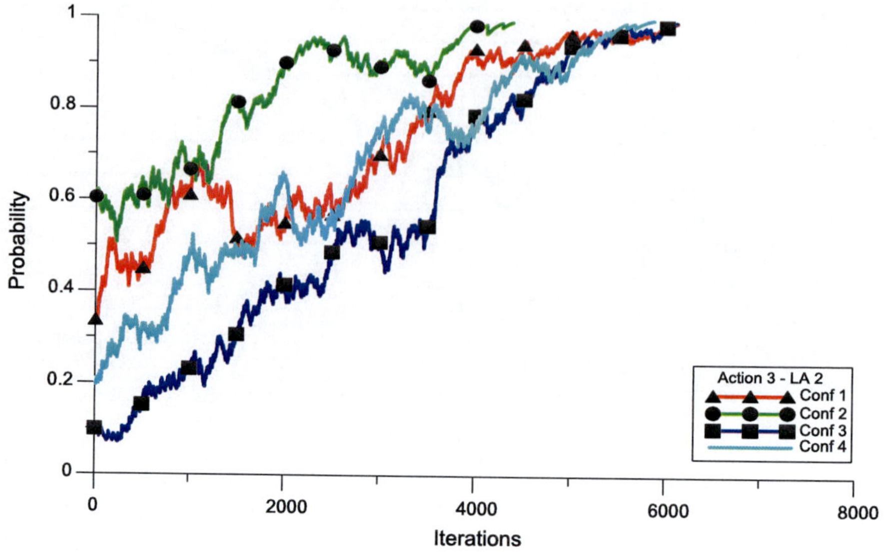

Fig. 7.18 Evolution of probability of action 3 of LA2 by starting from different initial configurations when (Normal, High, High), which is equivalent to configuration [(0,1,0), (0,0,1), (0,0,1)], is a compatible point

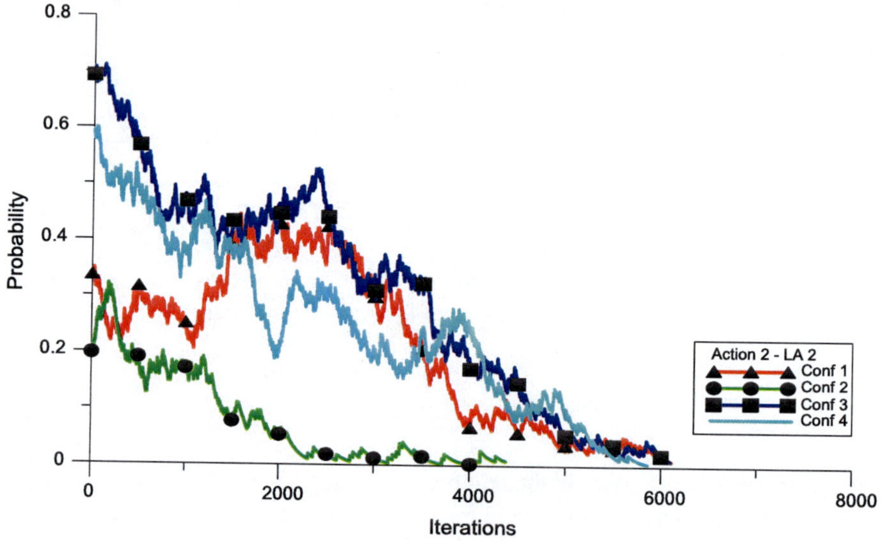

Fig. 7.19 Evolution of probability of action 2 of LA 2 by starting from different initial configurations when (Normal, High, High), which is equivalent to configuration [(0,1,0), (0,0,1), (0,0,1)], is a compatible point

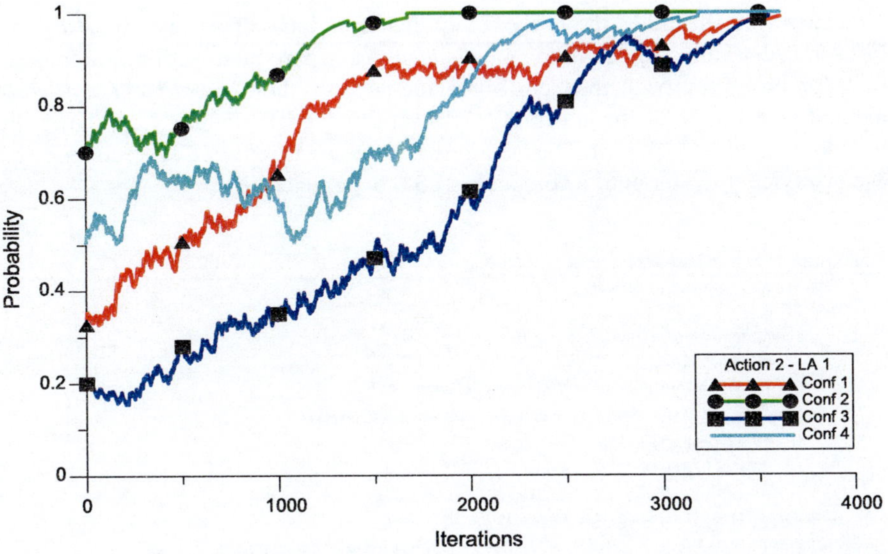

Fig. 7.20 Evolution of probability of action 2 of LA 1 by starting from different initial configurations when (Normal, High, High), which is equivalent to configuration [(0,1,0), (0,0,1), (0,0,1)], is a compatible point

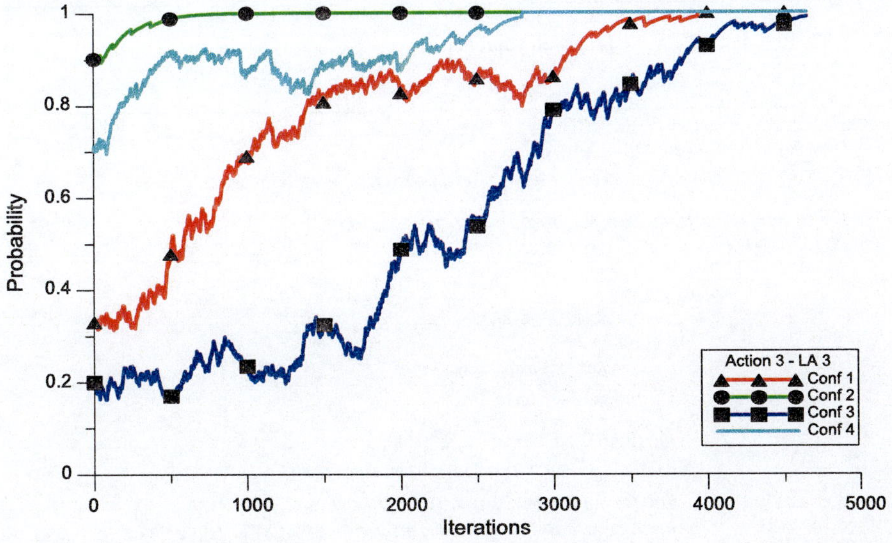

Fig. 7.21 Evolution of probability of action 3 of LA 3 by starting from different initial configurations when (Normal, High, High), which is equivalent to configuration [(0,1,0), (0,0,1), (0,0,1)], is a compatible point

Figure 7.22 illustrates the proposed approach in detail. There is a local rule in ICLA that determines the reinforcement signal to any particular learning automaton. In the proposed approach, the GenRSignal method uses the *Compatible_Local_Rule* method (see Fig. 7.24) to generate reinforcement signals. Figure 7.23 shows the pseudo-code of *GenRSignal*. The mentioned procedure is repeated every iteration by every service provider until a compatible LSA to be reached.

Algorithm 7-3. ICLA-based Competitive Loss Sharing

For each service provider, i in iteration t do {

For each $SP_j \in N_i$ *{*

  SP_j *at a random time asks* SP_i *about* $RP_i(t)$.

  SP_i *sends* $RP_i(t)$ *to* SP_j.

  SP_j *rescales actions of* LA_{ji} *according to the response of* SP_i.

  Using the rescaled actions, LA_{ji} *selects one action to specify* $\alpha_i^j(t)$.

  SP_j *announces* $\alpha_i^j(t)$ *to* SP_i.

  SP_i *updates* $RP_i(t)$.

}

SP_i *calculates* $X_i(t)$ *and announces it to all the* SP_j *that* $\alpha_i^j(t) \neq 0$.

SP_i *calculates* $PI_i(t) = I_i(t) - \sum_{j \in N_i}(\alpha_i^j(t) \times X_j(t)) + \sum_{j \in N_i} pr_i^j(t)$ *using all the announced* $X_j(t)$ *s.*

SP_i *calculates satisfaction signal (*$\beta_i(t)$*) and announces it to all the* SP_j *that* $\alpha_i^j(t) \neq 0$.

SP_i *receives satisfaction signals (*$\beta_j(t)$*) of the other* SP_j *that* $\alpha_i^j(t) \neq 0$.

SP_i *generates reinforcement signals for* LA_{ij} *using GenRSignal(*$\{(\alpha_i^j(t), X_j(t), pr_i^j(t)) | j \in N_i\}, \beta_j(t)$*).*

Probability vector of LA_{ij} *is updated using the generate reinforcement signal.*

} While (Threshold_Satisfied())

Fig. 7.22 Pseudocode for the proposed approach

Algorithm 7-4. generating reinforcement signal

GenRSignal($\{(\alpha_i^j(t), X_j(t), pr_i^j(t)) | j \in N_i\}, \beta_j(t)$*). {*

$$PI_i(t) = I_i(t) - \sum_{j \in N_i}(\alpha_i^j(t) \times X_j(t)) + \sum_{j \in N_i} pr_i^j(t)$$

  Rand = Generate_Random_Number(0,1);   // 0 < Rand < 1

  If (t==1) {

    $u_i^{max_{ii}}$;

    If (Rand > 0.5) {

      Reinforcement_Signal =1;

    } Else {

      Reinforcement_Signal =0;

    }

  } else {

    If ($u_i(PI_i(t)) > u_i^{max}$*) {* $u_i^{max_{ii}}$; *}*

    $\beta_i(t) = \frac{u_i(PI_i(t))}{u_i^{max}}$; // $\beta_i(t)$ is a satisfaction signal of SP_i

    $Er_i(t) = w_{ij}\beta_i(t) + (1 - w_{ij})\beta_j(t)$; // $0.5 \leq w_{ij} \leq 1$

    Reinforcement_Signal = Compatible_Local_Rule ($Er_i(t)$, $\{\alpha_i^j(t) | j \in N_i\}$*);*

  }

}

Fig. 7.23 Pseudocode for generating reinforcement signal

Algorithm 7-5. local rule of ICLA

Compatible_Local_Rule(Er_i, $\{\alpha_i^j | j \in N_i\}$) {

$$\underline{p}_i^{br}(\hat{\underline{p}}^t, Q_t^i) = \arg\max_{\underline{p}_i} \sum_{a^1}\sum_{a^2}\cdots\sum_{a^{\overline{m}_i}} \underline{p}_i(\hat{\underline{p}}^t, Q_t^i)(\alpha_i^i)\prod_{\substack{j=1\\ j\neq i}}^{\overline{m}_i}\hat{\underline{p}}_j^t(\alpha_i^j).Q_t^i(\{\alpha_i^j \mid j \in N_i\});$$

$$\Delta\underline{p}_i^{RS} = \underline{p}_i^{br} - \underline{p}_i^{c};$$

$$j_{\max} = \arg\max_j \Delta\underline{p}_i^{RS}(j);$$

$$j_{\min} = \arg\min_j \Delta\underline{p}_i^{RS}(j);$$

$$d_i = \left(\frac{\Delta\underline{p}_i^{RS}(1) - \Delta\underline{p}_i^{RS}(j_{\min})}{\Delta\underline{p}_i^{RS}(j_{\max}) - \Delta\underline{p}_i^{RS}(j_{\min})}, \ldots, \frac{\Delta\underline{p}_i^{RS}(\overline{m}_i) - \Delta\underline{p}_i^{RS}(j_{\min})}{\Delta\underline{p}_i^{RS}(j_{\max}) - \Delta\underline{p}_i^{RS}(j_{\min})}\right)$$

if (selected action $== a^k \in \{a^1, \ldots, a^{\overline{m}_i}\}$) then

  if ($\Delta\underline{p}_i^{RS}(k) \geq 0$) then

    set $\beta = 1$ with probability $d_i(k)$ and $\beta = 0$ with probability $(1 - d_i(k))$;

  else if ($\Delta\underline{p}_i^{RS}(k) < 0$) then

    set $\beta = 0$ with probability $d_i(k)$ and $\beta = 1$ with probability $(1 - d_i(k))$;

$$Q_{t+1}^i(\{\alpha_i^j \mid j \in N_i\}) = (1-\phi) \times Q_t^i(\{\alpha_i^j \mid j \in N_i\}) + \phi \times [Er_i^t +$$

$$\gamma.\sum_{a^1}\sum_{a^2}\cdots\sum_{a^{\overline{m}_i}} \underline{p}_i^{br}(\hat{\underline{p}}^t, Q_t^i)(\alpha_i^i)\prod_{\substack{j=1\\ j\neq i}}^{\overline{m}_i}\hat{\underline{p}}_j^t(\alpha_i^j).Q_t^i(\{\alpha_i^j \mid j \in N_i\})]$$

  return β;

}

Fig. 7.24 Pseudocode for the local rule of ICLA

7.4.1 Experiments

This section contains the results of the conducted experiments. First, the evaluation approach is described, and then and the results of experiments are reported in Sect. 7.4.1.2.

7.4.1.1 Evaluation Approach

To evaluate the proposed approach, the utility of service providers is used as a measure to check the usefulness of loss sharing. First of all, the utility of a service provider is compared in two different cases: With a loss-sharing agreement case (LSA) vs. lacking loss-sharing agreement case (NO-LSA). Then compatibility of the obtained LSA is checked. For this checking, the utility of the service providers is compared with the case when one of the service providers unilaterally deviates from the accepted LSA by changing its share ($\alpha_i^j(t)$). Due to the randomness of the environment, even with the same LSA, service providers may experience different utilities in two different iterations. Therefore, in experiments, the expected utility (Eu) is used instead of the utility function (u).

However, the proposed approach can operate under conditions with unknown utility functions and loss distributions. However, to measure the utility of service providers, exact forms for utility functions and loss distributions are provided in the experiments. For utility functions, exponential form functions similarly to Eq. (7.36) is used to describe the utility function of the service provider. This function form satisfies the conditions needed for the utility function of a risk-averse service provider.

$$u_i(PI_i) = c_i e^{-a_i PI_i} - c_i (c_i \langle 0, a_i \rangle 0) \tag{7.36}$$

Because it is common in risk management literature to model random losses with exponential distributions, it is assumed that random variable X_i has an exponential distribution with parameter θ_i .

7.4.1.2 Experiments

This section presents details of the experiments (e.g., settings) and results. In the experiments, there are 5 and 8 service providers in three different cases: cases A, B, and C. Figure 7.5 shows the neighboring relation (a tendency for cooperation in loss sharing) for each case. There are six actions for each learning automaton, which are labeled with action 0, action 1, …., and action 5. The convergence of learning automata to k^{th} action (action k) means that learning automaton has decided to put $\alpha_i^j(t)$ equal to $(0.2 \times k)$. The L_{R-I} the learning algorithm is used to update the probability vector of learning automata. The results of experiments are discussed from different viewpoints: capability of ICLA in convergence to a compatible LSA and usefulness of a compatible LSA due to the increasing utility of service providers as much as possible. Three different settings are used for risk aversion of service providers. For each setting, a_i is set according to Table 7.5. Here a_i is the risk aversion of a service provider. For all service providers, c_i in Eq. (7.36) and I_i are equal to -5 and 2000, respectively. The premium $(pr_i^j(t))$ is equal to $0.05 \times \alpha_i^j(t) \times X_i(t)$.

Table 7.6 illustrates the shares (α_i^j) in an obtained compatible LSA, learned by the proposed approach, in case A. In this table value of α_i^j is located in column ij. The shares in cases B and C are also illustrated in Tables 7.7 and 7.8, respectively. Diagrams of Fig. 7.25 illustrate the evolution of the probability vector of the learning automata located in cells of ICLA. Figure 7.25a–c shows these diagrams for LA_{12} in case A for setting 1, 2, and 3, respectively. Figure 7.25d–f show the same diagrams for case C. Comparing diagrams of Fig. 7.25. It can be seen that while risk aversion of a service provider (different settings of Table 7.5) has no significant impact on the convergence rate of learning automata, the structure of ICLA (neighboring relations) has a direct effect on the convergence rate. In other words, comparing Fig. 7.25a–f illustrate that a higher number of neighbors of service provider 2 causes more iterations to be needed for convergence LA_{12} . More number of neighbors for service

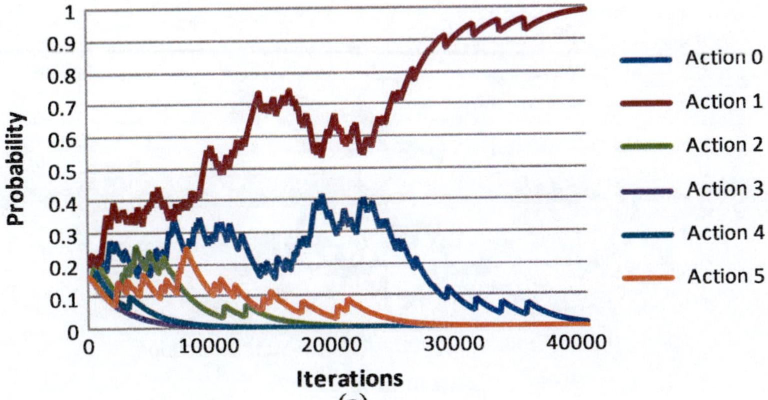

(a)

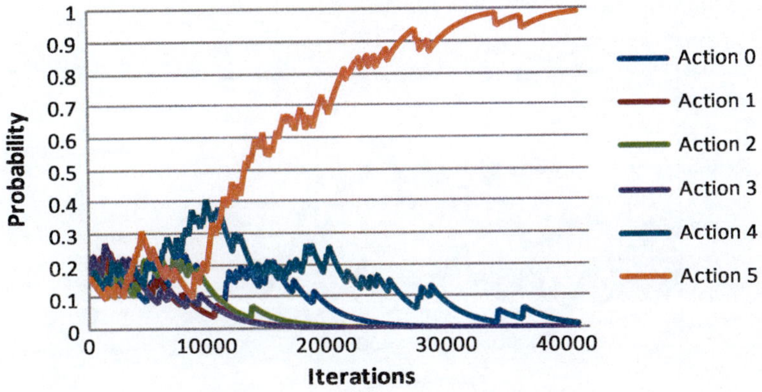

(b)

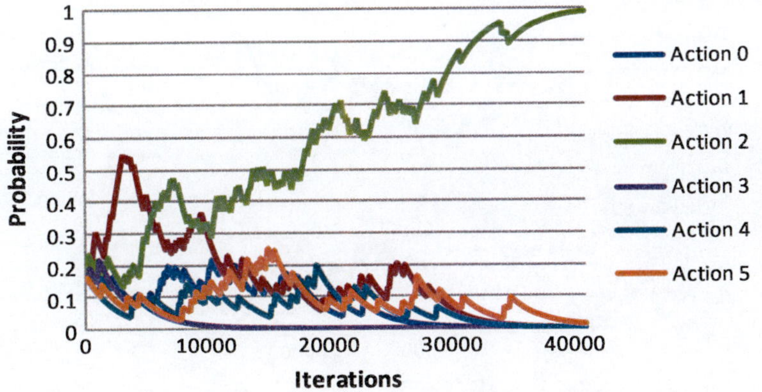

(c)

Fig. 7.25 Evolution of probability vector of **a** SP1in case A under setting 1 **b** SP1 in case A under setting 2 **c** SP1 in case A under setting 3 **d** SP1 in case of C under setting 1 **e** SP1 in case of C under setting 2 **f** SP1 in case of C under setting 3

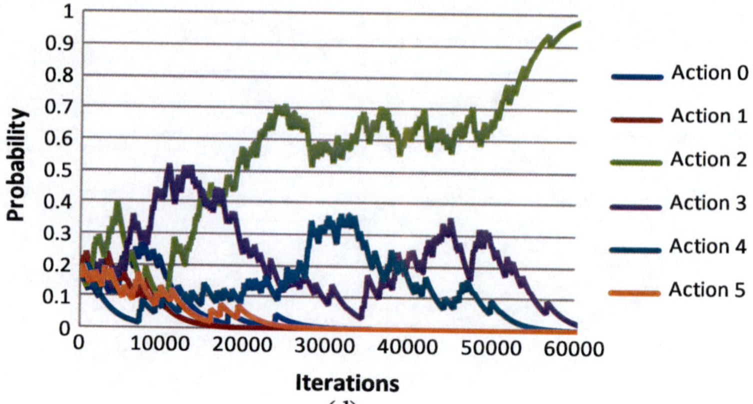

(d)

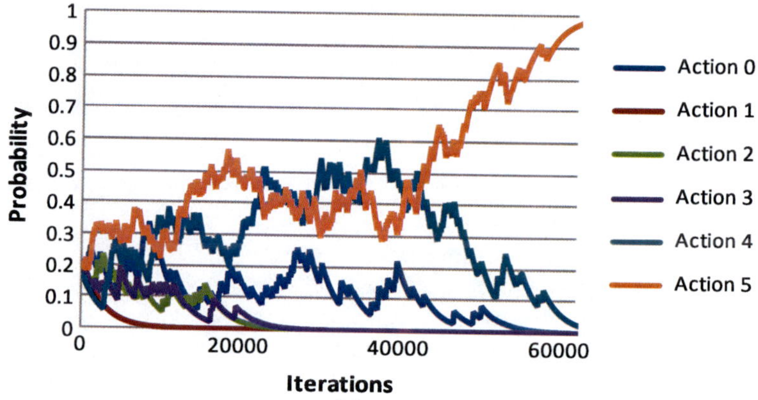

(e)

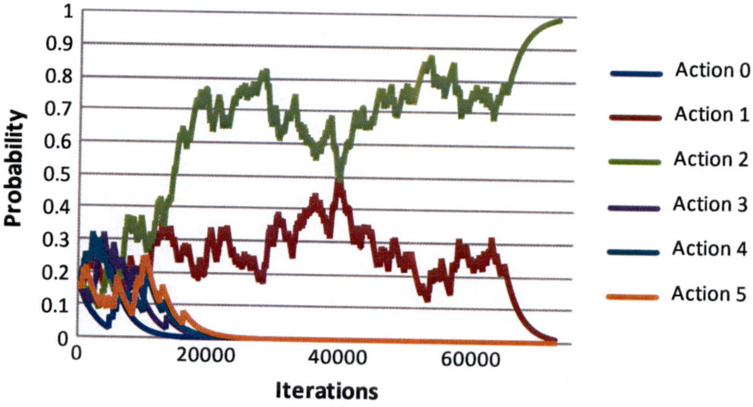

(f)

Fig. 7.25 (continued)

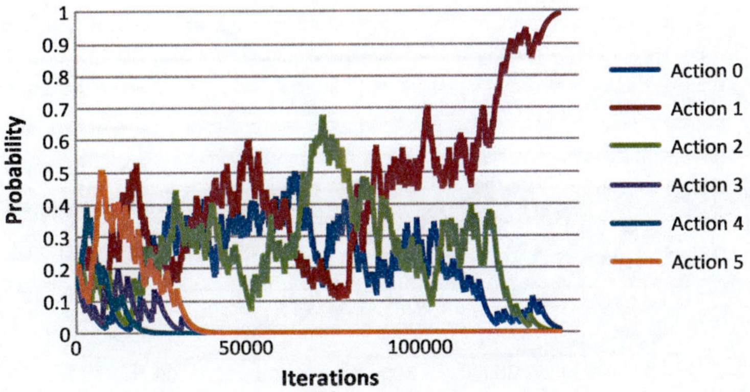

(a)

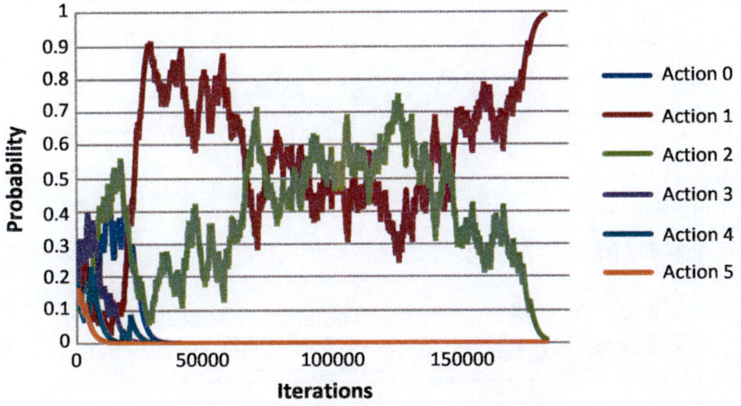

(b)

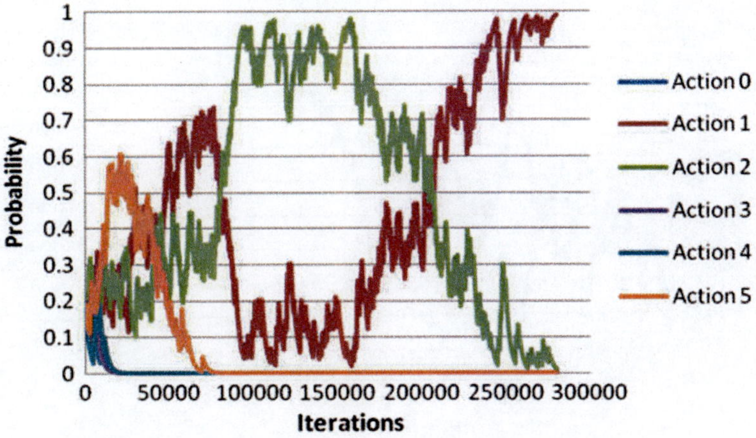

(c)

Fig. 7.26 Evolution of Probability Vector of **a** SP2 in case A under setting 2 **b** SP2 in case of B under setting 2 **c** SP2 in case of C under setting 2 **d** SP5 in case A under setting 2 **e** SP5 in case of B under setting 2 **f** SP5 in case of C under setting 2

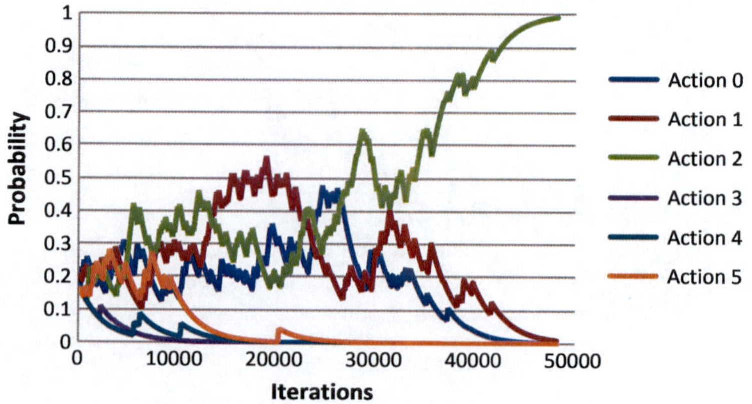

(d)

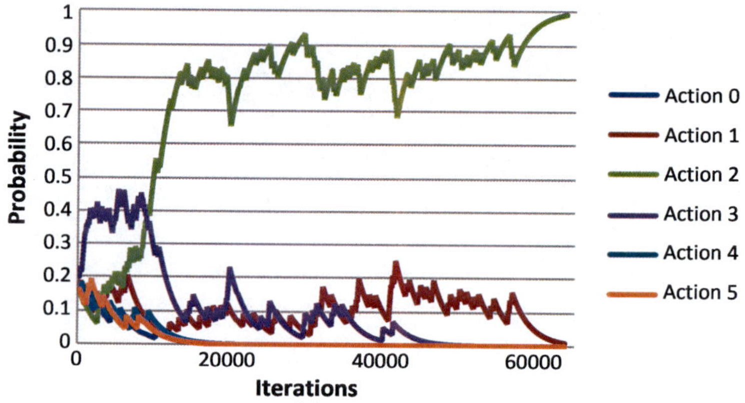

(e)

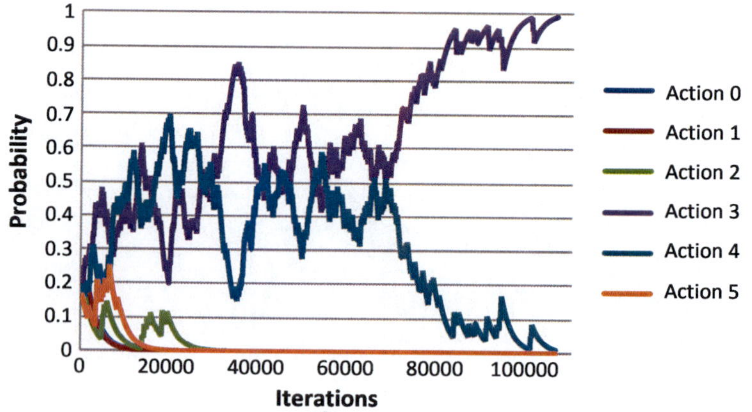

(f)

Fig. 7.26 (continued)

provider 2 is more effective on the convergence rate of learning automata located on cell 2. Figure 7.26a–e illustrates this effect. Adding a new neighbor increases the number of iterations, which is needed for convergence of a learning automaton.

Table 7.9 compares the variance of the utility of service providers when they operate under an LSA versus the case they operate independently. The reported result of Table 7.9 is obtained under settings 2 of Table 7.5. It can be seen that under an LSA agreement, service providers can reach more stable conditions with a lower variance of utility. According to the insurance literature, a lower variance is more desirable for risk-averse service providers. Beside lower variance, the utility of service providers increases by reaching an LSA (Table 7.10).

Figure 7.27 compares the utility of the service providers by operating under an LSA versus operating independently. Each point in the horizontal axis of diagrams of this figure shows a randomly sampled value of utility of a service provider. Generally, 300 random samples are taken. As shown in these diagrams, regardless of the few

Table 7.6 Risk aversion of service providers

Risk aversion	SP1	SP2	SP3	SP4	SP5	SP6	SP7	SP8
Setting 1	0.0005	0.0005	0.0005	0.0005	0.0005	0.0005	0.0005	0.0005
Setting 2	0.0001	0.0002	0.0003	0.0004	0.0005	0.0006	0.0007	0.0008
Setting 3	0.0008	0.0007	0.0006	0.0005	0.0004	0.0003	0.0002	0.0001

Table 7.7 The proposed shares by the ICLA-based approach in case A

Case A	SP1 (α)		SP2 (α)					SP3 (α)		SP4 (α)		SP5 (α)	
	11	12	21	22	23	24	25	32	33	42	44	52	55
Setting 1	0.4	0.6	0.2	0.2	0.2	0.2	0.2	0.4	0.6	0.4	0.6	0.4	0.6
Setting 2	0.8	0.2	0.4	0.2	0.2	0.2	0.0	0.6	0.4	0.8	0.2	0.8	0.2
Setting 3	0.4	0.6	0.0	0.2	0.2	0.2	0.4	0.4	0.6	0.4	0.6	0.4	0.6

Table 7.8 The proposed shares by the ICLA-based approach in case B

Case B	SP1 (α)		SP2 (α)					SP5 (α)		SP6 (α)	
	11	12	21	22	23	25	26	52	55	62	66
Setting 1	0.4	0.6	0.2	0.2	0.2	0.2	0.2	0.4	0.6	0.4	0.6
Setting 2	0.6	0.4	0.4	0.2	0.2	0.2	0.0	0.8	0.2	0.8	0.2
Setting 3	0.4	0.6	0.0	0.2	0.2	0.2	0.4	0.4	0.6	0.4	0.6
	SP4 (α)		SP3 (α)					SP7 (α)		SP8 (α)	
	43	44	32	33	34	37	38	73	77	83	88
Setting 1	0.6	0.4	0.2	0.2	0.2	0.2	0.2	0.4	0.6	0.6	0.4
Setting 2	0.6	0.4	0.4	0.2	0.2	0.2	0.0	0.8	0.2	0.8	0.2
Setting 3	0.4	0.6	0.0	0.2	0.2	0.2	0.4	0.4	0.6	0.4	0.6

Table 7.9 The proposed shares by the ICLA-based approach in case C

Case C	SP1 (α)		SP2 (α)							SP5 (α)			SP6 (α)		
	11	12	21	22	23	25	26	27	28	52	53	55	62	63	66
Setting 1	0.4	0.6	0.0	0.2	0.2	0.2	0.2	0.2	0.0	0.2	0.2	0.6	0.2	0.2	0.6
Setting 2	0.8	0.2	0.4	0.2	0.2	0.2	0.0	0.0	0.0	0.4	0.4	0.2	0.4	0.4	0.2
Setting 3	0.4	0.6	0.0	0.0	0.0	0.2	0.2	0.2	0.4	0.2	0.4	0.4	0.2	0.4	0.4
	SP4 (α)		SP3 (α)							SP7 (α)			SP8 (α)		
	43	44	32	33	34	35	36	37	38	72	73	77	82	83	88
Setting 1	0.4	0.6	0.0	0.2	0.2	0.2	0.2	0.0	0.2	0.2	0.2	0.6	0.2	0.2	0.6
Setting 2	0.8	0.2	0.4	0.2	0.2	0.2	0.0	0.0	0.0	0.4	0.4	0.2	0.4	0.4	0.2
Setting 3	0.4	0.6	0.0	0.0	0.0	0.2	0.2	0.2	0.4	0.2	0.4	0.4	0.2	0.4	0.4

cases, most of the time, the service providers reach better utility when they operate under an LSA agreement. Figure 7.28 illustrates how many times service providers have reached better utility without LSA. For example, service provider one has reached better utility without LSA just in 5.3% of the times, and 94.7% of times, it has reached better utility by operating under an LSA. To check the compatibility of the obtained LSA, the procedure of Fig. 7.29 is applied to the LSA, and the utility of the service providers is measured. Diagrams of Fig. 7.30 illustrate the average utility of SP1 when the selected action of LA_{12} changes and the other service providers are committed to the LSA.

As can be seen in these diagrams, when SP1 unilaterally changes its action, the change is not profitable for SP1, and its utility decreases. Diagrams of Fig. 7.31 illustrates that the unilateral deviation of SP2 is not profitable for SP2 as well. There are similar situations for the other six service providers. This shows that the obtained LSA by the proposed approach is a compatible LSA.

7.5 Conclusion

Many of the risks involved in cloud environments may lead to a lot of simultaneous service level agreement (SLA) violations and big losses. Offering insurance coverage to all users in such situations needs service providers with high financial capabilities. In this chapter, a new irregular cellular learning automaton (ICLA)-based loss-sharing approach is presented to make the big losses tolerable for cloud service providers. The base idea of the proposed approach is loss sharing among the service providers. Due to heterogeneity and dynamicity of cloud environments, an appropriate approach for cloud environments must have the ability to adapt to

Table 7.10 Comparing utility variance of service providers with/without LSA

Case	Mode	SP1	SP2	SP3	SP4	SP5	SP6	SP7	SP8
A	No-LSA	0.0024151	0.0022444	0.0031789	0.0030288	0.0027222	–	–	–
	LSA	0.0001034	5.698E−05	0.0001352	0.000101	4.268E−05	–	–	–
B	No-LSA	0.0024132	0.0022195	0.0032807	0.0030064	0.0027081	0.003286	0.0032954	0.0034784
	LSA	0.0001124	5.712E−05	0.0001302	0.000094	4.212E−05	5.786E−05	0.0001371	0.000493
C	No-LSA	0.0024395	0.0022102	0.0031421	0.0030387	0.0026987	0.003188	0.0032398	0.0034254
	LSA	0.0001098	4.987E−05	0.0001089	0.000091	3.728E−05	4.827E−05	0.0001187	0.000408

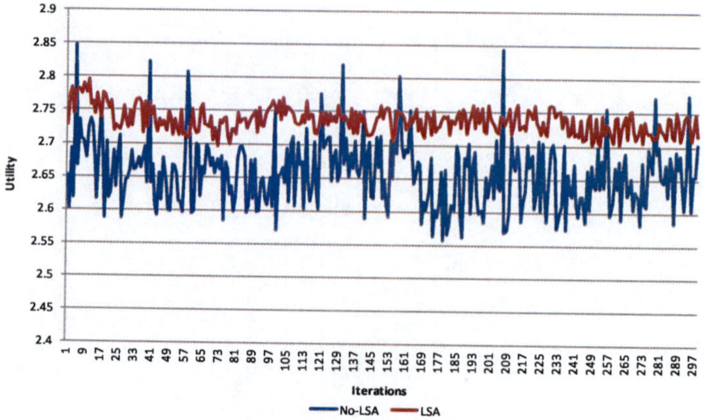

(a)

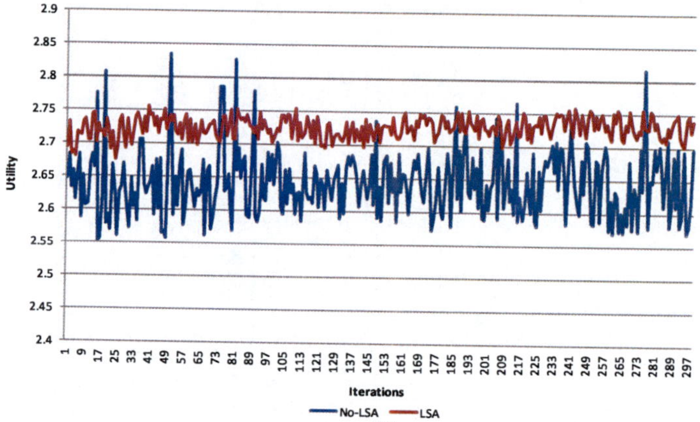

(b)

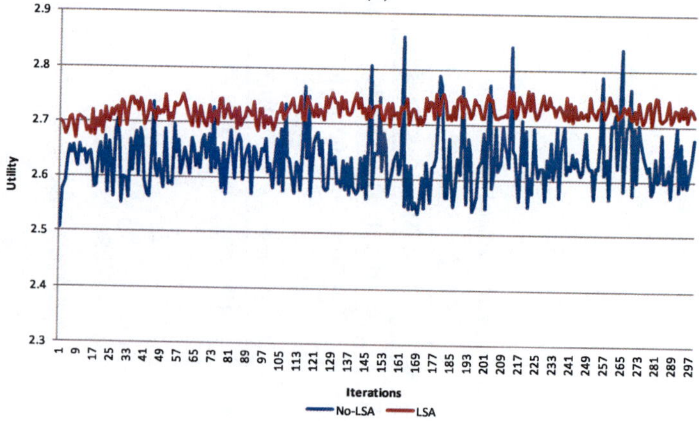

(c)

Fig. 7.27 Comparing average utility of a service provider in LSA and No-LSA modes for **a** SP1 **b** SP2 **c** SP3 **d** SP4 **e** SP5 **f** SP6 **g** SP7 **h** SP8

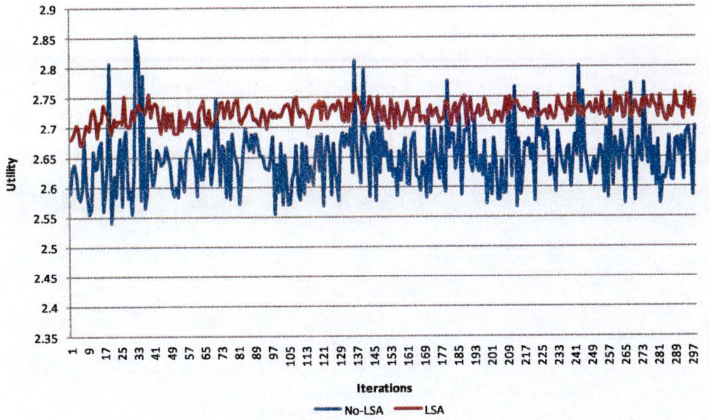

(d)

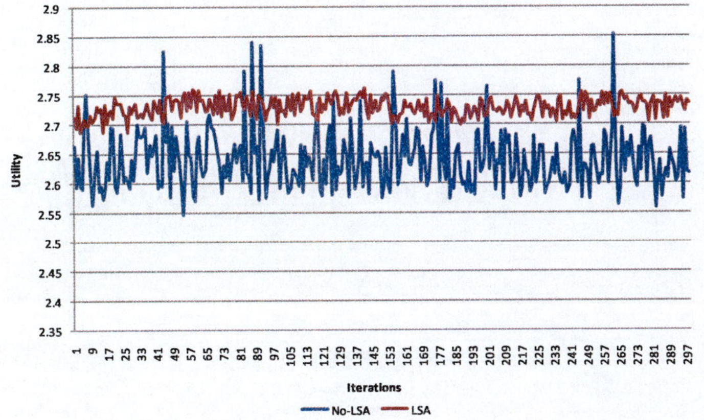

(e)

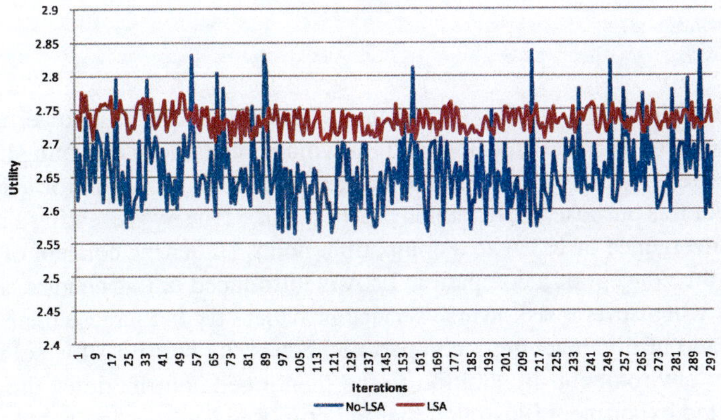

(f)

Fig. 7.27 (continued)

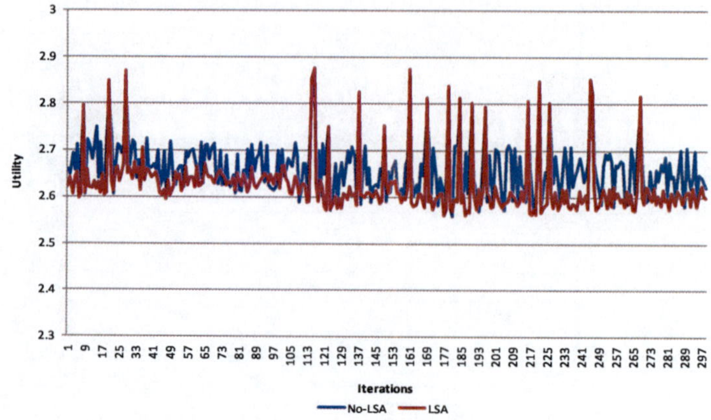

(g)

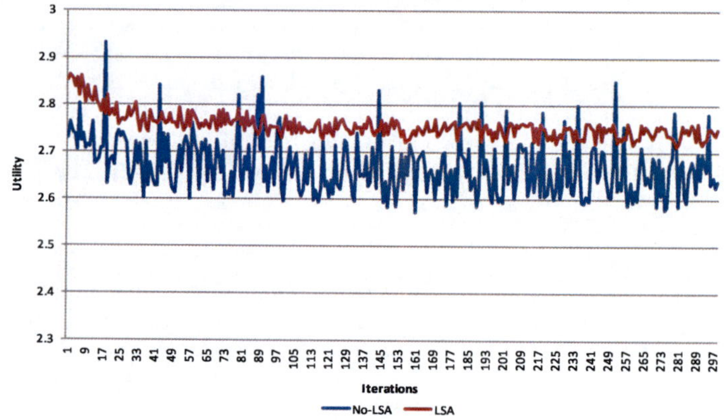

(h)

Fig. 7.27 (continued)

dynamic conditions, e.g., variable loss rates. Therefore ICLA, as a powerful mathematical model for decentralized applications which is capable of operating in random and stochastic environments, is used to present an efficient solution for loss sharing problem. For this purpose, we extended the theory of ICLA by presenting a new local rule for convergence of ICLA to a compatible point. Under the concept of compatible point, the concept of a compatible LSA is introduced in this chapter, and it has shown that when several risk-averse service providers try to increase their financial capability and utility, reaching a compatible LSA is the best possible solution in a competitive environment. In addition to the mentioned contributions, the result of the conducted experiment illustrated that the proposed loss-sharing approach could decrease the variance of the big losses. The low variance of losses is a highly favorable condition for risk-averse service providers and improve their perceived utility.

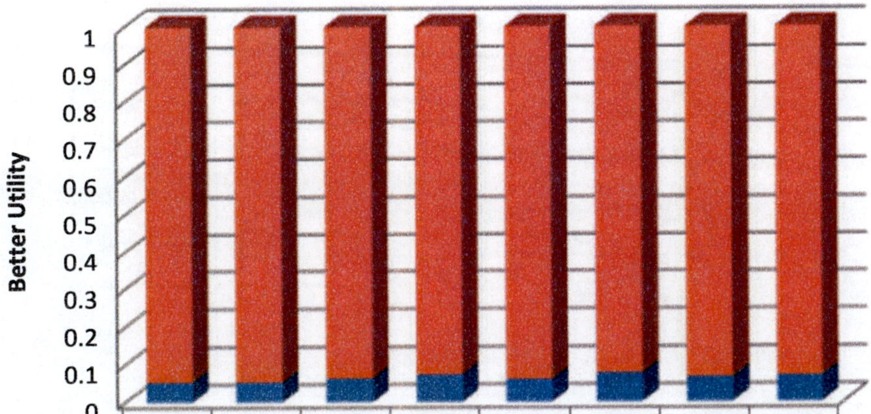

	SP1	SP2	SP3	SP4	SP5	SP6	SP7	SP8
LSA	0.947	0.947	0.937	0.927	0.94	0.923	0.934	0.93
No-LSA	0.053	0.053	0.063	0.073	0.06	0.077	0.066	0.07

Fig. 7.28 The proportion of the times that a service provider has reached better utility in LSA and No-LSA modes

Algorithm 7-6. checking compatibility of an LSA

```
CheckCompatibility(LSA)  {

    for 200 times do {

        Set the selected actions of all learning automata according to LSA
        for (i=1 to Number_of_SP )

            for(j=1 to Number_of_LAs_in_Cell_i)

                for (k=1 to 6)  //  Number of actions of a LA
                {
                    α_i^j(t) = Equvalent_Value_of_k_th_Action();

                    for(iteration=1 to 1000)
                        Sample_Utility_of_All_SP();
                    Calculate__AND_Log_Average_Utility_ of_All_SP();
                }
            }
        }
}
```

Fig. 7.29 Pseudo Code for checking compatibility of an LSA

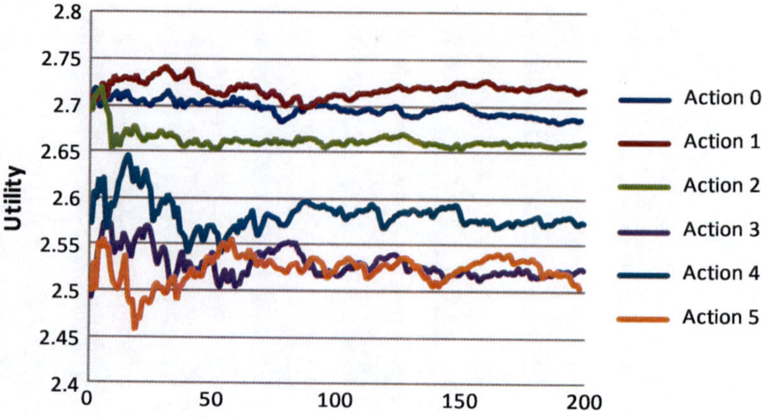

Fig. 7.30 Effect of unilateral deviation of SP1 on its utility

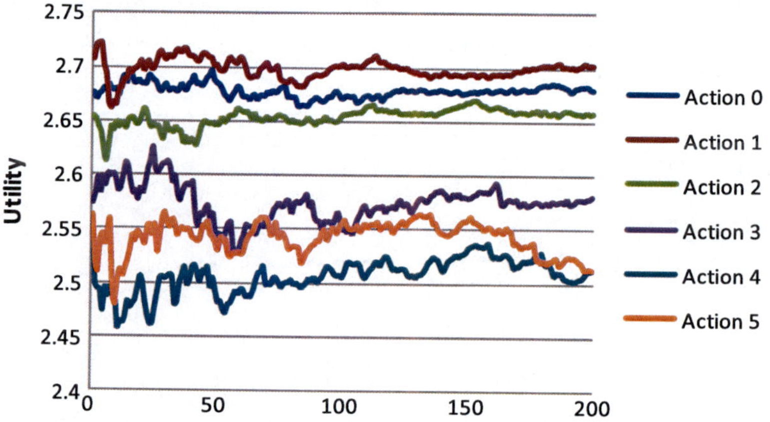

Fig. 7.31 Effect of unilateral deviation of SP2 on its utility

References

Ahmadinia, M., Meybodi, M., Esnaashari, M., Alinejad-Rokny, H.: Energy-efficient and multi-stage clustering algorithm in wireless sensor networks using cellular learning automata. IETE J. Res. **59**, 774 (2013). https://doi.org/10.4103/0377-2063.126958

Akbari Torkestani, J., Meybodi, M.R.: Clustering the wireless Ad Hoc networks: A distributed learning automata approach. J. Parallel Distrib. Comput. **70**, 394–405 (2010). https://doi.org/10.1016/j.jpdc.2009.10.002

Akbari Torkestani, J., Meybodi, M.R.: A cellular learning automata-based algorithm for solving the vertex coloring problem. Expert Syst. Appl. **8**, 9237–9247 (2011)

Beigy, H., Meybodi, M.R.: A self-organizing channel assignment algorithm: a cellular learning automata approach. In: Springer-Verlag Lecture Notes in Computer Science, pp. 119–126, Springer (2003)

Beigy, H., Meybodi, M.R.: A mathematical framework for cellular learning automata. Adv. Complex Syst. **07**, 295–319 (2004). https://doi.org/10.1142/S0219525904000202

Beigy, H., Meybodi, M.R.: Cellular learning automata based dynamic channel assignment algorithms. Int. J. Comput. Intell. Appl. **8**, 287–314 (2009)

Beigy, H., Meybodi, M.R.R.: Cellular learning automata with multiple learning automata in each cell and its applications. IEEE Trans. Syst. Man, Cybern. Part B **40**, 54–65 (2010). https://doi.org/10.1109/TSMCB.2009.2030786

Billard, E. A.: Asymmetry in learning automata playing multi-level games. In: SMC'98 Conference Proceedings. In: 1998 IEEE International Conference on Systems, Man, and Cybernetics (Cat. No.98CH36218). IEEE, pp. 2202–2206 (1998)

Billard, E. A.: Instabilities in learning automata playing games with delayed information. In: Proceedings of IEEE International Conference on Systems, Man and Cybernetics, IEEE, pp. 1160–1165

Billard, E.A.: Chaotic behavior of learning automata in multi-level games under\ndelayed information. In: 1997 IEEE International Conference on Systems, Man, and Cybernetics. Computational Cybernetics and Simulation. IEEE, pp. 1412–1417. Orlando, USA

Eraghi, A.E., Torkestani, J.A., Meybodi, M.R.: Cellular learning automata-based graph coloring problem. Comput. Eng., pp. 10–12. Perth, Australia (2009)

Esnaashari, M., Meybodi, M.R.: A cellular learning automata based clustering algorithm for wireless sensor networks. Sens. Lett. **6**, 723–735 (2008)

Esnaashari, M., Meybodi, M.R.: Deployment of a mobile wireless sensor network with k-coverage constraint: a cellular learning automata approach. Wirel. Networks. **19**, 945–968 (2013). https://doi.org/10.1007/s11276-012-0511-7

Esnaashari, M., Meybodi, M.R.: Irregular cellular learning automata. IEEE Trans. Cybern **45**, 1622–1632 (2018). https://doi.org/10.1016/j.jocs.2017.08.012

Gharehchopogh, F.S., Ebrahimi, S.: A novel approach for edge detection in images based on cellular learning automata. Int. J. Comput. Vis. Image Process. **2**, 51–61 (2012). https://doi.org/10.4018/ijcvip.2012100105

Ghavipour, M., Meybodi, M.R.: Irregular cellular learning automata-based algorithm for sampling social networks. Eng. Appl. Artif. Intell. **59**, 244–259 (2017). https://doi.org/10.1016/j.engappai.2017.01.004

Hasanzadeh Mofrad, M., Sadeghi, S., Rezvanian, A., Meybodi, M.R.: Cellular edge detection: Combining cellular automata and cellular learning automata. AEU – Int. J. Electron. Commun. **69**, 1282–1290 (2015). https://doi.org/10.1016/j.aeue.2015.05.010

Kashki, M., Abido, M.A., Abdel-Magid, Y.L.: Pole placement approach for robust optimum design of PSS and TCSC-based stabilizers using reinforcement learning automata. Electr. Eng. **91**, 383–394 (2010). https://doi.org/10.1007/s00202-010-0147-5

Khomami, M.M.D., Rezvanian, A., Meybodi, M.R.: A new cellular learning automata-based algorithm for community detection in complex social networks. J. Comput. Sci. **24**, 413–426 (2018). https://doi.org/10.1016/j.jocs.2017.10.009

Kumpati, S., Narendra, M.A.L.T.: Learning Automata: An Introduction. Prentice-Hall (1989)

Lakshmivarahan, S.: Learning Algorithms Theory And Applications. Springer-Verlag, New York (1981)

Liu, L., Hu, G., Xu, M., Peng, Y.: Learning Automata based spectrum allocation in cognitive networks. Wcnis, pp. 503–508. IEEE, Beijing, China (2010)

Masoumi, B., Meybodi, M.R.: Learning automata based multi-agent system algorithms for finding optimal policies in Markov games. Asian J. Control **14**, 137–152 (2012). https://doi.org/10.1002/asjc.315

Meybodi, M.R., Khojasteh, M.R.: Application of cellular learning automata in modelling of commerce networks. In: Proceedings of 6th annual international computer society of iran computer conference CSICC-2001, pp. 284–295

Misra, S., Tiwari, V., Obaidat, M.S.: Lacas: learning automata-based congestion avoidance scheme for healthcare wireless sensor networks. IEEE J. Sel. Areas Commun. **27**, 466–479 (2009). https://doi.org/10.1109/JSAC.2009.090510

Morshedlou, H., Meybodi, M.R.: A new local rule for convergence of ICLA to a compatible point. IEEE Trans. Syst. Man, Cybern. Syst. **47**, 3233–3244 (2017). https://doi.org/10.1109/TSMC.2016.2569464

Mousavian, A., Rezvanian, A., Meybodi, M.R.: Cellular learning automata based algorithm for solving minimum vertex cover problem. In: 2014 22nd Iranian conference on electrical engineering (ICEE). IEEE, pp. 996–1000 (2014)

Nejad, H.C., Azadbakht, B., Adenihvand, K., et al.: Fuzzy cellular learning automata for lesion detection in retina images. J. Intell. Fuzzy Syst. **27**, 2297–2303 (2014). https://doi.org/10.3233/IFS-141194

Nicopolitidis, P., Papadimitriou, G.I., Pomportsis, A.S.: Learning automata-based polling protocols for wireless LANs. IEEE Trans. Commun. **51**, 453–463 (2003). https://doi.org/10.1109/TCOMM.2003.809788

Nisan, N., Schapira, M., Valiant, G.Z.A.: Best-response mechanisms. In: ICS. pp. 155–165 (2011)

Nowé. A., Verbeeck. K., Peeters, M.: Learning automata as a basis for multi agent reinforcement learning, pp 71–85 (2006)

Rezapoor Mirsaleh, M., Meybodi, M.R.: A new memetic algorithm based on cellular learning automata for solving the vertex coloring problem. Memetic. Comput. **8**, 211–222 (2016). https://doi.org/10.1007/s12293-016-0183-4

Rezvanian, A., Saghiri, A.M., Vahidipour, S.M., et al.: Recent advances in learning automata. Springer (2018a)

Rezvanian, A., Saghiri, A.M., Vahidipour, S.M., et al.: Cellular learning automata, pp. 21–88 (2018b)

Rezvanian, A., Saghiri, A.M., Vahidipour, S.M., et al.: Learning automata for wireless sensor networks. In: Recent advances in learning automata, pp. 91–219 (2018c)

Rezvanian, A., Saghiri, A.M., Vahidipour, S.M., et al.: Learning automata for cognitive peer-to-peer networks. In: Recent advances in learning automata, pp. 221–278 (2018d)

Rezvanian, A., Moradabadi, B., Ghavipour, M., et al.: Social network sampling. In: Learning Automata Approach for Social Networks. Springer (2019a)

Rezvanian, A., Moradabadi, B., Ghavipour, M., et al.: Social link prediction. In: Learning Automata Approach for Social Networks. Springer, pp 169–239 (2019b)

Rodríguez, A., Vrancx, P., Grau, R., Nowé, A.: An RL approach to common-interest continuous action games. In: Proceedings of the 11th International Conference on Autonomous Agents and Multiagent Systems, pp. 1401–1402 (2012)

Saghiri, A.M., Meybodi, M.R.: An adaptive super-peer selection algorithm considering peers capacity utilizing asynchronous dynamic cellular learning automata. Appl. Intell. **48**, 271–299 (2018). https://doi.org/10.1007/s10489-017-0946-8

Sastry, P.S., Phansalkar, V.V., Thathachar, M.A.L.: Decentralized learning of Nash equilibria in multi-person stochastic games with incomplete information. IEEE Trans. Syst. Man Cybern. **24**, 769–777 (1994). https://doi.org/10.1109/21.293490

Szepesvári, C., Littman, M.L.: A Unified Analysis of Value-Function-based reinforcement-learning algorithms. Neural. Comput. **11**, 2017–2060 (1999). https://doi.org/10.1162/089976699300016070

Thierens. D.: An adaptive pursuit strategy for allocating operator probabilities. In: Proceedings of the 2005 Conference on Genetic and Evolutionary Eomputation—GECCO '05. ACM Press, New York, New York, USA, p. 1539

Tilak, O., Martin, R., Mukhopadhyay, S.: Decentralized indirect methods for learning automata games. IEEE Trans. Syst. Man, Cybern. Part B **41**, 1213–1223 (2011). https://doi.org/10.1109/TSMCB.2011.2118749

Tilak, O., Mukhopadhyay, S., Tuceryan, M., Raje, R.: A novel reinforcement learning framework for sensor subset selection. In: 2010 International Conference on Networking, Sensing and Control (ICNSC). IEEE, pp. 95–100 (2010)

Torkestani, J.A., Meybodi, M.R.: An intelligent backbone formation algorithm for wireless ad hoc networks based on distributed learning automata. Comput. Networks **54**, 826–843 (2009)

Tuan, T.A., Tong, L.C., Premkumar, A.B.: An adaptive learning automata algorithm for channel selection in cognitive radio network. In: 2010 International Conference on Communications and Mobile Computing. IEEE, pp. 159–163 (2010)

Vahidipour, S.M., Meybodi, M.R., Esnaashari, M.: Cellular adaptive Petri net based on learning automata and its application to the vertex coloring problem. Discret Event. Dyn. Syst. **27**, 609–640 (2017a). https://doi.org/10.1007/s10626-017-0251-z

Vahidipour, S.M., Meybodi, M.R., Esnaashari, M.: Adaptive Petri net based on irregular cellular learning automata with an application to vertex coloring problem. Appl. Intell. **46**, 272–284 (2017b). https://doi.org/10.1007/s10489-016-0831-x

Vrancx, P., Verbeeck, K., Nowe, A.: Decentralized learning in markov games. IEEE Trans. Syst. Man, Cybern Part B **38**, 976–981 (2008a). https://doi.org/10.1109/TSMCB.2008.920998

Vrancx, P., Verbeeck, K., Nowé, A.: Networks of learning automata and limiting games. Adaptive agents and multi-agent systems III, pp. 224–238. Adaptation and Multi-Agent Learning. Springer, Berlin Heidelberg, Berlin, Heidelberg (2008b)

Wheeler, R., Narendra, K.: Decentralized learning in finite Markov chains. IEEE Trans. Automat. Contr. **31**, 519–526 (1986). https://doi.org/10.1109/TAC.1986.1104342

Wolfram, S.: Theory and Applications of Cellular Automata. World Scientific Publication (1986)

Zhong, W., Xu, Y., Tao, M.: Precoding strategy selection for cognitive MIMO multiple access channels using learning automata. In: IEEE International Conference on Communications, pp. 1–5 (2010)

Zweifel, P., Eisen, R.: Insurance economics. Springer, Berlin Heidelberg, Berlin, Heidelberg (2012)

Chapter 8
Cellular Learning Automata Versus Multi-agent Reinforcement Learning

8.1 Introduction

Convergence toward Nash Equilibrium point in non-cooperative distributed systems with autonomous elements or agents is a very important issue because it provides efficient solutions for various problems in such systems. By introducing Reinforcement Learning (RL) as an efficient approach in game theory (Erev and Roth 1998), increasing literature is concerned with the theoretical convergence of RL-based approaches towards Nash equilibrium. Many attempts such as Q-learning based (Hu and Wellman 2003; Tesauro 2003; Leslie and Collins 2005) and MARL approaches (Hu and Wellman 1998; Shoham et al. 2003; Nowé et al. 2006; Rodríguez et al. 2012) are carried out to offer approaches for learning of Nash Equilibrium points. The present works cover the problems in which strategy of all system elements or players have direct effects on each other. This is even though many decentralized and distributed systems in the modern world have just local interactions. Elements of such systems have no direct effect on the other elements in the system except the local ones. Finding an equilibrium point equivalent to NE point in such systems can also provide valuable solutions for challenges and problems in such systems. Although the existing approaches in literature can be applied to such systems with local interaction, the time complexity of such approaches commonly are $|A|^n$, where n is the number of system elements or players, and $|A|$ is the size of possible strategy set.

In this chapter, a new approach is presented for learning of Nash equivalent equilibrium point in systems with local interactions such that it reduces the time complexity of learning drastically. For this purpose, Irregular Cellular Learning Automata (ICLA) (Esnaashari and Meybodi 2018) is used to represent the system. Each element or player of the system is modeled using a cell of ICLA. Each cell equipped with at least one learning automaton which receives reinforcement signals

R. Vafashoar et al., *Cellular Learning Automata: Theory and Applications*, Studies in Systems, Decision and Control 307,
https://doi.org/10.1007/978-3-030-53141-6_8

generated by the local rule of ICLA. A new local rule is proposed in Chap. 7, which guarantees to learn of Nash equivalent equilibrium point. Formal proofs for this learning are provided as well, and the results of the conducted experiments support our theoretical findings.

8.2 ICLA Game

ICLA game is an extension of a stochastic game of learning automata (Thathachar and Sastry 2004; Rezvanian et al. 2018a). Consider a game with N players where there is a set of actions for each player to choose. Over an iteration of the game, the players select a specific action from their respective action sets. Then they receive stochastic payoffs from the environment. The payoff each player gets from the environment depends on the joint actions of the players (The selected own action of the player and selected actions of the other players). These payoffs may be different for different players. The probability distributions that establish payoff of joint actions are unknown but stationary for all the players. Now assume a new game that payoffs of players i and j are independent of the selected actions of each other, and they are indifferent about each other's' choice. There exists a sequence of players $i, i_1, \ldots, i_m, j$ for $m \geq 1$ such that every two successive players in the mentioned sequence care about the choice of each other. Notice at the instant, the payoffs of players i and j do not depend on the selected action of each other, but in the long term, they would be influenced by each other's choices indirectly and through other players who are in the sequence. Now, if in the former game the latter game, we represent each player using a learning automaton (LA), then we have a game of learning automata in the former case and an ICLA game in the latter case. Actions of a player constitute the action set of LA. Hence at any given instant, the probability distribution vector of a LA is equivalent to a mixed strategy of a player. The action set of each LA determines its state set. All the LAs that LA i care about their choices are in LA i's neighborhood vector. Local rule of ICLA game establishes the stochastic payoff of each learning automaton.

Local rules of ICLA play an essential role in the convergence of ICLA game to compatible points. Simply rewarding or punishing LAs, just based on payoff each LA earned, does not guarantee the convergence of ICLA game to a compatible point when LAs use the $L_{R\text{-}I}$ learning algorithm. (Beigy and Meybodi 2004) has provided some conditions for the local rules under which, CLA will converge to a compatible point. ICLA games are too appropriate choices for modeling social interactions in non-centralized systems.

Proposition 8.1 A compatible configuration of an ICLA game is equivalent to Nash point of a stochastic normal form game.

Proof: For proof, assume, there is an ICLA game in which all players (LAs) care about the choice of each other. This ICLA game is equivalent to a normal form stochastic game. Now according to this ICLA game $d_{ir}\left(\underline{p}\right)$ is defined as

$$d_{ir}\left(\underline{p}\right) = \sum_{\alpha_2} \cdots \sum_{\alpha_{\overline{m}}} F^i(r, \alpha_2, \ldots, \alpha_{\overline{m}}) \prod_{l \neq i} p_{l\alpha_l} \tag{8.1}$$

A configuration $\underline{p}$ is compatible if and only if

$$\sum_r \left(\sum_{\alpha_2} \cdots \sum_{\alpha_{\bar{m}}} F^i(r, \alpha_2, \ldots, \alpha_{\bar{m}}) \prod_{l \neq i} p_{l\alpha_l} \right) \times p_{ir}$$
$$\geq \sum_r \left(\sum_{\alpha_2} \cdots \sum_{\alpha_{\bar{m}}} F^i(r, \alpha_2, \ldots, \alpha_{\bar{m}}) \prod_{l \neq i} p_{l\alpha_l} \right) \times q_{ir} \tag{8.2}$$

for all configurations $\underline{q} \in \mathrm{K}$ and all learning automaton A_i. Equation (8.2) is the definition of a Nash point in stochastic games and $\underline{p}$ determines the strategy of players.

8.3 Comparing CLA with Multi-agent Reinforcement Learning (MARL)

Multi-agent systems are employed in various fields and domains such as resource management, data mining, distributed control, collaborative decision support, economics, etc. A major part of the research on multi-agent learning concerns reinforcement learning techniques (MARL). This section provides a comparison between MARL and CLA for problems in distributed systems, but the first classification of CLAs is presented.

8.3.1 CLA Classification

There are two classes of CLAs in literature: static and dynamic CLAs. In static CLAs, the main structure of CLA remains unchanged during the evolution of CLA (Beigy and Meybodi 2007, 2008, 2010; Mousavian et al. 2014; Mozafari et al. 2015; Zhao et al. 2015; Rezvanian et al. 2018b; Khomami et al. 2018; Esnaashari and Meybodi 2018) but in dynamic class one of the aspects of CLA such as a local rule, structure, neighboring definition may change during evolution (Esnaashari and Meybodi 2011, 2013; Rezvanian et al. 2018c). Also, in a static class, depending on the structure, CLAs can be classified again as regular (Beigy and Meybodi 2004) or irregular (Rezvanian et al. 2018d; Esnaashari and Meybodi 2018). In Irregular class, the structure regularity assumption is removed. Each one of static and dynamic classes contains open or closed models. In closed models, the states of neighboring cells, which is called the local environment, affect the action selection process of the LA of that cell.

In contrast, in open models, each cell has its local environment, own exclusive environment, and one global environment defined for the whole CLA, which effect on the action selection process. CLA classes can be further classified into synchronous or asynchronous models. In a synchronous case, it is assumed that there is an external clock which triggers synchronous events for the cells and all the cells perform their local rules at the same time (Beigy and Meybodi 2007), but in asynchronous case, at a given instance part of the cells are activated. Their state may changes while and the state of the rest of the cells remains unchanged (Beigy and Meybodi 2008).

Moreover, Asynchronous CLAs in dynamic class can be either time-driven or step-driven (Esnaashari and Meybodi 2013b). In a time-driven model, each cell is assumed to have an internal clock that wakes up the LA associated with that cell while in step-driven mode, a cell is selected in a fixed or random sequence. Please note that the problem of the definition of asynchronous dynamic CLAs depends on its application, and all the reported dynamic CLAs in literature are closed and asynchronous (Esnaashari and Meybodi 2011, 2013). In (Beigy and Meybodi 2010), a model of static CLA with multiple LAs in each cell is introduced. In this model, the set of LAs of a cell is fixed during the evolution of CLA.

8.3.2 MARL and CLA

Table 8.1 provides a brief comparison between MARL techniques and CLA based approaches. Different opinions on the goal of MARL have led to the proposal of many different goals, among which two major goals can be distinguished: stability of the agents' learning dynamics, and adaptation to the changing behavior of the other agents. Both these goals are concerned with CLA as well. In CLA literature (Beigy and Meybodi 2004; Rezvanian et al. 2018b; Esnaashari and Meybodi 2018), the convergence of learning automata, which is an equivalent concept to stability, is an ultimate target. Moreover, considering the selected actions by the neighbors to specify the reward/punish signals shows that adaptation is a focal point in the theory of CLA. When one or more agents fail in a multi-agent system, the remaining agents can take over some of their tasks. This implies that MARL is inherently robust. However, in CLA agent are simple learning automata which make it more reliable, but by the failure of a learning automaton, CLA can continue its task without problem. Therefore, CLA is robust, as well. About scalability, by design, most multi-agent systems and also CLAs allow the easy insertion of new agents or learning automata into the system, leading to a high degree of scalability. Experience sharing can help to learn automata or agents with similar tasks to learn faster and better. However, there is no existent work on experience sharing in CLA literature. Speed-up of CLA and MARL can be realized thanks to parallel computation when the LAs or agents exploit the

Table 8.1 Comparison of MARL techniques and CLA-based approaches

Measure	MARL Techniques	CLA based Approaches
Goal	Stability—Adaptation	Convergence (Convergence needs stability and adaptation)
Robust	Yes	Yes
Scalability	Scalable	Scalable
Experience Sharing	Possible	Possible
Parallel Computation	Yes	Yes
Handling Nonstationary	Yes	Yes
Exploration-exploitation trade-off	Yes	Yes
Coordination	Yes	Yes
Applicability in Games	Yes	Yes
One-Shot Games	No	No
Static Games	Yes	Yes
Dynamic Games	Yes	Yes (Dynamic Class of CLAs with changing local rule during evolution)
Stage Games	Yes	Yes (Static class of CLAs can be employed in a single-stage or multi-stage with similar policy but for multi-stage games with different policy dynamic class of CLAs are required)
Common Payoff Games	Yes	Yes (Open Models)
Zero-sum Games	Yes	Yes (Open Models)
Opponent independent	Possible	Possible
Opponent-aware	Possible	Possible
Homogeneity	Homogeneous-Heterogeneous	Homogeneous-Heterogeneous
Learning Model	Model-based, Model-free	Model-free
Input to agents or LAs	The actions of the other agents, their actions and rewards	States of neighboring cells (local environment), own exclusive environment (not necessary), global environment (not necessary)
Complexity	Exponential in the number of agents	In Systems with local interaction, complexity is exponential in the number of neighbors (neighbor agents), which is too smaller than the number of all cells or agents. With global interactions, complexity is exponential in the number of cells (agents)

decentralized structure of the task. Nonstationary refers to changing the best policy by changing the policy of the other agents' policies. Policies in CLA are probability vectors of learning automata. Some of the existent MARL techniques and also CLA-based approaches handle the nonstationary problems. The exploration-exploitation trade-off needs MARL techniques to make a balance between the exploitation of current knowledge, and exploratory, information-gathering actions taken to improve that knowledge. The exploitation in CLA is choosing action using the probability vector of learning automata, and exploration is in the form of updating probability vector based on reinforcement signals that learning automata receive. The need for coordination comes from the fact that the effect of any agent's action on the environment also depends on the actions taken by the other agents. Hence, the agents' choices of actions must be mutually consistent in order to achieve their intended effect. In CLA coordination problem must be handled using local rules which control input reinforcement signals to learning automata.

CLA-based approaches and MARL techniques are employed in different research fields that show their power and abilities to find efficient solutions. For example, evolutionary computation concerns principles of biological evolution to locate solutions of the given problems (Bäck 1996; De Jong 2006). Populations of candidates (here agent behaviors) are maintained. Using a fitness function, candidates are evaluated and selected for mutation or breeding base on their fitness. Panait and Luke (2005) offer a comprehensive survey of evolutionary learning, including MARL techniques, for cooperative multi-agent systems. For the interested reader, examples of co-evolution techniques, where the behaviors of the agents evolve in parallel, can be found in (Potter and Jong 1994; Ficici and Pollack 2000; Panait et al. 2003). The power of CLA in the evolutionary field encouraged researchers to utilize CLA for presenting Evolutionary Computing Models and approaches (Rastegar and Meybodi 2004; Rastegar et al. 2006; Sharifi et al. 2012) based on CLA. MARL also has tight connections with game theory. Game theory, which is the study of interacting rational players or agents who try to maximize their utilities (Başar and Olsder 1998), and especially the theory of learning in games (Fudenberg and Levine 1998), make an essential contribution to MARL. Bowling and Veloso (Bowling and Veloso 2002) argue numerous MARL algorithms, showing that these algorithms combine temporal-difference RL with game-theoretic concepts for the static games arising in each state of the dynamic environment. A static game is a stateless stochastic game that there is no state signal, and the rewards depend only on the joint actions of agents. Because in CLA, direct interaction among all agents (learning automaton) is not necessary, therefore new solution concepts for equilibrium points are defined instead of points such as Nash. Beigy and Meybodi (2004) proposed an equilibrium point for CLA, which is called a compatible point. When instead of local interactions of CLA, each learning automaton has direct interactions with all the learning automata in CLA; the concept of compatible point will be equivalent to Nash point concept.

CLA based approaches cannot be utilized in one-shot games. However, it seems these approaches to be good candidates for repeated games (The static game is called a repeated game when it repeatedly played by the same agents). The main difference from a one-shot game is that the agents can use some of the game iterations to gather information about the other agents or the reward functions and make more informed decisions after that. The gathered information in CLA is represented in the form of probability vectors of learning automata. MARL algorithms learn strategies separately for every state of a game. The agent's overall policy is then the aggregate of these strategies. In CLA, because strategies are probability vectors, they can be compared easily with each other to compare different states of the system. For similar states, additional strategies can be removed to decrease the complexity of learning in the environment. The curse of dimensionality encompasses the exponential growth of the discrete state-action space in the number of state and action variables (dimensions). The complexity of MARL is exponential also in the number of agents because each agent adds its variables to the joint state-action space. This is while the time complexity of the CLA based approach is exponential in the number of neighbor cells, not all the cells.

8.3.3 Complexity of Learning Compatible Point in ICLA Game

In Sect. 7.3.1 from Chap. 7, existence of a compatible point is proved by lemma 7-1 in the ICLA game and then a new local rule presented. It is shown by Theorem 7.1 proved that using the proposed local rule of Fig. 7.2, ICLA converges to a compatible point. As described before, complexity analysis of the proposed local rule illustrated that the proposed local rule in terms of space complexity is linear in the number of learning automata (size of ICLA), polynomial in the number of actions of learning automata, but exponential in the number of neighbors of learning automata. Since reinforcement learning algorithms, like Q-learning, compute values for each joint-action or state-action pair, this leads to exponential computational complexity. Also, the computational complexity of the proposed approach is polynomial in the number of actions of learning automata but exponential in the number of neighbors of learning automata.

8.4 Experiments

In this section, we first compare the result of an ICLA-based approach using the proposed local rule with a well-known MARL approach, called Nash-Q, and then to illustrate the superiority and effectiveness of the proposed local on the convergence of ICLA to a compatible point, some numerical experiments are conducted.

8.4.1 CLA Versus Nash-Q

In this experiment, we aim to compare ICLA based approaches for learning equilibrium point with Nash-Q (Hu and Wellman 2003), which is a well-known MARL approach. The first capability of a simple Q-learner versus a single LA is evaluated in a learning process. For this purpose, a simple Q-learner is employed for the learning of M_{LT}. The M_{LT} using a single LA is defined as Eq. (8.3) and using a single Q-learner as Eq. (8.4).

$$M_{LT} = \sum_{i=1}^{num\_of\_actions} p_i \times associated\_value(action(i)) \tag{8.3}$$

$$M_{LT} = \sum_{i=1}^{num\_of\_actions} Q_i \times associated\_value(action(i)) \tag{8.4}$$

Figure 8.2a illustrates the learned values of M_{LT} for different user types using a LA and Q-learner. As illustrated in this figure, the output of both learners is similar. Now capability of a group of Q-learners (in the form of a multi-agent system) and a group of learning automata (in the form of a cellular learning automat) are evaluated in the following experiments. Hu and Wellman (2003) has extended Q-learning to a non-cooperative multi-agent context, using the framework of general-sum stochastic games. We aim to compare the result of Nash-Q versus ICLA using the proposed local rule. For this evaluation, graphs of Fig. 8.1 (a, c, e, and g) are used. In each node of the graphs, there is an agent in the Nash-Q approach and a cell in ICLA. The edges of graphs illustrate the relation between agents in Nash-Q and neighboring relations in ICLA. Because the Nash-Q approach cannot be applied to incomplete graphs. Therefore complete forms of these graphs are used for the Nash-Q approach. Figure 8.1 (b, d, f, and h) illustrate the complete forms. In following the graphs of Fig. 8.1 are called G_a, G_b, G_c, G_d, G_e, G_f, G_g and G_h.

Here response of the environment is generated using a payoff matrix. This matrix contains the probability of rewarding the selected actions of agents in Nash-Q and learning automata an ICLA. For example, when selected joint actions are (Low, High, High) and the corresponding entry in the payoff matrix is (0.91, 0.88, 0.91) (see, e.g., Table 8.2) then the reinforcement signals to $Agent_A$ (LA_A), $Agent_B$ (LA_B) and $Agent_C$ (LA_C) is reward with probability 0.91, 0.88, and 0.91, respectively. Tables 8.2, 8.3, 8.4 and 8.5 show the payoff matrices for the graphs G_a, G_b, G_c and G_d with two and three actions. For the other graphs and these graphs with more actions, the payoff matrix is too large. Therefore it must be generated using a computer program (Fig. 8.2).

Figure 8.3a shows convergence rate with different amounts of reward parameter (an in L_{R-I} learning algorithms and β in Nash-Q) in G_a and G_b with two actions. The results are obtained from 1000 times the running of the learning process. Figures 8.3b, c show the same diagrams G_a and G_b with three and five actions. As illustrated in

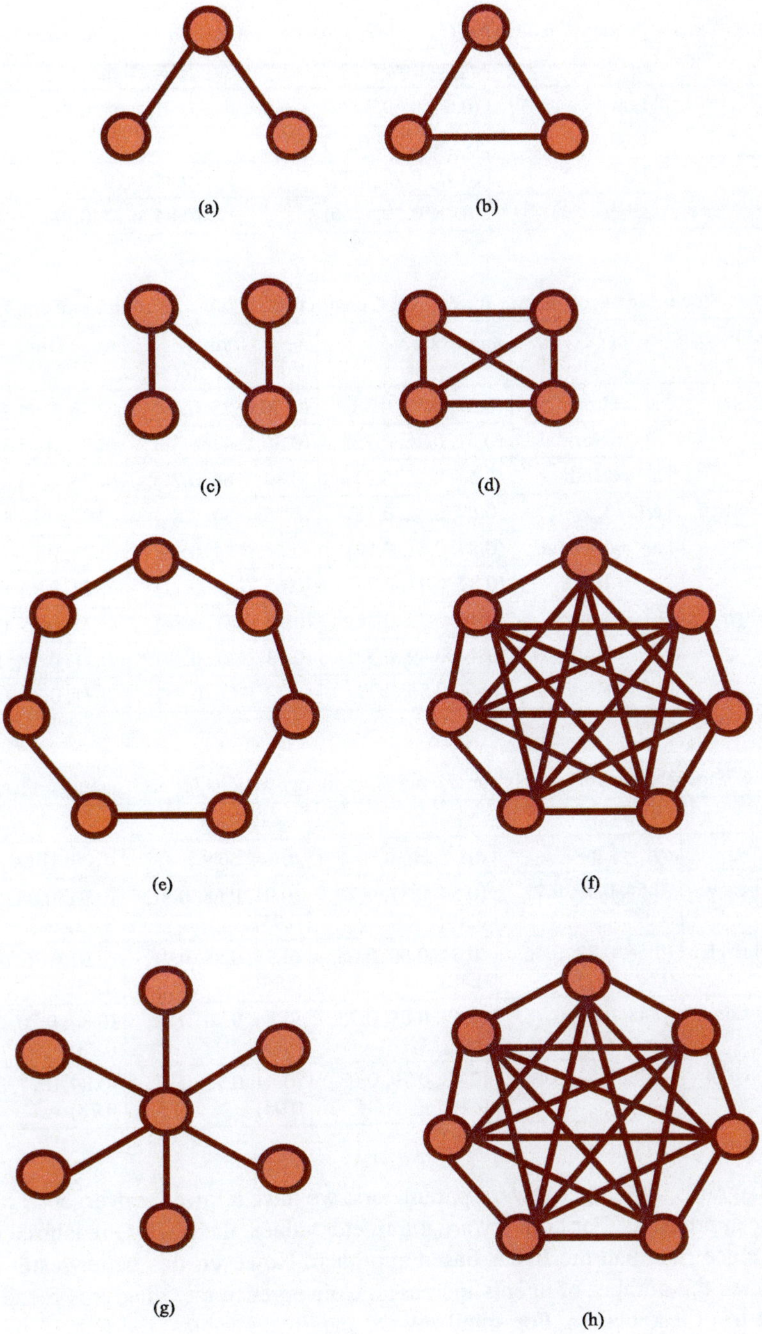

Fig. 8.1 Visual graphs of **a** ICLA-a **b** ICLA-b **c** ICLA-c **d** ICLA-d **e** ICLA-e **f** ICLA-f **g** ICLA-g **h** ICLA-h

Table 8.2 Stochastic payoff matrix for G_a and G_b with two actions for each agent and LA

a_A	a_C	a_B = Low	a_B = High
Low	Low	(0.54, 0.59, 0.72)	(0.91, 0.68, 0.81)
	High	(0.54, 0.82, 0.66)	**(0.91,0.88, 0.91)**
High	Low	(0.48, 0.75, 0.72)	(0.84, 0.70, 0.81)
	High	(0.48, 0.59, 0.66)	(0.84, 0.77, 0.91)

Table 8.3 Stochastic payoff matrix for G_a and G_b with three actions for each agent and LA

Reward Probability for (a_A, a_B, a_C)		a_B = Low	a_B = Normal	a_B = High
a_A = Low	a_C = Low	(0.51, 0.81, 0.38)	(0.36, 0.25, 0.84)	(0.53, 0.44, 0.32)
	a_C = Normal	(0.51, 0.76, 0.52)	(0.36, 0.47, 0.39)	(0.53, 0.25, 0.57)
	a_C = High	(0.51, 0.45, 0.77)	(0.36, 0.66, 0.71)	(0.53, 0.36, 0.90)
a_A = Normal	a_C = Low	(0.87, 0.32, 0.38)	(0.65, 0.90, 0.84)	(0.92, 0.51, 0.32)
	a_C = Normal	(0.87, 0.61, 0.52)	(0.65, 0.54, 0.39)	(0.92, 0.75, 0.57)
	a_C = High	(0.87, 0.71, 0.77)	(0.65, 0.82, 0.71)	**(0.92, 0.95, 0.90)**
a_A = High	a_C = Low	(0.44, 0.37, 0.38)	(0.73, 0.61, 0.84)	(0.71, 0.65, 0.32)
	a_C = Normal	(0.44, 0.49, 0.52)	(0.73, 0.37, 0.39)	(0.71, 0.51, 0.57)
	a_C = High	(0.44, 0.56, 0.77)	(0.73, 0.80, 0.71)	(0.71, 0.73, 0.90)

Table 8.4 Stochastic payoff matrix for G_c and G_d with two actions for each agent and LA

		a_B = Low		a_B = High	
a_A	a_C	a_D = Low	a_D = High	a_D = Low	a_D = High
Low	Low	(0.54, 0.59, 0.72, 0.62)	(0.54, 0.71, 0.75, 0.48)	(0.91, 0.68, 0.81, 0.95)	(0.91, 0.68, 0.82, 0.78)
	High	(0.54, 0.82, 0.66, 0.62)	(0.54, 0.69, 0.68, 0.48)	**(0.91, 0.88, 0.91, 0.95)**	(0.91,0.76, 0.89, 0.78)
High	Low	(0.48, 0.75, 0.72, 0.62)	(0.48, 0.61, 0.75, 0.48)	(0.84, 0.70, 0.81, 0.95)	(0.84, 0.70, 0.82, 0.78)
	High	(0.48, 0.59, 0.66, 0.62)	(0.48, 0.79, 0.68, 0.48)	(0.84, 0.77, 0.91, 0.95)	(0.84, 0.77, 0.89, 0.78)

these diagrams, by bigger reward parameters, we have a lower convergence rate by both the approaches. For high reward parameter values, the Nash-Q reaches a better convergence rate than the ICLA-based approach. However, this better result is not valid when the number of agents increases. Convergence rate diagrams for G_g and G_h confirm these claims. For small reward parameter values, the ICLA approach outperforms the Nash-Q. Although the better convergence rate is obtained using the ICLA-based approach, by increasing the reward parameter, the drop-in convergence rate using the ICLA-based approach is more sensible than drop using the Nash-Q

Table 8.5 Stochastic payoff matrix for G_c and G_d with three actions for each agent and LA

Reward Probability for (a_A, a_B, a_C, a_D)			a_B = Low	a_B = Normal	a_B = High
a_D = Low	a_A = Low	a_C = Low	(0.51, 0.81, 0.38, 0.38)	(0.36, 0.25, 0.84, 0.38)	(0.53, 0.44, 0.32, 0.38)
		a_C = Normal	(0.51, 0.76, 0.52, 0.42)	(0.36, 0.47, 0.39, 0.42)	(0.53, 0.25, 0.57, 0.42)
		a_C = High	(0.51, 0.45, 0.77, 0.12)	(0.36, 0.66, 0.71, 0.12)	(0.53, 0.36, 0.90, 0.12)
	a_A = Normal	a_C = Low	(0.87, 0.32, 0.38, 0.38)	(0.65, 0.90, 0.84, 0.38)	(0.92, 0.51, 0.32, 0.38)
		a_C = Normal	(0.87, 0.61, 0.52, 0.42)	(0.65, 0.54, 0.39, 0.42)	(0.92, 0.75, 0.57, 0.42)
		a_C = High	(0.87, 0.71, 0.77, 0.12)	(0.65, 0.82, 0.71, 0.12)	(0.92, 0.95, 0.90, 0.12)
	a_A = High	a_C = Low	(0.44, 0.37, 0.38, 0.38)	(0.73, 0.61, 0.84, 0.38)	(0.71, 0.65, 0.32, 0.38)
		a_C = Normal	(0.44, 0.49, 0.52, 0.42)	(0.73, 0.37, 0.39, 0.42)	(0.71, 0.51, 0.57, 0.42
		a_C = High	(0.44, 0.56, 0.77, 0.12)	(0.73, 0.80, 0.71, 0.12)	(0.71, 0.73, 0.90, 0.12)
a_D = Normal	a_A = Low	a_C = Low	(0.51, 0.81, 0.38, 0.54)	(0.36, 0.25, 0.84, 0.54)	(0.53, 0.44, 0.32, 0.54)
		a_C = Normal	(0.51, 0.76, 0.52, 0.47)	(0.36, 0.47, 0.39, 0.47)	(0.53, 0.25, 0.57, 0.47)
		a_C = High	(0.51, 0.45, 0.77, 0.78)	(0.36, 0.66, 0.71, 0.78)	(0.53, 0.36, 0.90, 0.78)
	a_A = Normal	a_C = Low	(0.87, 0.32, 0.38, 0.54)	(0.65, 0.90, 0.84, 0.54)	(0.92, 0.51, 0.32, 0.54)
		a_C = Normal	(0.87, 0.61, 0.52, 0.47)	(0.65, 0.54, 0.39, 0.47)	(0.92, 0.75, 0.57, 0.47)
		a_C = High	(0.87, 0.71, 0.77, 0.78)	(0.65, 0.82, 0.71, 0.78)	**(0.92, 0.95, 0.90, 0.78)**
	a_A = High	a_C = Low	(0.44, 0.37, 0.38, 0.54)	(0.73, 0.61, 0.84, 0.54)	(0.71, 0.65, 0.32, 0.54)
		a_C = Normal	(0.44, 0.49, 0.52, 0.47)	(0.73, 0.37, 0.39, 0.47)	(0.71, 0.51, 0.57, 0.47)
		a_C = High	(0.44, 0.56, 0.77, 0.78)	(0.73, 0.80, 0.71, 0.78)	(0.71, 0.73, 0.90, 0.78)
a_D = High	a_A = Low	a_C = Low	(0.51, 0.81, 0.38, 0.29)	(0.36, 0.25, 0.84, 0.29)	(0.53, 0.44, 0.32, 0.29)
		a_C = Normal	(0.51, 0.76, 0.52, 0.62)	(0.36, 0.47, 0.39, 0.62)	(0.53, 0.25, 0.57, 0.62)

(continued)

Table 8.5 (continued)

Reward Probability for (a_A, a_B, a_C, a_D)		a_B = Low	a_B = Normal	a_B = High
	a_C = High	(0.51, 0.45, 0.77, 0.41)	(0.36, 0.66, 0.71, 0.41)	(0.53, 0.36, 0.90, 0.41)
a_A = Normal	a_C = Low	(0.87, 0.32, 0.38, 0.29)	(0.65, 0.90, 0.84, 0.29)	(0.92, 0.51, 0.32, 0.29)
	a_C = Normal	(0.87, 0.61, 0.52, 0.62)	(0.65, 0.54, 0.39, 0.62)	(0.92, 0.75, 0.57, 0.62)
	a_C = High	(0.87, 0.71, 0.77, 0.41)	(0.65, 0.82, 0.71, 0.41)	(0.92, 0.95, 0.90, 0.41)
a_A = High	a_C = Low	(0.44, 0.37, 0.38, 0.29)	(0.73, 0.61, 0.84, 0.29)	(0.71, 0.65, 0.32, 0.29)
	a_C = Normal	(0.44, 0.49, 0.52, 0.62)	(0.73, 0.37, 0.39, 0.62)	(0.71, 0.51, 0.57, 0.62)
	a_C = High	(0.44, 0.56, 0.77, 0.41)	(0.73, 0.80, 0.71, 0.41)	(0.71, 0.73, 0.90, 0.41)

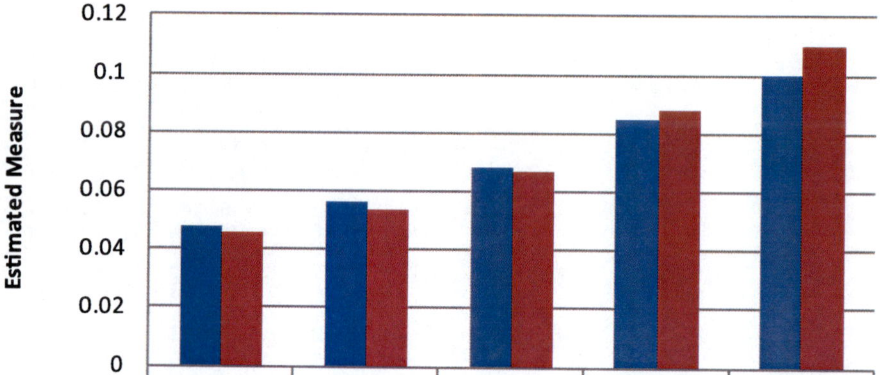

	User Type 1	User Type 2	User Type 3	User Type 4	User Type 5
LA	0.048	0.056	0.068	0.085	0.1
Q-Learner	0.046	0.054	0.067	0.088	0.11

Fig. 8.2 The learned values of M_{LT} for different user type using a LA and Q-learner

approach. This is for the reason that changes in ICLA configuration will not be negligible in two successive rounds using high reward parameter values. This causes drop in convergence rate because the algorithm 2 (see Fig. 8.2) assumes the best response in iteration t (see $\underline{p}_i^{br}$ in algorithm 2) is too close to the best response in iteration t + 1. Using Table 8.2 as the payoff matrix to generate reinforcement signals, learners converge to (Low, High, High), which is a Nash equilibrium point.

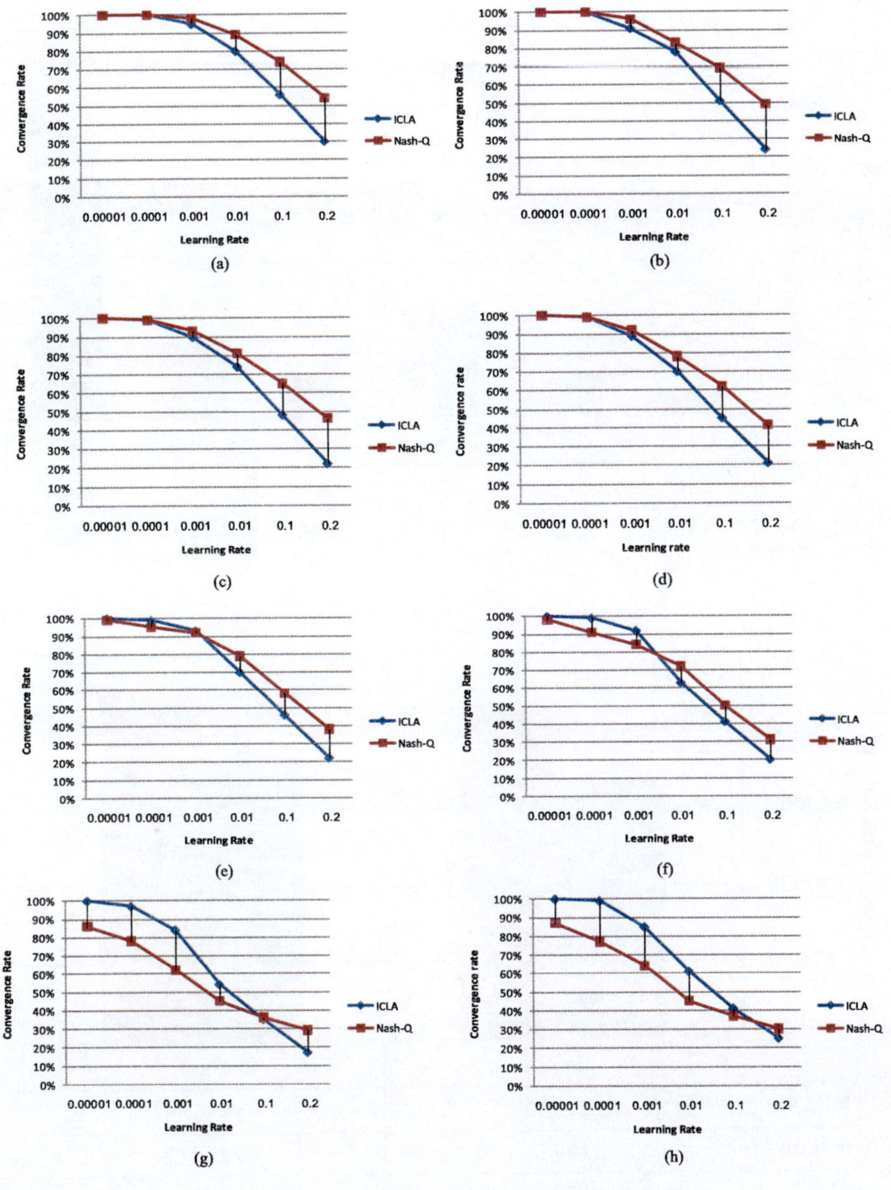

Fig. 8.3 Convergence Rate of ICLA and Nash-Q using different learning rate

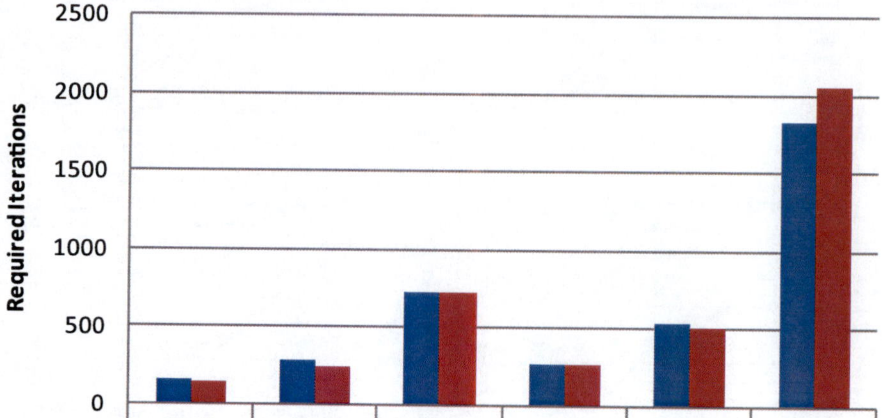

	Ga-2	Ga-3	Ga-5	Gc-2	Gc-3	Gc-5
ICLA	156	283	720	274	541	1824
Nash-Q	140	242	725	269	512	2054

(a)

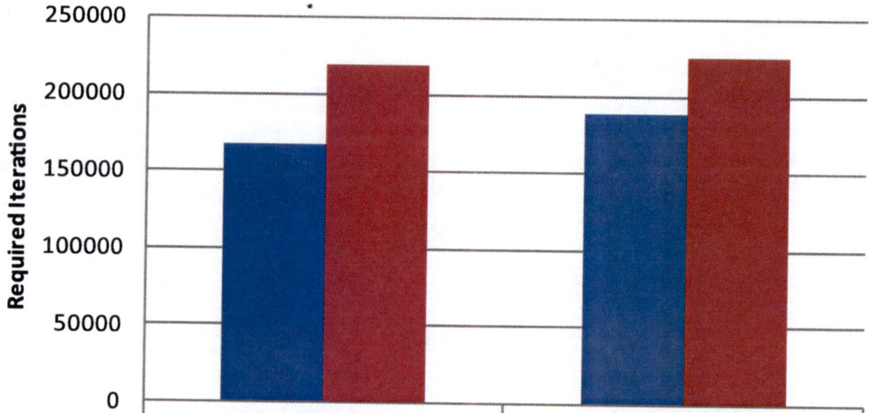

	Ge-5	Gg-5
ICLA	167589	189638
Nash-Q	219754	225412

(b)

Fig. 8.4 Comparing the numbers of required iterations

Figure 8.4a compare the number of required iterations for convergence in G_a and G_c with two, three, and five actions using the ICLA based approach with Nash-Q. Figure 8.4b makes the same comparison in G_e and G_g with five actions. These diagrams illustrate that there is no notable difference in the number of required iterations for convergence between the ICLA-based approach and Nash-Q. Since the number of update actions in each iteration differs for ICLA based approach and Nash-Q, therefore, the number of required update actions is also compared by Fig. 8.5a, b. This diagram shows the superiority of the ICLA-based approach over Nash-Q in terms of speed. These diagrams illustrate that the superiority of the ICLA-based approach over Nash-Q depends on the maximum degree of the graph. For complete graphs, there is no notable difference in convergence speed, but for graphs with a low maximum degree ICLA-based approach has better performance. This demonstrates that for the applications that can be modeled using the graphs, the ICLA-based approach is a better choice for learning an equilibrium point.

8.4.2 CLA and Convergence to Equilibrium Points

In this section, to illustrate the superiority and effectiveness of the proposed local on the convergence of ICLA to a compatible point, some numerical experiments are conducted. The results are compared with the results of the approach proposed in (Sastry et al. 1994) for convergence to a Nash point. To apply the proposed approach of (Sastry et al. 1994), the neighboring relation among learning automatons should be extended to include neighboring of all learning automata with each other. The ICLAs in Fig. 8.1b, d, f, and h illustrate the extended neighboring relation for ICLAs of Fig. 8.1a, c, e and g respectively. For clarity, first, it is explained how input reinforcement signals, which are used by the proposed local rule, are generated using the utility functions of players (Learning Automata). In a stochastic game, the expected utility of each player is determined according to the selected actions of all players. In the conducted experiments, for simplicity purposes, a simple utility function form as Eq. (8.5) is used to evaluate the proposed approach. The introduced utility function here is just for test and evaluation purposes. In real applications, utility functions vary depending on the application-dependent configurations. In Eq. (8.5), w_{ij}^k determines the effect of choosing action k by LA_j on the utility of LA_i. The amount of w_{ij}^k can be positive or negative. The value of b is one if the selected action of LA_j is action k otherwise b = 0. N_{ij} is one if LA_j and LA_i are neighbors; otherwise, $N_{ij} = 0$. u_0 is a constant utility value.

$$u_i = u_0 + \sum_{\substack{j=1 \\ j \neq i}}^{n} \sum_{k=1}^{5} b \times w_{ij}^k \times N_{ij} \tag{8.5}$$

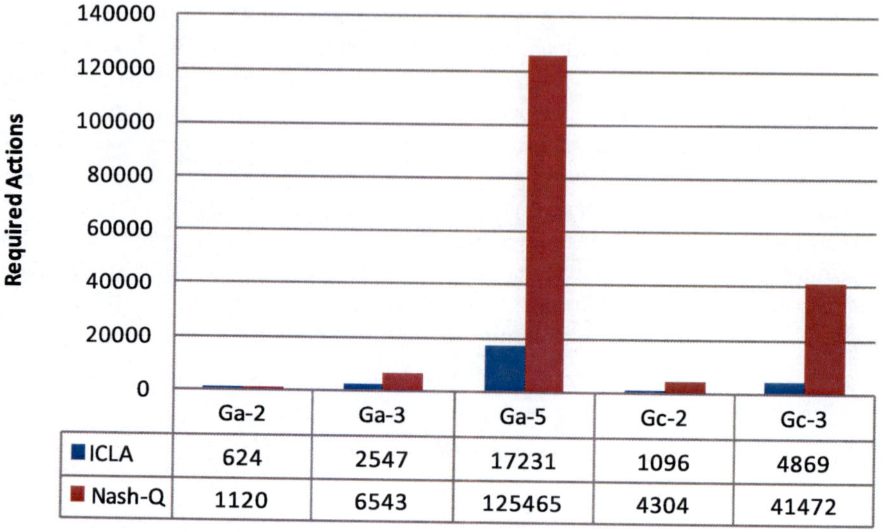

(a)

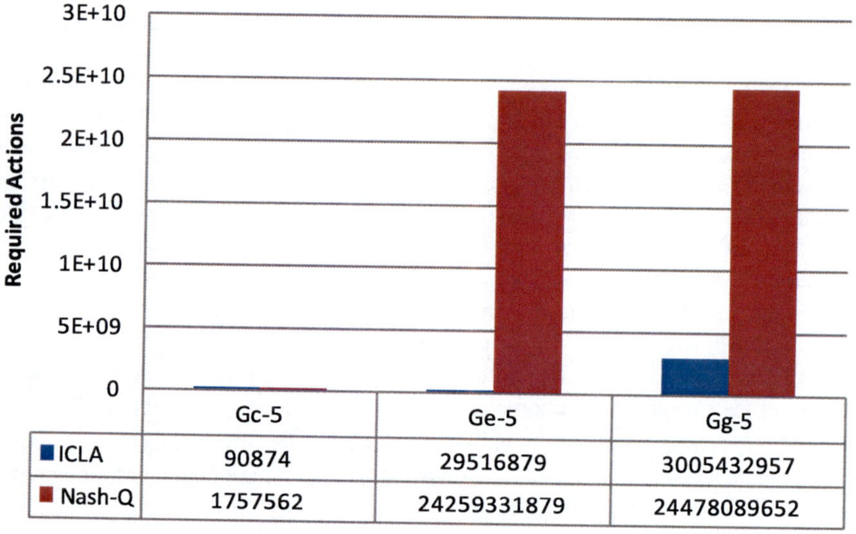

(b)

Fig. 8.5 Comparing the numbers of required actions (updates)

To have a stochastic input reinforcement signal, the maximum possible value of $u_i(u_{\max})$ is calculated as well. The input reinforcement signal will be 1 with probability $u_i/u_{\max}$ and 0 with probability $1-(u_i/u_{\max})$.

Now the results obtained by applying the proposed local rule on ICLAs of Fig. 8.1a, c, e, and g is compared by the results obtained by applying the approach of

(Sastry et al. 1994) on ICLAs of Fig. 8.1b, d, f, and h. Figure 8.6a illustrates the evolution of the probability vector of the learning automaton, which is located on cell B of ICLA-a. Figure 8.6b illustrates the same diagram for ICLA-b. As shown in diagrams of Fig. 8.6a, b, almost the same number of iterations are required for convergence to a compatible point in both cases. Figure 8.6c, d, which illustrate the evolution of

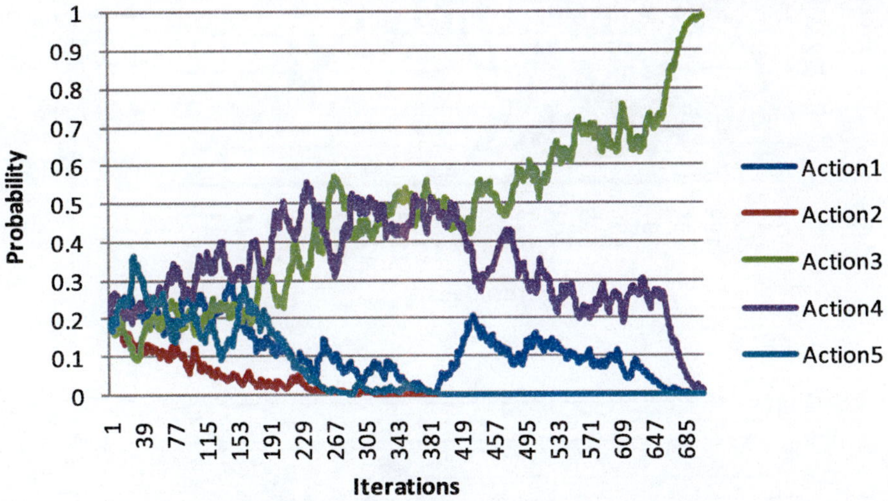

(a)

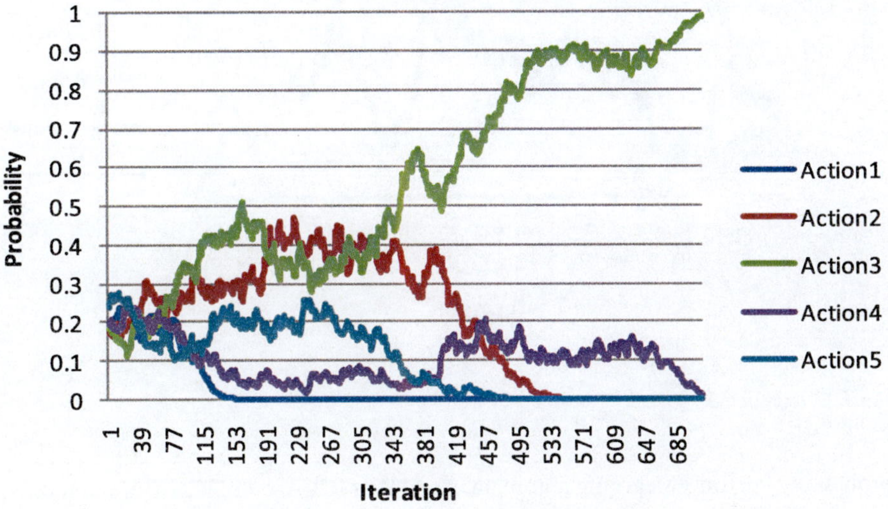

(b)

Fig. 8.6 The evolution of learning automaton probability vector in **a** ICLA-a **b** ICLA-b **c** ICLA-c **d** ICLA-d

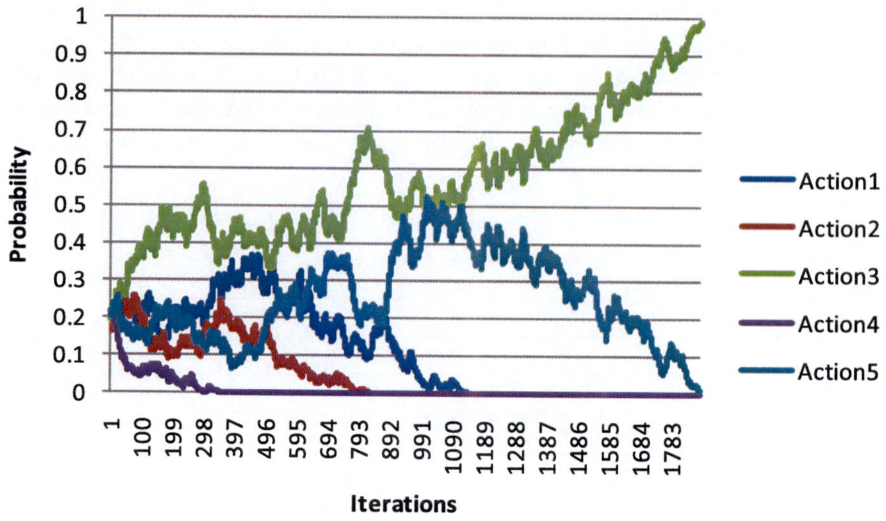

(c)

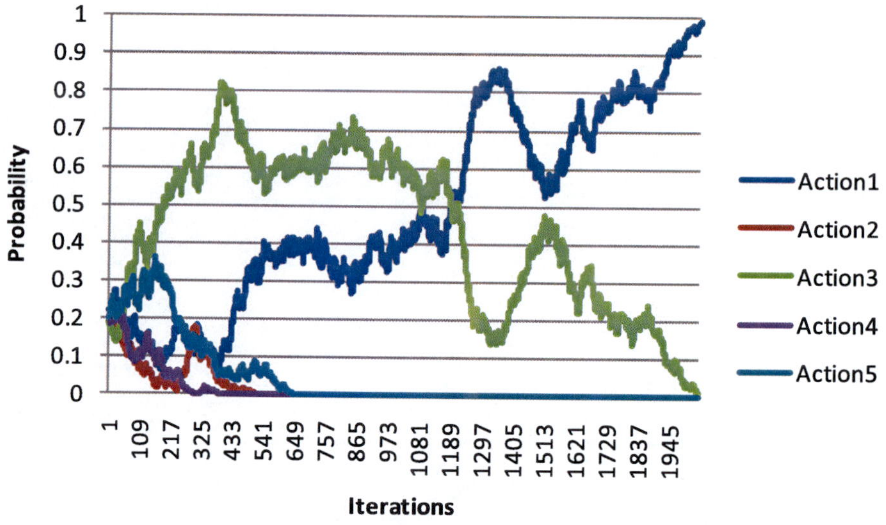

(d)

Fig. 8.6 (continued)

probability vectors of learning automata in cell B of ICLA-c and ICLA-d, support the idea that there is no significant difference between the number of required iterations for convergence using the proposed approach of Chap. 6 and the approach of (Sastry et al. 1994). Comparing the results for ICLA-e versus ICLA-f and ICLA-g versus ICLA-h confirm this idea as well. Although, almost the same numbers of iterations are required for convergence by both the approaches, the number of updates per iteration in two approaches has significant differences. Since running time directly depends on the number of updates. Therefore the proposed approach has a significant advantage in terms of speed over the approach of (Sastry et al. 1994). Figure 8.7a, c illustrates the number of required iterations for convergence in ICLA-a, ICLA-b, ICLA-c, and ICLA-d. In Fig. 8.7b, d, the number of required actions for convergence is illustrated. These diagrams show the superiority of the proposed approach over the approach of (Sastry et al. 1994) in terms of speed. Another important factor in the speed of convergence is the learning rate of the learning algorithms. As illustrated in Fig. 8.8a, b, a smaller value for the learning rate causes more iterations to be needed for convergence. However, the size of the learning rate parameter has a reverse effect on the convergence rate. With smaller values, a higher convergence rate can be achieved. Diagrams of Fig. 8.9 illustrate the effect of learning rate size on convergence rate. Also, these diagrams show that using the proposed approach higher convergence rate can be obtained in comparison to the approach of (Sastry et al. 1994). To study the convergence of an ICLA to a compatible point when it starts learning from different initial configurations, two numerical experiments are conducted using ICLA-a and ICLA-c. As illustrated in Fig. 8.10a, b, the convergence of ICLA-a does not depend on the preliminary probability vector of learning automata. Diagrams of Fig. 8.11a, b show the same result for ICLA-c. By starting from various configurations, the probability of an action that should be selected to reach a compatible point grows up to 1 regardless of the initial probability of that action (see the diagrams of part (a)). In the meantime, the probabilities of the other actions decrease until they reach zero (see the diagrams of part (b)). Different initial probability vectors just influence on the number of required iterations for convergence. Using the proposed local rule:

- ICLA-a has converged to ($LA_A = action3$, $LA_B = action2$, $LA_C = action1$)
- ICLA-c has converged to ($LA_A = action2$, $LA_B = action3$, $LA_C = action3$, $LA_D = action3$)
- ICLA-e has converged to ($LA_A = action5$, $LA_B = action5$, $LA_C = action1$, $LA_D = action4$, $LA_E = action1$, $LA_F = action4$, $LA_G = action2$)
- ICLA-g has converged to ($LA_A = action5$, $LA_B = action2$, $LA_C = action2$, $LA_D = action4$, $LA_E = action1$, $LA_F = action2$, $LA_G = action2$)

To verify the point or configuration, which ICLA has converged to it, is compatible or not, the expected utility of each learning automaton of ICLA should be maximum value among all points, which can be obtained by the unilateral deviation of that learning automaton. For example, in ICLA-a, ($LA_A = action3$, $LA_B = action2$, $LA_C = action1$) is a compatible point, if:

- The expected utility of LA_A in this point is the maximum utility among all the expected utilities of LA_A in ($LA_A = *$, $LA_B = action2$, $LA_C = action1$) points.
- The expected utility of LA_B in this point is the maximum value among all the expected utilities of LA_B in ($LA_A = action3$, $LA_B = *$, $LA_C = action1$) points.

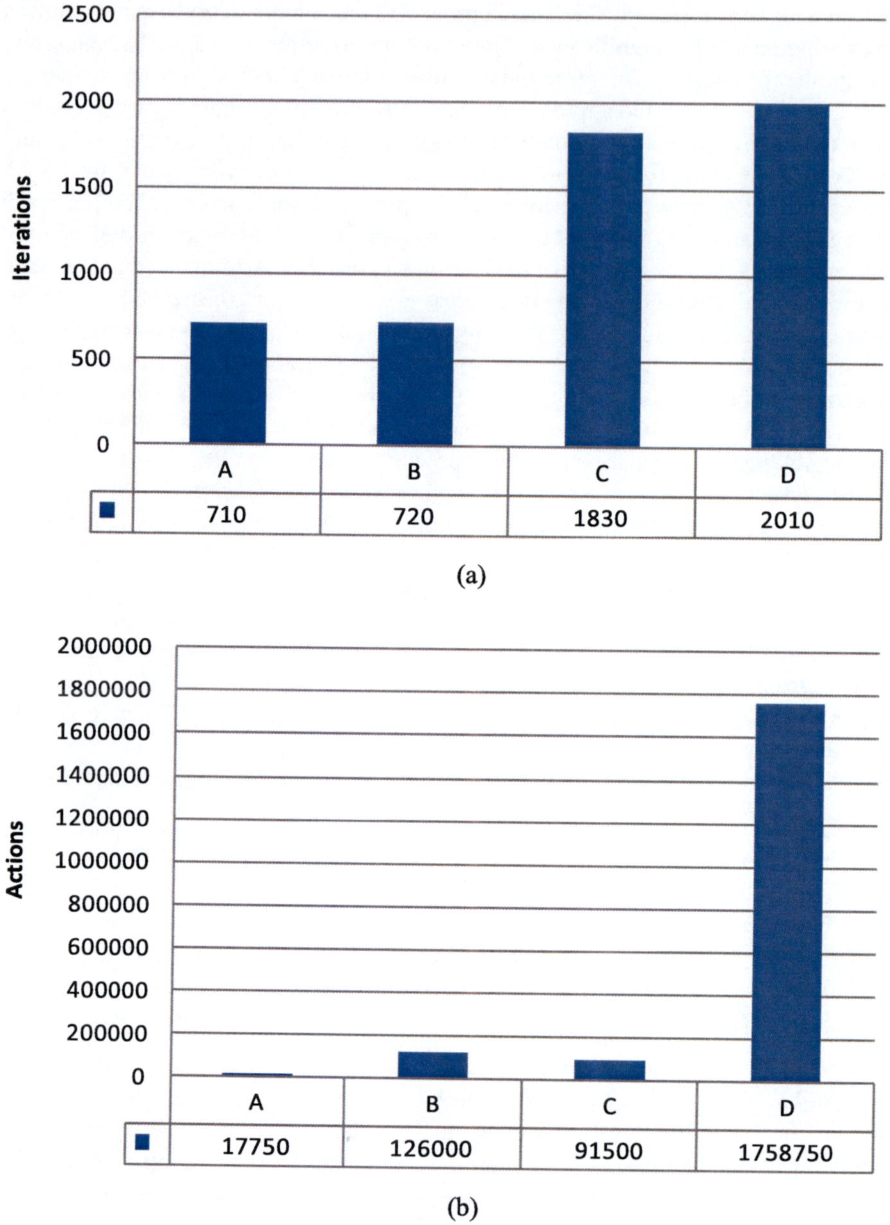

Fig. 8.7 Comparing the numbers of iterations and action (updates)

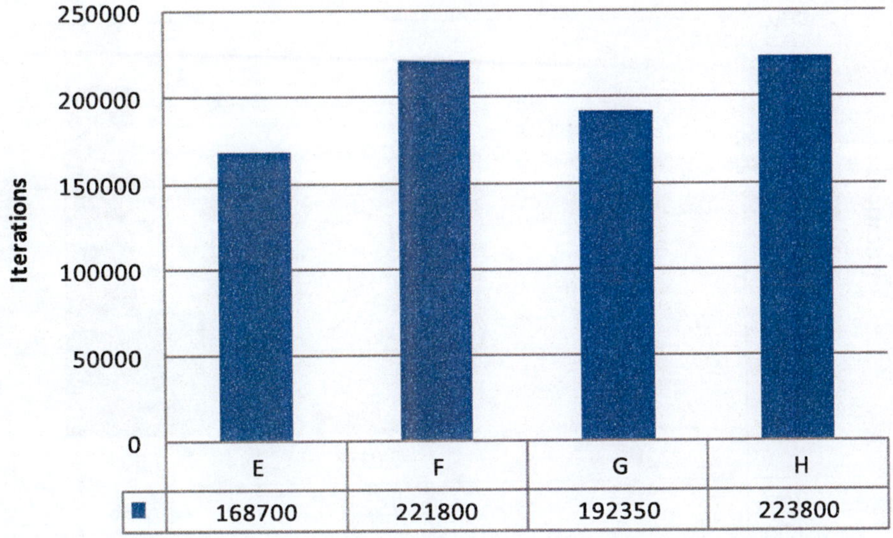

(c)

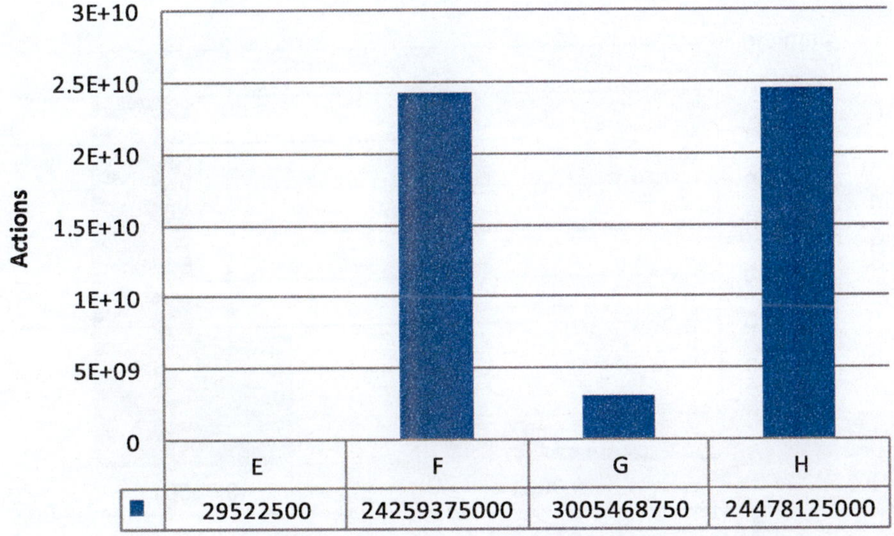

(d)

Fig. 8.7 (continued)

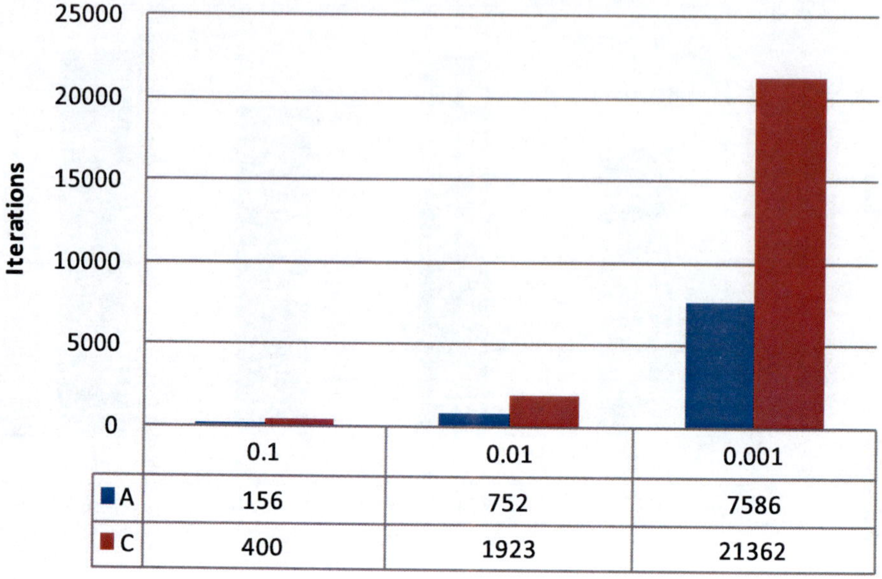

(a)

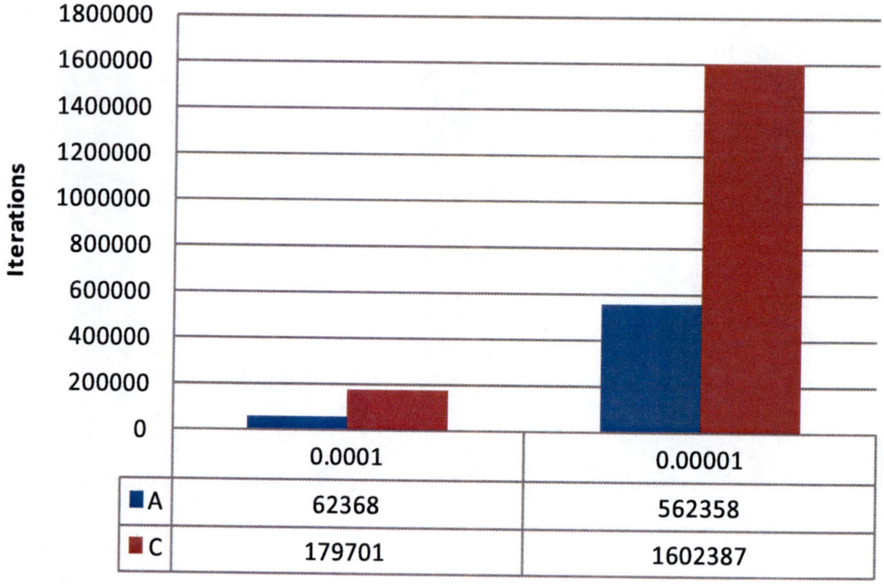

(b)

Fig. 8.8 Number of required iterations based on learning rate

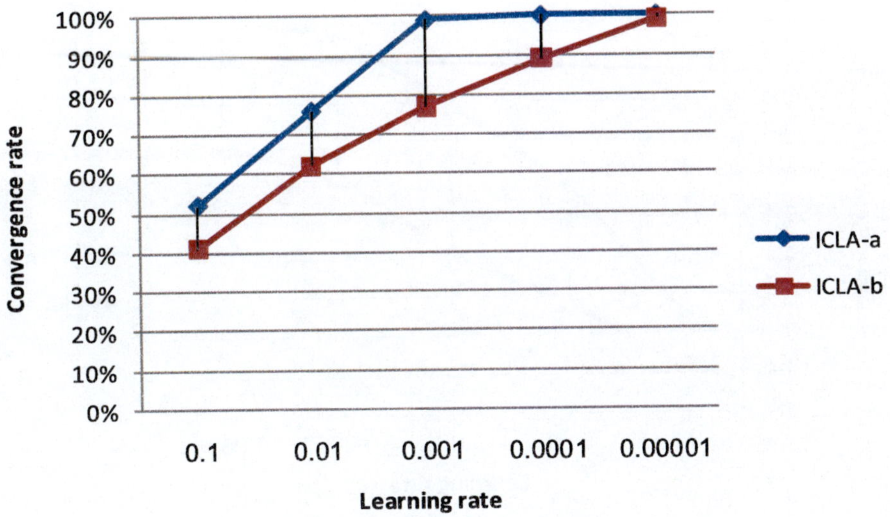

(a)

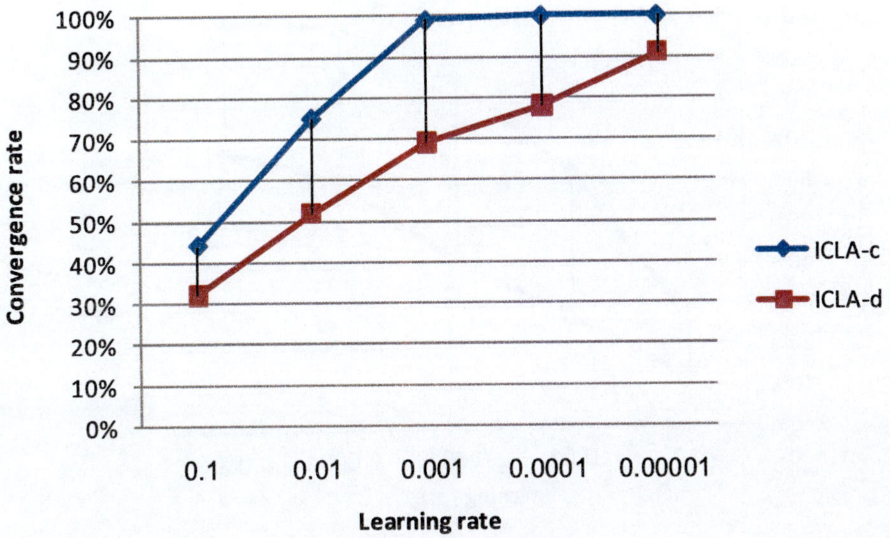

(b)

Fig. 8.9 Effect of learning rate size on the convergence rate

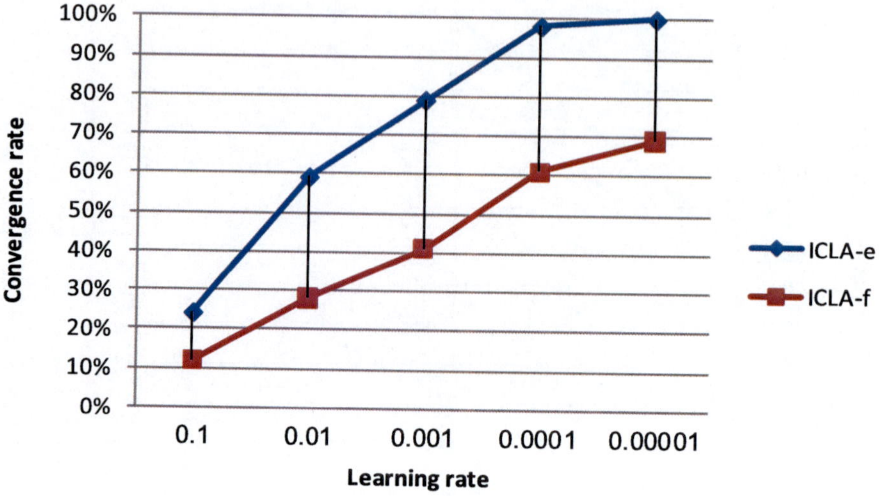

(c)

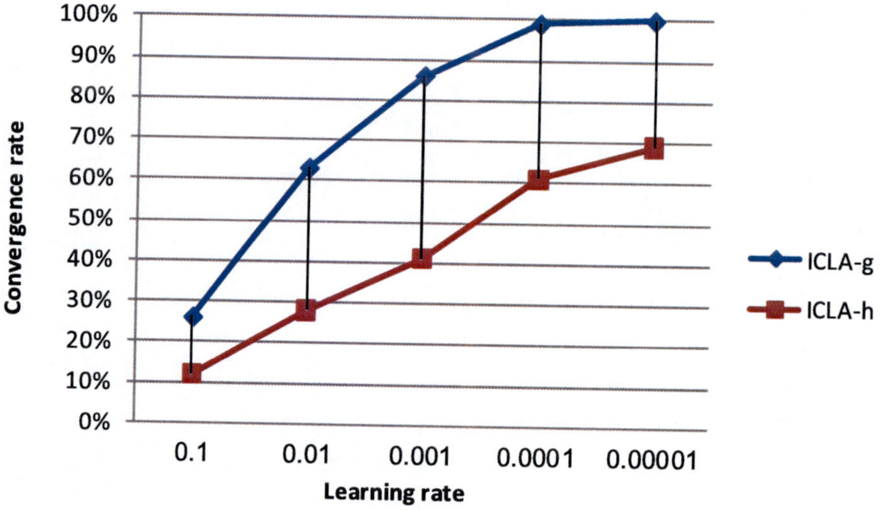

(d)

Fig. 8.9 (continued)

- The expected utility of LA_C in this point is the maximum value among all the expected utilities of LA_C in ($LA_A = action3$, $LA_B = action2$, $LA_C = *$) points.

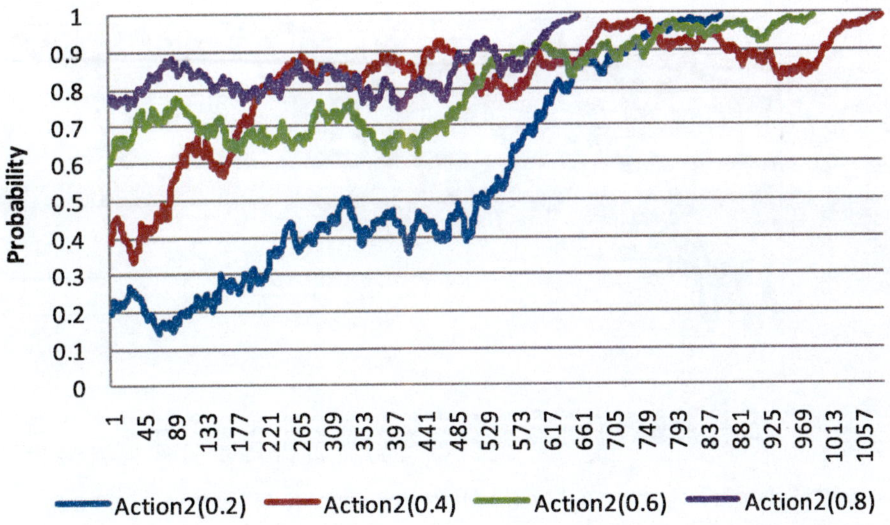

(a)

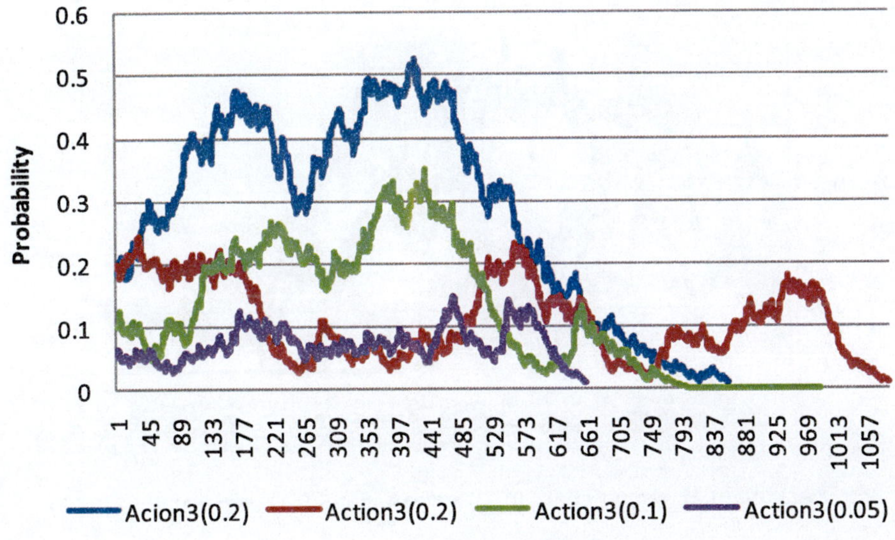

(b)

Fig. 8.10 Convergence of ICLA to a compatible point by starting from different initial configurations

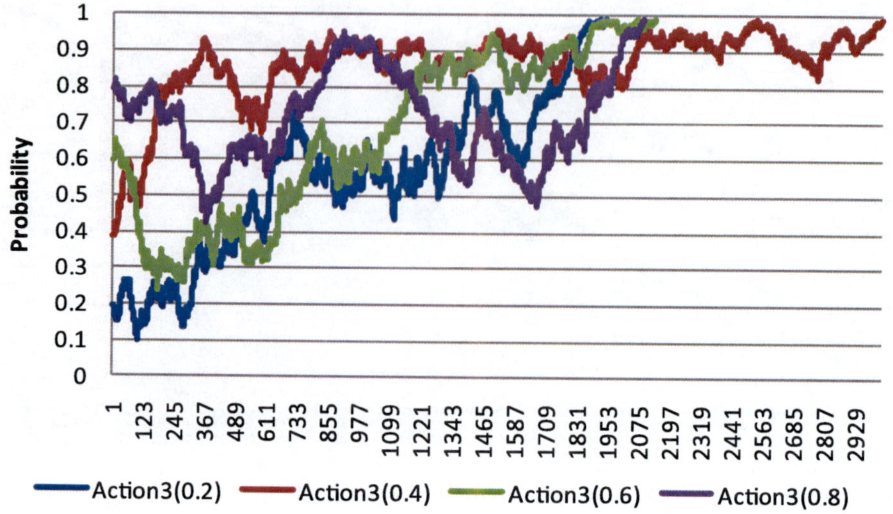

(a)

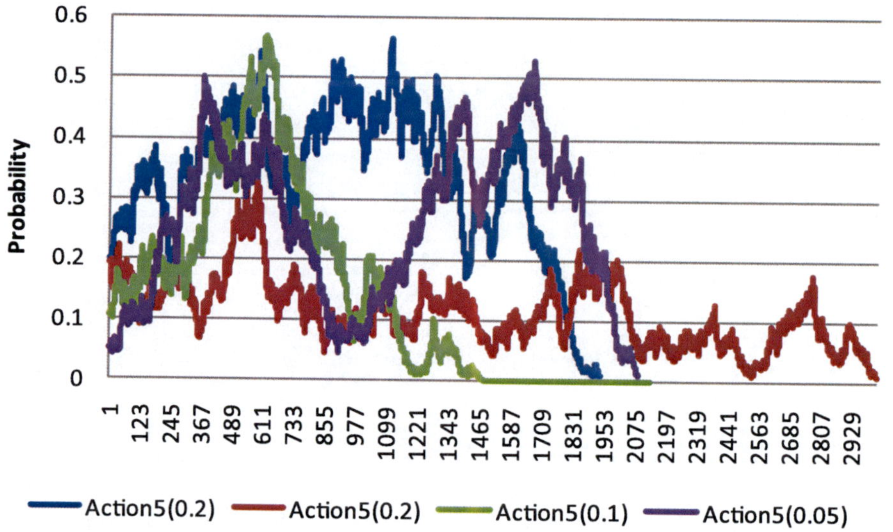

(b)

Fig. 8.11 Convergence of ICLA to a compatible point by starting from different initial configurations

Figures 8.12, 8.13, 8.14 and 8.15 compare the utility of each learning automaton at the convergence point and all the points which can be reached by unilateral deviation. As illustrated in these figures, the results indicate that the convergence points are compatible points.

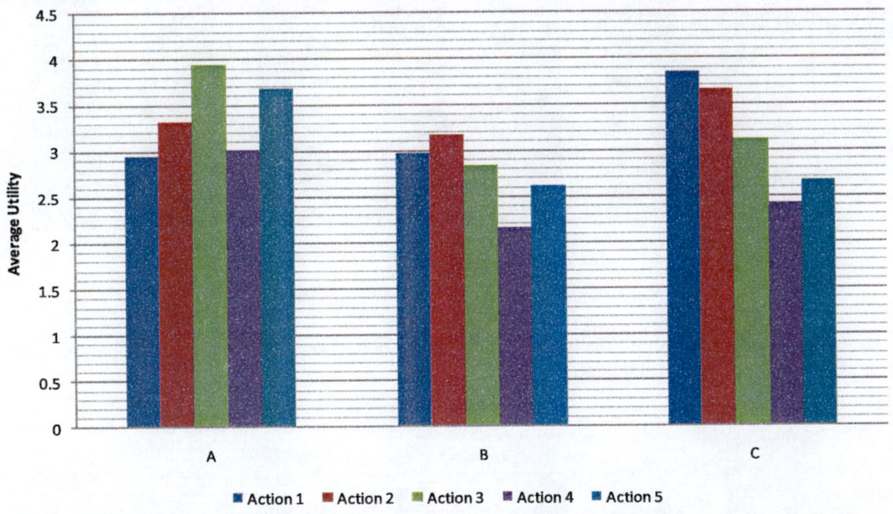

Fig. 8.12 The utility of each learning automaton at the convergence point and all the points which can be reached by unilateral deviation in ICLA-a

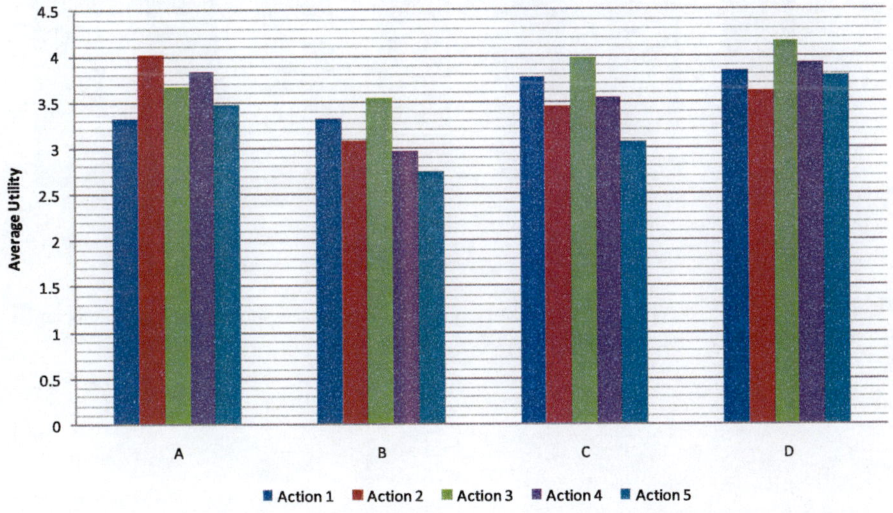

Fig. 8.13 The utility of each learning automaton at the convergence point and all the points which can be reached by unilateral deviation in ICLA-c

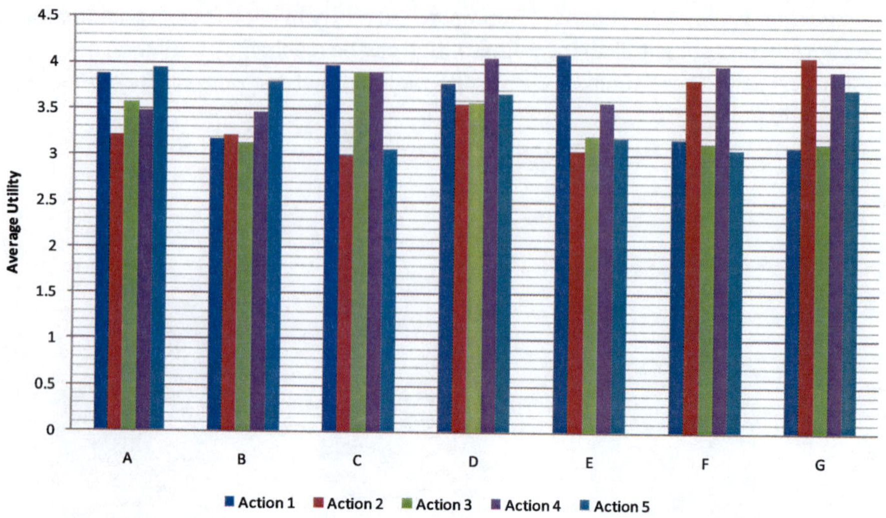

Fig. 8.14 The utility of each learning automaton at the convergence point and all the points which can be reached by unilateral deviation in ICLA-e

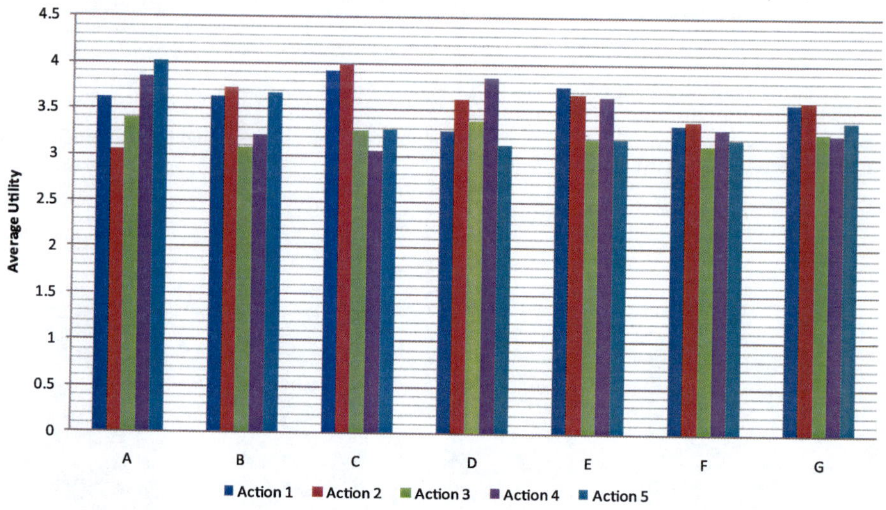

Fig. 8.15 The utility of each learning automaton at the convergence point and all the points which can be reached by unilateral deviation in ICLA-g

8.5 Conclusion

In this chapter, we employed irregular cellular learning automaton (ICLA) in the field of MARL and especially the learning of equilibrium points. We extended the theory of ICLA by presenting a new local rule which makes ICLA capable of convergence to Nash-equivalent points, which are called compatible points. We discovered that the ICLA model matches perfectly with the MARL paradigm. From the MARL point of view, this means that the theory of ICLA can serve as a theoretical analysis tool for the MARL algorithms. Increasing literature in the field of MARL is concerned with the theoretical convergence of RL-based approaches towards Nash equilibrium. Convergence toward Nash Equilibrium points in multi-agent systems with autonomous agents is a very important issue because it provides efficient solutions for various problems in such systems. RL, or more precisely, the technique of Q-learning, has already been used for learning in stochastic games; however, the proposed solutions are limited by some conditions. At present, great attention is going to the Markov game model for learning in MAS. Although, these approaches cover the MAS problems in which strategy of all agents or players has direct effects on each other. This is although there many decentralized and distributed systems in the modern world have just local interactions. Elements of such systems have no direct effect on the other elements in the system except the local ones. Learning of an equilibrium point equivalent to NE point in such systems can also provide valuable solutions for challenges and problems in such systems. Applying the existing RL approaches in the literature to such systems with local interaction, if possible, leads to high time and space costs. The time complexity of current RL approaches commonly are $|A|^n$ where n is the number of agent or players and $|A|$ is the size of the possible strategy set. As shown in this chapter, learning of Nash equivalent equilibrium point in systems with local interactions using the ICLA model can reduce the time complexity of learning drastically.

References

Bäck, T.: Evolutionary Algorithms in Theory and Practice: Evolution Strategies, Evolutionary Programming, Genetic Algorithms. Oxford University Press (1996)

Başar, T., Olsder, G.J.: Dynamic Noncooperative Game Theory, 2nd edn. Society for Industrial and Applied Mathematics (1998)

Beigy, H., Meybodi, M.R.: Open synchronous cellular learning automata. Adv. Complex Syst. **10**, 527–556 (2007)

Beigy, H., Meybodi, M.R.: Asynchronous cellular learning automata. Automatica **44**, 1350–1357 (2008)

Beigy, H., Meybodi, M.R.: A mathematical framework for cellular learning automata. Adv. Complex Syst. **07**, 295–319 (2004). https://doi.org/10.1142/S0219525904000202

Beigy, H., Meybodi, M.R.R.: Cellular learning automata with multiple learning automata in each cell and its applications. IEEE Trans. Syst. Man Cybern. Part B **40**, 54–65 (2010). https://doi.org/10.1109/TSMCB.2009.2030786

Bowling, M., Veloso, M.: Multiagent learning using a variable learning rate. Artif. Intell. **136**, 215–250 (2002). https://doi.org/10.1016/S0004-3702(02)00121-2

De Jong, K.: Evolutionary Computation: A Unified Approach. MIT Press (2006)

Erev, I., Roth, A.E.: Predicting how people play games: reinforcement learning in experimental games with unique, mixed strategy equilibria. Am. Econ. Rev. **88**, 848–881 (1998)

Esnaashari, M., Meybodi, M.R.: Deployment of a mobile wireless sensor network with k-coverage constraint: a cellular learning automata approach. Wirel. Netw. **19**, 945–968 (2013). https://doi.org/10.1007/s11276-012-0511-7

Esnaashari, M., Meybodi, M.R.: Irregular cellular learning automata. IEEE Trans. Cybern. **45**, 1622–1632 (2018). https://doi.org/10.1016/j.jocs.2017.08.012

Esnaashari, M., Meybodi, M.R.M.: A cellular learning automata-based deployment strategy for mobile wireless sensor networks. J. Parallel Distrib. Comput. **71**, 988–1001 (2011)

Ficici, S.G., Pollack, J.B.: A Game-Theoretic Approach to the Simple Coevolutionary Algorithm. pp. 467–476 (2000)

Fudenberg, D., Levine, D.: The Theory of Learning in Games. MIT Press (1998)

Hu, J., Wellman, M.P.: Nash Q-learning for general-sum stochastic games. J. Mach. Learn. Res. 1039–1069 (2003)

Hu, J., Wellman, M.P.: Multiagent reinforcement learning: theoretical framework and an algorithm. In: Proceedings of the Fifteenth International Conference on Machine Learning (ICML98), pp. 242–250 (1998)

Khomami, M.M.D., Rezvanian, A., Meybodi, M.R.: A new cellular learning automata-based algorithm for community detection in complex social networks. J. Comput. Sci. **24**, 413–426 (2018). https://doi.org/10.1016/j.jocs.2017.10.009

Leslie, D.S., Collins, E.J.: Individual Q-learning in normal form games. SIAM J. Control Optim. **44**, 495–514 (2005). https://doi.org/10.1137/S0363012903437976

Mousavian, A., Rezvanian, A., Meybodi, M.R.: Cellular learning automata based algorithm for solving minimum vertex cover problem. In: 2014 22nd Iranian Conference on Electrical Engineering (ICEE). IEEE, pp. 996–1000 (2014)

Mozafari, M., Shiri, M.E., Beigy, H.: A cooperative learning method based on cellular learning automata and its application in optimization problems. J. Comput. Sci. **11**, 279–288 (2015). https://doi.org/10.1016/j.jocs.2015.08.002

Nowé, A., Verbeeck, K., Peeters, M. Learning Automata as a Basis for Multi Agent Reinforcement Learning, pp. 71–85 (2006)

Panait, L., Wiegand R.P., Luke, S.: Improving coevolutionary search for optimal multiagent behaviors. In: Proceedings of the 18th International Joint Conference on Artificial Intelligence. Acapulco, Mexico, pp. 653–658 (2003)

Panait, L., Luke, S.: Cooperative multi-agent learning: the state of the art. Auton. Agent Multi Agent Syst. **11**, 387–434 (2005). https://doi.org/10.1007/s10458-005-2631-2

Potter, M.A., Jong, K.A.: A cooperative coevolutionary approach to function optimization. In: Parallel Problem Solving from Nature—PPSN III, pp. 249–257 (1994)

Rastegar, R., Meybodi, M.R.: A new evolutionary computing model based on cellular learning automata. In: IEEE Conference on Cybernetics and Intelligent Systems, IEEE, pp. 433–438 (2004)

Rastegar, R., Meybodi, M.R., Hariri, A.: A new fine-grained evolutionary algorithm based on cellular learning automata. Int. J. Hybrid Intell. Syst. **3**, 83–98 (2006). https://doi.org/10.3233/HIS-2006-3202

Rezvanian, A., Saghiri, A.M., Vahidipour, S.M., et al.: Learning Automata Theory. pp. 3–19 (2018a)

Rezvanian, A., Saghiri, A.M., Vahidipour, S.M., et al.: Cellular Learning Automata, pp. 21–88 (2018b)

Rezvanian, A., Saghiri, A.M., Vahidipour, S.M., et al.: Learning automata for wireless sensor networks. In: Recent Advances in Learning Automata, pp. 91–219 (2018c)

Rezvanian, A., Saghiri, A.M., Vahidipour, S.M., et al.: Learning automata for cognitive peer-to-peer networks. In: Recent Advances in Learning Automata, pp. 221–278 (2018d)

Rodríguez, A., Vrancx, P., Grau, R., Nowé, A.: An RL approach to common-interest continuous action games. In: Proceedings of the 11th International Conference on Autonomous Agents and Multiagent Systems, pp. 1401–1402 (2012)
Sastry, P.S., Phansalkar, V.V., Thathachar, M.A.L.: Decentralized learning of nash equilibria in multi-person stochastic games with incomplete information. IEEE Trans. Syst. Man Cybern. **24**, 769–777 (1994). https://doi.org/10.1109/21.293490
Sharifi, A., Noroozi, V., Bashiri, M., et al.: Two phased cellular PSO: a new collaborative cellular algorithm for optimization in dynamic environments. In: 2012 IEEE Congress on Evolutionary Computation, CEC 2012, pp. 1–8 (2012)
Shoham, Y., Powers, R., Grenager, T.: Multi-Agent Reinforcement Learning: A Critical Survey (2003)
Tesauro, G.: Extending Q-learning to general adaptive multi-agent systems. In: Proceedings of the 16th International Conference on Neural Information Processing (NIPS03), pp. 871–878 (2003)
Thathachar, M.A.L., Sastry, P.S.: Networks of Learning Automata. Springer US, Boston, MA (2004)
Zhao, Y., Jiang, W., Li, S., et al.: A cellular learning automata based algorithm for detecting community structure in complex networks. Neurocomputing **151**, 1216–1226 (2015). https://doi.org/10.1016/j.neucom.2014.04.087